冯康文集

(第二卷)

中国科学院数学与系统科学研究院
计算数学与科学工程计算研究所　编

曹礼群　唐贻发　审校

科学出版社
北京

内 容 简 介

《冯康文集》第一卷，主要收集了冯康教授关于广义函数、有限元法、边界元方法和弹性力学等方面的论文.

本书是《冯康文集》第二卷，主要收集了冯康教授关于数学物理反演问题，辛几何与流体动力学中的数值方法，线性哈密尔顿系统的辛差分格式，辛算法、切触算法和保体积算法，常微分方程多步法的步进算子，欧拉型差分格式，动力系统的保结构算法，哈密尔顿系统的辛算法等方面的论文.

本书可作为计算数学及其相关专业的教师和科研人员的参考用书，也可供高年级本科生和研究生阅读参考.

图书在版编目(CIP)数据

冯康文集 / 中国科学院数学与系统科学研究院计算数学与科学工程计算研究所编.—北京: 科学出版社, 2020.7

ISBN 978-7-03-065620-9

Ⅰ. ①冯… Ⅱ. ①中… Ⅲ. ①数学-文集 Ⅳ. O1-53

中国版本图书馆 CIP 数据核字 (2020) 第 119153 号

责任编辑：李静科 / 责任校对：彭珍珍
责任印制：吴兆东 / 封面设计：无极书装

科学出版社 出版
北京东黄城根北街 16 号
邮政编码：100717
http://www.sciencep.com
北京建宏印刷有限公司 印刷
科学出版社发行 各地新华书店经销
*
2020 年 7 月第 一 版 开本: 787 × 1092 1/16
2021 年 5 月第二次印刷 印张: 27 1/4
字数: 643 000

定价：468.00 元(含 2 卷)
(如有印装质量问题，我社负责调换)

再版前言

《冯康文集》共两卷, 由中国科学院计算数学与科学工程计算研究所前身中国科学院计算中心整理编辑. 第一、二卷由国防工业出版社分别于 1994 年和 1995 年首次出版. 《冯康文集》收录了冯先生公开出版的研究工作, 是非常珍贵的文献.

时逢冯康先生诞辰 100 周年, 为缅怀冯康先生为中国计算数学事业作出的巨大贡献, 计算数学与科学工程计算研究所再版《冯康文集》.

中国科学院数学与系统科学研究院
计算数学与科学工程计算研究所
2020 年 7 月

前　　言

冯康教授于 1993 年 8 月 17 日与世长辞了. 他作为著名的数学和物理学家, 作为中国计算数学的奠基人和开拓者, 为发展我国计算数学和科学与工程计算事业, 倾注了毕生心血, 建立了不朽的功勋; 他培养了一批计算数学人才, 创建了我国现代计算数学队伍; 他开辟了一个个新兴的学科方向, 给我们留下了极为丰富和宝贵的科学遗产; 他发表了许多论文、专著, 并留下了大量珍贵的手稿.

为了使更多的学者, 特别是年轻的计算数学家, 能够分享到冯康教授的科学遗产, 促进计算数学的发展, 我们除了把冯康教授的手稿整理出来, 陆续在有关杂志上发表外, 决定收集出版《冯康文集》.

《冯康文集》第一卷已于 1994 年出版发行. 本书是第二卷, 主要收集了冯康教授关于数学物理反演问题, 哈密尔顿系统的辛算法, 动力系统的保结构算法等方面的论文.

在此, 我们敬告读者, 如果您处珍藏有冯康教授尚未发表的论文或报告手稿, 请复印一份寄给我们, 并欢迎您协助整理, 尽早发表.

让我们以冯康教授的科学成就为起点, 开创我国计算数学繁荣昌盛的未来.

在此, 我们向为《冯康文集》的出版做出了贡献的所有人员表示衷心的感谢, 感谢他们辛勤和卓有成效的工作.

《冯康文集》整理编辑组

1995 年 5 月

《冯康文集》整理编辑组

组　长: 石钟慈

副组长: 崔俊芝

王烈衡　　余德浩　　秦孟兆

汪道柳　　李旺尧

冯康教授生平

冯康, 中国科学院学部委员, 中国科学院计算中心名誉主任, 数学和物理学家, 计算数学家, 中国计算数学的奠基人和开拓者; 1920 年 9 月 9 日出生于江苏省南京市; 因患脑蛛网膜下腔出血, 经多方抢救无效, 于 1993 年 8 月 17 日 13 时 45 分逝世, 享年 73 岁.

冯康于 1939 年春考入福建协和学院数理系; 同年秋又考入重庆中央大学电机工程系, 两年后转物理系学习直到 1944 年毕业. 1945 年到 1951 年, 他先后在复旦大学物理系、清华大学物理系和数学系任助教; 1951 年调到刚组建的中国科学院数学研究所任助理研究员; 1951 年到 1953 年在苏联斯捷克洛夫数学研究所工作. 从 1945 年到 1953 年他曾先后在当代著名数学大师陈省身、华罗庚和庞特利亚金等人指导下工作. 1957 年根据国家十二年科学发展计划, 他受命调到中国科学院计算技术研究所, 参加了我国计算技术和计算数学的创建工作, 成为我国计算数学和科学工程计算学科的奠基者和学术带头人. 1978 年调到中国科学院计算中心任中心主任, 1987 年改任计算中心名誉主任直到逝世. 冯康教授事业心极强, 刻苦工作, 成就卓著, 受到了党和人民的尊敬, 以及国内外学者的赞誉. 1959 年被评为全国先进工作者, 1964 年被选为第三届全国人大代表,1979 年被评为全国劳动模范, 1980 年当选为中国科学院学部委员. 曾任全国计算机学会副主任委员；全国计算数学会理事长、名誉理事长, 国际计算力学会创始理事, 英国伦敦凯莱计算与信息力学研究所科技顾问、国际力学与数学交互协会名誉成员、英国爱丁堡国际数学研究中心科学顾问等多个学会、协会职务. 他是全国四种计算数学杂志的主编, 先后担任美国《计算物理》、日本《应用数学》、荷兰《应用力学与工程的计算方法》、美国《科学与工程计算》、《中国科学》等杂志的编委, 并任《中国大百科全书 · 数学卷》副主编.

冯康的科学成就是多方面的和非常杰出的, 1957 年前他主要从事基础数学研究, 在拓扑群和广义函数理论方面取得了卓越的成就. 1957 年以后他转向应用数学和计算数学研究, 由于其具有广博而扎实的数学、物理基础, 使得他在计算数学这门新兴学科上做出了一系列开创性和历史性的贡献.

50 年代末与 60 年代初, 冯康在解决大型水坝计算问题的集体研究实践的基础上, 独立于西方创造了一套求解偏微分方程问题的系统化、现代化的计算方法, 当时命名为基于变分原理的差分方法, 即现时国际通称的有限元方法. 有限元方法的创立是计算数学的一项划时代成就, 它已得到国际上的公认.

70 年代, 冯康建立了间断有限元函数空间的嵌入理论, 并将椭圆方程的经典理论推广到具有不同维数的组合流形, 为弹性组合结构提供了严密的数学基础, 在国际上为首创.

与此同时, 冯康对传统的椭圆方程归化为边界积分方程的理论作出了重要的贡献, 提出自然边界元方法, 这是当今国际上边界元方法的三大流派之一. 1978 年以来, 冯康先后应邀赴法国、意大利、日本、美国等十多所著名的科研机构及大学主讲有限元和自然边界元方法, 受到高度评价.

1984 年起冯康将其研究重点从以椭圆方程为主的稳态问题转向以哈密尔顿方程和波动方程为主的动态问题. 他于 1984 年首次提出基于辛几何计算哈密尔顿体系的方法, 即哈密顿体系的保结构算法, 从而开创了哈密尔顿体系计算方法这一富有活力及发展前景的新领域; 冯康指导和带领了中国科学院计算中心一个研究组投入了此领域的研究, 取得了一系列优秀成果. 新的算法解决了久悬未决的动力学长期预测计算方法问题, 正在促成天体轨道、高能加速器、分子动力学等领域计算的革新, 具有更为广阔的发展前景. 冯康多次应邀在国内及西欧、苏联、北美等多国讲学或参加国际会议作主题报告, 受到普遍欢迎及高度评价, 国际和国内已兴起了许多后继研究. 1995 年国际工业与应用数学大会已决定邀请冯康就此主题作一小时大会报告.

由于在科学上的突出贡献, 他曾先后获得 1978 年全国科学大会重大成果奖、国家自然科学二等奖、国家科技进步二等奖及科学院自然科学一等奖等.

冯康除了本人的研究工作外, 还承担了众多的行政工作. 他花了大量的心血做了大量学术指导工作. 早在 60 年代, 他亲自为当时中国科学院计算技术研究所三室 200 多人讲授现代计算方法和具体指导科学研究, 他们中的许多人已成为我国计算数学的业务骨干. 冯康费尽心血, 大力培养年轻优秀人才, 他亲自培养的研究生目前已遍布国内外, 有的已成为国际知名学者.

冯康非常关心全国计算数学学科的发展及队伍的建设, 多次提出重要的指导性意见. 他曾向中央领导建议并呼吁社会各方重视科学与工程计算, 倡议将科学与工程计算列入国家基础研究重点项目, 等等. 冯康用极大的热情, 从科学技术发展的战略高度上阐明了科学与工程计算的地位和作用, 有力地促进了计算数学在我国 “四化” 建设中发挥其应有的作用. “科学与工程计算的方法和理论” 已列为 “八五” 期间国家基础研究重大关键项目, 冯康任首席科学家.

冯康的一生, 是为科学事业奋斗不息的一生, 是为祖国繁荣昌盛无私奉献的一生. 他在研究工作中, 积极倡导理论联系实际, 并身体力行, 自觉运用辩证法, 把握住事物的本质, 成功地开创了科学的新方向、新道路、新领域, 带领一批又一批人在新方向上做出卓越的贡献. 他从不满足, 具有强烈的进取心和为国争光的使命感, 这使他一直走在世界计算数学队伍前列. 在他年已古稀之时, 仍经常废寝忘食、通宵达旦地工作. 就在冯康患病住院的前一个小时, 他还在为一项新的工作奔波、伏案疾书, 在他从昏迷中清醒的片刻, 首先询问的是 1993 年 “华人科学与工程计算青年学者会议” 的准备工作, 关心着下一代的成长. 他的心里只有科学事业. 他是 “将军”, 总在运筹帷幄; 他又是士兵, 一直在冲锋陷阵. 他是导师, 总是开辟方向, 指导我们前进; 他又是益友, 总和我们研究人员在一起. 他是老一代知识分子的优秀代表, 是我们学习的榜样.

中国科学院计算中心

1993 年 8 月 18 日

纪念冯康先生

P. Lax①

(原载 SIAM NEWS 26 卷 93 年 11 期)

Feng Kang, China's, leading applied mathematician, died suddenly on August 17, in his 73rd year, after a long and distinguished career that had shown no sign of slowing.

Feng's early education was in electrical engineering, physics, and mathematics, a background that subtly shaped his later interests. He spent the early 1950s at the Steklov lnstitute in Moscow. Under the influence of Pontryagin, he began by working on problems of topological groups and Lie groups. On his return to China, he was among the first to popularize the theory of distributions.

In the late 1950s, Feng turned his attention to applied mathematics, where his most important contributions lie. Independently of parallel developments in the West, he created a theory of the finite element method. He was instrumental in both the implementation of the method and the creation of its theoretical foundation using estimates in Sobolev spaces. He showed how to combine boundary and domain finite elements effectively, taking advantage of integral relations satisfied by solutions of partial differential equations. In particular, he showed how radiation conditions can be satisfied in this way. He oversaw the application of the method to problems in elasticity as they occur in structural problems of engineering.

In the late 1980s, Feng proposed and developed so-called symplectic algorithms for solving evolution equations in Hamiltonian form. Combining theoretical analysis and computer experimentation, he showed that such methods, over long times, are much superior to standard methods. At the time of his death, he was at work on extensions of this idea to other structures.

Feng's significance for the scientific development of China cannot be exaggerated. He not only put China on the map of applied and computational mathematics, through his own research and that of his students, but he also saw to it that the needed resources were made available. After the collapse of the Cultural Revolution, he was ready and able to help the country build again from the ashes of this selfinflicted conflagration. Visitors to China were deeply impressed by his familiarity with new developments everywhere.

① 美国科学院院士, 柯朗研究所教授, 原美国总统科学顾问, 原美国数学会会长, 原柯朗研究所所长.

Throughout his life, Feng was fiercely independent, utterly courageous, and unwilling to knuckle under to authority. That such a person did survive and thrive shows that even in the darkest days, the authorities were aware of how valuable and irreplaceable he was.

In Feng's maturity the well-deserved honors were bestowed upon him—membership in the Academia Sinica, the directorship of the Computing Center, the editorship of important journals, and other honors galore.

By that time his reputation had become international, many remember his small figure at international conferences, his eyes and mobile face radiating energy and intelligence. He will be greatly missed by the mathematical sciences and by his numerous friends.

目 录

1 数学物理中的反问题

Inverse Problems in Mathematical Physics

数学物理方程中的问题大致可以分为两类. 一类问题是在给定的方程模式下, 再给出规定具体环境的定解条件, 如方程的系数包括源项或边界条件等, 人们就可试图求解以便定出过程演化、联系影响的定量特征, 一般称为正问题或正演问题, 起着由因推果的作用, 它们的研究与应用都比较成熟, 迄今占着主导地位. 另一类问题则是在给定的方程模式下, 人们已知其解或解的某些部分, 要求反过来求该方程的系数, 源项或边界的形状等等, 这就是所谓反问题或反演问题, 起着倒果求因的作用. 反问题的研究发展虽然还未成熟, 但近年来由于应用上的需要, 特别是为要查明不可达、不可触之处的形貌性态, 为了由表及里、索隐探秘, 就得用特定的物理手段以取得反响信息, 并据此提成数学上的反问题来求解. 在经济、国防、科技以及生活领域的资源勘探、遥感遥测、无损探伤、诊断构象、目标侦察等等问题上都有重要应用, 工程技术中的定向设计和系统识别等问题也属于这类范畴.

下面根据历史的渊源, 扼要介绍一些在理论和方法上有代表性, 在实践应用上有重要意义的反问题.

§1 Huygens 等时摆

重力摆的运动方程是

$$m\frac{d^2s}{dt^2} = -\frac{\partial U}{\partial s}, \quad U = mgy(s) \tag{1}$$

此处 s 为弧长坐标, $y = y(s)$ 为轨线, 设 l 为摆长, 当取圆周轨道时, $s = l\theta$, $y = l(1 - \cos\theta)$, 运动方程成为

$$ml\frac{d^2\theta}{dt^2} = -mg\sin\theta \tag{2}$$

取小幅角近似, $0 \sim \theta \sim \sin\theta$, 方程简化为

$$\frac{d^2\theta}{dt^2} = -\frac{g}{l}\theta, \quad 周期\ T = 2\pi\sqrt{\frac{l}{g}} \tag{3}$$

这就是单摆的等时性, 但只是近似的. 圆周摆的严格解的周期则是

$$T = 4\sqrt{\frac{l}{g}}\int_0^{\theta_0}\frac{1}{\sqrt{\cos\theta - \cos\theta_0}}d\theta = 2\pi\sqrt{\frac{l}{g}}\left(1 + \frac{1}{16}\theta_0^2 + \cdots\right)$$

它是不等时的, 依赖于初始角 θ_0. Huygens(1673) 提出并解决了定长摆的轨线形状使其严格等时 [1].

仿照单摆方程取 $y = s^2/2l$ 即得

$$\frac{d^2 s}{dt^2} = -\frac{g}{l}s, \quad T = 2\pi\sqrt{\frac{l}{g}}$$

由 $ds^2 = dx^2 + dy^2$ 得轮半径为 $l/4$ 的旋轮线

$$x = \frac{l}{4}(\phi + \sin\phi), \quad y = \frac{l}{4}(1 - \cos\phi), \quad s = l\sin\phi/2$$

这是历史上最早的反问题雏型, 也是最早的定向设计.

§2　Abel 积分方程

Abel (1823) 把 Huygens 问题推广为: 已知从高度 $y = \eta$ 降至 $y = 0$ 的走时曲线 $\tau = \tau(\eta)$, 反求轨道曲线 $y = y(s)$. 当 $y(s)$ 为单调时导出下列积分方程及其反演公式:

$$\frac{1}{\sqrt{2g}}\int_0^\eta \frac{s'(y)dy}{\sqrt{\eta - y}} = \tau(\eta)$$
$$s(y) = \frac{\sqrt{2g}}{\pi}\int_0^y \frac{\tau(\eta)d\eta}{\sqrt{y - \eta}}$$

取 $\tau(\eta) =$ 常数即得旋轮线, 这是历史上最早的积分方程, 也是最早的完整形式的反问题. 经过四分之三个世纪以后, Hebglotz-Wiechebt (1905-07) 应用 Abel 型反演解决了在一定的对称条件下根据地震波的走时曲线反推地层内部形貌的方法, 据此 Mohobovic (1909) “发现” 地壳与地幔之间的断层, 自此以后 Abel 反演广泛应用于地震学、地球物理、空间探测、形貌重构、显微成像以及立体构成 (Stereology) 等众多领域 [2,3,4].

§3　Radon 变换

已知函数 $f(x), x \in R^n$, 可以算出它在超平面 $\alpha \cdot x = p$ 的积分值.

$$\hat{f}(\alpha, p) = g(\alpha, p) = \int_{\alpha \cdot x = p} f(x)dx^{n-1}, \quad \alpha \in S^{n-1}, \quad p \in R$$

能否从函数 $g(\alpha, p)$ 重构 $f(x)$? 在二维圆对称的情况 $(r^2 = x_1^2 + x_2^2)$, 问题归结为 Abel 反演

$$f(x_1, x_2) = F(r), \quad g(\alpha_1, \alpha_2, p) = G(p)$$
$$G(p) = 2\int_\rho^\infty \frac{rF(r)dr}{\sqrt{r^2 - \rho^2}}, \quad F(r) = -\frac{1}{\pi}\int_r^\infty \frac{G'(p)dp}{\sqrt{p^2 - r^2}}$$

Lorentz (1906) 首先解决了三维反演:

$$f(x)=-\frac{1}{8\pi^2}\int_{s^2}\left[g_p'(\alpha,p)\right]_{p=\alpha\cdot x}d\alpha$$

Radon 则系统提出并解决了一般 n 维反演 [5,6] 包括二维反演:

$$f(x)=-\frac{1}{4\pi^2}\int_{S}\left[\frac{1}{p}\cdot g_p'(\alpha,p)\right]_{p=\alpha\cdot x}d\alpha$$

Radon 变换与 Fourier 变换有密切联系, 即 n 维 F.T. 可以表为 n 维 R.T. 与对变量 p 的一维 F.T. 的组合, 但 R.T. 比 F.T. 有更突出的几何直观性, 并有与微分算子的互换性 $\widehat{\Delta u}=\frac{\partial^2}{\partial p^2}\hat{u}$, 从而使它成为微分方程理论中的有效工具.

Radon 变换自创始后又经历了四分之三个世纪才显示了实践应用上的重要性. 当前计算机化 X 光分层扫描构象 (Tomography[7]) 的数学基础就是二维 R.T., 这是 X 光自 Roentgen 发明 (1900 年诺贝尔奖) 以来在医疗诊断上的最大进展 (Hounsfield-Cormack, 1979 年诺贝尔奖). 基于 R.T. 的分层扫描也可借助于声波、光波、电磁波、地震波, 从而对于无损探伤、射电望远镜、电子显微镜、雷达目标重构、等离子体监测和地震勘探油气数据处理等多方面有广泛应用. 特别是基于三维 R.T. 的核磁共振分层构象, 预期在诊断效能及无伤害性方面均远优于 X 光, 正为众所瞩目.

§4 Gelfand-Levitan 散射反演

一维 Schrödinger 方程, 即 Sturm-Liouville 方程

$$\left(-\frac{d^2}{dx^2}+U(x)\right)\psi=E\psi$$

的正演问题是从给定的在量子力学中代表位势, 在波动传播中代表介质阻抗的 $U(x)$ 推算其谱, 即本征值 E 和本征函数的无穷远处的散射性态.

离散谱:

$$E=-\lambda_n<0,\quad \psi_n(x)\sim C_n e^{-\lambda_n x},\quad x\to+\infty$$

连续谱:

$$E=k^2>0,\quad \psi(k,x)\sim\begin{cases}e^{-ikx}+R(k)e^{ikx}, & \text{当 } x\to+\infty\\ e^{-ikx}, & \text{当 } x\to-\infty\end{cases}$$

所谓基于谱的散射反演问题要求从已知谱 $-\lambda_n$, k^2 及其相应的散射量 C_n, $R(k)$ 反推函数 $U(x)$. 为此 Gelfand-Levitan (1951) 提出了独特的方法, 建立积分方程 [8]

$$K(x,y)+A(x+y)=\int_x^\infty K(x,z)A(y+z)dz=0$$

该

$$A(x)=\sum_n C_n^2 e^{-\lambda_n x}+\frac{l}{2\pi}\int_{-\infty}^{+\infty} R(k)e^{ikx}dk$$

对于 $x\leqslant y$ 求解 $K(x,y)$, 然后

$$U(x)=2\frac{d}{dx}k(x,x)$$

当对 Schrödinger 方程应用几何渐近 (即 W.K.B. 方法) 时, 则上述反演问题简化为 Abel 型方程, 因此, 和 R.T. 相仿, G.L. 反演是 Abel 反演的另一发展, 这三者之间的理论联系尚待探索. G.L. 方程应用于量子力学的位势重构, 层状介质波动速度反演, 抛物型方程的扩散、传热系数反演, 自控系统最优设计以及孤立子理论等许多方面.

另一类与上述问题相平行的是所谓基于动态响应的散射反演, 特别是对于波动方程

$$\rho\frac{\partial^2\vec{u}}{\partial t^2}=\operatorname{grad}\rho c^2\operatorname{div}\vec{u}\quad (z<0)$$

在地表原点受激后震波在地下半空间传播, 并在地表 $(z=0)$ 收取散射回来的反响, 要求由表及里推出反映地下构造的介质参数分布, 如 c (波速), ρ (密度) 或其组合 ρc (阻抗). 这是地震探油技术中关键性数学问题. G-L 方法可以用于一维层状介质模型. 如何把它推广到实际的三维或探索其他的有效途径乃是有重大实用价值和理论意义的课题. 例如, 传统而较粗放的射线法即几何渐近法仍有很大的发展前景. 从地表各点间的沿短程线的走时来重构地下的黎曼度量的问题相当于一种广义的非线性的 R.T., 这里取的不是直线积分而是短程曲线积分, 这和 R.T. 一样都是所谓积分几何的问题 [5,9], 对此近年兴起的付氏积分算子以及奇点分类的拓扑理论是合适的工具 [10].

§5　几何的反演问题

类似于上述反谱反射散问题还有其他的几何性质的反演问题. 例如, 反谱构形的问题, 即已知 Laplace 算子在 Dirichlet 边界条件下的谱 $(\lambda_1,\lambda_2,\cdots)$, 能否反推区域 Ω 的形状, 即人们能否“以耳代目”, “听出”鼓的形状 [11]? 虽然问题的解答基本上是否定的, 但是根据谱确可推出一系列有关 Ω 的几何信息, 例如在二维情况有渐近展开 $(t\to 0)$

$$\sum_n e^{-\lambda_n t}=\frac{A}{4\pi t}-\frac{L}{4(4\pi t)^{\frac12}}+\frac16(1-H)+O(1),$$

这里 A 是 Ω 的面积, L 是 $\partial\Omega$ 的长度, H 是 Ω 内的孔穴数等 [12]. 由此渐近展开还可能推出更多的关于 Ω 的几何信息.

另一类是反射构形问题, 即入射平面波 $e^{i(\alpha\cdot x)k}$ 被障碍物 Ω 干扰产生散射波 ψ_α, 它是 Helmholtz 方程

$$-\Delta\psi_\alpha=k^2\psi_\alpha$$

在区域 Ω 外部的 Dirichlet (或 Neumann) 问题的解, 并且 (三维情况)

$$\psi_\alpha(\gamma,\beta)=\frac{e^{ikr}}{r}A(\alpha,\beta,k)+O\left(\frac{1}{r^2}\right),\quad r\to\infty$$

$|A|^2$ 为物理上可测量的散射截面, $\alpha,\beta\in S^2$ 分别是入射角和散射角. 能否根据散射截面数据反推 Ω 的形状? 在雷达目标的侦查以及反射器的综合设计等方面就需要解决这样的问题. 可以证明 [13], 对于凸体 Ω 在高频时,

$$|A(\alpha,\beta,k)|^2\to\frac{(R(\alpha,\beta))^2}{K(\alpha-\beta)},\quad k\to\infty,\quad \alpha\neq\beta$$

此处 $k(\alpha-\beta)$ 为 $\partial\Omega$ 上法向并行于 $\alpha-\beta$ 处的高斯曲率, $R(\alpha,\beta)=\dfrac{(\alpha,\beta)-1}{|\alpha-\beta|^2}$ 为反射系数, 因此足够多的散射截面数据可以定出 $\partial\Omega$ 上的曲率分布 $K(\alpha)$, 因此 $\partial\Omega$ 的重构归结于一个微分几何反演问题, 即 Minkowski 问题: 已知球 S^{n-1} 上正函数 $K(\alpha)$ 满足条件

$$\int_{S^{n-1}}\frac{\alpha d\alpha}{K(\alpha)}=0$$

能否在 R^n 中重构凸面 $\partial\Omega$ 使得对应的高斯曲率分布就是已知的 $K(\alpha)$? 答案是可能的, 而且除了刚性运动外是唯一的 [14].

以上的例子以及其他许多问题, 包括 Abel, Radon, Gelfand 反演都说明积分几何与反演问题有密切的背景联系, 特别在理论基础和算法发展方面有指导意义.

参考文献

[1] Sommerfeld A. Lectures on Theoretical Physics, Vol.1: Mechanics. Trans from 4th German ed. by Martin Stern O. Academic Press, 1952

[2] Keller J B. Inverse Problems. American Mathematical Monthly. 1976, 83(2): 107-118

[3] Landau L D, Lifshitz E M. Mechanics. 3rd ed., Tr. from the Russian by J.B. Sykes and J.S. Bell, Oxford, Pergamon, 1976

[4] Aki K, Richards P G. Quantitative Seismology: Theory and Methods. San Francisco: Freeman and Company, 1980

[5] Gelfand I M, Shilov G E. Generalized Functions, Vol 1: Properties and Operators. New York: Academic Press, 1966

[6] Helgason S. The Radon Transform. Boston: Birkhauser, 1980

[7] Herman G T, ed. Image Reconstruction from Projections, Implemantion and Applications. Berlin: Springer, 1979

[8] Chadan K, Sabatier P C. Inverse Problems in Quantum Scattering Theory. 2nd ed. New York: Springer-Verlag, 1989

[9] Romanov V G. Integral Geometry and Inverse Problems for Hyperbolic Equations. Berlin: Springer, 1974

[10] Dangelmayr G, Guttinger E T. Topological Approch to Remote Sensing. Geophys J of the Royal Astronomical Society. 1982, 7: 79-126

[11] Kac M. Can One Hear the Shape of a Drum? American Mathematical Monthly. 1966, 73(4): 1-23

[12] Singer I M. Eigenvalues of the Laplacian and Invariants of Manifolds. Proc of the Intern Congress of Mathematicians, Vol 1. Vancouver. 1974

[13] Majda A. High Frequency Asymptotics for the Scattering Matrix and the Inverse Problem of Acoustical Scattering. Communication on Pure and Applied Mathematics. 1976, 29(3): 261-291

[14] Chen S Y, Yau S T. On the Regularity of the Solution of the n-Dimentional Minkowski. Communications on Pure and Applied Mathematics. 1976, 29(5): 495-516

2 On Difference Schemes and Symplectic Geometry[①]

论差分格式与辛几何

§1 Introductory Remarks

In this paper we present some considerations and results of a preliminary study, specifically within the framework of symplectic geometry, of difference schemes for numerical solution of the canonical system of equations

$$\frac{dp_i}{dt} = -\frac{\partial H}{\partial q_i}, \quad \frac{dq_i}{dt} = \frac{\partial H}{\partial p_i}, \quad i = 1, \cdots, n \tag{1}$$

with given Hamiltonian function $H(p_1, \cdots, p_n, q_1, \cdots, q_n)$.

The canonical system (1) with remarkable elegance and symmetry was first introduced by Hamilton in 1824 as a general mathematical scheme for problems of geometrical optics. The success of this approach was evidenced by the subsequent theoretical prediction and experimental confirmation of the phenomenon of conical refraction. The approach was then successfully applied by Hamilton himself in 1834 to an entirely different area—analytical dynamics. It was immediately followed and analytically developed by Jacobi into a well-established mathematical formalism for mechanics, which is an alternative of, and equivalent to, the Newtonian and Lagrangian formalisms. The proper geometrization of Hamiltonian formalism was started by Poincaré in 1890's, his contributions, together with the later ones by Cartan, Birkhoff, Weyl, Siegel, et al., in the 20th century, gave rise to a new discipline, called symplectic geometry, which serves as the mathematical foundation of Hamiltonian formalism.

For a certain long period of time, however, Hamiltonian formalism and symplectic geometry had not attracted deserved attention from the general mathematical community, their theoretical as well as practical significance remained not fully recognized. A turn of interest was triggered first by Kolmogorov-Arnold-Moser's researches on the invariance of conditional periodicity under small perturbation of the Hamiltonian around the integrable system, which brought to light the potential of the symplectic approach.

① In Feng K, ed. Proc 1984 Bejing Symp Diff Geometry and Diff Equations. Beiing: Science Press, 1985: 42-58

This was followed by Keller-Maslov's contribution to the symplectic-geometrical foundation of the WKB asymptotic method for solving wave and Schrödinger equations and extending thereby the validity of the method beyond the caustic singularities. Since then, in the recent 2 decades there is an ever growing interest of research and realization of the importance of Hamiltonian formalism in many different areas of pure and applied mathematics. It is known that, Hamiltonian formalism, apart from its classical links with analytical mechanics, geometrical optics, calculus of variations and non-linear PDE of first order, has inherent connections also with unitary representations of Lie groups and geometric quantization (Kirillov, Kostant, et al.), with linear PDE and pseudodifferential operators (Hörmander, Egorov, et al.), with classification of singularities (Arnold, et al.), with integrability theory of non-linear evolution equations with soliton solutions, with optimal control theory, etc. It is also under extension to infinite dimensions for various field theories, including hydrodynamics, elasticity, electrodynamics, plasma physics, relativity, etc. Now it is almost certain that all real physical processes with negligible dissipation can be described, in some way or other, by Hamiltonian formalism, so the latter is becoming one of the most useful tools in the mathematical arsenal of physical and engineering sciences. In this way, a systematic study of physical methods of Hamiltonian systems is motivated and would eventually lead to more general applicability and more direct accessibility of the Hamiltonian formalism.

§2 Digressions on Hamiltonian Formalism

We give here a brief summary of the Hamiltonian formalism and its basic geometrical properties. For simplicity we use the usual coordinate description and consider only the classical phase space R^{2n} of a dynamical system with n degrees of freedom. For details, see, e.g., [1]. $R^{2n}=R_p^n\times R_q^n, z=[z_1,\cdots,z_{2n}]'\in R^{2n}$ splits into $z=\begin{bmatrix}p\\q\end{bmatrix}, q=[q_1,\cdots,q_n]'=[z_{n+1},\cdots,z_{2n}]'\in R_q^n$, R_q^n is the configuration space, whose "points" q represents positions of the system; $p=[p_1,\cdots,p_n]'=[z_1,\cdots,z_n]'\in R_p^n$, R_p^n is the momenta space, whose "vectors" p represents the momenta of the system.

The phase space R^{2n} is equipped with a standard symplectic structure defined by a "fundamental" differential 2-form on R^{2n}:

$$\omega_J=\sum_{i=1}^n dp_i\wedge dq_i=\sum_{i=1}^n dz_i\wedge dz_{n+i}, \tag{2}$$

i.e., to each $z\in R^{2n}$ a bilinear antisymmetric form

$$\omega_J(\xi,\eta)_z=\xi' J\eta \tag{3}$$

for each pair of tangent vectors $\xi = [\xi_1, \cdots, \xi_{2n}]'$, $\eta = [\eta_1, \cdots, \eta_{2n}]'$ at z, J is the standard antisymmetric matrix

$$J = \begin{bmatrix} 0 & I_n \\ -I_n & 0 \end{bmatrix}, \quad J' = -J = J^{-1}, \quad \det J = 1.$$

The fundamental 2-form ω_J is non-singular and closed, i.e., $d\omega_J = 0$.

Let $w : R^{2n} \to R^{2n}$ be a differential mapping, $z \in R^{2n} \to w(z) \in R^{2n}$, the corresponding Jacobi matrix is denoted by

$$\frac{\partial w}{\partial z} = \begin{bmatrix} \frac{\partial w_1}{\partial z_1} & \cdots & \frac{\partial w_1}{\partial z_{2n}} \\ \vdots & & \vdots \\ \frac{\partial w_{2n}}{\partial z_1} & \cdots & \frac{\partial w_{2n}}{\partial z_{2n}} \end{bmatrix}$$

mapping w induces, for each $z \in R^{2n}$, a linear mapping w_* of the tangent space at z into the tangent space at $w(z)$ by

$$\xi = [\xi_1, \cdots, \xi_{2n}]' \to w_*\xi = \frac{\partial w}{\partial z}\xi$$

w also induces, for each 2-form ω on R^{2n}, a 2-form $w^*\omega$ on R^{2n} by the formula

$$w^*\omega(\xi, \eta)_z = \omega\left(\frac{\partial w}{\partial z}\xi, \frac{\partial w}{\partial z}\eta\right)_{w(z)}.$$

If $\omega(\xi, \eta)_z = \xi A(z)\eta$, $A'(z) = -A(z)$, then $w^*\omega(\xi, \eta) = \xi B(z)\eta$, where

$$B(z) = \left(\frac{\partial w}{\partial z}\right)' A(w(z))\frac{\partial w}{\partial z}.$$

A diffeomorphism (differentiable, one-one, onto mapping) of R^{2n} is called a canonical transformation if w preserves the standard symplectic structure, i.e., $w^*\omega_J = \omega_J$, i.e.,

$$\left(\frac{\partial w}{\partial z}\right)' J \left(\frac{\partial w}{\partial z}\right) = J \tag{4}$$

i.e., the Jacobian $\frac{\partial w}{\partial z}$ is a symplectic matrix for each z.

For every pair of smooth functions $\phi(z), \psi(z)$ on R^{2n}, we associate a smooth function $\chi(z) = \{\phi, \psi\}$, called the Poisson bracket by

$$\{\phi, \psi\} = \phi'_z J^{-1} \psi_z,$$

where $\phi'_z = \left[\frac{\partial \phi}{\partial z_1}, \cdots, \frac{\partial \phi}{\partial z_n}\right]$. The Poisson brackets are anti-symmetric and satisfy Jacobi identity.

Choose a smooth function $H(z)=H(z_1,\cdots,z_{2n})=H(p_1,\cdots,p_n,q_1,\cdots,q_n)$. The equations (1), or written alternatively as

$$\frac{dz}{dt}=J^{-1}H_z, \tag{5}$$

is called the canonical system of equations with Hamiltonian $H(z)$. According to the general theory of ODE, for each Hamiltonian system (5), there corresponds a one-parameter group of diffeomorphisms g^t at least locally in t and z, of R^{2n} such that

$$g^0=\text{ identity},\quad g^{t_1+t_2}=g^{t_1}\cdot g^{t_2},$$

such that, if $z(0)$ is taken as the initial condition, then the solution of (5) is generated by

$$z(t)=g^t z(0).$$

The basic property of Hamiltonian system (1) is that g^t are canonical transformations

$$g^{t*}\omega_J=\omega_J, \tag{6}$$

for all t. This leads to the following class of phase-area conservation laws

$$\begin{aligned}
\int_{g^t\sigma^2}\omega_J&=\int_{\sigma^2}\omega_J,\quad \text{every 2-chain } \sigma^2\subset R^{2n},\\
\int_{g^t\sigma^4}\omega_J\wedge\omega_J&=\int_{\sigma^4}\omega_J\wedge\omega_J,\quad \text{every 4-chain } \sigma^4\subset R^{2n},\\
&\cdots\\
\int_{g^t\sigma^{2n}}\omega_J\wedge\cdots\wedge\omega_J&=\int_{\sigma^{2n}}\omega_J\wedge\cdots\wedge\omega_J,\quad \text{every } 2n\text{-chain } \sigma^{2n}\subset R^{2n},
\end{aligned} \tag{7}$$

the last one is the Liouville's phase-volume conservation law.

Another class of conservation law is related to energy and all the first integrals. A smooth function $\varphi(z)$ is said to be a first integral if $\varphi(g^t z)=\varphi(z)$ for all t, z, the latter is equivalent to the condition $\{\phi,H\}=0$. H, usually representing the energy, is itself a first integral.

The above situations can be generalized. A symplectic structure in R^{2n} is specified by a non-degenerate, closed 2 form $\omega_K=\sum k_{ij}(z)dz_i\wedge dz_j$

$$\omega_K(\xi,\eta)_z=\frac{1}{2}\xi' K(z)\eta,\quad K'(z)=-K(z),\quad \det K(z)\neq 0.$$

A differentiable mapping $w: R^{2n}\to R^{2n}$ is called K-canonical if $w^*\omega_K=\omega_K$, i.e.

$$\left(\frac{\partial w}{\partial z}\right)' K(w(z))\frac{\partial w}{\partial z}=K(z). \tag{8}$$

The Poisson bracket is defined as

$$\{\phi, \psi\}_K = \phi'_z K^{-1}(z)\psi_z,$$

which is anti-symmetric and satisfies the Jacobi identity. The equations of the form

$$\frac{dz}{dt} = K^{-1}H_z \tag{9}$$

is called the K-canonical system with Hamiltonian H, whose solutions are generated by a one-parameter group g^t, which consists of K-canonical transformations. The conservation laws of (7) remain true with J replaced by K, while the Liouville's theorem remains unchanged. The condition of first integrals for (9) is analogous, H is also among the first integrals.

Darboux theorem establishes the equivalence between all symplectic structures: Every non-singular closed 2-form ω_K can be brought to the standard form

$$\sum k_{ij}(z)dz_i \wedge dz_j = \sum dw_i \wedge dw_{n+1}$$

locally by suitable coordinate transformation $z \to w(z)$.

§3 Difference Schemes for Linear Canonical Systems

Take the Hamiltonian to be a quadratic form

$$H(z) = \frac{1}{2}z'Sz, \quad S' = S, \tag{10}$$

and K be an anti-symmetric non-singular constant matrix, then the K-canonical system (9) becomes linear

$$\frac{dz}{dt} = Bz, \quad B = K^{-1}S, \tag{11}$$

the generating one-parameter group is a group of linear transformations which coincides with their own Jacobians

$$z(t) = G(t)z(0), \quad G(t) = \exp tB. \tag{12}$$

The matrix B is infinitesimally K-symplectic

$$KB + B'K = 0, \tag{13}$$

its exponential transform exp tB is K-symplectic and S-orthogonal

$$(\exp tB)'K(\exp tB) = K, \quad (\exp tB)'S(\exp tB) = S, \tag{14}$$

i.e., both the symplectic structure and the energy are conserved.

In a wider context let $\psi(\lambda)$ be a meromorphic function, in case the matrix B has no eigenvalue at the poles of $\psi(\lambda)$ — we then say that B is non-exceptional—, the transform $\psi(B)$ is well-determined. It can be shown that, in order that $\psi(B)$ be K-symplectic for all non-exceptional infinitesimal K-symplectic matrices tB, it is necessary and sufficient that

$$\psi(\lambda)\psi(-\lambda)=1. \tag{15}$$

When this is satisfied, we have, for all integers $m \geqslant 0$,

$$(\psi(tB))'KB^m(\psi(tB))=KB^m. \tag{16}$$

Let

$$K_i=KB^{2i-2},\quad S_i=KB^{2i-1},\quad i=1,2,3,\cdots,n. \tag{17}$$

$K_1=K$, K_i are anti-symmetric, $S_1=S$, S_i are symmetric, then (14) is extended to

$$(\psi(tB))'K_i(\psi(tB))=K_i,\quad (\psi(tB))'S_i(\psi(tB))=S_i. \tag{18}$$

The independency of the sets K_i and S_i depends on the degree of the minimal polynomial of B. Thus the K-canonical transformation $\psi(tB)$ has many conservation laws of phase-areas and symplectic structures as well as many quadratic first integrals $\varphi_i(z)=\frac{1}{2}z'S_iz$. The exponential transform $\psi(\lambda)=\exp\lambda\rightarrow\exp B$ and Cayley transform $\psi(\lambda)=\frac{1+\lambda}{1-\lambda}\rightarrow\frac{I+B}{I-B}$ satisfy the condition (15).

In our study of numerical methods, we are interested in the Hamiltonian equations less as a system of ODE's per se, but rather as a specific system with Hamiltonian structure. It is natural to look forward to those discrete systems which preserve as much as possible the intrinsic properties of the continuous system. We hope this would lead to more satisfactory practical performance and theoretical foundation.

Consider now three kinds of difference schemes for linear Hamiltonian system (11). Let τ be the time-step, $z(n\tau)\sim z^n, n=0,1,2,\cdots$. Scheme I — Centered implicit Euler scheme

$$\frac{1}{\tau}\left(z^{m+1}-z^m\right)=B\frac{1}{2}\left(z^{m+1}+z^m\right),\quad B=K^{-1}S. \tag{19}$$

The transition $z^m\rightarrow z^{m+1}$ is given by the following linear transformation F_τ which coincides with its own Jacobian

$$z^{m+1}=F_\tau z^m,\quad F_\tau=\left(I-\frac{\tau}{2}B\right)^{-1}\left(I+\frac{\tau}{2}B\right)=\psi\left(-\frac{\tau}{2}B\right), \tag{20}$$

where $\psi(\lambda) = \dfrac{1-\lambda}{1+\lambda}$, the Cayley transform function. Note that the corresponding transition $z(m\tau) \to z((m+1)\tau)$ for the true solution at the same time-step τ is given by

$$z((m+1)\tau) = G_\tau z(m\tau), \quad G_\tau = \exp \tau B.$$

So G_τ and F_τ are the exponential and Cayley transforms respectively of the same B, so they have the same sets of invariant "symplectic structures" K_i and invariant "energies" S_i as given by (17). Even more than that, it can be easily proved that each function of anyone of the following types

(a) Quadratic form $f(z)$,

(b) Bilinear form $g(z, w)$,

(c) Linear form $\ell(z)$

is invariant under the differential equations (11) if and only if it is invariant under the difference equations (19). So the conservation properties of (11) and (19) are the same.

For the comparison of the stability properties of (11) and (19), we take, for simplicity, $K = J$ and the "separable" Hamiltonian

$$H(p, q) = U(p) + V(q) = \text{ kinetic energy } + \text{ potential energy},$$

where

$$U(p) = \frac{1}{2}p'Mp, \quad M = M', \quad \text{positive definite,}$$

$$V(q) = \frac{1}{2}q'Lq, \quad L = L', \quad \text{not necessarily positive definite,}$$

so that

$$S = \begin{bmatrix} M & 0 \\ 0 & L \end{bmatrix}, \quad B = J^{-1}S = \begin{bmatrix} 0 & -L \\ M & 0 \end{bmatrix},$$

and systems (11), (19) can be written as

$$\frac{dp}{dt} = -Lq, \quad \frac{dq}{dt} = Mp, \tag{21}$$

$$\frac{1}{\tau}\left(p^{m+1} - p^m\right) = -L\frac{1}{2}\left(q^{m+1} + q^m\right), \quad \frac{1}{\tau}\left(q^{m+1} - q^m\right) = M\frac{1}{2}\left(p^{m+1} + p^m\right). \tag{22}$$

The eigenvalue λ of B is related to the eigenvalue μ of the Pencil $L - \mu M^{-1}$ by $\lambda^2 = -\mu$, where μ is real, $\mu = 0$ or $-\omega^2$ or $+a^2$, where ω and a are positive. The Jordan normal form of the matrices B, G_τ, and F_τ consists of n diagonal blocks of order 2 of the following three possible types

	Type1	Type2	Type3
B	$\begin{bmatrix} 0 & 0 \\ 0 & 0 \end{bmatrix}$	$\begin{bmatrix} i\omega & 0 \\ 0 & -i\omega \end{bmatrix}$	$\begin{bmatrix} a & 0 \\ 0 & -a \end{bmatrix}$
G_τ	$\begin{bmatrix} 1 & 0 \\ 0 & 1 \end{bmatrix}$	$\begin{bmatrix} e^{i\omega\tau} & 0 \\ 0 & e^{-i\omega\tau} \end{bmatrix}$	$\begin{bmatrix} e^{a\tau} & 0 \\ 0 & e^{-a\tau} \end{bmatrix}$
F_τ	$\begin{bmatrix} 1 & 0 \\ 0 & 1 \end{bmatrix}$	$\begin{bmatrix} \dfrac{1+i\omega\tau/2}{1-i\omega\tau/2} & 0 \\ 0 & \dfrac{1-i\omega\tau/2}{1+i\omega\tau/2} \end{bmatrix}$	$\begin{bmatrix} \dfrac{1+a\tau/2}{1-a\tau/2} & 0 \\ 0 & \dfrac{1-a\tau/2}{1+a\tau/2} \end{bmatrix}$

When type 3 is missing in B, all eigenvalues of both (20) and (21) are unimodular with linear elementary divisors. Type 3 leads to instability for both (20) and (21).

Note that

$$G_\tau = \exp \tau \begin{bmatrix} 0 & -L \\ M & 0 \end{bmatrix} = \begin{bmatrix} G_{\tau,11} & G_{\tau,12} \\ G_{\tau,21} & G_{\tau,22} \end{bmatrix},$$

$$G_{\tau,11} = \sum_{m=0}^{\infty} \frac{(-1)^m}{(2m)!}(\tau)^{2m}(LM)^m, \quad G_{\tau,12} = -\sum_{m=0}^{\infty} \frac{(-1)^m}{(2m+1)!}(\tau)^{2m+1}(LM)^m L, \tag{23}$$

$$G_{\tau,21} = \sum_{m=0}^{\infty} \frac{(-1)^m}{(2m+1)!}(\tau)^{2m+1}(ML)^m M, \quad G_{\tau,22} = \sum_{m=0}^{\infty} \frac{(-1)^m}{(2m)!}(\tau)^{2m}(ML)^m.$$

Scheme Ⅱ. Staggered explicit scheme for separable Hamiltonian systems (20)

$$\begin{aligned} &\frac{1}{\tau}\left(p^{m+1} - p^m\right) = -Lq^{m+1/2}, \\ &\frac{1}{\tau}\left(q^{m+1+1/2} - q^{m+1/2}\right) = Mp^{m+1}. \end{aligned} \tag{24}$$

The p's are set at integer times $t = m\tau$, the q's at half-integer times $t = \left(m + \dfrac{1}{2}\right)\tau$.

In this case the transition

$$w^m = \begin{bmatrix} p^m \\ q^{m+1/2} \end{bmatrix} \to w^{m+1} = \begin{bmatrix} p^{m+1} \\ q^{m+1+1/2} \end{bmatrix}$$

is given by the linear transformation

$$w^{m+1} = F_\tau w^m, \quad F_\tau = \begin{bmatrix} I & 0 \\ -\tau M & I \end{bmatrix}^{-1} \begin{bmatrix} I & -\tau L \\ 0 & I \end{bmatrix}. \tag{25}$$

In order to analyze this scheme we introduce a linear transformation for the true solutions

$$z(t)=\left[\begin{array}{c}p(t)\\q(t)\end{array}\right]\rightarrow w(t)=\left[\begin{array}{c}p(t)\\q\left(t+\frac{\tau}{2}\right)\end{array}\right],$$

$$w(t)=Tz(t),\quad T=\begin{bmatrix}I & 0\\G_{\frac{\tau}{2},21} & G_{\frac{\tau}{2},22}\end{bmatrix},\quad z(t)=T^{-1}w(t).$$

Define

$$K=(T^{-1})'JT^{-1}=\left[k_{ij}(z)\right],\quad K'=-K.$$

It can be shown that $d\left(\sum k_{ij}(z)dz_i\wedge dz_j\right)=0$ and $\det K\neq 0$, so K actually defines a symplectic structure. Then $w(t)$ satisfies the K-canonical system

$$\frac{dw}{dt}=K^{-1}\widetilde{H}_w=K^{-1}\widetilde{S}w \tag{26}$$

with Hamiltonian $\widetilde{H}(w)=H\left(T^{-1}w\right)=\dfrac{1}{2}w'\widetilde{S}w$,

$$\widetilde{S}=T^{-1'}ST^{-1},$$

so the transition

$$\widetilde{G}_\tau: w(t)=\begin{bmatrix}p(t)\\q\left(t+\frac{\tau}{2}\right)\end{bmatrix}\rightarrow w(t+\tau)=\begin{bmatrix}p(t+\tau)\\q\left(t+\tau+\frac{\tau}{2}\right)\end{bmatrix}$$

is linear and K-symplectic, i.e., $\widetilde{G}_\tau'K\widetilde{G}_\tau=K$.

It can be proved that F_τ is also K-symplectic as is expected. However, the energy conservation properties are somehow different, F_τ does preserve $\tilde{H}(w)$, which is the true energy after synchronizing p^m, $q^{m+\frac{1}{2}}$ at staggered moments by T^{-1} to p^m,q^m at the same moment. F_τ preserves, instead, a modified Hamiltonian $\hat{H}(w)=\dfrac{1}{2}w'\hat{S}w, \hat{S}=$ $T^{-1'}JT^{-1}\begin{bmatrix}I & -\frac{\tau}{2}B\\-\frac{\tau}{2}A & I\end{bmatrix}^{-1}\begin{bmatrix}0 & -B\\A & 0\end{bmatrix}$. Note that in practical computation, the synchronization is done by defining $q^m=\dfrac{1}{2}\left(q^{m-\frac{1}{2}}+q^{m+\frac{1}{2}}\right)$, then all the first integrals $\varphi(p,q)$, including the Hamiltonian $H(p,q)$ are conserved approximately as

$$\phi\left(p^{m+1},q^{m+1}\right)=\phi\left(p^m,q^m\right)\mod O\left(\tau^3\right).$$

The eigenvalue λ of F_τ is related to the eigenvalue μ of the pencil $L-\mu M^{-1}$ by $\lambda^2+\lambda\left(\tau^2\mu-2\right)+1=0$, this leads to the Jordan normal form of F, consisting again of three possible types

$$
\begin{array}{ccc}
 & \text{Type 1} & \text{Type 2} \\
F_\tau & \begin{bmatrix} 1 & 0 \\ 0 & 1 \end{bmatrix} & \begin{bmatrix} 1+\dfrac{\omega^2\tau^2}{2}+\dfrac{i\omega\tau}{2}\sqrt{4-\omega^2\tau^2} & 0 \\ 0 & 1+\dfrac{\omega^2\tau^2}{2}-\dfrac{i\omega\tau}{2}\sqrt{4-\omega^2\tau^2} \end{bmatrix}
\end{array}
$$

$$
\text{Type 3} \qquad F_\tau \quad \begin{bmatrix} 1+\dfrac{a^2\tau^2}{2}+\dfrac{a\tau}{2}\sqrt{4+a^2\tau^2} & 0 \\ 0 & 1+\dfrac{a^2\tau^2}{2}+\dfrac{a\tau}{2}\sqrt{4+a^2\tau^2} \end{bmatrix}
$$

Type 2: When $\tau < \dfrac{2}{\omega}$, the two eigenvalues are unimodular, complex-conjugate, distinct. They collide at -1 when $\tau = \dfrac{2}{\omega}$. As $\tau > \dfrac{2}{\omega}$ they become distinct and real, one with modulus > 1 and other with modulus < 1. Type 3: the two eigenvalues are real and distinct, one with modulus > 1 and other with modulus < 1. In case L being non-negative definite, Type 3 is missing, then all eigenvalues of F_τ are unimodular and belonging to linear elementary divisors when $\tau < \dfrac{2}{\omega_{\max}}$.

We apply the above scheme to the 1-D wave equation

$$
\frac{\partial^2 u}{\partial t^2} = c^2\frac{\partial^2 u}{\partial x^2}, \quad 0 < x < 1, \quad u(0,t) = u(l,t) = 0
$$

with finite element semi-discretization

$$
\begin{aligned}
\frac{d^2 u_k}{dt^2} &= \frac{c^2}{h^2}\left[u_{k-1} - 2u_k + u_{k+1}\right], \quad k = 1, \cdots, n; \\
u_0 &= u_{n+1} = 0, \quad h = \frac{1}{n+1}.
\end{aligned}
$$

Let $q_k = u_k, p_k = \frac{\partial u_k}{\partial t}$, we get a canonical system (24) with

$$
M = I, \quad L = \frac{c^2}{h^2}\begin{bmatrix} 2 & -1 & & & \\ -1 & 2 & \ddots & & \\ & \ddots & \ddots & \ddots & \\ & & -1 & 2 & -1 \\ & & & -1 & 2 \end{bmatrix}.
$$

The types 1 and 3 are missing, $\omega_k = \dfrac{2c}{h}\cos\dfrac{k\pi}{2n+1} < \omega_1 = \dfrac{2c}{h}\cos\dfrac{\pi}{2n+1} < \dfrac{2c}{h}$. So the Courant condition $\tau \leqslant \dfrac{h}{c}$ ensures stability of (24). The scheme is in fact equivalent to the classical 5-point scheme for the wave equation, see [2]. There is an interesting study [3], with further references there, on computer simulation of fluids base on Hamiltonian formalism with spatiotemporal staggered scheme.

Schemes Ⅲ. Energy-conservative schemes by Hamiltonian differencing. For simplicity, we illustrate the cases only by $n=2$. Let $z=z^m, \tilde{z}=z^{m+1}$,

$$\begin{aligned}
\frac{1}{\tau}\left(\widetilde{p}_1-p_1\right) &= -\frac{1}{\widetilde{q}_1-q}\left\{H\left(p_1,p_2,\widetilde{q}_1,q_2\right)-H\left(p_1,p_2,q_1,q_2\right)\right\},\\
\frac{1}{\tau}\left(\widetilde{p}_2-p_2\right) &= -\frac{1}{\widetilde{q}_2-q_2}\left\{H\left(\widetilde{p}_1,p_2,\widetilde{q}_1,\widetilde{q}_2\right)-H\left(\widetilde{p}_1,p_2,\widetilde{q}_1,q_2\right)\right\},\\
\frac{1}{\tau}\left(\widetilde{q}_1-q_2\right) &= \frac{1}{\widetilde{p}_1-p_1}\left\{H\left(\widetilde{p}_1,p_2,\widetilde{q}_1,q_2\right)-H\left(p_1,p_2,\widetilde{q}_1,q_2\right)\right\},\\
\frac{1}{\tau}\left(\widetilde{q}_2-q_2\right) &= \frac{1}{\widetilde{p}_2-p_2}\left\{H\left(\widetilde{p}_1,\widetilde{p}_2,\widetilde{q}_1,\widetilde{q}_2\right)-H\left(\widetilde{p}_1,p_2,\widetilde{q}_1,\widetilde{q}_2\right)\right\}.
\end{aligned} \tag{27}$$

By addition and cancellation we have energy conservation for arbitrary Hamiltonian $H\left(\widetilde{p}_1,\widetilde{p}_2,\widetilde{q}_1,\widetilde{q}_2\right)=H\left(p_1,p_2,q_1,q_2\right)$.

For quadratic Hamiltonian, $H=\frac{1}{2}z'Sz$, we get

$$\frac{1}{\tau}\left(z^{m+1}-z^m\right)=J^{-1}S\frac{1}{2}\left(z^{m+1}+z^m\right)-\frac{1}{2}JR\left(z^{m+1}-z^m\right),$$

where

$$S=\begin{bmatrix} s_{11} & s_{12} & s_{13} & s_{14}\\ & s_{22} & s_{23} & s_{24}\\ & * & s_{33} & s_{34}\\ & & & s_{44}\end{bmatrix}=S',\quad R=\begin{bmatrix} 0 & -s_{12} & s_{13} & -s_{14}\\ & 0 & s_{23} & s_{24}\\ & * & 0 & -s_{34}\\ & & & 0\end{bmatrix}=-R',$$

and

$$z^{m+1}=F_\tau z^m,\quad F_\tau=\left(I+\frac{\tau}{2}JR-\frac{\tau}{2}J^{-1}S\right)^{-1}\left(I+\frac{\tau}{2}JR+\frac{\tau}{2}J^{-1}S\right).$$

Let $\bar{K}=J-\frac{\tau}{2}R, \bar{B}=\bar{K}^{-1}S$, we can prove that F_τ is the Cayley transform

$$F_\tau=\psi\left(-\frac{\tau}{2}\bar{B}\right),\quad \psi(\lambda)=\frac{1-\lambda}{1+\lambda},$$

so we have invariant "symplectic structures" $\bar{K}_1=\bar{K},\bar{K}_2,\bar{K}_3,\cdots$ and invariant "energies" $\bar{S}_1,=S,\bar{S}_2,\bar{S}_3,\cdots$ like (17), (18).

§4 Difference Schemes for General Canonical Systems

The three kinds of schemes for linear systems in the previous section can be generalized to the general non-linear case.

Scheme I. For the general canonical system (1), we put

$$\frac{1}{\tau}\left(z^{m+1}-z^{m}\right)=J^{-1}H_{z}\left(\frac{1}{2}z^{m+1}+\frac{1}{2}z^{m}\right). \tag{28}$$

The transition $z^m \to z^{m+1}$ is non-linear in general. By differentiation,

$$\frac{\partial z^{m+1}}{\partial z^{m}}-I=\tau J^{-1}H_{zz}\left(\frac{z^{m+1}+z^{m}}{2}\right)\left[\frac{1}{2}\frac{\partial z^{m+1}}{\partial z^{m}}+\frac{1}{2}I\right],$$

here $H_{zz}\left(\frac{z^{m+1}+z^m}{2}\right)$ is the Hessian matrix of the function $H(z)$, evaluated at $z=\dfrac{z^{m+1}+z^m}{2}$, $\dfrac{\partial z^{m+1}}{\partial z^m}$ is the Jacobian matrix F_τ, so

$$F_{\tau}=\left[I-\frac{\tau}{2}J^{-1}H_{zz}\left(\frac{z^{m+1}+z^{m}}{2}\right)\right]^{-1}\left[I+\frac{\tau}{2}J^{-1}H_{zz}\left(\frac{z^{m+1}+z^{m}}{2}\right)\right].$$

When z remains bounded and take τ sufficiently small we can keep the infinitesimally symplectic matrix $\frac{\tau}{2}J^{-1}H_{zz}\left(\frac{z^{m+1}+z^m}{2}\right)$ non-exceptional, then F_τ, as a Cayley transform, is symplectic. Thus all the conservation laws for phase areas remain true. However, unlike the linear case, the first integrals $\phi(z)$ including H itself are not conserved exactly. Instead, the approximate conservation

$$\phi\left(z^{m+1}\right)=\phi\left(z^{m}\right) \mod O\left(\tau^{3}\right)$$

can be shown.

We remark that the analogous averaged implicit Euler scheme

$$\frac{1}{\tau}\left(z^{m+1}-z^{m}\right)=J^{-1}\left[\frac{1}{2}H_{z}\left(z^{m+1}\right)+\frac{1}{2}H_{z}\left(z^{m}\right)\right], \tag{29}$$

which reduces, like (28), to the same symplectic scheme (19) with $K=J$ for linear problems.

$$F_{\tau}=\left[I-\frac{\tau}{2}J^{-1}H_{zz}\left(z^{m+1}\right)\right]\left[I+\frac{\tau}{2}J^{-1}H_{zz}\left(z^{m}\right)\right],$$

which is not symplectic in general.

Scheme II. For the canonical system with general separable Hamiltonian $H(p,q)=U(p)+V(q)$, we have

$$\begin{aligned}&\frac{1}{\tau}\left(p^{m+1}-p^{m}\right)=-V_{q}\left(q^{m+1/2}\right),\\&\frac{1}{\tau}\left(q^{m+1+1/2}-q^{m+1/2}\right)=U_{p}\left(p^{m+1}\right).\end{aligned} \tag{30}$$

The transition $\begin{bmatrix}p^m\\q^{m+1/2}\end{bmatrix}\to\begin{bmatrix}p^{m+1}\\q^{m+1+1/2}\end{bmatrix}$ has Jacobian

$$F_{\tau}=\begin{bmatrix}I&0\\-\tau M&I\end{bmatrix}^{-1}\begin{bmatrix}I&-\tau L\\0&I\end{bmatrix},$$

which can be shown to be K-symplectic as for (24), (26), but with

$$M=U_{pp}\left(p^{m+1}\right), \quad L=V_{qq}\left(q^{m+1/2}\right).$$

This leads to a class of modified conservation laws of phase areas, but with Liouville's theorem unchanged.

The first integrals $\phi(p,q)$, including $H(p,q)$, are approximately conserved as

$$\phi\left(p^{m+1},\frac{1}{2}\left(q^{m+1+1/2}+q^{m+1/2}\right)\right)=\phi\left(p^{m},\frac{1}{2}\left(q^{m+1/2}+q^{m-1/2}\right)\right), \quad \operatorname{mod}O\left(\tau^{3}\right).$$

Scheme Ⅲ. This has already been constructed for the nonlinear case in the previous section, the Hamiltonian $H(z)$ is always conserved exactly. However, the first integrals $\phi(z)$, other than the Hamiltonian, are approximately conserved to a lower order as

$$\phi\left(z^{m+1}\right)=\phi\left(z^{m}\right), \quad \operatorname{mod}O\left(\tau^{2}\right)$$

due to some kind of asymmetry in the algorithm. Moreover, except in the linear systems, symplectic properties for the Jacobian of transition could not be established in general.

The details of the results and some other developments will be published elsewhere.

References

[1] Arnold V I. Mathematical Methods of Classical Mechanics. New York: Springer. 1978

[2] Richtmyer R D, Morton K W. Difference Methods for Initial-Value Problems, 2nd edition. New York: Interscience. 1967

[3] Buneman O. Advantages of Hamiltonian Formulations in Computer Simulations. In Tabor, ed. Mathematical Methods in Hydrodynamics and Integrability of Dynamical Systems. Amer Inst Phys. USA. 1981

3 Difference Schemes for Hamiltonian Formalism and Symplectic Geometry①

哈密尔顿形式的差分格式与辛几何

§1 Introduction

The present program [1] that the author and his group have started is a systematic study of the numerical methods for the solution of differential equations of mathematical physics expressed in Hamiltonian formalism. As is well known, Hamiltonian canonical systems serve as the basic mathematical formalism, for diverse areas of physics, mechanics, engineering, as well as pure and applied mathematics, e.g., geometrical optics, analytical dynamics, nonlinear PDE's of first order, group representations, WKB asymptotics, pseudo-differential and Fourier integral operators, electrodynamics, plasma physics, elasticity, hydrodynamics, relativity, control theory, etc. It is generally accepted that all real physical processes with negligible dissipation could be expressed, in some way or other, in suitable Hamiltonian forms. So the general methods developed for the numerical solution of Hamiltonian equations, if good, would have wide applications.

Since symplectic geometry is the mathematical foundation of Hamiltonian formalism, a wealth of theoretical results is already accumulated which should be and could be explored for numerical purposes. So the proper mode of research in this area should be geometrical. We try to conceive, design, analyse and evaluate difference schemes and algorithms specifically within the framework of symplectic geometry. The approach proves to be quite successful as one might expect, and we actually derive in this way numerous "unconventional" difference schemes.

Due to historical reasons, classical symplectic geometry, however, lacks the "computational" component in the modern sense. Our present study might be considered as an attempt to fill the blank. We got a number of results (e.g. Th.1, §2) which are crucial for the construction of symplectic difference schemes on the one hand and which have independent theoretical interest in themselves on the other hand.

① J Comp Math. 1986 4(3): 279-289

In this paper, we consider the canonical system in finite dimensions

$$\frac{dp_i}{dt} = -H_{q_i}, \quad \frac{dq_i}{dt} = H_{p_i}, \quad i = 1, 2, \cdots, n, \tag{1.1}$$

with Hamiltonian $H(p_1, \cdots, p_n, q_1, \cdots, q_n)$.

In the following, vectors are always represented by column matrices, and matrix transpose is denoted by prime. Let $z = (z_1, \cdots, z_n, z_{n+1}, \cdots, z_{2n})' = (p_1, \cdots, p_n, q_1, \cdots, q_n)'$, $H_z = (H_{z_1}, \cdots, H_{z_{2n}})'$. (1.1) can be written as

$$\frac{dz}{dt} = J^{-1}H_z, \quad J = \begin{bmatrix} 0 & I_n \\ -I_n & 0 \end{bmatrix}, \tag{1.2}$$

defined in phase space R^{2n} with a standard symplectic structure given by the nonsingular anti-symmetric closed differential 2-form

$$\omega = \sum dz_i \wedge dz_{n+i} = \sum dp_i \wedge dq_i.$$

The Fundamental Theorem on Hamiltonian Formalism says that the solution of the canonical system (1.2) can be generated by a one-parameter group G_t of canonical transformations of R^{2n} (locally in t and z) such that

$$G_{t_1} G_{t_2} = G_{t_1+t_2},$$
$$z(t) = G_t z(0).$$

A transformation $z \to \hat{z}$ of R^{2n} is called canonical if it is a local diffeomorphism whose Jacobian $\frac{\partial \hat{z}}{\partial z} = M$ is everywhere symplectic, i.e.

$$M'JM = J, \quad \text{i.e.} \quad M \in Sp(2n).$$

Linear canonical transformations are simply symplectic transformations.

The canonicity of G_t implies the preservation of 2-form ω, 4-form $\omega \wedge \omega, \cdots$, $2n$-form $\omega \wedge \omega \wedge \cdots \wedge \omega$. They constitute the class of conservation laws of phase area of even dimensions for the Hamiltonian system (1.2).

Moreover, the Hamiltonian system possesses another class of conservation laws related to the energy $H(z)$. A function $\phi(z)$ is said to be an invariant integral of (1.2) if it is invariant under (1.2)

$$\phi(z(t)) \equiv \phi(z(0))$$

which is equivalent to

$$\{\phi, H\} = 0,$$

where the Poisson bracket for two functions $\phi(z)$, $\psi(z)$ are defined as

$$\{\phi, \psi\} = \phi_z' J^{-1} \psi_z.$$

H itself is always an invariant integral, see, e.g., [2].

For the numerical study, we are less interested in (1.2) as a general system of ODE per se, but rather as a specific system with Hamiltonian structure. It is natural to look for those discretization systems which preserve as many as possible the characteristic properties and inner symmetries of the original continuous systems. We hope that this might lead to more satisfactory theoretical foundation and practical performance.

The above digressions on Hamiltonian systems suggest the following guideline for difference schemes to be constructed. The transition from the k-th time step z^k to the next$(k+1)$-th time step z^{k+1} should be canonical for all k and, moreover, the invariant integrals of the original system should remain invariant under these transitions.

§2 Some Difference Schemes for Hamiltonian Systems

2.1 The centered Euler scheme and its generalizations

Consider first the case for which the Hamiltonian is a quadratic form

$$H(z) = \frac{1}{2} z' S z, \quad S' = S, \quad H_z = S(z). \tag{2.1}$$

Then the canonical system is linear

$$\frac{dz}{dt} = Lz, \tag{2.2}$$

where $L = J^{-1} S$ is infinitesimally symplectic, i.e.

$$L'J + JL = 0.$$

The solution of (2.2) is

$$z(t) = G_t z(0),$$

where $G(t) = \exp tL$, as the exponential transform of infinitesimally symplectic tL, is symplectic.

Proposition 1 *The weighted Euler scheme*

$$\frac{1}{\tau}(z^{k+1} - z^k) = L(\alpha z^{k+1} + (1-\alpha) z^k)$$

for the linear system (2.2) *is symplectic if and only if* $\alpha = \frac{1}{2}$, *i.e. it is the case of time-centered Euler scheme with the transition*

$$z^{k+1} = F_\tau z^k, \quad F_\tau = \phi(\tau L), \quad \phi(\lambda) = \frac{1 + \dfrac{\lambda}{2}}{1 - \dfrac{\lambda}{2}}, \tag{2.3}$$

where F_τ, *the Cayley transform of infinitesimally symplectic* τL, *is symplectic.*

In order to generalize the time-centered Euler scheme, we need, apart from the exponential or Cayley transforms, other matrix transforms carrying infinitesimally symplectic matrices into symplectic ones.

Theorem 1 *Let* $\psi(\lambda)$ *be a function of complex variable* λ *satisfying*

(I) $\psi(\lambda)$ *is analytic with real coefficients in a neighborhood* D *of* $\lambda = 0$,

(II) $\psi(\lambda)\psi(-\lambda) \equiv 1$ *in* D,

(III) $\psi_\lambda(0) \neq 0$.

A *is a matrix of order* $2n$, *then*

$$(\psi(\tau L))' A \psi(\tau L) = A$$

for all τ *with sufficiently small* $|\tau|$, *if and only if*

$$L'A + AL = 0.$$

If, moreover, $\exp \lambda - \psi(\lambda) = O(|\lambda|^{m+1})$, *then*

$$z^{k+1} = \psi(\tau L) z^k, \tag{2.4}$$

considered as an approximative scheme for the canonical system (2.2), *is symplectic, of m-th order accuracy and has the property that the bilinear form* $z'A\omega$ *is invariant under* $\psi(\tau L)$ *if and only if it is invariant under* G_τ *of* (2.2) ([5]).

Remark 1 The last property is remarkable in the sense that all the bilinear invariants of the system (2.2), no more and no less, are kept invariant under the scheme (2.4), in spite of the fact that the latter is only approximate.

Remark 2 The approximative scheme in Theorem 1 becomes difference scheme only when $\psi(\lambda)$ is a rational function. As a concrete application for the construction of symplectic difference schemes, take the diagonal Padé approximants to the exponential function

$$\exp \lambda - \frac{P_m(\lambda)}{P_m(-\lambda)} = O(|\lambda|^{2m+1}),$$

where

$$\begin{aligned}
&P_0(\lambda) = 1,\\
&P_1(\lambda) = 2 + \lambda,\\
&P_2(\lambda) = 12 + 6\lambda + \lambda^2,\\
&P_3(\lambda) = 120 + 60\lambda + 12\lambda^2 + \lambda^3,\\
&\cdots\\
&P_m(\lambda) = 2(2m-1)P_{m-1}(\lambda) + \lambda^2 P_{m-2}(\lambda), \text{etc.}
\end{aligned}$$

Theorem 2 *The difference schemes*

$$z^{k+1} = \frac{P_m(\tau L)}{P_m(-\tau L)} z^k, \quad m = 1, 2, \cdots \tag{2.5}$$

for the system (2.2) *are symplectic, A-stable, of* $2m$*-th order accuracy, and having the same set of bilinear invariants as that of system* (2.2). *The case* $m = 1$ *is the centered Euler scheme*[5].

For the comparison of stability properties of (2.4) and (2.2), we consider the simple case of separable Hamiltonian

$$S = \begin{bmatrix} S_1 & 0 \\ 0 & S_2 \end{bmatrix}, \quad L = J^{-1}S = \begin{bmatrix} 0 & -S_2 \\ S_1 & 0 \end{bmatrix},$$
$$S_1 = S_1' \text{ positive definite}; \quad S_2 = S_2'.$$

The eigenvalue λ of L is related to the eigenvalue μ of the pencil $S_2 - \mu S_1^{-1}$ by $\lambda^2 = -\mu$, where μ is real, $\mu = 0$ or $+\omega^2$ or $-a^2(\omega > 0, a > 0)$. The Jordan normal form of L, G_τ, F_τ consists of n diagonal blocks of order 2 of the following three possible types

	type1	type2	type3
L	$\begin{bmatrix} 0 & 0 \\ 0 & 0 \end{bmatrix}$	$\begin{bmatrix} i\omega & 0 \\ 0 & -i\omega \end{bmatrix}$	$\begin{bmatrix} a & 0 \\ 0 & -a \end{bmatrix}$
G_τ	$\begin{bmatrix} 1 & 0 \\ 0 & 1 \end{bmatrix}$	$\begin{bmatrix} e^{i\omega\tau} & 0 \\ 0 & e^{-i\omega\tau} \end{bmatrix}$	$\begin{bmatrix} e^{a\tau} & 0 \\ 0 & e^{-a\tau} \end{bmatrix}$
F_τ	$\begin{bmatrix} 1 & 0 \\ 0 & 1 \end{bmatrix}$	$\begin{bmatrix} \dfrac{P_m(i\omega\tau)}{P_m(-i\omega\tau)} & 0 \\ 0 & \dfrac{P_m(-i\omega\tau)}{P_m(\omega\tau)} \end{bmatrix}$	$\begin{bmatrix} \dfrac{P_m(a\tau)}{P_m(-a\tau)} & 0 \\ 0 & \dfrac{P_m(-a\tau)}{P_m(a\tau)} \end{bmatrix}$

When type 3 is missing, all eigenvalues of both G_τ and F_τ are unimodular with linear elementary divisors. Type 3 leads to instability for both G_τ and F_τ.

For the general non-linear canonical system (1.2), the time-centered Euler scheme is

$$\frac{1}{\tau}(z^{k+1}-z^k)=J^{-1}H_z\left(\frac{1}{2}(z^{k+1}+z^k)\right). \tag{2.6}$$

The transition $z^k \to z^{k+1}$ is canonical with Jacobian

$$F_\tau=\left[I-\frac{\tau}{2}J^{-1}H_{zz}\left(\frac{1}{2}(z^{k+1}+z^k)\right)\right]^{-1}\left[I+\frac{\tau}{2}J^{-1}H_{zz}\left(\frac{1}{2}(z^{k+1}+z^k)\right)\right]$$

everywhere symplectic. However, unlike the linear case the invariant integrals $\phi(z)$ of system (1.2), including $H(z)$, are conserved only approximately

$$\phi(z^{k+1})-\phi(z^k)=O(\tau^3).$$

The analogous averaged Euler scheme

$$\frac{1}{\tau}(z^{k+1}-z^k)=J^{-1}\left[\frac{1}{2}H_z(z^{k+1})+\frac{1}{2}H_z(z^k)\right]$$

which reduces, like (2.6), to the same symplectic scheme (2.4) for linear systems, is not canonical in general.

2.2 Staggered explicit schemes for separable Hamiltonians

For the non-linear separable system we have

$$\begin{aligned}&\frac{1}{\tau}(p^{k+1}-p^k)=-V_q(q^{k+1/2}),\\&\frac{1}{\tau}(q^{k+1+1/2}-q^{k+1/2})=U_p(p^{k+1}).\end{aligned}$$

The p's are set at integer times $t=k\tau$, q's at half-integer times $t=\left(k+\frac{1}{2}\right)\tau$. The transition

$$w^k=\begin{bmatrix}p^k\\q^{k+1/2}\end{bmatrix}\to\begin{bmatrix}p^{k+1}\\q^{k+1+1/2}\end{bmatrix}=w^{k+1}=F_\tau w^k,$$

$$F_\tau=\begin{bmatrix}I&0\\\tau S_1&I\end{bmatrix}^{-1}\begin{bmatrix}I&-\tau S_2\\0&I\end{bmatrix},\quad S_1=U_{pp}(p^{k+1}),\quad S_2=V_{qq}(q^{k+1/2}),$$

is symplectic, of 2nd order accuracy and practically explicit. Since p, q are computed at different times, we need synchronization, e.g. using

$$q^k=\frac{1}{2}(q^{k-1/2}+q^{k+1/2})$$

to compute the invariant integrals $\phi(p,q)$,

$$\phi(p^{k+1},q^{k+1})-\phi(p^k,q^k)=O(\tau^3).$$

For the comparison of stability for the linear system (2.2) with separable Hamiltonian with the staggered scheme and the application of the latter to the wave equation, see [1]. In [1], a class of energy conservative schemes was constructed using the differencing of the Hamiltonian function; the symplectic property is not satisfactory. The problem of compatibility of energy conservation with phase area conservation in difference schemes is solved successfully for linear canonical systems (Theorem 1); it seems difficult, however, for the general non-linear systems.

§3 A General Theory of Generating Functions and Hamilton-Jacobi Equations

Our approach in this part was inspired by the early works of Siegel [3] and Hua [4]. Every matrix

$$A=\begin{bmatrix}A_1\\A_2\end{bmatrix}\in M(4n,2n), A_1,A_2\in M(2n), \text{rank } A=2n$$

defines in R^{4n} a $2n$-dimensional subspace $\{A\}$ spanned by its column vectors. $\{A\}=\{B\}$ if and only if $A\sim B$, i.e.

$$AP=B, \text{i.e.}, \begin{bmatrix}A_1P\\A_2P\end{bmatrix}=\begin{bmatrix}B_1\\B_2\end{bmatrix}, \text{for some } P\in GL(2n).$$

The spaces of symmetric and symplectic matrices of order $2n$ will be denoted by $Sm(2n)$, $Sp(2n)$ respectively. Let

$$J_{4n}=\begin{bmatrix}0 & I_{2n}\\-I_{2n} & 0\end{bmatrix},\quad \widetilde{J}_{4n}=\begin{bmatrix}-J_{2n} & 0\\0 & J_{2n}\end{bmatrix},$$

$$X=\begin{bmatrix}X_1\\X_2\end{bmatrix},\quad Y=\begin{bmatrix}Y_1\\Y_2\end{bmatrix}\in M(4n,2n), \text{of rank } 2n.$$

Subspace $\{X\}\subset R^{4n}$ is called J_{4n}-Lagrangian (and $\begin{bmatrix}X_1\\X_2\end{bmatrix}$ is called a symmetric pair)if

$$X'J_{4n}X=O_{2n},\quad \text{i.e.}\quad X_1'X_2-X_2'X_1=O_{2n},$$

If, moreover, the transversality condition $|X_2|\neq 0$ holds, then $X_1X_2^{-1}=N\in Sm(2n)$ and $\begin{bmatrix}X_1\\X_2\end{bmatrix}\sim\begin{bmatrix}N\\I\end{bmatrix}$, where N is determined uniquely by the subspace $\{X\}$. Similarly, subspace $\{Y\} \subset R^{4n}$ is called $\widetilde{J}_{4n}$-Lagrangian (and $\begin{bmatrix} Y_1 \\ Y_2 \end{bmatrix}$ is called a symplectic pair) if

$$Y'\widetilde{J}_{4n}Y = O_{2n}, \quad \text{i.e.} \quad Y_1'J_{2n}Y_1 - Y_2'J_{2n}Y_2 = O_{2n}.$$

If moreover, $|Y_2| \neq 0$, then $Y_1Y_2^{-1} = M \in Sp(2n)$ and $\begin{bmatrix} Y_1 \\ Y_2 \end{bmatrix} \sim \begin{bmatrix} M \\ I \end{bmatrix}$, where M is determined uniquely by the subspace $\{Y\}$.

A $2n$-dimensional submanifold $U \subset R^{4n}$ is called J_{4n}-Lagrangian (respectively $\widetilde{J}_{4n}$-Lagrangian) if the tangent plane of U is a J_{4n}-Lagrangian (respectively $\widetilde{J}_{4n}$-Lagrangian) subspace of the tangent space at each point of U.

Let $z \to \hat{z} = g(z)$ be a canonical transformation in R^{2n}, with Jacobian $g_z = M(z) \in Sp(2n)$. The graph

$$V = \left\{ \begin{bmatrix} \hat{z} \\ z \end{bmatrix} \in R^{4n} | \hat{z} = g(z) \right\}$$

of g is a $\widetilde{J}_{4n}$-Lagrangian submanifold, whose tangent plane is spanned by the symplectic pair $\begin{bmatrix} M(z) \\ I \end{bmatrix}$.

Similarly, let $w \to \hat{w} = f(w)$ be a gradient transformation in R^{2n}, the Jacobian $f_w = N(w) \in Sm(2n)$. This is equivalent to the (local) existence of a scalar function $\phi(w)$ such that $f(w) = \phi_w(w)$. The graph

$$U = \left\{ \begin{bmatrix} \hat{w} \\ w \end{bmatrix} \in R^{4n} | \hat{w} = f(w) \right\}$$

of f is a J_{4n}-Lagrangian submanifold with tangent planes spanned by the symmetric pair $\begin{bmatrix} N(w) \\ I \end{bmatrix}$.

Theorem 3 *$T \in GL(4n)$ carries every $\widetilde{J}_{4n}$-Lagrangian submanifold into J_{4n}-Lagrangian submanifold if and only if*

$$T'J_{4n}T = \mu\widetilde{J}_{4n}, \textit{for some } \mu \neq 0,$$

i.e.

$$A_1 = -\mu^{-1}J_{2n}C', \quad B_1 = \mu^{-1}J_{2n}A',$$
$$C_1 = \mu^{-1}J_{2n}D', \quad D_1 = -\mu^{-1}J_{2n}B',$$
$$T = \begin{bmatrix} A & B \\ C & D \end{bmatrix}, \quad T^{-1} = \begin{bmatrix} A_1 & B_1 \\ C_1 & D_1 \end{bmatrix}.$$

The totality of T's in Theorem 3 will be denoted by $CSp(\widetilde{J}_{4n}, J_{4n})$, the subset with $\mu = 1$ by $Sp(\widetilde{J}_{4n}, J_{4n})$. The latter is not empty, since $\widetilde{J}_{4n}$ is congruent to J_{4n}. Fix $T_0 \in Sp(\widetilde{J}_{4n}, J_{4n})$; then every $T \in CSp(\widetilde{J}_{4n}, J_{4n})$ is a product

$$T = MT_0, \quad M \in CSp(4n) = \text{conformal symplectic group}.$$

T^{-1} for $T \in CSp(\widetilde{J}_{4n}, J_{4n})$ carries J_{4n}-Lagrangian submanifolds into $\widetilde{J}_{4n}$-Lagrangian submanifolds.

A major component of the transformation theory in symplectic geometry is the method of generating functions. Canonical transformations can in some way be expressed in implicit form, as gradient transformations with generating functions via suitable linear transformations. The graphs of canonical and gradient transformations in R^{4n} are $\widetilde{J}_{4n}$-Lagrangian and J_{4n}-Lagrangian submanifolds respectively. Theorem 3 leads to the existence and construction of the generating functions, under certain transversality conditions, for the canonical transformations.

Theorem 4 *Let* $T = \begin{bmatrix} A & B \\ C & D \end{bmatrix}, T^{-1} = \begin{bmatrix} A_1 & B_1 \\ C_1 & D_1 \end{bmatrix}, T \in CSp(\widetilde{J}_{4n}, J_{4n})$, *which define linear transformations*

$$\hat{w} = A\hat{z} + Bz, \quad \hat{z} = A_1\hat{w} + B_1 w,$$
$$w = C\hat{z} + Dz, \quad z = C_1\hat{w} + D_1 w.$$

Let $z \to \hat{z} = g(z)$ *be a canonical transformation in (some neighborhood of)* R^{2n}, *with Jacobian* $g_z = M(z) \in Sp(2n)$ *and graph*

$$V^{2n} = \left\{ \begin{bmatrix} \hat{z} \\ z \end{bmatrix} \in R^{4n} | \hat{z} - g(z) = 0 \right\}.$$

If (in some neighborhood of R^{4n}*)*

$$|CM + D| \neq 0, \tag{3.1}$$

then there exists in (some neighborhood of) R^{2n} *a gradient transformation* $w \to \hat{w} = f(w)$ *with Jacobian* $f_w = N(w) \in Sm(2n)$ *and graph*

$$U^{2n} = \left\{ \begin{bmatrix} \hat{w} \\ w \end{bmatrix} \in R^{4n} | \hat{w} - f(w) = 0 \right\}$$

and a scalar function—generating function—$\phi(w)$ *such that*

(1) $f(w) = \phi_w(w)$;

(2) $N = (AM + B)(CM + D)^{-1}, M = (NC - A)^{-1}(B - ND)$;

(3) $T(V^{2n}) = U^{2n}, V^{2n} = T^{-1}(U^{2n})$.

This corresponds to the fact that, under the transversality condition (3.1),

$$[\hat{w} - \phi_w(w)]_{\hat{w}=A\hat{z}+Bz, w=C\hat{z}+D_z} = 0$$

gives the implicit representation of the canonical transformation $\hat{z} = g(z)$ via linear transformation T and generating function ϕ.

For the time-dependent canonical transformation, related to the time-evolution of the solutions of a canonical system (1.2) with Hamiltonian function $H(z)$, we have the following general theorem on the existence and construction of the time-dependent generating function and Hamilton-Jacobi equation depending on T and H under transversality condition.

Theorem 5 *Let T be such as in Theorems 3 and 4. Let $z \to \hat{z} = g(z,t)$ be a time-dependent canonical transformation (in some neighborhood) of R^{2n} with Jacobian $g_z(z,t) = M(z,t) \in Sp(2n)$ such that*

(a) *$g(*,0)$ is a linear canonical transformation $M(z,0) = M_0$, independent of z.*

(b) *$g^{-1}(*,0)g(*,t)$ is the time-dependent canonical transformation carrying the solution $z(t)$ at moment t to $z(0)$ at moment $t = 0$ for the canonical system (1.2), so $\hat{z} = M_0 z_0$. If we have the transversality condition:*

$$|CM_0 + D| \neq 0, \tag{3.2}$$

then there exists, for sufficiently small $|t|$ and in (some neighborhood of) R^{2n}, a time-dependent gradient transformation $w \to \hat{w} = f(w,t)$ with Jacobian $f_w(w,t) = N(w,t) \in Sm(2n)$ and a time-dependent generating function $\phi(w,t)$ such that

(1) *$[\hat{w} - f(w,t)]_{\hat{w}=A\hat{z}+Bz, w=C\hat{z}+Dz} = 0$ is the implicit representation of the canonical transformation $\hat{z} = g(z,t)$;*

(2) $N = (AM + B)(CM + D)^{-1}, M = (NC - A)^{-1}(B - ND)$;

(3) $\phi_w(w,t) = f(w,t)$;

(4) $\phi_t(w,t) = -\mu H(C_1\phi_w(w,t) + D_1 w), w = C\hat{z} + Dz$.

Equation (4) is the most general Hamilton-Jacobi equation abbreviated as H.J. equation for the Hamiltonian canonical system (1.2) and linear transformation $T \in CSp(\widetilde{J}_{4n}, J_{4n})$. Here the generating function $\phi(w,t) = \phi_H(w,t)$ is determined by the Hamiltonian H and the choice of T and M_0.

Special types of generating functions:

$$\text{(I)}\quad T=\begin{bmatrix}-I_n & 0 & 0 & 0\\ 0 & 0 & I_n & 0\\ 0 & I_n & 0 & 0\\ 0 & 0 & 0 & I_n\end{bmatrix},\mu=1,M_0=J_{2n},|CM_0+D|\neq 0;$$

$$w=\begin{bmatrix}\hat{q}\\ q\end{bmatrix},\phi=\phi(\hat{q},q,t);$$

$$\hat{w}=\begin{bmatrix}-\hat{p}\\ p\end{bmatrix}=\begin{bmatrix}\phi_{\hat{q}}\\ \phi_q\end{bmatrix},\phi_t=-H(\phi_q,q).$$

This is the generating function and H. J. equation of the first kind [2].

$$\text{(II)}\quad T=\begin{bmatrix}-I_n & 0 & 0 & 0\\ 0 & 0 & 0 & -I_n\\ 0 & I_n & 0 & 0\\ 0 & 0 & I_n & 0\end{bmatrix},\mu=1,M_0=I_{2n},|CM_0+D|\neq 0;$$

$$w=\begin{bmatrix}\hat{q}\\ p\end{bmatrix},\phi=\phi(\hat{q},p,t);$$

$$\hat{w}=\begin{bmatrix}-\hat{p}\\ -q\end{bmatrix}=\begin{bmatrix}\phi_{\hat{q}}\\ \phi_p\end{bmatrix},\phi_t=-H(p,-\phi_p).$$

This is the generating function and H.J. equation of the second kind [2].

$$\text{(III)}\quad T=\begin{bmatrix}-J_{2n} & J_{2n}\\ \frac{1}{2}I_{2n} & \frac{1}{2}I_{2n}\end{bmatrix},\mu=-1,M_0=I_{2n},|CM_0+D|\neq 0;$$

$$w=\frac{1}{2}(z+\hat{z}),\phi=\phi(w,t);$$

$$\hat{w}=J(z-\hat{z})=\phi_w,\phi_t=H(w-\frac{1}{2}J\phi_w).$$

This is a new type of generating functions and H.J. equations, not encountered in the classical literature.

By recursions we can determine explicitly all possible time-dependent generating functions for analytic Hamiltonians [6].

Theorem 6 *Let $H(z)$ depend analytically on z. Then $\phi(w,t)$ in Theorem 5 is ex-*

pressible as convergent power series in t for sufficiently small $|t|$:

$$\begin{aligned}
\phi(w,t) =& \sum_{k=0}^{\infty} \phi^{(k)}(w)t^k, \\
\phi^{(0)}(w) =& \frac{1}{2}w'N_0w, \quad N_0=(AM_0+B)(CM_0+D)^{-1}, \\
\phi^{(1)}(w) =& -\mu H(E_0w), \quad E_0=(CM_0+D)^{-1}, \\
k \geqslant 1: \quad \phi^{(k+1)}(w) =& \frac{-1}{k+1}\sum_{m=1}^{k}\frac{\mu}{m!}\sum_{i_1,\cdots,i_m=1}^{2n} H_{z_{i_1}\cdots z_{i_m}}(E_0w) \\
&\times \sum_{\substack{k_1+\cdots+k_m=k \\ k_j\geqslant 1}} (C_1\phi_w^{(k_1)}(w))_{i_1}\cdots(C_1\phi_w^{(k_m)}(w))_{i_m}.
\end{aligned}$$

§4 Construction of Canonical Difference Schemes via Generating Functions

Generating functions play the central role for the construction of canonical difference schemes for Hamiltonian systems. The general methodology for the latter is as follows: Choose some suitable type of generating function (Theorem 5) with its explicit expression (Theorem 6). Truncate or approximate it in some way and take gradient of this approximate generating function. Then we get automatically the implicit representation of some canonical transformation for the transition of the difference scheme. In this way one can get an abundance of canonical difference schemes. This methodology is unconventional in the ordinary sense, but natural from the point of view of symplectic geometry [6]. As an illustration we construct a family of canonical difference schemes of arbitrary order from the truncations of the Taylor series of the generating functions for each choice of $T \in CSp(\widetilde{J}_{4n}, J_{4n})$. Since in any difference scheme the transitional transformation always tends to identity as the time step $\tau \to 0$, so we take $M_0 = I_{2n}$, $\hat{z}=z(0)=$ initial value and in Theorems 5, 6 we have specifically (3.2) in the form, transversality condition $|C+D| \neq 0$ and $N_0=(A+B)(C+D)^{-1}, E_0=(C+D)^{-1}, N_1=(C'+D')(A+B), F=C_1E_0'=\frac{1}{\mu}D'(C'+D')^{-1}$.

Theorem 7 *Using Theorems 5 and 6 with $M_0=I_{2n}$, for sufficiently small $\tau>0$ as the time-step, define*

$$\psi^{(m)}(w,\tau)=\sum_{k=0}^{m}\phi^{(k)}(w)\tau^k, \quad m=1,2,\cdots. \tag{4.1}$$

Then the gradient transformation

$$w \to \hat{w} = \psi_w^{(m)}(w,\tau) \tag{4.2}$$

with Jacobian $N^{(m)}(w,\tau)\in Sm(2m)$ *satisfies*

$$|N^{(m)}C-A|\neq 0 \tag{4.3}$$

and defines implicitly a canonical difference scheme $\hat{z}=z^k\to z^{k+1}=z$ *of m-th order accuracy upon substitution*

$$\hat{w}=Az^k+Bz^{k+1},\quad w=Cz^k+Dz^{k+1}. \tag{4.4}$$

For the special case of type (III), the generating function $\phi(w,t)$ is odd in t, so $\phi^{(2k)}(w)\equiv 0$. Then Theorem 7 leads to a family of canonical difference schemes of arbitrary even order accuracy, generalizing the centered Euler scheme.

Theorem 8 *Using Theorems* 5 *and* 6 *for the type* (III) *generating function and for sufficiently small* $\tau>0$ *as the time-step, define*

$$\Phi^{(2m)}(w,\tau)=\sum_{k=1}^{m}\phi^{2k-1}(w)\tau^{2k-1},\quad m=1,2,\cdots. \tag{4.5}$$

Then the gradient transformation

$$w\to\hat{w}=\Phi_w^{(2m)}(w,\tau) \tag{4.6}$$

with Jacobian $N^{(2m)}(w,\tau)\in Sm(2n)$ *satisfies*

$$|N^{(2m)}C-A|\neq 0$$

and defines implicitly a canonical difference scheme $\hat{z}=z^k\to z^{k+1}=z$ *of 2m-th order accuracy upon substitution* (4.4). *The case* $m=1$ *is the centered Euler scheme* (2.6).

For linear canonical system (2.1), (2.2) the type (III) generating function is the quadratic form

$$\phi(w,t)=\frac{1}{2}w'(2J\tanh\frac{\tau}{2}L)w,\quad L=J^{-1}S,\quad S'=S, \tag{4.7}$$

where

$$\tanh\lambda=\lambda-\frac{1}{3}\lambda^3+\frac{2}{15}\lambda^5-\frac{17}{312}\lambda^7+\cdots=\sum_{k=1}^{\infty}a_{2k-1}\lambda^{2k-1}.$$

$$a_{2k-1}=2^{2k}(2^{2k}-1)B_{2k}/(2k)!,\quad B_{2k}\text{ —Bernoulli numbers},$$

$$2J\tanh\frac{\tau}{2}L\in Sm(2n).$$

(4.6) becomes symplectic difference schemes

$$z^{k+1}-z^k=\left(\sum_{k=1}^{m}a_{2k-1}\left(\frac{\tau}{2}L\right)^{2k-1}\right)(z^{k+1}+z^k).$$

The case $m=1$ is the centered Euler scheme (2.3).

References

[1] Feng K. On difference schemes and symplectic geometry. In Feng K, ed. Proc 1984 Bejing Symp Diff Geometry and Diff Equations. Beijing: Science Press, 1985, 42-58

[2] Arnold V I. Mathematical Methods of Classical Mechanics. New York: Springer. 1978

[3] Siegel C L. Symplectic geometry. Amer J Math. 1943, 65: 1-86

[4] Hua L K. On the theory of automorphic functions of a matrix variable I, II. Amer J Math. 1944, 66: 470-488, 531-563

[5] Feng K, Wu H M, Qin M Z. Symplectic difference schemes for linear Hamiltonian canonical systems. J Comp Math. 1990, 8(4): 371-380

[6] Feng K, Wu H M, Qin M Z, Wang D L. Construction of canonical difference schemes for Hamiltonian formalism via generating functions. J Comp Math. 1989, 7(1): 71-96

4 Symplectic Geometry and Numerical Methods in Fluid Dynamics[①]

辛几何与流体动力学中的数值方法

§1 Introduction

It is an honor and a pleasure for me to present the inaugural talk at the Tenth International Conference on Numerical Methods in Fluid Dynamics in Beijing. I want to thank the Organizing Committee, its Secretory, Prof. H. Cabannes, the Conference Chairman, Prof. F.G. Zhuang, and the co-chairman, Prof. Y.L. Zhu for the kind invitation.

We present a brief survey of considerations and results of a study [1, 2, 3, 4, 6], undertaken by the author and his group, on the links between the *Hamiltonian formalism* and the *numerical methods* for solving dynamical problems expressed in the form of the *canonical system* of differential equations

$$\frac{dp_i}{dt} = -\frac{\partial H}{\partial q_i}, \quad \frac{dq_i}{dt} = \frac{\partial H}{\partial p_i}, \quad i = 1, \cdots, n \tag{1.1}$$

with given *Hamiltonian function* $H(p_1, \cdots, p_n, q_1, \cdots, q_n)$.

The canonical system (1.1) with remarkable elegance and symmetry was introduced by Hamilton as a general mathematical scheme, first for problems of geometrical optics in 1824, then for conservative dynamical problems in 1834. The approach was followed and developed further by Jacobi into a well-established mathematical formalism for analytical dynamics, which is an alternative of, and equivalent to, the Newtonian and Lagrangian formalisms. The geometrization of the Hamiltonian formalism was undertaken by Poincare in 1890's and by Cartan, Birkhoff, Weyl, Siegel, etc., in the 20th century; this gave rise a new discipline, called *symplectic geometry*, which serves as the mathematical foundation of the Hamiltonian formalism.

It is known that, Hamiltonian formalism, apart from its classical links with analytical mechanics, geometrical optics, calculus of variations and non-linear PDE of first

① In Zhuang F G, Zhu Y L, eds. Proc 10'th Inter Conf on Numerical Methods on Fluid Dynamics. Beijing. 1986. Lecture Notes in Physics. 1986 264: 1-7. Berlin: Speringer Verlag

order, has inherent connections also with unitary representations of Lie groups, geometric quantization, pseudo-differential and Fourier integral operators, classification of singularities, integrability of non-linear evolution equations, optimal control theory, etc.. It is also under extension to infinite dimensions for various field theories, including fluid dynamics, elasticity, electrodynamics, plasma physics, relativity, etc.. Now it is almost certain that all real physical processes with negligible dissipation can be described, in some way or other, by Hamiltonian formalism, so the latter is becoming one of the most useful tools in the mathematical arsenal of physical and engineering sciences. In this way, a systematic study of numerical methods of Hamiltonian systems is motivated and would eventually lead to more general applicability and more direct accessibility of the Hamiltonian formalism. We try to conceive, design, analyse and evaluate difference schemes and algorithms specifically within the framework of symplectic geometry. The approach proves to be quite successful as one might expect, we actually derive in this way numerous "unconventional" difference schemes. Due to historical reasons, classical symplectic geometry, however, lacks the "computational" component in the modern sense. Our present study might be considered as an attempt to fill the blank.

In the following, vectors are always represented by column matrices, matrix transpose is denoted by prime $'$. Let $z = (z_1, \cdots, z_n, z_{n+1}, \cdots, z_{2n})' = (p_1, \cdots, p_n, q_1, \cdots, q_n)'$,

$$H_z = \left[\frac{\partial H}{\partial p_1}, \cdots, \frac{\partial H}{\partial p_n}, \frac{\partial H}{\partial q_1}, \cdots, \frac{\partial H}{\partial q_n}\right]',$$

$$J_{2n} = J = \begin{bmatrix} 0 & I_n \\ -I_n & 0 \end{bmatrix}, \quad J' = J^{-1} = -J.$$

(1.1) can be written as

$$\frac{dz}{dt} = J^{-1} H_z, \tag{1.2}$$

defined in phase space R^{2n} with a standard *symplectic structure* given by the non-singular anti-symmetric closed differential 2-form

$$\omega = \sum dz_i \wedge dz_{n+i} = \sum dp_i \wedge dq_i.$$

According to *Darboux Theorem*, the symplectic structure given by any non-singular closed differential 2-form can be brought to the above standard form, at least locally, by suitable change of co-ordinates.

The *Fundamental Theorem on Hamiltonian Formalism* says that the solution $z(t)$ of the canonical system (1.2) can be generated by a *one-parameter group* $G(t)$, depending on the given Hamiltonian H, of *canonical transformations* of R^{2n} (locally in t and z) such that

$$z(t) = G(t)z(0).$$

A transformation $z \to \hat{z}$ of R^{2n} is called *canonical*, or *symplectic*, if it is a local diffeomorphism whose Jacobian $\frac{\partial \hat{z}}{\partial z} = M$ is everywhere symplectic, i.e.

$$M'JM = J, \quad \text{i.e.} \quad M \in Sp(2n).$$

The canonicity of $G(t)$ implies the preservation of 2-form ω, 4-form $\omega \wedge \omega, \cdots$, $2n$-form $\omega \wedge \omega \wedge \cdots \wedge \omega$. They constitute the class of *conservation laws of phase area* of even dimensions for the Hamiltonian system (1.2).

Moreover, the Hamiltonian system possesses another class of conservation laws related to the *energy* $H(z)$. A function $\varphi(z)$ is said to be an *invariant integral* of (1.2) if it is invariant under (1.2)

$$\varphi(z(t)) \equiv \varphi(z(0))$$

which is equivalent to

$$\{\varphi, H\} = 0,$$

where the *Poisson Bracket* for two functions $\varphi(z), \psi(z)$ are defined as

$$\{\varphi, \psi\} = \varphi_z' J^{-1} \psi_z,$$

H itself is always an invariant integral, see, e.g., [5].

The above digressions on Hamiltonian systems suggest the following *guidelines* for the numerical study of dynamical problems: The problem should be expressed in some suitable *Hamiltonian formalism*. The numerical schemes should preserve as much as possible the characteristic properties and inner symmetries of the original system. The transition from the k-th time step z^k to the next $(k+1)$-th time step z^{k+1} should be *canonical* for all k and, moreover, the invariant integrals of the original system should *remain invariant* under these transitions.

§2 Canonical Difference Schemes for Linear Canonical Systems

Consider the case for which the Hamiltonian is a quadratic form

$$H(z) = \frac{1}{2} z' S z, \quad S' = S, \quad H_z = Sz, \tag{2.1}$$

then the canonical system

$$\frac{dz}{dt} = Lz, \quad L = J^{-1}S \tag{2.2}$$

is *linear*, where L is *infinitesimally symplectic*, i.e. L satisfies $L'J + JL = 0$. The solution of (2.2) is

$$z(t) = G(t)z(0),$$

where $G(t) = \exp tL$, as the *exponential transform* of infinitesimally symplectic tL, is symplectic.

It is easily seen that the weighted Euler scheme

$$\frac{1}{\tau}(z^{k+1} - z^k) = L(\alpha z^{k+1} + (1-\alpha)z^k)$$

for the linear system (2.2) is *symplectic* if and only if $\alpha = \dfrac{1}{2}$, i.e. it is the case of *time-centered Euler Scheme* with the transition matrix F_τ,

$$z^{k+1} = F_\tau z^k, \quad F_\tau = \varphi(\tau L), \quad \varphi(\lambda) = \frac{1+\dfrac{\lambda}{2}}{1-\dfrac{\lambda}{2}}, \tag{2.3}$$

F_τ, as the *Cayley transform* of infinitesimally symplectic τL, is symplectic. The 2nd order canonical Euler scheme (2.3) can be generalized to canonical schemes of arbitrary high order [2, 3]. For example, by taking the matrix transform function $\varphi(\lambda)$ in (2.3) to be the diagonal *Padé approximants* $P_m(\lambda)/P_m(-\lambda)$ to the exponential function $\exp\lambda$, where

$$\begin{aligned}
P_0(\lambda) &= 1, \\
P_1(\lambda) &= 2 + \lambda, \\
P_2(\lambda) &= 12 + 6\lambda + \lambda^2, \\
&\cdots \\
P_m(\lambda) &= 2(2m-1)P_{m-1}(\lambda) + \lambda^2 P_{m-2}(\lambda),
\end{aligned}$$

we can prove that the difference schemes

$$z^{k+1} = \frac{P_m(\tau L)}{P_m(-\tau L)} z^k \quad m = 1, 2, \cdots \tag{2.4}$$

for the system (2.2) are symplectic, A-stable, of $2m$-th order of accuracy, and having the same set of quadratic invariant integrals including $H(z)$ as that of system (2.2). The case $m = 1$ is the time-centered Euler scheme (2.3).

For the general non-linear canonical system (1.2), the time-centered Euler scheme

$$\frac{1}{\tau}(z^{k+1} - z^k) = J^{-1}H_z\left(\frac{1}{2}(z^{k+1} + z^k)\right) \tag{2.5}$$

is *canonical*. However, unlike the linear case, the invariant integrals $\varphi(z)$ of system (1.2), including $H(z)$, are conserved only approximately

$$\varphi(z^{k+1}) - \varphi(z^k) = O(\tau^3).$$

the time-centered Euler schemes (2.3), (2.5) and their canonical generalizations (2.4) are all *implicit*. For the case of *separable* Hamiltonian

$$H(p,q) = U(p) + V(q),$$

one can construct *time-staggered* schemes which are *canonical*, of 2nd order accuracy and *practically explicit* [1, 2], e.g.

$$\begin{aligned} &\frac{1}{\tau}(p^{k+1} - p^k) = -V_q(q^{k+\frac{1}{2}}), \\ &\frac{1}{\tau}(q^{k+1+\frac{1}{2}} - q^{k+\frac{1}{2}}) = U_p(p^{k+1}). \end{aligned} \tag{2.6}$$

The $p's$ are set at integer times $t = k\tau$, $q's$ at half-integer times $t = \left(k+\frac{1}{2}\right)\tau$. We need averaging, e.g., using

$$q^k = \frac{1}{2}\left(q^{k-\frac{1}{2}} + q^{k+\frac{1}{2}}\right)$$

to compute the invariant integrals $\varphi(p,q)$ and get

$$\varphi(p^{k+1}, q^{k+1}) - \varphi(p^k, q^k) = O(\tau^3).$$

For the comparison of stability for the linear system (2.3) and the canonical schemes (2.4), (2.6) and the application of (2.6) to the wave equation, see [1].

§3 Construction of Canonical Difference Schemes via Generating Functions

A major component of the transformation theory in symplectic geometry is the method of *generating functions*, see, e.g., [5], which also plays a central role for the construction of canonical difference schemes. In [2, 4] a constructive general theory of generating functions is given, roughly as follows: Let

$$T = \begin{bmatrix} A & B \\ C & D \end{bmatrix}, \quad T^{-1} = \begin{bmatrix} A_1 & B_1 \\ C_1 & D_1 \end{bmatrix},$$

T be a non-singular real matrix of order $4n$ satisfying

$$T' = \begin{bmatrix} 0 & I_{2n} \\ -I_{2n} & 0 \end{bmatrix} T = \mu \begin{bmatrix} -J_{2n} & 0 \\ 0 & J_{2n} \end{bmatrix}, \quad \text{for some } \mu \neq 0. \tag{3.1}$$

T defines a linear transformation in product space $R^{2n} \times R^{2n}$ by

$$\begin{aligned} \hat{w} &= A\hat{z} + Bz, \\ w &= C\hat{z} + Dz, \end{aligned} \quad \begin{bmatrix} \hat{z} \\ z \end{bmatrix}, \begin{bmatrix} \hat{w} \\ w \end{bmatrix} \in R^{2n} \times R^{2n}. \tag{3.2}$$

Let $z \to \hat{z} = g(z,t)$ be a *time-dependent canonical transformation* defined by

$$g(z,t) = M_0 G(z, -t) \tag{3.3}$$

where $G(w,t)$ is the one-parameter group of canonical transformations for the canonical system (1.2) with given Hamiltonian $H(z)$; M_0 is a constant symplectic *matrix* $\in Sp(2n)$. The Jacobian

$$M(z,t) = \frac{\partial g(z,t)}{\partial z} \in Sp(2n), \quad M(z,0) = M_0.$$

If the transversality condition

$$|CM_0 + D| \neq 0 \tag{3.4}$$

holds, then there exist, for sufficiently small $|t|$ and in (some neighborhood of) R^{2n}, a *timedependent gradient transformation* $w \to \hat{w} = f(w,t)$ with Jacobian $\frac{\partial f(w,t)}{\partial w} = N(w,t) \in S_m(2n)$, i.e., everywhere *symmetric*, and a *time-dependent generating function* $\phi(w,t)$ with gradient $\phi_w(w,t) = f(w,t)$ such that

$$[\hat{w} - f(w,t)]_{\hat{w}=A\hat{z}+Bz, w=C\hat{z}+Dz} = 0$$

is an implicit representation of the canonical transformation $\hat{z} = g(z,t)$. The generating function $\phi(w,t)$ satisfies the *Hamilton-Jacobi equation*

$$\phi_t(w,t) = -\mu H(C_1\phi_w(w,t) + D_1 w), \quad w = C\hat{z} + Dz. \tag{3.5}$$

By recursions we can determine explicitly all possible time-dependent generating functions for Hamiltonians $H(z)$ analytic in z:

$$\begin{aligned} \phi(w,t) &= \sum_{k=0}^{\infty} \phi^{(k)}(w) t^k, \\ \phi^{(0)}(w) &= \frac{1}{2} w' N_0 w, \quad N_0 = (AM_0 + B)(CM_0 + D)^{-1}, \\ \phi^{(1)}(w) &= -\mu H(E_0 w), \quad E_0 = (CM_0 + D)^{-1}, \\ k \geqslant 1: \quad \phi^{(k+1)}(w) &= -\frac{1}{k+1} \sum_{m=1}^{k} \frac{\mu}{m!} \sum_{i_1,\cdots,i_m=1}^{2n} H_{z_{i_1},\cdots,z_{i_m}}(E_0 w). \end{aligned} \tag{3.6}$$

$$\cdot \sum_{\substack{k_1+\cdots+k_m=k \\ k_j \geqslant 1}} (C_1\phi_w^{(k_1)}(w))_{i_1} \cdots (C_1\phi_w^{(k_m)}(w))_{i_m}.$$

Choose

$$T = \begin{bmatrix} -I_n & 0 & 0 & 0 \\ 0 & 0 & I_n & 0 \\ 0 & I_n & 0 & 0 \\ 0 & 0 & 0 & I_n \end{bmatrix}, \quad \mu = 1, \quad M_0 = J_{2n},$$

we get the generating function of the 1st type for the case $|\frac{\partial \hat{q}}{\partial p}| \neq 0$:

$$\phi = \phi(\hat{q}, q, t), \quad -\hat{p} = \phi_{\hat{q}}, \quad p = \phi_q, \quad \phi_t + H(\phi_q, q) = 0.$$

Choose

$$T = \begin{bmatrix} -I_n & 0 & 0 & 0 \\ 0 & 0 & 0 & -I_n \\ 0 & I_n & 0 & 0 \\ 0 & 0 & I_n & 0 \end{bmatrix}, \quad \mu = 1, \quad M_0 = I_{2n},$$

we get the generating function of the 2nd type for the case $|\frac{\partial \hat{q}}{\partial q}| \neq 0$:

$$\phi = \phi(\hat{q}, p, t), \quad -\hat{p} = \phi_{\hat{q}}, \quad -q = \phi_p, \quad \phi_t + H(p, -\phi_p) = 0.$$

Choose

$$T = \begin{bmatrix} -J_{2n} & J_{2n} \\ \frac{1}{2}I_{2n} & \frac{1}{2}I_{2n} \end{bmatrix}, \quad \mu = -1, \quad M_0 = I_{2n},$$

we get the generating function of a new type — the *Euler type* —

$$\phi = \phi(w, t), \quad w = \frac{1}{2}(\hat{z} + z), \quad \hat{w} = J(\hat{z} - z) = \phi_w, \quad \phi_t - H(w - \frac{1}{2}J\phi_w) = 0.$$

The *general methodology* for the construction of canonical difference schemes is as follows: Choose some suitable type of generating function with its explicit expression (3.6) truncate or approximate it in some way and take gradient of this approximation, then we get automatically the implicit representation of some canonical transformation for the transition of the difference scheme. In this way one can get an abundance of canonical difference schemes. This methodology is *unconventional* in the ordinary sense, but natural from the point of view of symplectic geometry [4]. As an illustration we choose the *Euler type* generating function $\phi(w, t)$, which is odd in t:

Take sufficiently small $\tau > 0$ as the time-step, define

$$\psi^{(2m)}(w,\tau) = \sum_{k=1}^{m} \phi^{(2k-1)}(w)\tau^{2k-1}, \quad m = 1, 2, \cdots. \tag{3.7}$$

Then the gradient transformation

$$w \to \hat{w} = \psi_w^{(2m)}(w,\tau) \tag{3.8}$$

represents implicitly a canonical scheme $\hat{z} = z^k \to z^{k+1} = z$ of $2m$-th order accuracy upon substitution (3.2). The case $m = 1$ is the centered Euler scheme (2.5). For linear canonical system (2.2), the generating function is the quadratic form

$$\begin{aligned}
\phi(w,t) &= \frac{1}{2} w' \left(2J \tanh\left(\frac{\tau}{2}L\right)\right) w, \quad L = J^{-1}S, \quad S' = S, \\
\tanh\lambda &= \lambda - \frac{1}{3}\lambda^3 + \frac{2}{15}\lambda^5 - \frac{17}{312}\lambda^7 + \cdots = \sum_{k=1}^{\infty} a_{2k-1}\lambda^{2k-1},
\end{aligned}$$

(3.8) becomes symplectic difference schemes

$$z^{k+1} - z^k = \left(\sum_{k=1}^{m} a_{2k-1} \left(\frac{\tau}{2}L\right)^{2k-1} \right) (z^{k+1} + z^k).$$

The case $m = 1$ is the centered Euler scheme (2.3).

§4 Hamiltonian Formalism in Infinite Dimensions for Fluids

Physical processes in continuous media are dynamical systems of infinite dimensions, the corresponding symplectic geometry has not yet been fully developed theoretically. The constructive theory of generating functions [2, 4] of the 2nd type and of the Euler type and the corresponding construction of canonical difference schemes (involving time discretization only) has been generalized by Qing and Li to the case of phase space of infinite dimensions of the form $B^* \times B$, where B is a reflexive Banach space, B^* its dual [6], the Hamiltonian canonical systems are of the form

$$\frac{\partial p}{\partial t} = -\frac{\delta H}{\delta q}, \quad \frac{\partial q}{\partial t} = \frac{\delta H}{\delta p}$$

$H = H(p, q)$ is a functional, the right hand sides are variational derivatives. In case B is self-dual, the generalization is valid also for the generating functions of the 1st kind.

The problem of the Hamiltonian structure for the equations of ideal fluids has a long history, dated back to 1850's. There are several different approaches to its solution. We mention here only the oldest one — the representation of velocity by *Clebsch variables*.

Take, e.g., the case of compressible ideal fluid,

$$\vec{v} = \frac{\lambda}{\rho}\text{grad } \mu + \text{grad } \phi$$

where ρ is the density, λ, μ, ϕ are Clebsch potentials. Then the flow equations can be put in the canonical form

$$\begin{aligned}
&\frac{\partial \phi}{\partial t} = -\frac{\delta H}{\delta \rho}, \quad \frac{\partial \mu}{\partial t} = -\frac{\delta H}{\delta \lambda}, \\
&\frac{\partial \rho}{\partial t} = \frac{\delta H}{\delta \phi}, \quad \frac{\partial \lambda}{\partial t} = \frac{\delta H}{\delta \mu}, \\
&H = \int \{\frac{1}{2}\rho \vec{v}^2 + e(\rho)\} dr, \quad e(\rho) = \text{internal energy.}
\end{aligned}$$

Buneman used this formalism (modified) in computer simulation with a scheme *staggered both in space and time* and pointed out the inherent computational advantages of the canonical formalism [7].

References

[1] Feng K. On difference schemes and symplectic geometry. In Feng K, ed. Proc 1984 Beijing Symp Diff Geometry and Diff Equations. Beijing: Science Press, 1985: 42-58

[2] Feng K. Difference schemes for Hamiltonian formalism and symplectic geometry. J Comp Math. 1986, 4(3): 279-289

[3] Feng K, Wu H M, Qin M Z. Symplectic difference schemes for linear Hamiltonian canonical systems. J Comp Math. 1990, 8(4): 371-380

[4] Feng K, Wu H M, Qin M Z, Wang D L. Construction of canonical difference schemes for Hamiltonian formalism via generating functions. J Comp Math. 1989, 7(1): 71-96

[5] Arnold V I. Mathematical Methods of Classical Mechanics. New York: Springer. 1978

[6] Li C W, Qin M Z. Symplectic difference schemes for infinite dimensional Hamiltonian systems. J Comp Math. 1988, 6(2): 164-174

[7] Buneman O. Advantages of Hamiltonian Formulations in Computer Simulations. In Tabor, ed. Mathematical Methods in Hydrodynamics and Integrability of Dynamical Systems. Amer Inst Phys. USA. 1981

5 Canonical Difference Schemes for Hamiltonian Canonical Differential Equations①

哈密尔顿典则方程的典则差分格式

The present program[1] that the author and his group have started is a systematic study, within the framework of symplectic geometry, of the numerical methods for the solution of differential equations of mathematical physics expressed in Hamiltonian formalism. As is well known, Hamiltonian canonical systems serve as the basic mathematical formalism, for diverse areas of physics, mechanics, engineering, as well as pure and applied mathematics, e.g., geometrical optics, analytical dynamics, non-linear PDE's of first order, group representations, WKB asymptotics, pseudodifferential and Fourier integral operators, electrodynamics, plasma physics, elasticity, hydrodynamics, relativity, control theory, etc. It is generally accepted that all real physical processes with negligible dissipation could be expressed, in some way or other, in suitable Hamiltonian forms.

We consider the canonical system in finite dimensions

$$\frac{dp_i}{dt} = -H_{q_i}, \quad \frac{dq_i}{dt} = H_{p_i}, \quad i = 1, 2, \cdots, n, \tag{1}$$

with Hamiltonian $H(p_1, \cdots, p_n, q_1, \cdots, q_n,)$.

In the following, vectors are always represented by column matrices, matrix transpose is denoted by a prime $'$. Let $z = (z_1, \cdots, z_n, z_{n+1}, \cdots, z_{2n})' = (p_1, \cdots, p_n, q_1, \cdots, q_n)'$, (1) can be written as

$$\frac{dz}{dt} = J^{-1}H_z, \quad J = \begin{pmatrix} 0 & I_n \\ -I_n & 0 \end{pmatrix} \tag{2}$$

defined in phase space R^{2n} with a standard symplectic structure given by the non-singular anti-symmetric closed differential 2-form $\omega = \sum dz_i \wedge dz_{n+i} = \sum dp_i \wedge dq_i$. The Fundamental Theorem on Hamiltonian Formalism says that the solution of the canonical system (2) can be generated by a one-parameter group G_t of canonical transformations

① In Proc Inter Workshop on Applied Diff Equations. Singapore: World Scientific. 1986: 59-73

of R^{2n} (locally in t and z) such that

$$G_{t_1}G_{t_2} = G_{t_1+t_2}$$
$$z(t) = G_t z(0).$$

A transformation $z \to \hat{z}$ of R^{2n} is called canonical if it is a local diffeomorphism whose Jacobian $\frac{\partial \hat{z}}{\partial z} = M$ is symplectic everywhere, i.e.

$$M'JM = J, \quad \text{i.e.} \quad M \in Sp(2n).$$

Linear canonical transformations are simply symplectic transformations.

The canonicity of G_t implies the preservation of 2-form ω, 4-form $\omega \wedge \omega, \cdots$, $2n$-form $\omega \wedge \omega \cdots \wedge \omega$. They constitute the class of conservation laws of phase area of even dimensions for the Hamiltonian system (2).

Moreover, the Hamiltonian systems possesses another class of conservation laws related to the energy $H(z)$. A function $\phi(z)$ is said to be an invariant integral of (2) if it is invariant under (2)

$$\phi(z(t)) \equiv \phi(z(0))$$

which is equivalent to

$$\{\phi, H\} = 0,$$

where the Poisson Bracket for two functions $\phi(z), \psi(z)$ are defined as

$$\{\phi, \varphi\} = \phi_z' J^{-1} \varphi_z.$$

H itself is always an invariant integral. See, e.g., Refs. 5 and 6.

For the numerical study, we are less interested in (2) as a general system of ODE per se, but rather as a specific system with Hamiltonian structure. It is natural to look for those discretization systems which preserve as much as possible the characteristic properties and inner symmetries of the original continuous systems. To this end the transition $\hat{z} \to z$ from the k-th time step $z^k = \hat{z}$ to the next $(k+1)$-th time step $z^{k+1} = z$ should be canonical for all k and, moreover, the invariant integrals of the original system should remain invariant under these transitions. We try to conceive, design, analyse and evaluate difference schemes and algorithms specifically within the framework of symplectic geometry. The approach proves to be quite successful as one might expect, we actually derive in this way numerous "unconventional" difference schemes.

We give a brief survey of some of our preliminary results[1-4].

Consider at first the case for which the Hamiltonian is a quadratic form

$$H(z) = \frac{1}{2}z'Sz, \quad S' = S, \quad H_z = S(z). \tag{3}$$

Then the canonical system is linear

$$\frac{dz}{dt} = Bz, \tag{4}$$

where $B = J^{-1}S$ is infinitesimally symplectic, i.e.

$$B'J + JB = 0.$$

The solution is

$$z(t) = G_t z(0),$$

where $G(t) = \exp tB$, as the exponential transform of infinitesimally symplectic tB, is symplectic.

Theorem 0 *The weighted Euler scheme*

$$\frac{z^{k+1} - z^k}{\tau} = B(\alpha z^{k+1} + (1-\alpha)z^k) \tag{5}$$

for the linear system (4) *is symplectic if and only if* $\alpha = 1/2$, *i.e. it is the case of time-centered Euler Scheme with the transition.*

For the time-centered case, the transition matrix

$$z^{k+1} = F_\tau z^k, \quad F_\tau = \phi(\tau B), \quad \phi(\lambda) = \frac{1 + \dfrac{\lambda}{2}}{1 - \dfrac{\lambda}{2}} \tag{6}$$

F_τ, as Cayley transform of infinitesimally symplectic τB, is symplectic.

In order to generalize the time-centered Euler scheme, we need, apart from the exponential or Cayley transforms, other matrix transforms carrying infinitesimally symplectic matrices into symplectic ones.

Theorem 1 *Let* $\psi(\lambda)$ *be a function of complex variable* λ *satisfying*

(I) $\psi(\lambda)$ *is analytic with real coefficients in a neighborhood* D *of* $\lambda = 0$,

(II) $\psi(\lambda)\psi(-\lambda) \equiv 1$ *in* D,

(III) $\psi(0) \neq 0$.

A *is a matrix of order* $2n$, *then* $(\psi(\tau B))'A\psi(\tau B) = A$ *for all* τ *with sufficiently small* $|\tau|$, *if and only if*

$$B'A + AB = 0.$$

If, moreover, $\exp\lambda - \psi(\lambda) = O(|\lambda|^{m+1})$, *then*

$$z^{k+1} = \psi(\tau B)z^k, \tag{7}$$

considered as an approximative scheme for the canonical system (4), *is symplectic, of m-th order of accuracy and has the property that the bilinear form* $z'Aw$ *is invariant under* $\psi(\tau B)$ *if and only if it is invariant under* G_τ *of* (4).

Remark 1 The last property is remarkable in the sense that, all the bilinear invariants of the system (4), no more and no less, are kept invariant under the scheme (7), in spite of the fact that the latter is only approximate.

Remark 2 The approximative scheme in Theorem 1 becomes difference schemes only when $\psi(\lambda)$ is a rational function. As a concrete application for the construction of symplectic difference schemes, take the diagonal Padé approximants to the exponential function

$$\exp\lambda - \frac{P_m(\lambda)}{P_m(-\lambda)} = O(|\lambda|^{2m+1})$$

Theorem 2 *The difference schemes*

$$z^{k+1} = \frac{P_m(\tau B)}{P_m(-\tau B)} z^k, \quad m = 1, 2, \cdots \tag{8}$$

for the system (4) *are symplectic, A-stable, of 2m-th order of accuracy, and having the same set of bilinear invariants as that of system* (4). *The case* $m = 1$ *is the centered Euler scheme.*

For the general non-linear canonical system (2), the time-centered Euler scheme is

$$\frac{1}{\tau}(z^{k+1} - z^k) = J^{-1}H_z\left(\frac{1}{2}(z^{k+1} + z^k)\right). \tag{9}$$

The transition $z^{k+1} \to z^k$ is canonical with Jacobian

$$F_\tau = \left[I - \frac{\tau}{2}J^{-1}H_{zz}\left(\frac{1}{2}(z^{k+1} + z^k)\right)\right]^{-1}\left[I + \frac{\tau}{2}J^{-1}H_{zz}\left(\frac{1}{2}(z^{k+1} + z^k)\right)\right]$$

symplectic everywhere. However, unlike the linear case the invariant integrals $\varphi(z)$ of system (2), including $H(z)$, are conserved only approximately

$$\varphi(z^{k+1}) - \varphi(z^k) = O(\tau^3).$$

For non-linear separable system for which $H(p, q) = U(p) + V(q)$ where U, V are kinetic and potential energies respectively, we define the staggered scheme

$$\begin{aligned} &\frac{1}{\tau}\left(p^{k+1} - p^k\right) = -V_q\left(q^{k+\frac{1}{2}}\right), \\ &\frac{1}{\tau}\left(q^{k+1+\frac{1}{2}} - q^{k+\frac{1}{2}}\right) = U_p\left(p^{k+1}\right). \end{aligned} \tag{10}$$

The $p's$ are set at integer times $t = k\tau$, $q's$ at half-integer times $t = (k+\frac{1}{2})\tau$. The transition

$$w^k = \begin{pmatrix} p^k \\ q^{k+\frac{1}{2}} \end{pmatrix} \to \begin{pmatrix} p^{k+1} \\ q^{k+1+\frac{1}{2}} \end{pmatrix} = w^{k+1} = F_\tau w^k,$$

$$F_\tau = \begin{pmatrix} I & 0 \\ -\tau M & I \end{pmatrix}^{-1} \begin{pmatrix} I & -\tau L \\ 0 & I \end{pmatrix},$$

$$M = U_{pp}(p^{k+1}), \quad L = V_{qq}(q^{k+\frac{1}{2}}),$$

is symplectic, of 2nd order of accuracy and practically explicit. Since p, q are computed at different times, we need synchronization, e.g., using

$$q^k = \frac{1}{2}(q^{k-\frac{1}{2}} + q^{k+\frac{1}{2}})$$

to compute the invariant integrals $\varphi(p, q)$,

$$\varphi(p^{k+1}, q^{k+1}) - \varphi(p^k, q^k) = O(\tau^3).$$

For the comparison of stability for the linear system (4) with separable Hamiltonian with the canonical schemes in Theorems 1, 2 the staggered scheme and the application of the latter to the wave equation see Ref 1. In Ref 1, a class of energy conservative schemes was constructed using the differencing of the Hamiltonian function, the symplectic property is not satisfactory. The problem of compatibility of energy conservation with phase area conservation in difference schemes is solved successfully for linear canonical systems (Theorem 1), it seems to be difficult, however, for the general non-linear systems.

In order to develop a general method of construction of canonical difference schemes we first give a constructive generalization of the classical theory of generating functions and Hamilton-Jacobi equations. Our approach is in part inspired by the early works of Siegel[7] and Hua[8]. Every $4n \times 2n$ matrix

$$A = \begin{pmatrix} A_1 \\ A_2 \end{pmatrix} \in M(4n, 2n), \quad A_1, A_2 \in M(2n), \quad \text{rank} \quad A = 2n$$

defines in R^{4n} a $2n$-dimensional subspace $\{A\}$ spanned by its column vectors. $\{A\} = \{B\}$ if and only if $A \sim B$, i.e.

$$AP = B, \quad \text{i.e.,} \quad \begin{pmatrix} A_1 P \\ A_2 P \end{pmatrix} = \begin{pmatrix} B_1 \\ B_2 \end{pmatrix}, \quad \text{for some } P \in GL(2n).$$

The spaces of symmetric and symplectic matrices of order $2n$ will be denoted by $Sm(2n)$, $Sp(2n)$ resp. Let

$$J_{4n} = \begin{pmatrix} 0 & I_{2n} \\ -I_{2n} & 0 \end{pmatrix}, \quad \tilde{J}_{4n} = \begin{pmatrix} -J_{2n} & 0 \\ 0 & +J_{2n} \end{pmatrix}$$

$$X = \begin{pmatrix} X_1 \\ X_2 \end{pmatrix}, \quad Y = \begin{pmatrix} Y_1 \\ Y_2 \end{pmatrix} \in M(4n, 2n), \quad \text{of rank} \quad 2n.$$

Subspace $\{X\} \subset R^{4n}$ is called J_{4n}-Lagrangian (and $\begin{pmatrix} X_1 \\ X_2 \end{pmatrix}$, is called a symmetric pair) if

$$X' J_{4n} X = 0_{2n}, \quad \text{i.e.} \quad X_1' X_2 - X_2' X_1 = 0_{2n}.$$

If, moreover $|X_2| \neq 0$, then $X_1 X_2^{-1} = N \in Sm(2n)$ and $\begin{pmatrix} X_1 \\ X_2 \end{pmatrix} \sim \begin{pmatrix} N \\ I \end{pmatrix}$, N is determined uniquely by the subspace $\{X\}$. Similarly, subspace $\{Y\} \subset R^{4n}$ is called $\tilde{J}_{4n}$-Lagrangian (and $\begin{pmatrix} Y_1 \\ Y_2 \end{pmatrix}$, is called a symplectic pair) if

$$Y' \tilde{J}_{4n} Y = 0_{2n}, \quad \text{i.e.} \quad Y_1' J_{2n} Y_1 - Y_2' J_{2n} Y_2 = 0_{2n}.$$

If, moreover, $|Y_2| \neq 0$, then $Y_1 Y_2^{-1} = N \in Sp(2n)$ and $\begin{pmatrix} Y_1 \\ Y_2 \end{pmatrix} \sim \begin{pmatrix} M \\ I \end{pmatrix}$, M is determined uniquely by the subspace $\{Y\}$.

A $2n$-dimensional submanifold $U \subset R^{4n}$ is called J_{4n}-Lagrangian (resp. $\tilde{J}_{4n}$-Lagrangian) if the tangent plane of U is a J_{4n}-Lagrangian (resp. $\tilde{J}_{4n}$-Lagrangian) subspace of the tangent space at each point of U.

Let $z \to \hat{z} = g(z)$ be a canonical transformation in R^{2n}, with Jacobian $g_z = M(z) \in Sp(2n)$. The graph of g,

$$V = \left\{ \begin{pmatrix} \hat{z} \\ z \end{pmatrix} \in R^{4n} | \hat{z} = g(z) \right\},$$

is a $\tilde{J}_{4n}$-Lagrangian submanifold, whose tangent plane is spanned by the symplectic pair $\begin{pmatrix} M(z) \\ I \end{pmatrix}$.

Similarly, let $w \to \hat{w} = f(w)$ be a gradient transformation in R^{2n}, the Jacobian $f_w = N(w) \in Sm(2n)$. This is equivalent to the (local) existence of a scalar function $\phi(w)$ such that $f(w) = \phi_w(w)$. The graph of f,

$$U = \left\{ \begin{pmatrix} \hat{w} \\ w \end{pmatrix} \in R^{4n} | \hat{w} = f(w) \right\},$$

is a J_{4n}-Lagrangian submanifold, whose tangent plane is spanned by the symmetric pair $\begin{pmatrix} N(w) \\ I \end{pmatrix}$. The following algebraic theorem is crucial for our construction.

Theorem 3 *$T \in GL(4n)$ carries every $\tilde{J}_{4n}$-Lagrangian submanifolds into J_{4n}-Lagrangian submanifolds if and only if*

$$T' J_{4n} T = \mu \tilde{J}_{4n}, \quad for\ some \quad \mu \neq 0,$$

i.e.

$$A_1 = -\mu^{-1} J_{2n} C', \quad B_1 = \mu^{-1} J_{2n} A', \quad T = \begin{pmatrix} A & B \\ C & D \end{pmatrix},$$

$$T^{-1} = \begin{pmatrix} A_1 & B_1 \\ C_1 & D_1 \end{pmatrix}, \quad C_1 = \mu^{-1} J_{2n} D', \quad D_1 = -\mu^{-1} J_{2n} B'.$$

The totality of $T's$ in Theorem 3 will be denoted by $CSp(\tilde{J}_{4n}, J_{4n})$. Canonical transformations can in some way be expressed in implicit form, as gradient transformations with generating functions via suitable linear transformations. The graphs of canonical and gradient transformations in R^{4n} are $\tilde{J}_{4n}$-Lagrangian and J_{4n}-Lagrangian submanifolds respectively. Theorem 3 leads to the existence and construction of the generating functions, under certain transversality condition, for the canonical transformations.

Theorem 4 *Let $T = \begin{pmatrix} A & B \\ C & D \end{pmatrix}$, $T^{-1} = \begin{pmatrix} A_1 & B_1 \\ C_1 & D_1 \end{pmatrix}$, $T \in CSp(\tilde{J}_{4n}, J_{4n})$, they define linear transformations*

$$\hat{w} = A\hat{z} + Bz, \quad \hat{z} = A_1 \hat{w} + B_1 w,$$

$$w = C\hat{z} + Dz, \quad z = C_1 \hat{w} + D_1 w.$$

Let $z \to \hat{z} = g(z)$ be a canonical transformation in (some neighborhood of) R^{2n}, with Jacobian $g_z = M(z) \in Sp(2n)$ and graph

$$V^{2n} = \left\{ \begin{pmatrix} \hat{z} \\ z \end{pmatrix} \in R^{4n} | \hat{z} - g(z) = 0 \right\}.$$

If (in some neighborhood of R^{4n}) the transversality condition

$$|CM + D| \neq 0, \tag{11}$$

holds, then there exists in (some neighborhood of) R^{4n} a gradient transformation $w \to \hat{w} = f(w)$ with Jacobian $f_w = N(w) \subset Sm(2n)$ and graph

$$U^{2n} = \left\{ \begin{pmatrix} \hat{w} \\ w \end{pmatrix} \in R^{4n} | \hat{w} - f(w) = 0 \right\}$$

and a scalar function—generating function—$\phi(w)$ *such that*

(1) $f(w) = \phi_w(w)$,

(2) $N = (AM + B)(CM + D)^{-1}, M = (NC - A)^{-1}(B - ND)$,

(3) $T(V^{2n}) = U^{2n}, V^{2n} = T^{-1}(U^{2n})$.

This corresponds to the fact that, under the transversality condition (11)

$$[\hat{w} - \phi_w(w)]_{\hat{w}=A\hat{z}+Bz, w=C\hat{z}+Dz=0}$$

gives the implicit representation of the canonical transformation $\hat{z} = g(z)$ *via linear transformation* T *and generating function* ϕ.

For the time-dependent canonical transformation, related to the time-evolution of the solutions of a canonical system (2) with Hamiltonian function $H(z)$, we have the following general theorem on the existence and construction of time-dependent generating function and Hamilton-Jacobi equation depending on T and H under the transversality condition.

Theorem 5 *Let* T *be such as in Theorems* 3 *and* 4. *Let* $z \to \hat{z} = g(z,t)$ *be a time-dependent canonical transformation (in some neighborhood) of* R^{2n} *with Jacobian* $g_z(z,t) = M(z,t) \in Sp(2n)$ *such that*

(a) $g(*,0)$ *is a linear canonical transformation,* $M(z,0) = M_0$ *be independent of* z.

(b) $g^{-1}(*,0)g(*,t)$ *is the time-dependent canonical transformation carrying the solution* $z(t)$ *at moment to* $z(0)$ *at moment* $t = 0$ *for the canonical system* (2). *If*

$$|CM_0 + D| \neq 0.$$

then for sufficiently small $|t|$, *the transversality condition* (11) *holds and there exists in (some neighborhood of)* R^{2n} *a time-dependent gradient transformation* $w \to \hat{w} = f(w,t)$ *with Jacobian* $f_w(w,t) = N(w,t) \in Sm(2n)$ *and a time-dependent generating function* $\phi(w,t)$ *such that*

(1) $[\hat{w} - f(w,t)]_{\hat{w}=A\hat{z}+Bz, w=C\hat{z}+Dz} = 0$ *is the implicit representation of the canonical transformation* $\hat{z} = g(z,t)$

(2) $N = (AM + B)(CM + D)^{-1}, M = (NC - A)^{-1}(B - ND)$

(3) $\phi_w(w,t) = f(w,t)$

(4) $\phi_t(w,t) = -\mu H(C_1\phi_w(w,t) + D_1 w), w = C\hat{z} + D_z$.

The last equation is the most general Hamilton-Jacobi equation for the canonical system with Hamiltonian $H(z)$ *and linear transformation* $T \in CS_p(\tilde{J}_{4n}, J_{4n})$.

By recursions we can determine explicitly the power series expansions for all possible time-dependent generating functions for analytic Hamiltonians.

Theorem 6 *Let $H(z)$ depends analytically on z, then $\phi(w,t)$ in Theorem 5 is expressible as convergent power series in t for sufficiently small $|t|$:*

$$\begin{aligned}
\phi(w,t) &= \sum_{k=0}^{\infty} \phi^{(k)}(w)t^k, \\
\phi^{(0)}(w) &= \frac{1}{2}w' N_0 w, \quad N_0 = (AM_0 + B)(CM_0 + D)^{-1} \\
\phi^{(1)}(w) &= -\mu H(E_0 w), \quad E_0 = (CM_0 + D)^{-1} \\
\phi^{(k+1)}(w) &= -\frac{\mu}{k+1}\sum_{m=1}^{k}\frac{1}{m!}\sum_{i_1,\cdots,i_m=1}^{2n} H_{z_{i_1},\cdots,z_{i_m}}(E_0 w) \\
&\quad \times \sum_{\substack{k_1+\cdots+k_m=k \\ k_j \geqslant 1}} \left(C_1\phi_w^{(k_1)}(w)\right)_{i_1}\cdots\left(C_1\phi_w^{(k_m)}(w)\right)_{i_m}, \quad k \geqslant 1.
\end{aligned}$$

Examples of special types of generating functions:

$$\text{(I)} \quad T = \begin{pmatrix} -I_n & 0 & 0 & 0 \\ 0 & 0 & I_n & 0 \\ 0 & I_n & 0 & 0 \\ 0 & 0 & 0 & I_n \end{pmatrix}, \quad \mu = +1, \quad M_0 = J_{2n}, \quad |CM_0 + D| \neq 0$$

$$w = \begin{pmatrix} \hat{q} \\ q \end{pmatrix}, \quad \phi = \phi(\hat{q}, q, t),$$

$$\hat{w} = \begin{pmatrix} -\hat{p} \\ p \end{pmatrix} = \begin{pmatrix} \phi_{\hat{q}} \\ \phi_q \end{pmatrix}, \quad \phi_t = -H(\phi_q, q).$$

This is the generating function and H.J. equation of the first kind for the “free” canonical transformations[5] .

$$\text{(II)} \quad T = \begin{pmatrix} I_n & 0 & 0 & 0 \\ 0 & 0 & 0 & -I_n \\ 0 & I_n & 0 & 0 \\ 0 & 0 & I_n & 0 \end{pmatrix}, \quad \mu = +1, \quad M_0 = I_{2n}, \quad |CM_0 + D| \neq 0$$

$$w = \begin{pmatrix} \hat{q} \\ p \end{pmatrix}, \quad \phi = \phi(\hat{q}, p, t),$$

$$w = \begin{pmatrix} -\hat{p} \\ -q \end{pmatrix} = \begin{pmatrix} \phi_{\hat{q}} \\ \phi_p \end{pmatrix}, \quad \phi_t = -H(p, -\phi_p).$$

This is the generating function and H.J. equation of the second kind[5].

$$(\text{III})\quad T=\begin{pmatrix} J_{2n} & -J_{2n} \\ \frac{1}{2}I_{2n} & \frac{1}{2}I_{2n} \end{pmatrix},\quad \mu=+1,\quad M_0=I_{2n},\quad |CM_0+D|\neq 0$$

$$w=\begin{pmatrix} \frac{1}{2}(\hat{p}+p) \\ \frac{1}{2}(\hat{q}+q) \end{pmatrix}=\begin{pmatrix} \bar{p} \\ \bar{p} \end{pmatrix}=\text{mean values},\quad \phi=\phi(\bar{p},\bar{q},t),$$

$$\hat{w}=\begin{pmatrix} q-\hat{q} \\ \hat{p}-p \end{pmatrix}=\begin{pmatrix} \phi_{\bar{p}} \\ \phi_{\bar{q}} \end{pmatrix},\quad \phi_t=-H(\bar{p}+\frac{1}{2}\phi_{\bar{q}},\bar{q}-\frac{1}{2}\phi_{\bar{p}})$$

This is a new type of generating functions and H.J. equations, not encountered in the classical literature. $\phi(\bar{p},\bar{q},t)$ is an odd function in t.

Generating functions play the central role for the construction of canonical difference schemes for Hamiltonian systems. The general methodology for the latter is as follows: Choose some suitable type of generating function (Theorem 5) with its explicit expression (Theorem 6). Truncate it or approximate it in some way, take the gradient of this approximate generating function, then we get automatically the implicit representation of some canonical transformation for the transition of the difference scheme. In this way one can get an abundance of canonical difference schemes. This methodology is unconventional in the ordinary sense, but natural from the point of view of symplectic geometry. As an illustration we construct a generalization of centered Euler scheme with arbitrarily high order of accuracy.

Theorem 7 *Choose the generating function of Example* (III) *with its m-th truncation*

$$\phi(\bar{p},\bar{q},\tau)=\sum_{k=0}^{\infty}\phi^{(2k+1)}(\bar{p},\bar{q},)\tau^{2k+1},$$

$$\phi_m(\bar{p},\bar{q},\tau)=\sum_{k=0}^{m-1}\phi^{(2k+1)}(\bar{p},\bar{q},)\tau^{2k+1},\quad m=1,2,\cdots,$$

with $\tau>0$ *small enough as the time-step,* $\phi(\tau)-\phi_m(\tau)=O(\tau^{2n})$. *Let* $\hat{z}=z^{k+1}, z=z^k$, $\bar{z}=\frac{1}{2}(z^k+z^{k+1})$, *then*

$$\hat{q}-q=\sum_{k=0}^{m-1}\phi_{\bar{p}}^{(2k+1)}(\bar{p},\bar{q})\tau^{2k+1},$$

$$p-\hat{p}=\sum_{k=0}^{m-1}\phi_{\bar{q}}^{(2k+1)}(\bar{p},\bar{q})\tau^{2k+1}$$

is a canonical difference scheme of $2m$*-th order of accuracy,* $m=1$ *is the centered Euler scheme* (9).

References

[1] Feng K. On difference schemes and symplectic geometry. In Feng K, ed. Proc 1984 Beijing Symp Diff Geometry and Diff Equations. Beijing: Science Press, 1985: 42-58

[2] Feng K. Difference schemes for Hamiltonian formalism and symplectic geometry. J Comp Math. 1986, 4(3): 279-289

[3] Feng K, Wu H M, Qin M Z. Symplectic difference schemes for linear Hamiltonian canonical systems. J Comp Math. 1990, 8(4): 371-380

[4] Feng K, Wu H M, Qin M Z, Wang D L. Construction of canonical difference schemes for Hamiltonian formalism via generating functions. J Comp Math. 1989, 7(1); 71-96

[5] Arnold V I. Mathematical Methods of Classical Mechanics. New York: Springer. 1978

[6] Abrabam R, Marsden J. Foundations of Mechanics. 2nd ed. Mass: Addison-Wesley, Reading. 1978

[7] Siegel C L. Symplectic geometry. Amer J Math. 1943, 65: 1-86

[8] Hua L K. On the theory of automorphic functions of a matrix variable I, II. Amer J Math. 1944, 66: 470-488, 531-563

6 The Symplectic Methods for the Computation of Hamiltonian Equations①②

计算哈密尔顿方程的辛方法

Abstract

The present paper gives a brief survey of results from a systematic study, undertaken by the authors and their colleagues, on the symplectic approach to the numerical computation of Hamiltonian dynamical systems in finite and infinite dimensions. Both theoretical and practical aspects of the symplectic methods are considered. Almost all the real conservative physical processes can be cast in suitable Hamiltonian formulation in phase spaces with symplectic structure, which has the advantages to make the intrinsic properties and symmetries of the underlying processes more explicit than in other mathematically equivalent formulations, so we choose the Hamiltonian formalism as the basis, together with the mathematical and physical motivations of our symplectic approach for the purpose of numerical simulation of dynamical evolutions. We give some symplectic difference schemes and related general concepts for linear and nonlinear canonical systems in finite dimensions. The analysis confirms the expectation for them to behave more satisfactorily, especially in the desirable conservation properties, than the conventional schemes. We outline a general and constructive theory of generating functions and a general method of construction of symplectic difference schemes based on all possible generating functions. This is crucial for the developments of the symplectic methods. A generalization of the above theory and method to the canonical Hamiltonian eqs. in infinite dimensions is also given. The multi-level schemes, including the leapfrog one, are studied from the symplectic point of view. We give an application of symplectic schemes, with some indications of their potential usefulness, to the computation of chaos.

CONTENTS

---

① Joint with Qin M Z

② In Zhu Y L, Guo Ben-yu, ed. Proc Conf on Numerical Methods for PDE's. Berlin: Springer, 1987. 1-37. Lect Notes in Math 1297

§1 Introduction

Recently it is evident that Hamiltonian formalism plays a fundamental role in the diverse areas of physics, mechanics, engineering, pure and applied mathematics, e.g. geometrical optics, analytical dynamics, nonlinear PDE's of first order, group representations, WKB asymptotics, pseudodifferential and Fourier integral operators, classification of singularities, integrability of non-linear evolution equations, optimal control theory, etc. It is also under extension to infinite dimensions for various field theories, including electrodynamics, plasma physics, elasticity, hydrodynamics etc. It is generally accepted that all real physical processes with negligible dissipation could be expressed, in some way or other, by Hamiltonian formalism, so the latter is becoming one of the most useful tools in the mathematical physics and engineering sciences.

Hamiltonian formalism has the important property of being area-preserving (symplectic) i.e. the sum of the areas of canonical variable pairs, projected on any two-dimensional surface in phase space, is time invariant. In numerically solving these equations one hopes that the approximating equation will hold this property.

In DD-5 Beijing Conference the first author [1] propose an approach for computing Hamiltonian equation from the viewpoint of symplectic geometry. This paper is a brief survey of considerations and developments [1-11, 15], obtained by the first author and his group, on the links between the Hamiltonian formalism and the numerical methods.

Now we will give a review of some facts from Hamiltonian mechanics which are fundamental to what follows. We consider the following canonical system of ordinary first order differential equations on R^{2n}

$$\frac{dp_i}{dt}=-\frac{\partial H}{\partial q_i},\quad \frac{dq_i}{dt}=\frac{\partial H}{\partial p_i},\quad i=1,2,\cdots,n, \tag{1.1}$$

where $H(p,q)$ is some real valued function. We call (1.1) a Hamiltonian system of differential equations (H-system). In the following, vectors are always represented by column matrices, matrix transpose is denoted by prime $'$. Let $z=(z_1,\cdots,z_n,z_{n+1},\cdots,z_{2n})'=(p_1,\cdots,p_n,q_1,\cdots,q_n)'$, $H_z=\left[\frac{\partial H}{\partial p_1},\cdots,\frac{\partial H}{\partial p_n},\frac{\partial H}{\partial q_1},\cdots,\frac{\partial H}{\partial q_n}\right]'$, $J_{2n}=J=\begin{bmatrix}\mathbf{0} & I_n\\ -I_n & 0\end{bmatrix}$, $J'=J^{-1}=-J$, where I_n is the $n\times n$ identity matrix. (1.1) can be written as

$$\frac{dz}{dt}=J^{-1}H_z, \tag{1.2}$$

defined in phase space R^{2n} with a standard symplectic structure given by the nonsingular anti-symmetric closed differential 2-form

$$\omega = \sum dz_i \wedge dz_{n+i} = \sum dp_i \wedge dq_i.$$

According to Darboux Theorem, the symplectic structure given by any non-singular anti-symmetric closed differential 2-form can be brought to the above standard form, at least locally, by suitable change of coordinates.

The right side of equation (1.2) gives a vector field. At each point (p, q) of the phase space, there is a $2n$-dimensional vector $(-H_q, H_p)$.

The fundamental theorem on Hamiltonian Formalism says that the solution $z(t)$ of the canonical system (1.2) can be generated by a one-parameter group $G(t)$, depending on given Hamiltonian H, of canonical transformations of R^{2n} (locally in t and z) such that

$$z(t) = G(t)z(0).$$

This group is also called the phase flow

$$G(t) : (p(0), q(0)) \to (p(t), q(t))$$

where $p(t)$, $q(t)$ are the solution of Hamilton's system of equations (1.1).

A transformation $z \to \hat{z}$ of R^{2n} is called canonical, or symplectic, if it is a local diffeomorphism whose Jacobian $\dfrac{\partial \hat{z}}{\partial z} = M$ is every-where symplectic, i.e.

$$M'JM = J, \quad \text{i.e.} \quad M \in Sp(2n).$$

Linear canonical transformation is special symplectic transformation.

The canonicity of $G(t)$ implies the preservation of 2-form ω, 4-form $\omega \wedge \omega, \cdots$, $2n$-form $\omega \wedge \omega \wedge \cdots \wedge \omega$. They constitute the class of conservation laws of phase area of even dimensions for the Hamiltonian system (1.2).

Moreover, the Hamiltonian system possesses another class of conservation laws related to the energy $H(z)$. A function $\psi(z)$ is said to be an invariant integral of (1.2) if it is invariant under (1.2)

$$\psi(z(t)) \equiv \psi(z(0))$$

which is equivalent to

$$\{\psi, H\} = 0,$$

where the Poisson Brackets for any pair of differentiable functions ϕ and ψ are defined as

$$\{\phi, \psi\} = \phi_z' J^{-1} \psi_z.$$

H itself is always an invariant integral, see, e.g. [12].

For the numerical study, we are less interested in (1.2) as a general system of ODE per se, but rather as a specific system with Hamiltonian structure. It is natural to look for those discretization systems which preserve as much as possible the characteristic properties and inner symmetries of the original continuous systems. To this end the transition $\hat{z} \to z$ from the k-th time step $z^k = \hat{z}$ to the next $(k+l)$-th time step $z^{k+1} = z$ should be canonical for all k and, moreover, the invariant integrals of the original system should remain invariant under these transitions.

Thus, a difference scheme may be regarded as a transformation from time t^k to time t^{k+1}. We have the following

Definition 1 A difference scheme may be called symplectic or canonical scheme if its transitional transformation is symplectic.

We try to conceive, design, analyse and evaluate difference schemes and algorithms specifically within the framework of symplectic geometry. The approach proves to be quite successful as one might expect, we actually derive in this way numerous "unconventional" difference schemes.

An outline of the paper is as follows. In section 2 we review some symplectic difference schemes (S-schemes) for linear Hamiltonian system (LH-system) and nonlinear Hamiltonian system (NLH-system) and its related properties are given. In section 3, we systematically outline the general method of construction of S-schemes with any order accuracy via generating function. The constructive theory of generating function and the corresponding construction of S-schemes have been generalized to the case of phase space of infinite dimensions of the form $B^\star \times B$, where B is a reflexive Banach space, B^* its dual [3] [8]. Section 4 contains the main idea. The multi-level difference S-schemes of Hamiltonian type are described in §5. In §6 we show some computational results and comparison with R-K method. The last section is S-schemes and chaos. It is well known that canonical transformation is an area preserving mapping. Therefore S-schemes are suitable tools for studying chaotic behavior in Hamiltonian mechanics.

§2 S-schemes for Linear and Nonlinear Hamiltonian Systems

Consider the case for which the Hamiltonian is a quadratic form

$$H(z) = \frac{1}{2} z' S z, \quad S' = S, \quad H_z = Sz. \tag{2.1}$$

Then the canonical system

$$\frac{dz}{dt} = Lz, \quad L = J^{-1} S \tag{2.2}$$

is linear, where L is infinitesimally symplectic, i.e. L satisfies $L'J + JL = 0$.

The solution of (2.2) is

$$z(t) = G(t)z(0)$$

where $G(t) = \exp tL$, as the exponential transform of infinitesimally symplectic tL, is symplectic.

It is easily seen that the weighted Euler scheme

$$\frac{1}{\tau}\left(z^{k+1} - z^k\right) = L\left(\alpha z^{k+1} + (1-\alpha)z^k\right)$$

for the linear system (2.2) is symplectic iff $\alpha = \frac{1}{2}$, i.e. it is the case of time centered Euler scheme with the transition matrix F_τ,

$$z^{k+1} = F_\tau z^k, \quad F_\tau = \psi(\tau L), \quad \psi(\lambda) = \frac{1+\frac{\lambda}{2}}{1-\frac{\lambda}{2}}, \tag{2.3}$$

F_τ, as the Cayley transform of infinitesimally symplectic τL, is symplectic.

In order to generalize the time-centered Euler scheme, we need, apart from the exponential or Cayley transforms, other matrix transforms carrying infinitesimally symplectic matrices into symplectic ones.

Theorem 1 *Let $\psi(\lambda)$ be a function of complex variable λ satisfying*

(I) *$\psi(\lambda)$ is analytic with real coefficients in a neighborhood D of $\lambda = 0$,*

(II) *$\psi(\lambda)\psi(-\lambda) \equiv 1$ in D,*

(III) *$\psi'_\lambda(0) \neq 0$.*

A is a matrix of order $2n$, then $(\psi(\tau L))'A\psi(\tau L) = A$ for all τ with sufficiently small $|\tau|$, iff

$$L'A + AL = 0.$$

If, more, $\exp\lambda - \psi(\lambda) = O\left(|\lambda|^{m+1}\right)$, then

$$z^{k+1} = \psi(\tau L)z^k, \tag{2.4}$$

considered as an approximative scheme for the canonical system (2.2), is symplectic, of m-th order of accuracy and has the property that the bilinear form $z'Aw$ is invariant under $\psi(\tau L)$ iff it is invariant under G_t of (2.2).

Remark 1 The last property is remarkable in the sense that, all the bilinear invariants of the system (2.2), no more and no less, are kept invariant under the scheme (2.4), in spite of the fact that the latter is only approximate.

Remark 2 The approximative scheme in Theorem 1 becomes difference schemes only when $\psi(\lambda)$ is a rational function. As a concrete application to the construction of symplectic difference schemes, we take the diagonal Padè approximants to the exponential function

$$\exp \lambda - \frac{P_m(\lambda)}{P_m(-\lambda)} = O\left(|\lambda|^{2m+1}\right)$$

where $P_0(\lambda) = 1, P_1(\lambda) = 2+\lambda, P_2(\lambda) = 12+16\lambda+\lambda^2, \cdots, P_m(\lambda) = 2(2m-1)P_{m-1}(\lambda)+\lambda^2 P_{m-2}(\lambda)$.

Theorem 1′ *Difference schemes*

$$z^{k+1} = \frac{P_m(\tau L)}{P_m(-\tau L)} z^k, \quad m = 1, 2, \cdots \tag{2.5}$$

for the eq. (2.2) *are symplectic, A-stable, of* 2*m-th order of accuracy, and having the same set of bilinear invariants as that of eq.* (2.2), *and the case* $m = 1$ *is the centered Euler scheme* [5].

For the general non-linear canonical system (1.2), the time-centered Euler scheme is

$$\frac{1}{\tau}\left(z^{k+1} - z^k\right) = J^{-1} H_z\left(\frac{1}{2}\left(z^{k+1} + z^k\right)\right). \tag{2.6}$$

The transition $z^{k+1} \to z^k$ is canonical with Jacobian

$$F_\tau = \left[I - \frac{\tau}{2} J^{-1} H_{zz}\left(\frac{1}{2}\left(z^{k+1} + z^k\right)\right)\right]^{-1}\left[I + \frac{\tau}{2} J^{-1} H_{zz}\left(\frac{1}{2}\left(z^{k+1} + z^k\right)\right)\right]$$

symplectic everywhere. However, unlike the linear case, the invariant integrals ψ, not quadratic in z, including $H(z)$, are conserved only approximately:

$$\psi\left(z^{k+1}\right) - \psi\left(z^k\right) = O\left(\tau^3\right).$$

For the general non-linear separable system for which $H(p, q) = U(p) + V(q), U, V$, being kinetic and potential energies respectively, we define the staggered scheme

$$\begin{aligned} \frac{1}{\tau}\left(p^{k+1} - p^k\right) &= -V_q\left(q^{k+\frac{1}{2}}\right), \\ \frac{1}{\tau}\left(q^{k+1+\frac{1}{2}} - q^{k+\frac{1}{2}}\right) &= U_p\left(p^{k+1}\right). \end{aligned} \tag{2.7}$$

The p's are set at integer times $t = k\tau$, q's at half-integer times $t = \left(k + \frac{1}{2}\right)\tau$. The transition

$$w^k = \begin{bmatrix} p^k \\ q^{k+\frac{1}{2}} \end{bmatrix} \to \begin{bmatrix} p^{k+1} \\ q^{k+1+\frac{1}{2}} \end{bmatrix} = w^{k+1} = F_\tau w^k,$$

$$F_\tau = \begin{bmatrix} I & 0 \\ -\tau M & I \end{bmatrix}^{-1} \begin{bmatrix} I & -\tau L \\ 0 & I \end{bmatrix},$$

$$M = U_{pp}\left(p^{k+1}\right), L = V_{qq}\left(q^{k+\frac{1}{2}}\right),$$

is symplectic of 2nd order accuracy and practically explicit, since p, q are computed at different times, we have to use certain synchronization, e.g. using

$$q^k = \frac{1}{2}\left(q^{k-\frac{1}{2}} + q^{k+\frac{1}{2}}\right)$$

to compute the invariant integrals $\psi(p, q)$

$$\psi\left(p^{k+1}, q^{k+1}\right) - \psi\left(p^k, q^k\right) = O\left(\tau^3\right).$$

In [1, 9], a class of energy conservative schemes was constructed using the differencing of the Hamiltonian function, however, the symplectic property was not satisfactory. The problem of compatibility of energy conservation with phase area conservative in difference schemes is solved successfully for linear canonical system (Theorem 1). It seems difficult, however, for the general nonlinear system.

Suppose that the Hamiltonian function $H(p, q)$ does not depend on time

$$H + h = 0, \tag{2.8}$$

where h is a constant, is the integral of energy of the system.

We assume that (in some region) the equation (2.8) can be solved for some variable (no loss of generality) for example p_1, $H_{p_1} = \dfrac{dq_1}{dt} \neq 0$, so (2.8) is solved by

$$K\left(p_2, p_3, \cdots, p_n, q_1, \cdots, q_n, h\right) + p_1 = 0 \tag{2.9}$$

and $t \to q_1(t)$ has an inverse $q_1 \to t\left(q_1\right)$, so we can use q_1 instead of t as "time". Equations (1.1) can be written as

$$\frac{dq_r}{dq_1} = \frac{\partial K}{\partial p_r}, \quad \frac{dp_r}{dq_1} = -\frac{\partial K}{\partial q_r}, \quad (r = 2, 3, \cdots, n), \tag{2.10}$$

$$\frac{dt}{dq_1} = \frac{\partial K}{\partial h}, \quad \frac{dh}{dq_1} = 0 = -\frac{\partial K}{\partial t}. \tag{2.11}$$

The last two equations can be separated from the rest of the system, since the first $(2n-2)$ equations do not involve t, and h is a constant. The phase trajectories of (1.1) on the energy surface $H(z)+h=0$ satisfy (2.10). Moreover (2.11), (2.10) combine into a new H-system $dw/dq_1 = J^{-1}K_w$ in "time" q_1 in the modified phase space with new canonical variables $w=(h,p_2,\cdots,p_n;t,q_2,\cdots,q_n)$ with "time"-dependent Hamiltonian $K(w;q_1)$. Since K is independent of $w_{n+1}=t$, its conjugate variable $w_1=h$ is constant. We use the Euler scheme

$$\frac{1}{\Delta q_1}\left(w^{m+1}-w^m\right)=J^{-1}K_w\left(\frac{1}{2}\left(w^{m+1}+w^m\right);\frac{1}{2}\left(q_1^{m+1}+q_1^m\right)\right),\quad q_1^{m+1}=q_1^m+\Delta q_1.$$

The transition $w^m\rightarrow w^{m+1}$ is symplectic and preserves $w_1=h$, i.e. $h^m=h^{m+1}$. We compute, in addition, $K^m=K\left(w^m;q_1^m\right), p_1^m=-K^m$. Owing to the identity

$$H\left(-K\left(p_2,\cdots,p_n,q_1,q_2,\cdots,q_n,h\right),p_2,\cdots,p_n,q_1,\cdots,q_n\right)+h\equiv 0,$$

we have $H^m = H\left(p_1^m,p_2^m,\cdots,q_n^m\right) = -h^m = -h^{m+1} = H^{m+1}$. So we get energy conservative scheme which is also symplectic in a modified sense. Here the computed time steps $t^{m+1}-t^m$ are in general variable under the fixed "time" step Δq_1. This agrees with an idea of T.D. Lee [16], where time steps are to be solved to make energy conservative. The above approach to make S-schemes energy conservative is due to Qin Meng-zhao.

The problem of preservation of first integrals of system (1.2) under a S-scheme

$$z^{m+1}=S_H^{\tau}\left(z^m\right),$$

where S_H^{τ} is a symplectic transformation depending on $H(z)$ and step τ, is closely related to the invariance properties of S_H^{τ} under groups of symplectic transformations [7]. The scheme S_H^{τ} is said to be invariant under a group G of symplectic transformations if

$$g^{-1}\circ S_H^{\tau}\circ g=S_{H\circ g},\quad \forall g\in G.$$

Theorem 2 *Let F be a first integral of system $dz/dt=J^{-1}H_z$. Then F is preserved up to a constant by the scheme S_H^{τ}, i.e.*

$$F(z)\equiv F\left(S_H^{\tau}(z)\right)+c,\quad c=\ \textit{const}.$$

iff S_H^{τ} is invariant under the 1-parameter group G_F^t of the phase flow of the system $dz/dt=J^{-1}\ F_z$. The constant $c=0$ if S_H^{τ} has a fixed point.

It is known that all linear first integrals (e.g., linear momenta) of H-system are preserved by any compatible difference scheme, symplectic or not. However for quadratic first integrals (e.g., angular momenta), this is by far not case. In this aspect the symplectic schemes are distinguished as shown above at least for linear H-systems.

For general non-linear H-systems and for symplectic Euler schemes of arbitrary $2m$-th order (the case $m=1$ is (2.6), for higher order schemes see Theorem 8, §3), Ge Zhong, Wang Dao-liu, and Wu Yu-hua have proved the physically significant property of preservation of all the quadratic first integrals. This is easily seen for the case $m=1$: Let $F(z)=\frac{1}{2}z'Bz$, $B'=B$, be a first integral of (1.2), then $\{f,H\}=(Bz)'J^{-1}H_z(z)=0$ for all z. Multiply (2.6) by $\left(B\frac{1}{2}\left(z^{k+1}+z^k\right)\right)'$, we get $\frac{1}{\tau}\frac{1}{2}\left(z^{k+1}+z^k\right)'B\left(z^{k+1}-z^k\right)=0$, then $F\left(z^{k+1}\right)=F\left(z^k\right)$.

§3 Constructive Theory of Generating Functions and S-schemes

In this section we reproduce almost verbally the results from [4]. The generalization to Poisson maps was done in [11]. The generalization to infinite dimensions was done in [8] and will be outlined in §4.

In order to develop a general method of construction of canonical difference schemes we first give a constructive generalization of the classical theory of generating function and Hamilton-Jacobi equations. Our approach in this part was inspired by the early works of Siegel [13] and Hua [14]. Every matrix

$$A=\begin{bmatrix}A_1\\A_2\end{bmatrix}\in M(4n,2n),\quad A_1,A_2\in M(2n),\quad \operatorname{rank}A=2n$$

defines in R^{4n} a $2n$-dimensional subspace $\{A\}$ spanned by its column vectors. $\{A\}=\{B\}$ iff $A\sim B$, i.e.

$$AP=B,\quad \text{i.e.}\quad \begin{bmatrix}A_1P\\A_2P\end{bmatrix}=\begin{bmatrix}B_1\\B_2\end{bmatrix},\quad \text{for some } P\in GL(2n).$$

The spaces of symmetric and symplectic matrices of order $2n$ will be denoted by $Sm(2n)$, $Sp(2n)$ respectively. Let

$$J_{4n}=\begin{bmatrix}0&I_{2n}\\-I_{2n}&0\end{bmatrix},\quad \widetilde{J}_{4n}=\begin{bmatrix}-J_{2n}&0\\0&J_{2n}\end{bmatrix}$$

$$X=\begin{bmatrix}X_1\\X_2\end{bmatrix},\quad Y=\begin{bmatrix}Y_1\\Y_2\end{bmatrix}\in M(4n,2n),\ \text{of rank } 2n.$$

Subspace $\{X\}\subset R^{4n}$ is called J_{4n}-Lagrangian (and $\begin{bmatrix}X_1\\X_2\end{bmatrix}$ is called a symmetric pair) if

$$X'J_{4n}X=O_{2n},\quad \text{i.e.}\quad X_1'X_2-X_2'X_1=O_{2n}.$$

If, moreover, $|X_2| \neq 0$, then $X_1 X_2^{-1} = N \in Sm(2n)$ and $\begin{bmatrix} X_1 \\ X_2 \end{bmatrix} \sim \begin{bmatrix} N \\ I \end{bmatrix}$, where N is determined uniquely by the subspace $\{X\}$. Similarly, subspace $\{Y\} \subset R^{4n}$ is called $\widetilde{J}_{4n}$-Lagrangian $\left(\text{and } \begin{bmatrix} Y_1 \\ Y_2 \end{bmatrix} \text{ is called a symplectic pair}\right)$ if

$$Y' \widetilde{J}_{4n} Y = O_{2n}, \text{ i.e.,} \quad Y_1' J_{2n} Y_1 - Y_2' J_{2n} Y_2 = O_{2n}.$$

If, moreover, $|Y_2| \neq 0$, then $Y_1 Y_2^{-1} = M \in Sp(2n)$ and $\begin{bmatrix} Y_1 \\ Y_2 \end{bmatrix} \sim \begin{bmatrix} M \\ I \end{bmatrix}$, where M is determined uniquely by the subspace $\{Y\}$.

A $2n$-dimensional submanifold $U \subset R^{4n}$ is called J_{4n}-Lagrangian (respectively $\widetilde{J}_{4n}$-Lagrangian) if the tangent plane of U is a J_{4n}-Lagrangian (respectively $\widetilde{J}_{4n}$-Lagrangian) subspace of the tangent space at each point of U.

Let $z \to \hat{z} = g(z)$ be a canonical transformation in R^{2n}, with Jacobian $g_z = M(z) \in Sp(2n)$. The graph of g,

$$V = \left\{ \begin{bmatrix} \hat{z} \\ z \end{bmatrix} \in R^{4n} | \hat{z} = g(z) \right\},$$

is a $\widetilde{J}_{4n}$-Lagrangian submanifold, whose tangent plane is spanned by the symplectic pair $\begin{bmatrix} M(z) \\ I \end{bmatrix}$.

Similarly, let $w \to \hat{w} = f(w)$ be a gradient transformation in R^{2n}, the Jacobian $f_w = N(w) \in Sm(2n)$. This is equivalent to the (local) existence of a scalar function $\phi(w)$ such that $f(w) = \phi_w(w)$. The graph of f,

$$U = \left\{ \begin{bmatrix} \hat{w} \\ w \end{bmatrix} \in R^{4n} | \hat{w} = f(w) \right\},$$

is a J_{4n}-Lagrangian submanifold with tangent planes spanned by the symmetric pair $\begin{bmatrix} N(w) \\ I \end{bmatrix}$.

Theorem 3 *$T \in GL(4n)$ carries every J_{4n}-Lagrangian submanifold into J_{4n}-Lagrangian submanifold if and only if*

$$T' J_{4n} T = \mu \widetilde{J}_{4n}, \quad \textit{for some } \mu \neq 0,$$

i.e.

$$\begin{aligned} A_1 &= -\mu^{-1} J_{2n} C', \quad B_1 = \mu^{-1} J_{2n} A', \\ C_1 &= \mu^{-1} J_{2n} D', \quad D_1 = -\mu^{-1} J_{2n} B', \\ A &= \begin{bmatrix} A & B \\ C & D \end{bmatrix}, \quad T^{-1} = \begin{bmatrix} A_1 & B_1 \\ C_1 & D_1 \end{bmatrix}. \end{aligned} \tag{3.1}$$

The totality of T's in Theorem 3 will be denoted by $CSp\left(\tilde{J}_{4n}, J_{4n}\right)$, the subset with $\mu = 1$ by $Sp\left(\tilde{J}_{4n}, J_{4n}\right)$. The latter is not empty since $\tilde{J}_{4n}$ is congruent to J_{4n}. Fix $T_0 \in Sp\left(\tilde{J}_{4n}, J_{4n}\right)$; then every $T \in CSp\left(\tilde{J}_{4n}, J_{4n}\right)$ is a product

$$T = MT_0, \quad M \in CSp(4n) = \text{ conformal symplectic group.}$$

T^{-1} for $T \in CSp\left(\widetilde{J}_{4n}, J_{4n}\right)$ carries J_{4n}-Lagrangian submanifolds into $\widetilde{J}_{4n}$-Lagrangian submanifolds.

A major component of the transformation theory in symplectic geometry is the method of generating functions. Canonical transformations can in some way be expressed in implicit form, as gradient transformations with generating functions via suitable linear transformations. The graphs of canonical and gradient transformations in R^{4n} are $\widetilde{J}_{4n}$-Lagrangian and J_{4n}-Lagrangian submanifolds respectively. Theorem 3 leads to the existence and construction of the generating functions, under some transversality conditions, for the canonical transformations.

Theorem 4 *Let* $T = \begin{bmatrix} A & B \\ C & D \end{bmatrix}, T^{-1} = \begin{bmatrix} A_1 & B_1 \\ C_1 & D_1 \end{bmatrix}, T \in CSp\left(\widetilde{J}_{4n}, J_{4n}\right)$, *which define linear transformations*

$$\begin{aligned} \hat{w} &= A\hat{z} + Bz, \; \hat{z} = A_1 \hat{w} + B_1 w, \\ w &= C\hat{z} + Cz, \; z = C_1 \hat{w} + D_1 w. \end{aligned}$$

Let $z \to \hat{z} = g(z)$ *be a canonical transformation in (some neighborhood of)* R^{2n}, *with Jacobian* $g_z = M(z) \in Sp(2n)$ *and graph*

$$V^{2n} = \left\{ \begin{bmatrix} \hat{z} \\ z \end{bmatrix} \in R^{4n} | \hat{z} - g(z) = 0 \right\}.$$

If (in some neighborhood of R^{4n}*)*

$$|CM + D| \neq 0, \tag{3.2}$$

then there exists in (some neighborhood of) R^{2n} a gradient transformation $w \to \hat{w} = f(w)$ with Jacobian $f_w = N(w) \in Sm(2n)$ and graph

$$U^{2n} = \left\{ \begin{bmatrix} \hat{w} \\ w \end{bmatrix} \in R^{4n} | \hat{w} - f(w) = 0 \right\}$$

and a scalar function—generating function—$\phi(w)$ such that

(1) $f(w) = \phi_w(w)$;

(2) $N = (AM + B)(CM + D)^{-1}, M = (NC - A)^{-1}(B - ND)$;

(3) $T(V^{2n}) = U^{2n}, V^{2n} = T^{-1}(U^{2n})$.

This corresponds to the fact that, under the transversality condition (3.2),

$$[\hat{w} - \phi_w(w)]_{\hat{w}=A\hat{z}+Bz, w=C\hat{z}+Dz} = 0$$

gives the implicit representation of the canonical transformation $\hat{z} = g(z)$ via linear transformation T and generating function ϕ.

For the time-dependent canonical transformation, related to the time-evolution of the solutions of a canonical system (1.2) with Hamiltonian function $H(z)$, we have the following general theorem on the existence and construction of the time-dependent generating function and Hamilton-Jacobi equation depending on T and H under some transversality condition.

Theorem 5 *Let T be such as in Theorems 3 and 4. Let $z \to \hat{z} = g(z, t)$ be a time-dependent canonical transformation (in some neighborhood) of R^{2n} with Jacobian $g_z(z, t) = M(z, t) \in Sp(2n)$ such that*

(a) *$g(\star, 0)$ is a linear canonical transformation $M(z, 0) = M_0$, independent of z,*

(b) *$g^{-1}(\star, 0)g(\star, t)$ is the time-dependent canonical transformation carrying the solution $z(t)$ at moment t to $z(0)$ at moment $t = 0$ for the canonical system. If*

$$|CM_0 + D| \neq 0, \tag{3.3}$$

then there exists, for sufficiently small $|t|$ and in (some neighborhood of) R^{2n}, a time-dependent gradient transformation $w \to \hat{w} = f(w, t)$ with Jacobian $f_w(w, t) = N(w, t) \in Sm(2n)$ and a time-dependent generating function $\phi(w, t)$ such that

(1) *$[\hat{w} - f(w, t)]_{\hat{w}=A\hat{z}+Bz, w=C\hat{z}+Dz} = 0$ is the implicit representation of the canonical transformation $\hat{z} = g(z, t)$;*

(2) $N = (AM + B)(CM + D)^{-1}, M = (NC - A)^{-1}(B - ND)$;

(3) $\phi_w(w, t) = f(w, t)$;

(4) $\phi_t(w, t) = -\mu H(C_1\phi_w(w, t) + D_1 w), w = C\hat{z} + Dz$.

Equation (4) is the most general Hamilton-Jacobi equation abbreviated as H.J. equation for the Hamiltonian canonical system (1.2) and linear transformation $T \in CSp\left(\tilde{J}_{4n}, J_{4n}\right)$.

Special types of generating functions:

$$(\mathrm{I}) \quad \mathrm{T} = \begin{bmatrix} -I_n & 0 & 0 & 0 \\ 0 & 0 & I_n & 0 \\ 0 & I_n & 0 & 0 \\ 0 & 0 & 0 & I_n \end{bmatrix}, \ \mu = 1, \ M_0 = J_{2n}, \ |CM_0 + D| \neq 0;$$

$$w = \begin{bmatrix} \hat{q} \\ q \end{bmatrix}, \quad \phi = \phi(\hat{q}, q, t);$$

$$\hat{w} = \begin{bmatrix} -\hat{p} \\ p \end{bmatrix} = \begin{bmatrix} \phi_{\hat{q}} \\ \phi_q \end{bmatrix}, \quad \phi_t = -H\left(\phi_q, q\right)$$

are the generating function and H-J equation of the first kind [12].

$$(\mathrm{II}) \quad T = \begin{bmatrix} -I_n & 0 & 0 & 0 \\ 0 & 0 & 0 & -I_n \\ 0 & I_n & 0 & 0 \\ 0 & 0 & I_n & 0 \end{bmatrix}, \ \mu = 1, \ M_0 = I_{2n}, \ |CM_0 + D| \neq 0;$$

$$w = \begin{bmatrix} \hat{q} \\ p \end{bmatrix}, \quad \phi = \phi(\hat{q}, p, t);$$

$$\hat{w} = \begin{bmatrix} -\hat{p} \\ -q \end{bmatrix} = \begin{bmatrix} \phi_{\hat{q}} \\ \phi_p \end{bmatrix}, \quad \phi_t = -H\left(p, -\phi_p\right)$$

are the generating function and H-J equation of the second kind [2].

$$(\mathrm{III}) \quad T = \begin{bmatrix} -J_{2n} & J_{2n} \\ \frac{1}{2}I_{2n} & \frac{1}{2}I_{2n} \end{bmatrix}, \ \mu = -1, \ M_0 = I_{2n}, \ |CM_0 + D| \neq 0;$$

$$w = \frac{1}{2}(z + \hat{z}), \quad \phi = \phi'(w, t);$$

$$\hat{w} = J(z - \hat{z}) = \phi_w, \quad \phi_t = H\left(w - \frac{1}{2}J\phi_w\right)$$

are a new type of generating functions and H-J equations, not encountered in the classical literature.

By recursions we can determine explicitly all possible time-dependent generating functions for analytic Hamiltonians [6].

Theorem 6 *Let $H(z)$ depend analytically on z. Then $\phi(w,t)$ in Theorem 5 can be* *expressed as convergent power series in t for sufficiently small $|t|$:*

$$\begin{aligned}
&\phi(w,t) = \sum_{k=0}^{\infty} \phi^{(k)}(w)t^k, \\
&\phi^{(0)}(w) = \frac{1}{2}w'N_0 w, \quad N_0 = (AM_0 + B)(CM_0 + D)^{-1}, \\
&\phi^{(1)}(w) = -\mu H(E_0 w), \quad E_0 = (CM_0 + D)^{-1}, \\
k \geqslant 1: \quad &\phi^{(k+1)}(w) = \frac{-\mu}{k+1} \sum_{m=1}^{k} \frac{1}{m!} \sum_{i_1,\cdots,i_m=1}^{2n} H_{z_{i_1},\cdots,z_{i_m}}(E_0 w) \\
&\times \sum_{\substack{k_1+\cdots+k_m=k \\ k_j \geqslant 1}} \left(C_1 \phi_w^{(k_1)}(w)\right)_{i_1} \cdots \left(C_1 \phi_w^{(k_m)}(w)\right)_{i_m}.
\end{aligned} \tag{3.4}$$

Generating functions play the central role for the construction of canonical difference schemes for Hamiltonian systems. The general methodology for the latter is as follows: Choose some suitable type of generating function (Theorem 5) with its explicit expression (Theorem 6). Truncate or approximate it in some way and take gradient of this approximate generating function. Then we get automatically the implicit representation of some canonical transformation for the transition of the difference scheme. In this way one can get an abundance of canonical difference schemes. This methodology is unconventional in the ordinary sense, but natural from the point of view of symplectic geometry [6]. As an illustration we construct a family of canonical difference schemes of arbitrary order from the truncations of the Taylor series of the generating functions for each choice of $T \in CSp\left(\tilde{J}_{4n}, J_{4n}\right)$ and $M_0 \in Sp(2n)$ satisfying (3.3).

Theorem 7 *Using Theorems 5 and 6, for sufficiently small $\tau > 0$ as the time-step, we define*

$$\psi^{(m)}(w,\tau) = \sum_{k=0}^{m} \phi^{(k)}(w)\tau^k, \quad m = 1, 2, \cdots. \tag{3.5}$$

Then the gradient transformation

$$w \to \hat{w} = \psi_w^{(m)}(w,\tau) \tag{3.6}$$

with Jacobian $N^{(m)}(w,\tau) \in Sm(2m)$ satisfies

$$\left|N^{(m)}C - A\right| \neq 0 \tag{3.7}$$

and defines implicitly a canonical difference scheme $\hat{z} = z^k \to z^{k+1} = z$ of m-th order accuracy upon substitution

$$\hat{w} = Az^k + Bz^{k+1}, \quad w = Cz^k + Dz^{k+1}. \tag{3.8}$$

For the special case of type (Ⅲ), the generating function $\phi(w,t)$ is odd in t. Then Theorem 7 leads to a family of canonical difference schemes of arbitrary even order accuracy, generalizing the centered Euler scheme, as follows.

Theorem 8 *Using Theorems* 5 *and* 6, *for sufficiently small* $\tau>0$ *as the time-step, we define*

$$\psi^{(2m)}(w,\tau)=\sum_{k=1}^{m}\phi^{(2k-1)}(w)\tau^{2k-1},\quad m=1,2,\cdots. \tag{3.9}$$

Then the gradient transformation

$$w\to\hat{w}=\psi_w^{(2m)}(w,\tau) \tag{3.10}$$

with Jacobian $N^{(2m)}(w,\tau)\in Sm(2n)$ *satisfies*

$$\left|N^{(2m)}C-A\right|\neq 0$$

and defines implicitly a canonical difference scheme $\hat{z}=z^k\to z^{k+1}=z$ *of* $2m$*-th order accuracy upon substitution* (3.8). *The case* $m=1$ *is the centered Euler scheme* (2.6).

For linear canonical system (2.1), (2.2) the type (Ⅲ) generating function is the quadratic form

$$\phi(w,\tau)=\frac{1}{2}w'\left(2J\tanh\left(\frac{\tau}{2}L\right)\right), L=J^{-1}S, S'=S, \tag{3.11}$$

where

$$\tanh\lambda=\lambda-\frac{1}{3}\lambda^3+\frac{2}{15}\lambda^5-\frac{17}{312}\lambda^7+\cdots=\sum_{k=1}^{\infty}a_{2k-1}\lambda^{2k-1}.$$

$$a_{2k-1}=2^{2k}\left(2^{2k}-1\right)B_{2n}/(2k)!,\quad B_{2k}\text{ —Bernoulli numbers,}$$

$$2J\tanh\left(\frac{\tau}{2}L\right)\in Sm(2n).$$

(3.10) becomes symplectic difference schemes

$$z^{k+1}-z^k=\left(\sum_{k=1}^{m}a_{2k-1}\left(\frac{\tau}{2}L\right)^{2k-1}\right)\left(z^{k+1}+z^k\right). \tag{3.12}$$

The case $m=1$ is the centered Euler scheme (2.3).

Several specific S-schemes up to fourth order accuracy are given in Appendix 1.

§4 S-schemes for Infinite Dimensional Hamiltonian Systems

4.1 An infinite dimensional Hamiltonian equation

Suppose B is a reflexive Banach space and $B^\star$ its dual. E^n is an Euclidean space and n its dimension. The generalized coordinate in the Banach space is function $q(r,t)$:

$E^n \times R \to R, \forall t \in R$. We have $q(r,t) \in B$. B corresponds to the configuration space. We introduce $p(r,t)$, generalized momentum, where $r \in E^n, t \in R$. For $\forall t \in R, p(r,t) \in B^\star$. $B^\star$ corresponds to momentum space. $B \times B^\star$ is the phase space.

Let H be a functional in Hamiltonian mechanics. We have the Hamiltonian equation in $B \times B^\star$:

$$\begin{aligned} \frac{dp}{dt} &= -\frac{\delta H}{\delta q}(q,p,t), \\ \frac{dq}{dt} &= \frac{\delta H}{\delta p}(q,p,t), \end{aligned} \tag{4.1}$$

where $\dfrac{\delta}{\delta q}, \dfrac{\delta}{\delta p}$ are functional derivatives.

We denote $z = (q,p) \in B \times B^\star$, then we get formally the form of (4.1)

$$\frac{dz}{dt} = \begin{bmatrix} 0 & I \\ -I & 0 \end{bmatrix} H_z.$$

We define the operator J as

$$\begin{aligned} J: \quad & B \times B^\star \to (B \times B^\star)^\star = B^\star \times B \\ & J(q,p) = (-p,q) \end{aligned}$$

then it is easy to prove that $J \in GL\left((B \times B^\star), (B \times B^\star)^\star\right)$.

$$J = \begin{bmatrix} 0 & -I_{B^\star} \\ I_B & 0 \end{bmatrix}, \quad J^{-1} = \begin{bmatrix} 0 & I_B \\ -I_{B^\star} & 0 \end{bmatrix},$$

$$I_B : B \to B \text{ identity operator,}$$

$$I_{B^\star} : B^\star \to B^\star \text{ identity operator,}$$

we write them briefly as I.

Then the H-system has the form

$$\frac{dz}{dt} = J^{-1} H_z, \tag{4.2}$$

where $H_z = \left(\dfrac{\delta H}{\delta q}, \dfrac{\delta H}{\delta p}\right)$, with canonical 2-form:

$$\omega(\alpha_1, \alpha_2) = \langle J\alpha_1, \alpha_2 \rangle$$

where $\alpha_1, \alpha_2 \in B \times B^\star$, and $\langle , \rangle$ is the dual product.

4.2 Generating functional and difference schemes

Suppose $f : B \times B^\star \to (B \times B^\star)^\star$

$$\hat{w} = f(w).$$

We call f a gradient transformation, or potential operator, if there is $\phi : B \times B^\star \to R$ such that:

$$\hat{w} = \phi_w(w).$$

Let $g : B \times B^\star \to B \times B^\star$ be a canonical transformation

$$z \to \hat{z} = g(z)$$

$T \in GL\left((B \times B^\star) \times (B \times B^\star), (B \times B^\star)^\star \times (B \times B^\star)\right)$

$$T = \begin{bmatrix} A & B \\ C & D \end{bmatrix}$$

in which $A, B \in GL\left(B \times B^\star, (B \times B^\star)^\star\right), C, D \in GL(B \times B^\star, B \times B^\star)$.

Lemma 1 ([17]). *If the following conditions are satisfied:*

1. *F is an operator: $E \to E^\star, E$ is a reflexive Banach space,*

2. *F has Gateaux differential $DF(x, h)$ in U which belongs to E and is defined by $|x - x_0| < r$,*

3. *functional $\langle DF(x, h_1), h_2\rangle$ is continuous in U, then F is a potential operator iff $\langle DF(x, h_1), h_2\rangle$ is symmetric for $\forall x \in U$:*

$$\langle DF(x, h_1), h_2\rangle = \langle DF(x, h_2), h_1\rangle, \quad \forall h_1, h_2 \in E.$$

Theorem 3′ *T carries every symplectic transformation into gradient transformation, if:*

$$T^\star \begin{bmatrix} 0 & I \\ -I & 0 \end{bmatrix} T = \mu \begin{bmatrix} -J & 0 \\ 0 & J \end{bmatrix}$$

i.e.,

$$\begin{aligned} &A_1 = -\mu^{-1} J C^\star, \quad B_1 = \mu^{-1} J A^\star, \\ &C_1 = \mu^{-1} J A^\star, \quad D_1 = -\mu^{-1} J B^\star, \\ &T = \begin{bmatrix} A & B \\ C & D \end{bmatrix}, \quad T^{-1} = \begin{bmatrix} A_1 & B_1 \\ C_1 & D_1 \end{bmatrix}. \end{aligned} \tag{4.3}$$

With the aid of the above theorem, like the finite dimensional case, we have

Theorem 5′ *Let T be defined as above. $z \to \hat{z} = g(z,t)$ is a canonical transformation. $z(t)$ is the solution of* (4.2).

$$M(z,t) = g_z(z,t).$$

1) *$g(\star,0)$ is a linear canonical transformation, $M(z,0) = M_0$ is independent of z.*

2) *$g^{-1}(z,0)g(\star,t)$ is a canonical transformation dependent on t, and carries $z(t)$ into $z(0)$.*

3) *$CM_0 + D$ is nonsingular, then there is a gradient transformation dependent on $t : w \to \hat{w} = f(w,t)$ and a generating functional $\phi(w,t)$ for sufficiently small $|t|$, such that*:

(1) $[\hat{w} - f(w,t)]_{\hat{w}=A\hat{z}+Bz, w=C\hat{z}+Dz} = 0.$

(2) $\phi_t(w,t) = -\mu H\left(C_1\phi_w(w,t) + D_1 w\right)|_{w=C\hat{z}+Dz}\,.$ (4.4)

Like the finite dimensional case, we can determine by recursion all possible time-dependent generating functionals for Hamiltonians $H(z)$ analytic in z.

$$\begin{aligned}
\phi(w,t) &= \sum_{k=0}^{\infty}\phi^{(k)}(w)t^k,\\
\phi^{(0)}(w,t) &= \frac{1}{2}\langle w^\star, N_0 w\rangle, \quad N_0 = (AM_0+B)(CM_0+D)^{-1},\\
\phi^{(1)}(w) &= -\mu H(E_0 w), \quad E_0 = (CM_0+D)^{-1},\\
k \geqslant 1: \quad \phi^{(k+1)}(w) &= -\frac{1}{k+1}\sum_{m=1}^{k}\frac{\mu}{m!}\sum_{k_1+k_2+\cdots+k_m=k} H^{(m)}(E_0 w)\\
&\quad \times \left(C_1\phi_w^{(k_1)}(w)\right)\cdots\left(C_1\phi_w^{(k_m)}(w)\right).
\end{aligned} \tag{4.5}$$

Remark All conclusion concerning construction of S-scheme via generating functional is true when the J-symplectic form is replaced by K-symplectic form in finite or infinite dimensional case.

Remark When we take $B = R^n$, $q = (q_1, \cdots, q_n) \in R^n, p = (p_1, \cdots, p_n) \in (R^n)^\star = R^n$, equations (4.1) become the canonical Hamiltonian equations (1.1). All the conclusion of the finite dimensional case can be regarded as a special case of infinite dimensional case.

Remark Using above method, we get a semi-discrete S-scheme (involving time discretization only), in other words, we obtain an ∞-dim symplectic transformation in space $B \times B^\star$. In order to obtain a fully discrete S-scheme in $R^n \times R^n$, we must approximate spatial derivative by central difference. As we have done in §6, a fully discrete S-scheme for wave equation is obtained.

Another way is a method of lines: discretizing spatial derivatives, then we get a system of Hamiltonian ODE's. For fully discrete S-scheme, we use methods of constructing S-schemes for finite dimensional H-system $[1, 25]$.

§5 Multi-level S-schemes

We first consider an LH-system

$$\frac{du}{dt} = J^{-1}Au, \tag{5.1}$$

where A is a symmetric matrix. For this system we consider a three-level difference scheme

$$u^{n+1} - \phi_1 u^n - \phi_2 u^{n-1} = 0. \tag{5.2}$$

Introducing a new variable $v^n = u^{n-1}$, we obtain an equivalent two-level scheme

$$\begin{bmatrix} u^{n+1} \\ v^{n+1} \end{bmatrix} = \begin{bmatrix} \phi_1 & \phi_2 \\ I & 0 \end{bmatrix} \begin{bmatrix} u^n \\ v^n \end{bmatrix}. \tag{5.3}$$

In this case an equivalent H-system is given by

$$\frac{d}{dt}\begin{bmatrix} u \\ v \end{bmatrix} = \begin{bmatrix} 0 & J^{-1} \\ J^{-1} & 0 \end{bmatrix} \begin{bmatrix} 0 & A \\ A & 0 \end{bmatrix} \begin{bmatrix} u \\ v \end{bmatrix}. \tag{5.4}$$

Rewrite it in the form

$$\frac{dz}{dt} = K^{-1}Az \tag{5.5}$$

where

$$K^{-1} = \begin{bmatrix} 0 & J^{-1} \\ J^{-1} & 0 \end{bmatrix}.$$

We call scheme (5.2) an S-scheme, if scheme (5.3) is K-symplectic, i.e.

$$\begin{bmatrix} \phi_1^T & I_n \\ \phi_2^T & 0 \end{bmatrix} \begin{bmatrix} 0 & J \\ J & 0 \end{bmatrix} \begin{bmatrix} \phi_1 & \phi_2 \\ I_n & 0 \end{bmatrix} = \begin{bmatrix} 0 & J \\ J & 0 \end{bmatrix}. \tag{5.6}$$

After a short computation, we can prove

Lemma 2 ([7]) *Scheme* (5.3) *is an S-scheme, iff* ϕ_1 *is an infinitesimal symplectic matrix,* $\phi_2 = I$.

Theorem 9 *Scheme*

$$u^{n+1}=u^{n-1}+2\,\mathrm{sh}\left(\tau J^{-1}A\right)u^{n} \tag{5.7}$$

is an S-scheme with any order of accuracy, where

$$\mathrm{sh}(\lambda)=\lambda+\frac{\lambda^3}{3!}+\frac{\lambda^5}{5!}+\cdots+\frac{\lambda^{2m-1}}{(2m-1)!}+\cdots.$$

$m=1$: It is the usual leap-frog scheme of second order

$$u^{n+1}=u^{n-1}+2\tau J^{-1}Au^{n}. \tag{5.8}$$

$m=2$: It is the generalized leap-frog scheme of fourth order

$$u^{n+1}=u^{n-1}+2\tau J^{-1}Au^{n}+\frac{\tau^3}{3!}\left(J^{-1}A\right)^3u^{n}. \tag{5.9}$$

Leap-frog scheme for LH-systems can be generalized to the NLH-System (1.2). Consider a three-level scheme

$$Z^{n+1}-\phi_1\left(Z^{n}\right)-\phi_2\left(Z^{n-1}\right)=0 \tag{5.10}$$

Introducing a new variable $V^n=Z^{n-1}$, then

$$T:\quad\begin{aligned}Z^{n+1}&=\phi_1\left(Z^{n}\right)+\phi_2\left(V^{n}\right)\\ V^{n+1}&=Z^{n}\end{aligned} \tag{5.10'}$$

We say (5.10) is an S-scheme if the Jacobian

$$dT=\frac{\partial\left(Z^{n+1},V^{n+1}\right)}{\partial\left(Z^{n},V^{n}\right)}\quad\text{satisfies}$$

$$(dT)'\begin{bmatrix}0 & J^{-1}\\ J^{-1} & 0\end{bmatrix}(dT)=\begin{bmatrix}0 & J^{-1}\\ J^{-1} & 0\end{bmatrix}.$$

Lemma 2′ *Scheme* (5.10) *is an S-scheme, iff* $\dfrac{\partial\phi_1}{\partial Z^n}$ *is an infinitesimal symplectic matrix,* $\dfrac{\partial\phi_2}{\partial V^n}=I$.

For example, the leap-frog scheme is an S-scheme:

$$Z^{n+1}=Z^{n-1}-2\tau J^{-1}H_Z\left(Z^{n}\right).$$

As an application we consider two forms of H-system for wave equation $w_{tt}=w_{xx}$. The first form is to take its H-Functional as

$$H(v,u)=\frac{1}{2}\int\left(v^2+u_x^2\right)dx,\ \text{where } u=w, v=w_t.$$

We write the H-system

$$\frac{dz}{dt} = J^{-1}Az, \tag{5.11}$$

where

$$J^{-1}A = \begin{bmatrix} 0 & \frac{\partial^2}{\partial x^2} \\ 1 & 0 \end{bmatrix}, \quad z = \begin{bmatrix} v \\ u \end{bmatrix}, \quad A = \begin{bmatrix} 1 & 0 \\ 0 & -\frac{\partial^2}{\partial x^2} \end{bmatrix}.$$

Another form is to put $v = w_t, u = w_x$,

$$\frac{dz}{dt} = \mathcal{D}\frac{\delta H}{\delta z}, \quad H(v,u) = \frac{1}{2}\int \left(v^2 + u^2\right) dx, \tag{5.12}$$

where $\mathcal{D} = \begin{bmatrix} 0 & \frac{\partial}{\partial x} \\ \frac{\partial}{\partial x} & 0 \end{bmatrix}$, obviously it is a skew-adjoint operator. Rewrite it in form

$$\frac{dz}{dt} = K^{-1}Az, \tag{5.13}$$

where

$$K^{-1}A = \begin{bmatrix} 0 & \frac{\partial}{\partial x} \\ \frac{\partial}{\partial x} & 0 \end{bmatrix}, \quad A = \begin{bmatrix} I & 0 \\ 0 & I \end{bmatrix}, \quad K^{-1} = \mathcal{D}.$$

Let

$$\Delta_2 u_m = \frac{u(m+1) - 2u(m) + u(m-1)}{\Delta x^2} \tag{5.14}$$

$$\Delta_4 u_m = \frac{-u(m+2) + 16u(m+1) - 30u(m) + 16u(m-1) - u(m-2)}{12\Delta x^2} \tag{5.15}$$

$$\nabla_2 u_m = \frac{u(m+1) - u(m-1)}{2\Delta x} \tag{5.16}$$

$$\nabla_4 u_m = \frac{-u(m+2) + 8u(m+1) - 8u(m-1) + u(m-2)}{12\Delta x} \tag{5.17}$$

For the first kind of equation (5.11), the schemes $(5,8), (5.9)$ have forms respectively

$$\begin{bmatrix} v^{n+1} \\ u^{n+1} \end{bmatrix} = \begin{bmatrix} v^{n-1} \\ u^{n-1} \end{bmatrix} + 2\Delta t \begin{bmatrix} 0 & M \\ I & 0 \end{bmatrix} \begin{bmatrix} v^n \\ u^n \end{bmatrix}, \tag{5.18}$$

$$\begin{bmatrix} v^{n+1} \\ u^{n+1} \end{bmatrix} = \begin{bmatrix} v^{n-1} \\ u^{n-1} \end{bmatrix} + 2\Delta t \begin{bmatrix} 0 & M \\ I & 0 \end{bmatrix} \begin{bmatrix} v^n \\ u^n \end{bmatrix} + \frac{\Delta t^3}{3} \begin{bmatrix} 0 & M \\ I & 0 \end{bmatrix}^3 \begin{bmatrix} v^n \\ u^n \end{bmatrix}. \tag{5.19}$$

Here I is $n \times n$ identity matrix, M is $n \times n$ matrix. According to (5.14), (5.15) it may be put in two forms

$$M_1 : \frac{1}{\Delta x^2}\begin{bmatrix} -2 & 1 & 0 & \cdots & 1 \\ 1 & -2 & 1 & \cdots & 0 \\ \cdots & \cdots & \cdots & \cdots & \cdots \\ \cdots & \cdots & \cdots & \cdots & \cdots \\ 0 & 0 & \cdots & -2 & 1 \\ 1 & 0 & \cdots & 1 & -2 \end{bmatrix}$$

$$M_2 : \frac{1}{12\Delta x^2}\begin{bmatrix} -30 & 16 & -1 & 0 & \cdots & -1 & 16 \\ 16 & -30 & 16 & -1 & \cdots & 0 & -1 \\ -1 & 16 & -30 & 16 & -1 & \cdots & 0 \\ \cdots & \cdots & \cdots & \cdots & \cdots & \cdots & \cdots \\ \cdots & \cdots & \cdots & \cdots & \cdots & \cdots & \cdots \\ -1 & 0 & 0 & \cdots & 16 & -30 & 16 \\ 16 & -1 & 0 & \cdots & -1 & 16 & -30 \end{bmatrix}$$

In scheme (5.18) if $M = M_1$ we express this scheme with accuracy $O\left(\Delta t^2 + \Delta x^2\right)$ by SLFM $(2, 2)$ and if $M = M_2$ we express it by SLFM$(2, 4)$.

In scheme (5.19), if $M = M_1$ we express it by SLFM $(4, 2)$ and if $M = M_2$ we express it by SLFM $(4, 4)$.

For the second kind of equation (5.13), the schemes (5.8), (5.9) become

$$\begin{bmatrix} v^{n+1} \\ u^{n+1} \end{bmatrix} = \begin{bmatrix} v^{n-1} \\ u^{n-1} \end{bmatrix} + 2\Delta t \begin{bmatrix} 0 & M \\ M & 0 \end{bmatrix} \begin{bmatrix} v^n \\ u^n \end{bmatrix},$$

$$\begin{bmatrix} v^{n+1} \\ u^{n+1} \end{bmatrix} = \begin{bmatrix} v^{n-1} \\ u^{n-1} \end{bmatrix} + 2\Delta t \begin{bmatrix} 0 & M \\ M & 0 \end{bmatrix} \begin{bmatrix} v^n \\ u^n \end{bmatrix} + \frac{\Delta t^3}{3} \begin{bmatrix} 0 & M \\ M & 0 \end{bmatrix}^3 \begin{bmatrix} v^n \\ u^n \end{bmatrix}.$$

Here M is $n \times n$ matrix, according to (5.16), (5.17) they are

$$M_3 : \frac{1}{2\Delta x}\begin{bmatrix} 0 & 1 & 0 & \cdots & -1 \\ -1 & 0 & 1 & \cdots & 0 \\ \cdots & \cdots & \cdots & \cdots & \cdots \\ \cdots & \cdots & \cdots & \cdots & \cdots \\ 0 & \cdots & \cdots & 0 & 1 \\ 1 & 0 & \cdots & -1 & 0 \end{bmatrix}$$

$$M_4: \frac{1}{12\Delta x}\begin{bmatrix} 0 & 8 & -1 & 0 & \cdots & 1 & -8 \\ -8 & 0 & 8 & -1 & \cdots & 0 & 1 \\ 1 & -8 & 0 & 8 & -1 & \cdots & 0 \\ \cdots & \cdots & \cdots & \cdots & \cdots & \cdots & \cdots \\ \cdots & \cdots & \cdots & \cdots & \cdots & \cdots & \cdots \\ -1 & 0 & \cdots & \cdots & -8 & 0 & 8 \\ 8 & -1 & 0 & \cdots & 1 & -8 & 0 \end{bmatrix}$$

In scheme (5.18), if $M = M_3$ we express it by KLFM $(2,2)$ and if $M = M_4$ we express it by KLFM $(2,2)$, and if $M = M_4$ we express it by KLFM $(2,4)$.

In scheme (5.19), if $M = M_3$ we express it by KLFM $(4,2)$ and if $M = M_4$ we express it by KLFM $(4,4)$.

Lemma 3 *Eigenvalues of matrices*

$$\begin{bmatrix} 0 & M_1 \\ I & 0 \end{bmatrix}, \begin{bmatrix} 0 & M_2 \\ I & 0 \end{bmatrix}, \begin{bmatrix} 0 & M_3 \\ M_3 & 0 \end{bmatrix}, \begin{bmatrix} 0 & M_4 \\ M_4 & 0 \end{bmatrix}$$

are

$$\mu_k^{(1)} = \pm\frac{2i}{\Delta x}\sin\left(\frac{\pi k}{2N}\right), \quad \mu_k^{(2)} = \pm\frac{i}{\Delta x}\sqrt{4\sin^2\frac{\pi k}{2N} + \frac{4}{3}\sin^4\frac{\pi k}{2N}}$$

$$\mu_k^{(3)} = \pm\frac{i}{\Delta x}\sin\left(\frac{\pi k}{2N}\right), \quad \mu_k^{(4)} = \pm\frac{i}{\Delta x}\left(\frac{1}{6}\sin\frac{2\pi k}{2N} - \frac{8}{6}\sin\frac{\pi k}{2N}\right)$$

$$k = 0, 1, \cdots, 2N-1,$$

respectively.

Above eigenvalues and criterions in [20] are used to derive the stability conditions in the following Table 1.

Table 1

H-eq.	scheme	error of approximation	stability condition
$\frac{d}{dt}\begin{bmatrix} v \\ u \end{bmatrix} = \begin{bmatrix} 0 & -1 \\ 1 & 0 \end{bmatrix}\begin{bmatrix} v \\ -u_{xx} \end{bmatrix}$	SLFM(2,2)	$o\left(\Delta t^2 + \Delta x^2\right)$	$\Delta t/\Delta x \leqslant 1/2$
	SLFM(2,4)	$o\left(\Delta t^2 + \Delta x^4\right)$	$\Delta t/\Delta x \leqslant 0.4330$
	SLFM(4,2)	$o\left(\Delta t^4 + \Delta x^2\right)$	$\Delta t/\Delta x \leqslant 1.4237$
	SLFM(4,4)	$o\left(\Delta t^4 + \Delta x^4\right)$	$\Delta t/\Delta x \leqslant 1.2330$
$\frac{d}{dt}\begin{bmatrix} v \\ u \end{bmatrix} = \begin{bmatrix} 0 & \frac{\partial}{\partial x} \\ \frac{\partial}{\partial x} & 0 \end{bmatrix}\begin{bmatrix} v \\ u \end{bmatrix}$	KLFM(2,2)	$o\left(\Delta t^2 + \Delta x^2\right)$	$\Delta t/\Delta x \leqslant 1/2$
	KLFM(2,4)	$o\left(\Delta t^2 + \Delta x^4\right)$	$\Delta t/\Delta x \leqslant 0.7287$
	KLFM(4,2)	$o\left(\Delta t^4 + \Delta x^2\right)$	$\Delta t/\Delta x \leqslant 2.8473$
	KLFM(4,4)	$o\left(\Delta t^4 + \Delta x^4\right)$	$\Delta t/\Delta x \leqslant 2.0750$

Using above scheme, numerical solutions of problem (6.4) have been done, we will list the numerical results and test of stability in Appendix 2 Table (2-5).

Remark Under the condition that the eigenvalues of infinitesimally symplectic matrices J^{-1} A are purely imaginary, the region of stability of the generalized leap-frog scheme (5.6) tends to the full real axis as the orders of accuracy increase to infinity [10].

§6 Numerical Examples

6.1 Two-body problem

In celestial mechanics, the simplest and famous example is the two-body problem. We outline its arguments here

(i) $M = R^2 \times \{R^2 - (0)\}$, phase space.

(ii) $m \in M$, initial data.

(iii) Hamiltonian, $H(p, q) = \|p\|^2/2 - 1/\|q\|$.

It has been proved that if angular momentum $G \neq 0$ the path is an ellipse, parabola, or hyperbola according to $E < 0, E = 0, E > 0$. The case $G = 0$ is degenerate, the path is a straight line.

We choose (A-III-4)

$$z^{n+1} = z^n + \tau J^{-1} H_z \left(\frac{z^{n+1}+z^n}{2}\right) + \frac{\tau^3}{24}\left(\left(J^{-1}H_z\right)' H_{zz}\left(J^{-1}H_z\right)\right)_z \left(\frac{z^{n+1}+z^n}{2}\right), \quad (6.1)$$

where $z^n = \begin{pmatrix} p^n \\ q^n \end{pmatrix}$, τ is step length. This scheme is fourth order. Because the scheme is symplectic, we keep $G = q_1 p_2 - p_1 q_2$ constant, i.e., keep the path's curvature unchanged. We have computed all the kinds of paths. The paths are the same as the theory predicts. In elliptic case, the R-K method cannot be carried through. As a matter of fact, it will overflow after short time. Because scheme (6.1) is A-stable, it gives good result in all cases. In Appendix 2 Table 6 we will list comparison data between R-K method and S-schemes (fourth order and second order).

6.2 Wave equation

Let $B = H^1, P = H^1 \times H^{-1}$. In $H^2 \times H^1$, a subspace of P, we have the Hamiltonian functional

$$H(u, v) = \int_{R^2} \frac{1}{2}\left(v^2 + (\nabla u)^2\right) dx.$$

The corresponding Hamiltonian vector field is $(v, \Delta u)$. The associated H-equation is

$$\begin{aligned} \frac{dv}{dt} &= -\frac{\delta H}{\delta u} = \Delta u, \\ \frac{du}{dt} &= \frac{\delta H}{\delta v} = v. \end{aligned} \quad (6.2)$$

Choosing $T = \begin{bmatrix} -J & +J \\ \frac{1}{2}I & \frac{1}{2}I \end{bmatrix}$, we get a S-scheme (A-III-4) of fourth order for wave equation [8]

$$z^{n+1} = z^n + \tau J^{-1} H_z - \frac{2\tau^3}{4!} J^{-1} H_{zz} J^{-1} H_{zz} J^{-1} H_z, \tag{6.3}$$

e.g.

$$\begin{bmatrix} v \\ u \end{bmatrix}^{n+1} = \begin{bmatrix} v \\ u \end{bmatrix}^n + \tau \begin{bmatrix} 0 & \Delta \\ I & 0 \end{bmatrix} \begin{bmatrix} \bar{v} \\ \bar{u} \end{bmatrix} - \frac{\tau^3}{12} \begin{bmatrix} 0 & \Delta \\ I & 0 \end{bmatrix}^3 \begin{bmatrix} \bar{v} \\ \bar{u} \end{bmatrix},$$

where $\overline{v} = \frac{1}{2}\left(v^{n+1} + v^n\right), \overline{u} = \frac{1}{2}\left(u^{n+1} + u^n\right)$.

If we take Δ as the fourth-order centered difference approximation, using periodic boundary condition we get

$$z^{n+1} = z^n + \tau B\bar{z} - \frac{\tau^3}{12} B^3 \bar{z}.$$

where

$$B = \begin{bmatrix} 0 & M \\ I & 0 \end{bmatrix},$$

$$M = \frac{1}{12h^2} \begin{bmatrix} -30 & 16 & -1 & 0 & 0 & \cdots & -1 & 16 \\ 16 & -30 & 16 & -1 & 0 & \cdots & 0 & -1 \\ -1 & 16 & -30 & 16 & -1 & \cdots & 0 & 0 \\ \cdots & \cdots & \cdots & \cdots & \cdots & \cdots & \cdots & \cdots \\ \cdots & \cdots & \cdots & \cdots & \cdots & \cdots & \cdots & \cdots \\ -1 & 0 & 0 & \cdots & \cdots & 16 & -30 & 16 \\ 16 & -1 & 0 & \cdots & 0 & -1 & 16 & -30 \end{bmatrix},$$

$$\bar{z} = \frac{z^{n+1} + z^n}{2}.$$

τ is the time step length, h is the space step length, M is an $N \times N$ matrix, I is the $N \times N$ unit matrix, $h = \frac{2\pi}{N}$.

Computations were done. The initial and boundary conditions are taken to be

$$\begin{aligned} &u(0,x) = \sin x \quad \text{for } 0 \leqslant x \leqslant 2\pi, \\ &u_t(0,x) = \cos x \quad \text{for } 0 \leqslant x \leqslant 2\pi, \\ &u(t,0) = u(t,2\pi). \end{aligned} \tag{6.4}$$

The exact solution is $u(t,x) = \sin(x+t)$.

We list the numerical and corresponding exact results in Appendix 2 table 7.

§7 S-schemes and Chaos

Most systems with two or more degrees of freedom have some chaotic motions and some regular motions. The regular motions occur when the system is far away from resonances of low order. This is in accord with the conclusion of the KAM theorem. On the other hand, when the system is near such resonances, chaotic motion occurs.

Hénon and Heiles [21] considered a system with two degrees of freedom with coordinates $q_1(t)$ and $q_2(t)$ and conjugate momenta $p_1(t)$ and $p_2(t)$. Their Hamiltonian is

$$H(q_1, q_2, p_1, p_2) = \frac{1}{2}\left(p_1^2 + p_2^2\right) + \frac{1}{2}\left(q_1^2 + q_2^2 + 2q_1^2 q_2 - \frac{2}{3}q_2^3\right). \tag{7.1}$$

They found that for total energy E less than $1/12$, the solutions were regular for most initial conditions. As E increased the fraction of initial condition leading to chaotic solutions increased.

For (7.1), its canonical equations are

$$\frac{d}{dt}\begin{bmatrix} p_1 \\ p_2 \\ q_1 \\ q_2 \end{bmatrix} = \begin{bmatrix} 0 & 0 & 1 & 0 \\ 0 & 0 & 0 & 1 \\ -1 & 0 & 0 & 0 \\ 0 & -1 & 0 & 0 \end{bmatrix}^{-1} \begin{bmatrix} p_1 \\ p_2 \\ q_1 + 2q_1 q_2 \\ q_2 + q_1^2 - q_2^2 \end{bmatrix}. \tag{7.2}$$

Using the explicit scheme (A-II-1), we obtain

$$\begin{aligned} p_1^{n+1} &= p_1^n - \tau\left(q_1^n + 2q_1^n q_2^n\right), \\ p_2^{n+1} &= p_2^n - \tau\left(q_2^n + (q_1^2)^n - (q_2^2)^n\right), \\ q_1^{n+1} &= q_1^n + \tau p_1^{n+1}, \\ q_2^{n+1} &= q_2^n + \tau p_2^{n+1}. \end{aligned} \tag{7.3}$$

Obviously this scheme is a symplectic transformation, approximating original system in second order precision, of course, it is regarded as an area-preserving mapping. As the time step we selected $\tau = 0.0005$ for 20000 cycles, the CPU time was 10 minutes (IBM-4341). The results obtained are in agreement with the results of Hénon-Heiles (1964).

In Appendix 2 Fig. 1, we list the graphics of surfaces of section computed with scheme (7.3) for $E(\text{or}H) = 1/24, 1/12, 1/8, 1/6$. Good results are also obtained by the implicit S-scheme (fourth order) (A-III-4) .

Hénon [22] accepted Kruskal's suggestion [21] to study an area-preserving mapping directly.

$$\begin{aligned} T: x_{n+1} &= 1 - ax_n^2 + y_n \\ y_{n+1} &= bx_n \end{aligned} \tag{7.4}$$

describes conservative system, dissipative system and logistic system when $|b| = 1, |b| < 1$ and $|b| = 0$ respectively.

In the case $b = \pm 1, T$ is an area-preserving mapping and may be thought of as the Poincare section for a H-system with two degrees of freedom.

Remark The research of dynamical system with three degrees of freedom can be reduced to the study of a four dimensional mapping. Scheme (7.3) is an area preserving mapping, therefore it is a good tool for studying dynamical system with three degrees of freedom.

Now we consider the equation

$$\frac{du}{dt} = u(1-u) \tag{7.5}$$

with initial condition $u(0) = u_0$. Using leap-frog scheme (5.8), we get a difference equation

$$\frac{u_{n+1} - u_{n-1}}{2h} = u_n (1 - u_n), \tag{7.6}$$

with initial condition $u(0) = u_0$ and $u_1 = u_0 + hu_0 (1 - u_0)$ computed by Euler's forward difference scheme. Using leap-frog method described in (5.10), we obtain

$$\begin{aligned} u_{n+1} &= 2hu_n (1 - u_n) + v_n \\ v_{n+1} &= u_n. \end{aligned} \tag{7.7}$$

Thus we have a mapping $\bar{T} : R^2 \to R^2$ defined by

$$\begin{aligned} X &= 2hx(1-x) + y \\ Y &= x \end{aligned} \tag{7.8}$$

Scheme (7.6), in the sense of (5.6), is a S-scheme. The Jacobian matrix T is given by

$$d\bar{T} = \frac{\partial(X,Y)}{\partial(x,y)} = \begin{bmatrix} 2h(1-2x) & 1 \\ 1 & 0 \end{bmatrix}$$

Obviously $|d\bar{T}| = -1$, therefore it is an area-preserving scheme.

Let $u_{2m} = p_m, u_{2m-1} = q_m$. Rewrite the equation in the following form [9][23]

$$S: \begin{aligned} \frac{p_m - p_{m-1}}{2h} &= q_m (1 - q_m) \\ \frac{q_{m+1} - q_m}{2h} &= p_m (1 - p_m) \end{aligned} \tag{7.9}$$

The above system is H-system with $H(p,q) = \frac{1}{2}(p^2 - q^2) - \frac{1}{3}(p^3 - q^3)$. Symplectic transformation (7.9) can be regarded as a composition transformation $\bar{T} \cdot \bar{T}$.

In general we must use the central difference for conservative systems. On the contrary, in order to describe a dissipative system we must employ mixed difference [24]

$$(1-\theta)\frac{u_{n+1}-u_{n-1}}{2h}+\theta\frac{u_{n+1}-u_n}{h}=u_n\left(1-u_n\right), \quad 0\leqslant\theta\leqslant 1, \tag{7.10}$$

or equivalently, a mapping

$$\begin{aligned} T_\theta : u_{n+1} &= \frac{1-\theta}{1+\theta}v_n+\frac{2}{1+\theta}\left[(\theta+h)u_n-hu_n^2\right] \\ v_{n+1} &= u_n. \end{aligned} \tag{7.11}$$

Its Jacobian is

$$dT_\theta=\begin{bmatrix} 2\left(\theta+h-2hu_n\right)/(1+\theta) & (1-\theta)/(1+\theta) \\ 1 & 0 \end{bmatrix}.$$

Transform the mapping (7.11) by an affine transformation

$$\begin{aligned} u_n &= \frac{\theta+h}{2h}\left[\left(\frac{h-\theta}{1+\theta}\right)x_n+1\right] \\ v_n &= \frac{\theta+h}{2h}\left[\left(\frac{h-\theta}{1+\theta}\right)y_n+1\right] \end{aligned}$$

into the form (7.4) with $a=\left(h^2-\theta^2\right)/(1+\theta)^2, b=(1-\theta)/(1+\theta)$. In particular, we select $h=1.898, \theta=0.538$ then the corresponding map (7.4) with $a=1.4$ and $b=1.3$ is a Hénon's strange attractor [24], see Fig .3(c) in Appendix 2.

The step length $h=1.898$ is too large as time step for numerical integration. But it is still significant in studying an area mapping.

Appendix 1

In this Appendix we collect S-schemes for H-system (1.2) with precision up to fourth order. According to method of construction of S-schemes described in §3, we have

$$\begin{aligned} \phi^{(0)}(w) &= \frac{1}{2}w'N_0 w, \\ \phi^{(1)}(w) &= -\mu H\left(E_0 w\right), \\ \phi^{(2)}(w) &= -\frac{\mu}{2}H_{z_i}\left(C_1\phi_w^{(1)}\right)_i, \\ \phi^{(3)}(w) &= -\frac{\mu}{3}\left(\frac{1}{2!}H_{z_{i_1}z_{i_2}}\left(C_1\phi_w^{(1)}\right)_{i_1}\left(C_1\phi_w^{(1)}\right)_{i_2}+H_{z_i}\left(C_1\phi_w^{(2)}\right)_i\right), \\ \phi^{(4)}(w) &= -\frac{\mu}{4}\left(\frac{1}{3!}H_{z_{i_1}z_{i_2}z_{i_3}}\left(C_1\phi_w^{(1)}\right)_{i_1}\left(C_1\phi_w^{(1)}\right)_{i_2}\left(C_1\phi_w^{(1)}\right)_{i_3}\right. \\ &\quad \left.+H_{z_{i_1}z_{i_2}}\left(C_1\phi_w^{(2)}\right)_{i_1}\left(C_1\phi_w^{(1)}\right)_{i_2}+H_{z_1}\left(C_1\phi_w^{(3)}\right)_i\right). \end{aligned} \tag{A-0}$$

In §3 case II, $M_0 = I_{2n}, \mu = 1$, by (3.1), we have

$$w = \begin{vmatrix} \hat{q} \\ p \end{vmatrix}, \quad C_1 = \begin{bmatrix} 0 & 0 \\ 0 & -I \end{bmatrix}, \quad D_1 = \begin{bmatrix} 0 & I \\ 0 & 0 \end{bmatrix}, \quad \text{H-J equation is } \phi_t = -H\left(p, -\phi_p\right).$$

As matrices in theorem 6, N_0, E_0 are

$$N_0 = \begin{bmatrix} 0 & I \\ I & 0 \end{bmatrix}, \quad E_0 = \begin{bmatrix} 0 & I \\ I & 0 \end{bmatrix}, \quad E_0 w = \begin{bmatrix} p \\ \hat{q} \end{bmatrix}.$$

Substituting μ, N_0, E_0, C_1 into (A-0) yields

$$\phi^{(0)}(w) = \frac{1}{2} w' N_0 w = \frac{1}{2} \hat{q}_i p_i,$$

where repeated subscripts i subject to Einstein summation convention.

$$\begin{aligned}
\phi^{(1)}(w) = & -H(p\hat{q}), \\
\phi^{(2)}(w) = & -\frac{1}{2!} H_{q_i} H_{p_i}, \\
\phi^{(3)}(w) = & -\frac{1}{3!}\left(H_{q_i q_j} H_{p_i} H_{p_j} + H_{q_i p_j} H_{p_i} H_{q_j} + H_{p_i p_j} H_{q_i} H_{q_j}\right), \\
\phi^{(4)}(w) = & -\frac{1}{4!}\left(H_{q_i q_j q_k} H_{p_i} H_{p_j} H_{p_k} + H_{q_i q_j p_k} H_{p_i} H_{p_j} H_{q_k}\right. \\
& + H_{q_i p_j p_k} H_{p_i} H_{q_j} H_{q_k} + H_{p_i p_j p_k} H_{q_i} H_{q_j} H_{q_k} \\
& + 5 H_{q_i q_j} H_{p_j p_k} H_{p_i} H_{q_k} + H_{q_i p_j} H_{q_i p_k} H_{p_i} H_{q_k} \\
& \left. + 3 H_{q_i p_j} H_{p_i p_k} H_{q_j} H_{q_k} + 3 H_{q_i q_j} H_{q_k p_i} H_{p_j} H_{p_k}\right).
\end{aligned}$$

Scheme of first order: Using (3.5), (3.6), we have

$$\begin{aligned}
&\psi^{(1)}(w, \tau) = \phi^{(0)}(w) + \phi^{(1)}(w)\tau \\
&\hat{w} = \psi_w^{(1)}(w, \tau) = \phi_w^{(0)}(w) + \tau \phi_w^{(1)}(w) \\
&\begin{bmatrix} -\hat{p} \\ -q \end{bmatrix} = -\begin{bmatrix} p \\ \hat{q} \end{bmatrix} + \tau \begin{bmatrix} -H_q(p, \hat{q}) \\ -H_p(p, \hat{q}) \end{bmatrix},
\end{aligned}$$

where $\hat{p} = p^0, \hat{q} = q^0$. Thus we obtain scheme of first order accuracy

$$\begin{aligned}
p &= p^0 - \tau H_q\left(p, q^0\right), \\
q &= q^0 + \tau H_p\left(p, q^0\right).
\end{aligned} \tag{A-II-1}$$

Remark This scheme will have 2nd-order accuracy when computing p in the $(n+1)$-th time step and q in the $\left(n+1+\dfrac{1}{2}\right)$-th step. Especially, for separable H-system it is the staggered explicit scheme (2.7).

By similar derivation, we have scheme of second order:

$$\psi^{(2)}(w,\tau)=\psi^{(1)}(w,\tau)+\tau^2\phi^{(2)}(w),$$

$$\hat{w}=\psi_w^{(2)}(w,\tau)=\psi_w^{(1)}(w,\tau)+\tau^2\phi_w^{(2)}(w).$$

$$p_i=p_i^0-\tau H_{q_i}\left(p,q^0\right)-\frac{\tau^2}{2}\left[H_{q_jq_i}H_{p_j}+H_{q_j}H_{p_jq_i}\right],$$

$$q_i=q_i^0+\tau H_{p_i}\left(p,q^0\right)+\frac{\tau^2}{2}\left[H_{q_jp_i}H_{p_j}+H_{q_j}H_{p_jp_i}\right]. \tag{A-II-2}$$

Scheme of third order:

$$\psi^{(3)}(w,\tau)=\psi^{(2)}(w,\tau)+\tau^3\phi^{(3)}(w),$$

$$\hat{w}=\psi_w^{(3)}(w,\tau)=\psi_w^{(2)}(w,\tau)+\tau^3\phi_w^{(3)}(w).$$

$$\begin{aligned}p_i=&\,p_i^0-\tau H_{q_i}\left(p,q^0\right)-\frac{\tau^2}{2!}\left(H_{q_jq_i}H_{p_j}+H_{p_jq_i}H_{p_j}\right)\\&-\frac{\tau^3}{3!}\left[H_{q_kq_iq_i}H_{p_k}H_{p_j}+H_{q_kp_jq_i}H_{p_k}H_{q_j}+H_{p_kp_jq_i}H_{q_k}H_{q_j}\right.\\&+H_{q_jp_k}\left(H_{p_jq_i}H_{q_k}+H_{p_j}H_{q_kq_i}\right)+2H_{p_jp_k}H_{q_jq_i}H_{q_k}\\&\left.+2H_{q_jq_k}H_{p_jq_i}H_{p_k}\right],\end{aligned} \tag{A-II-3}$$

$$\begin{aligned}q_i=&\,q_i^0+\tau H_{p_i}\left(p,q^0\right)+\frac{\tau^2}{2!}\left(H_{q_jp_i}H_{p_j}+H_{p_jp_i}H_{q_j}\right)\\&+\frac{\tau^3}{3!}\left[H_{q_jp_kp_iH_{p_j}H_{q_k}}+H_{p_jp_kp_i}H_{q_j}H_{q_k}+H_{q_jq_kq_i}H_{p_j}H_{p_k}\right.\\&+H_{q_jp_k}\left(H_{p_jp_i}H_{q_k}+H_{p_j}H_{q_kp_i}\right)+2H_{p_jp_k}H_{q_jp_i}H_{q_k}\\&\left.+2H_{q_jq_k}H_{p_jp_i}H_{p_k}\right].\end{aligned}$$

Scheme of fourth order:

$$\psi^{(4)}(w,\tau)=\psi^{(3)}(w,\tau)+\tau^4\phi^{(4)}(w),$$

$$\hat{w}=\psi_w^{(4)}(w,\tau)=\psi_w^{(3)}(w,\tau)+\tau^4\phi_w^{(4)}(w). \tag{A-II-4}$$

In §3, case Ⅲ, $\mu=-1, M_0=I, w=\dfrac{1}{2}(z+\hat{z}), \hat{w}=J(z-\hat{z})$, by (3.1) we have $C_1=\dfrac{1}{2}J^{-1}=\dfrac{1}{2}\begin{bmatrix}0&-I\\I&0\end{bmatrix}, D_1=\begin{bmatrix}I&0\\0&I\end{bmatrix}$. H-J equation is $\phi_t=H\left(w-\dfrac{1}{2}J\phi_w\right)$,

$$N_0=\begin{bmatrix}0&0\\0&0\end{bmatrix}, E_0=\begin{bmatrix}I&0\\0&I\end{bmatrix}, E_0w=\frac{1}{2}(z+\hat{z})=\frac{1}{2}\left(z+z^0\right).$$

Substituting μ, N_0, E_0, C_1 into (A-0) yields

$$\phi^{(0)}(w)=0,$$

$$\phi^{(1)}(w)=H(w),$$

$$
\begin{aligned}
\phi^{(2)}(w) &= \frac{1}{2} H_z' \left(\frac{1}{2} J^{-1} H_z\right) = 0, \\
\phi^{(3)}(w) &= \frac{1}{3} H_{z_i}(w) \left(C_1 \phi_w^{(2)}\right)_i + \frac{1}{6} H_{z_i z_j} \left(C_1 \phi_w^{(1)}\right)_i \left(C_1 \phi_w^{(1)}\right)_j \\
&= \frac{1}{6} H_{z_i z_j} \left(C_1 \phi_w^{(1)}\right)_i \left(C_1 \phi_w^{(1)}\right)_j \\
&= \frac{1}{4!} \left\{ H_{p_i p_j} H_{q_i} H_{q_j} - 2 H_{p_i q_j} H_{p_j} H_{q_i} + H_{q_i q_j} H_{p_i} H_{p_j} \right\}, \quad \phi^{(4)}(w) = 0.
\end{aligned}
$$

By similar derivation as above, we have scheme of second order:

$$
\begin{aligned}
\psi^{(2)}(w,\tau) = \psi^{(1)}(w,\tau) + \tau^2 \phi^{(2)}(w) = \psi^{(1)}(w,\tau) = \phi^{(0)}(w) + \tau \phi^{(1)}(w), \\
\hat{w} = \psi^{(2)}(w,\tau) = \tau \phi_w^{(1)}(w) = \tau H_z(w).
\end{aligned}
$$

i.e.

$$
\begin{aligned}
J(z - \hat{z}) &= \tau H_z \left(\frac{z + \hat{z}}{2}\right), \\
z &= z^0 + \tau J^{-1} H_z \left(\frac{z + z^0}{2}\right),
\end{aligned}
\tag{A-III-1}
$$

i.e.

$$
\begin{aligned}
p &= p^0 - \tau H_q \left(\frac{p + p^0}{2}, \frac{q + q^0}{2}\right), \\
q &= q^0 + \tau H_p \left(\frac{p + p^0}{2}, \frac{q + q^0}{2}\right).
\end{aligned}
\tag{A-III-2}
$$

Scheme of fourth order:

$$
\begin{aligned}
\hat{w} &= \psi_w^{(4)}(w,\tau) = \tau \phi_w^{(1)}(w) + \tau^3 \phi_w^{(3)}(w) \\
&= \tau H_z(w) + \frac{\tau^3}{4!} \left[\left(J^{-1} H_z\right)' H_{zz} \left(J^{-1} H_z\right) \right]_z (w),
\end{aligned}
\tag{A-III-3}
$$

$$
\begin{aligned}
p_i = p_i^0 - \tau H_{q_i} \left(\frac{p + p^0}{2}, \frac{q + q^0}{2}\right) - \frac{\tau^3}{4!} &\left\{ H_{p_j p_k q_i} H_{q_j} H_{q_k} + 2 H_{p_j p_k} H_{q_j q_i} H_{q_k} \right. \\
&- 2 H_{p_j q_k q_i} H_{p_k} H_{q_j} - 2 H_{p_k q_j} H_{p_j q_i} H_{q_k} - 2 H_{p_k q_j} H_{p_j} H_{q_k q_i} \\
&\left. + 2 H_{q_j q_k} H_{p_j q_i} H_{p_k} + H_{q_j q_k q_i} H_{p_j} H_{p_k} \right\}, \\
q_i = q_i^0 + \tau H_{p_i} \left(\frac{p + p^0}{2}, \frac{q + q^0}{2}\right) + \frac{\tau^3}{4!} &\left\{ H_{p_j p_k p_i} H_{q_j} H_{q_k} + 2 H_{p_j p_k} H_{q_j p_i} H_{q_k} \right. \\
&- 2 H_{p_j q_k p_i} H_{p_k} H_{q_j} - 2 H_{p_k q_j} H_{p_j p_i} H_{q_k} - 2 H_{p_k q_j} H_{p_j} H_{q_k p_i} \\
&\left. + 2 H_{q_j q_k p_i} H_{p_j} H_{p_k} + 2 H_{q_j q_k} H_{p_j p_i} H_{p_k} \right\}.
\end{aligned}
\tag{A-III-4}
$$

Appendix 2

This Appendix consists of Tables 2-7 and Figures 1-3, which give some computer results.

Table 2. Comparison of the exact solutions with the numerical solutions of eq. (5.11) with (6.4) and of (5.12) with initial data $u(0,x) = v(0,x) = \sin x$, using different leap-frog schemes, where $u(n,m)$ denotes value of $u(t,x)$ at $t = nDt, x = mDx$. Courant number $C = Dt/Dx = 0.4, Dx = \frac{2}{40}$.

Table 2

$Dt = 0.06283$	Dx=0.15708		$C = Dt/Dx = 0.4$	
SCHEME	u(99,0)	u(99,10)	u(99,20)	u(99,30)
KLFM(4,2)	0.06279	0.99777	−0.06778	−0.99777
KLFM(4,4)	0.06279	0.99815	−0.06279	−0.99815
KLFM(2,2)	0.04109	0.99149	−0.04109	−0.99149
KLFM(2,4)	0.03701	0.99930	−0.03701	−0.99930
SLFM(2,2)	0.06047	0.99817	−0.06048	−0.99817
SLFM(2,4)	0.05636	0.99841	−0.05635	−0.99841
SLFM(4,2)	0.06277	0.99803	−0.06277	−0.99803
SLFM(4,4)	0.06277	0.99803	−0.06279	−0.99803
Analytical Solution	0.06279	0.99803	−0.06279	−0.99803

Table 3. Similar results as in Table 2, but only with schemes, SLFM (4,4) and KLFM $(4,4)$ under $C = 1.1$.

Table 3

$Dt = 0.17279$	Dx=0.15708		$C = Dt/Dx = 1.1$	
SCHEME	u(99,0)	u(99,10)	u(99,20)	u(99,30)
KLFM(4,4)	−0.98519	0.17145	0.98519	−0.17145
SLFM(4,4)	−0.98520	0.17111	0.98532	−0.17109
Analytical Solution	−0.98511	0.17191	0.98512	−0.17192

Table 4. Numerical tests of stability of scheme KLFM$(4,4)$ with $C = 2.09$ and 2.07, (critical $C = 2.075$).

Table 4

	$Dt = 0.15708$		KLFM(4,4)	
C-number	u(30,0)	u(99,0)	u(30,10)	u(99,10)
2.09	−0.68332	overflow	−0.73430	overflow
2.07	−0.63119	0.99853	−0.77565	0.05482
Analytical Solution	−0.63208	0.99870	−0.77490	0.05089

Table 5. Numerical tests of stability of scheme SLFM $(4,4)$ with $C = 1.23$ and 1.24,

(critical $C = 1.233$).

Table 5

	$Dt = 0.15708$		SLFM(4,4)	
C-number	$u(30,0)$	$u(99,0)$	$u(30,10)$	$u(99,10)$
1.23	−0.23288	0.75542	−0.97253	0.65509
1.24	0.09666	overflow	−0.55163	overflow
Analytical Solution	−0.23344	0.75630	−0.97237	0.65423

Table 6. Comparison of numerical results for 2 -body problem by 5-6- th order R-K scheme and by 4-th and 2-nd order S-schemes. This shows that lower order S-schemes are comparable in precision with higher order R-K schemes.

Table 6

Runge-Kutta-Verner method (5 or 6 order)

Initial data 1, 1, 1, 1. Step length: 0.05

	p^1	p^2	q^1	q^2
$t = 0.5$	0.88041	0.88041	1.46630	1.46630
$t = 2.0$	0.74699	0.74699	2.66716	2.66716
$t = 4.0$	0.68254	0.68254	4.08757	4.08757
$t = 7.0$	0.63992	0.63992	6.06290	6.06290
$t = 10.0$	0.61794	0.61794	7.94673	7.94673

S-Scheme 4 order (A-III-4)

Initial data 1, 1, 1, 1. Step length: 0.05

	p^1	p^2	q^1	q^2
$t = 0.5$	0.88041	0.88041	1.46629	1.46629
$t = 2.0$	0.74699	0.74699	2.66713	2.66713
$t = 4.0$	0.68254	0.68254	4.08750	4.08750
$t = 7.0$	0.63991	0.63991	6.06281	6.06281
$t = 10.0$	0.61793	0.61793	7.94660	7.94660

S-Scheme 2 order (A-III-2)

Initial data 1, 1, 1, 1. Step length: 0.05

	p^1	p^2	q^1	q^2
$t = 0.5$	0.88045	0.88045	1.46635	1.46635
$t = 2.0$	0.74706	0.74706	2.66732	2.66732
$t = 4.0$	0.68262	0.68262	4.08789	4.08789
$t = 7.0$	0.64001	0.64001	6.06348	6.06348
$t = 10.0$	0.61803	0.61803	7.94756	7.94756

Table 7. Comparison of the exact solution with the numerical results of eq. (5.11) with (6.4) using the scheme (6.3) adapted from A-III- 4 scheme.

Table 7

$h = 2 * 3.14159/40 \quad t = 1.0$

$X = M * h$	$\sin(x+i)$	A-III-4 $Dt = 0.15$	A-III-4 $Dt = 0.3$
$M = 0$	−0.54402	−0.54419	−0.54470
$M = 2$	−0.77668	−0.77679	−0.77717
$M = 4$	−0.93332	−0.93335	−0.93358
$M = 6$	−0.99859	−0.99855	−0.99861
$M = 8$	−0.96611	−0.96600	−0.96591
$M = 10$	−0.83907	−0.83890	−0.83870
$M = 12$	−0.62989	−0.62967	−0.62940
$M = 14$	−0.35906	−0.35878	−0.35850
$M = 16$	−0.05307	−0.05279	−0.05251
$M = 18$	0.25811	0.25835	0.25657
$M = 20$	0.54402	0.54422	0.54434
$M = 22$	0.77668	0.77682	0.77680
$M = 24$	0.93332	0.93337	0.93320
$M = 26$	0.99886	0.99857	0.99823
$M = 28$	0.96611	0.96601	0.96552
$M = 30$	0.83907	0.83916	0.83829
$M = 32$	0.62989	0.62968	0.62900
$M = 34$	0.35906	0.35881	0.35809
$M = 36$	0.05307	0.05282	0.05212
$M = 38$	−0.25811	−0.25832	−0.25896
$M = 40$	−0.54402	−0.54419	−0.54471

Fig. 1. Intersections of the orbits of Hénon-Heiles system with the Poincare section, computed by 2-nd order S-schemes. (A) : $H = \frac{1}{24}$, all motions are regular. (B): $H = \frac{1}{12}$, all motions are regular. (C): $H = \frac{1}{8}$, most of the motions are regular but chaotic motions appear. (D): $H = \frac{1}{6}$, almost all motions are chaotic.

Fig. 2. Orbit of the conservative system (7.11) with $\theta = 0$, (A) : $h = 0.4, u_0 = 0.58$. There are 2 rings, each ring contains 11 archipelagoes, each archipelago consists basically of 5 isles, (B) : $h = 0.1, u_0 = 0.54$. There are 2 rings with simple, regular structure. (C): $h = 0.4, u_0 = 0.5$. (D): $h = 0.4, u_0 = 0.6$.In (C), (D) the structure are grossly similar to that of (A), but more irregular, especially in (D).

Fig. 3. Orbit of dissipative system (7.11) with $\theta > 0$, arising from mixed leap-frog schemes (7.10) for system (7.5). (A) : $h = 1.898$, $\theta = 0.538$, $u_0 = 0.58$. This is the strange attractor of the Hénon map. (B) : $h = 0.1$, $\theta = 0.01$, $u_0 = 0.6$. (C) : $h = 0.4$, $\theta = 0.01$, $u_0 = 0.56$. (D) : $h = 0.5$, $\theta = 0.01$, $u_0 = 0.56$. The last 3 graphs show interesting spiral structures. Here (B) is a dissipative counterpart of Fig. 2(B), with 2 rings changed into 2 contracting spirals. In (C), (D), the spirals have more complicated

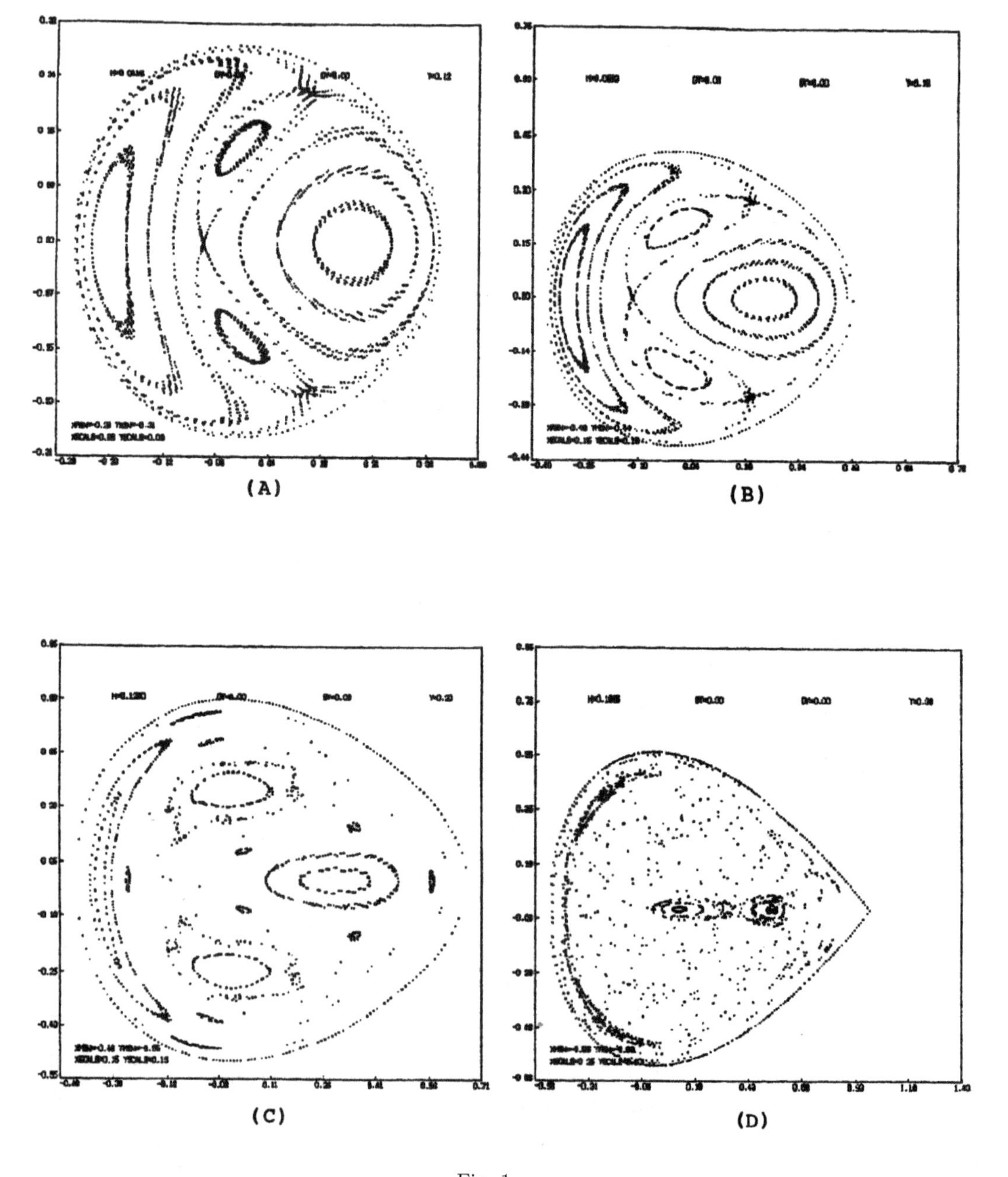

Fig. 1

structures than that of (B). Note that the number of branches of the contracting spiral is 8 in (C), 6 in (D).

Fig. 2 (A, C, D) and Fig. 3 (B, C, D) contain some interesting features, which seems, to the authors' knowledge, to be new.

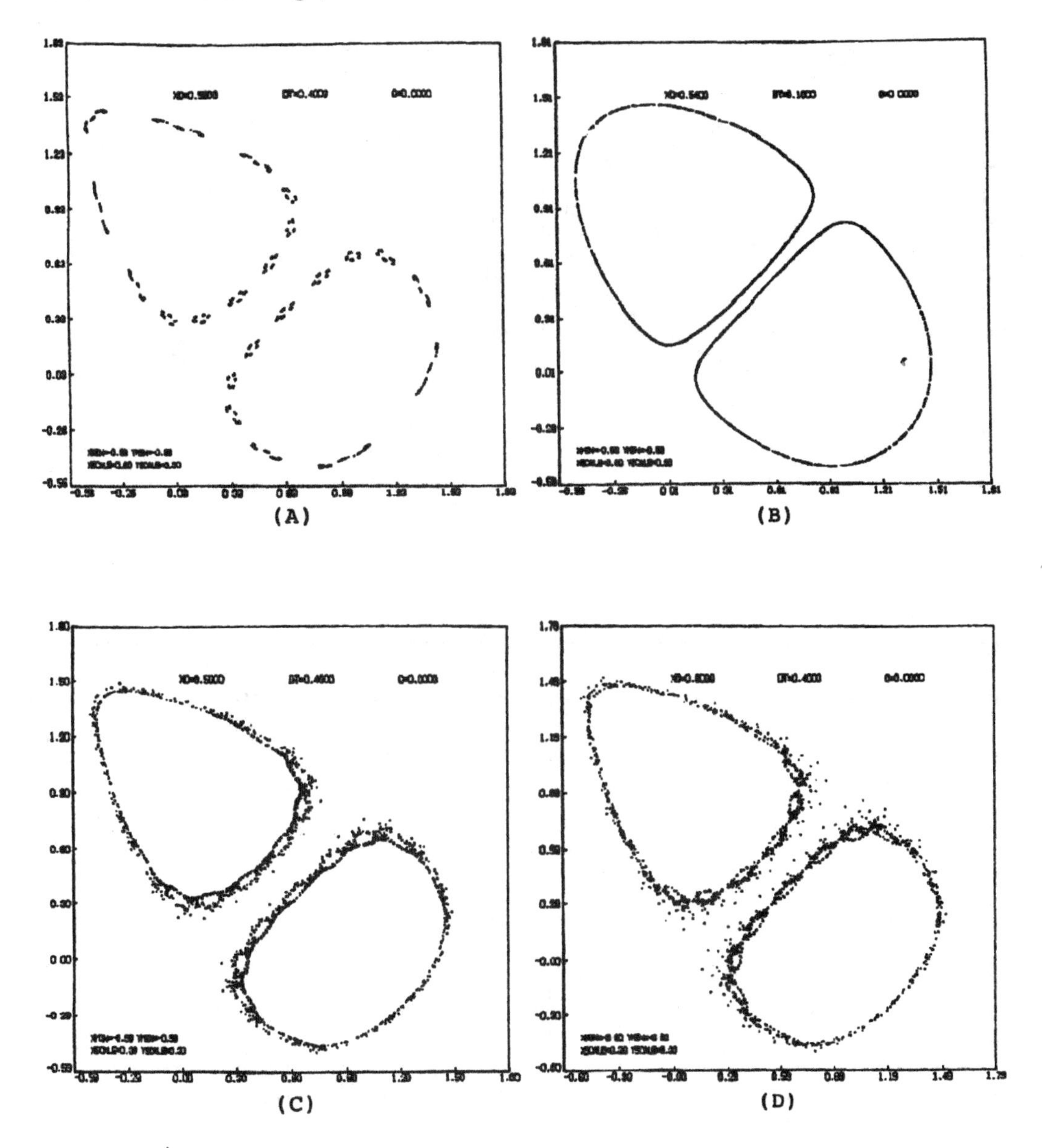

Fig. 2

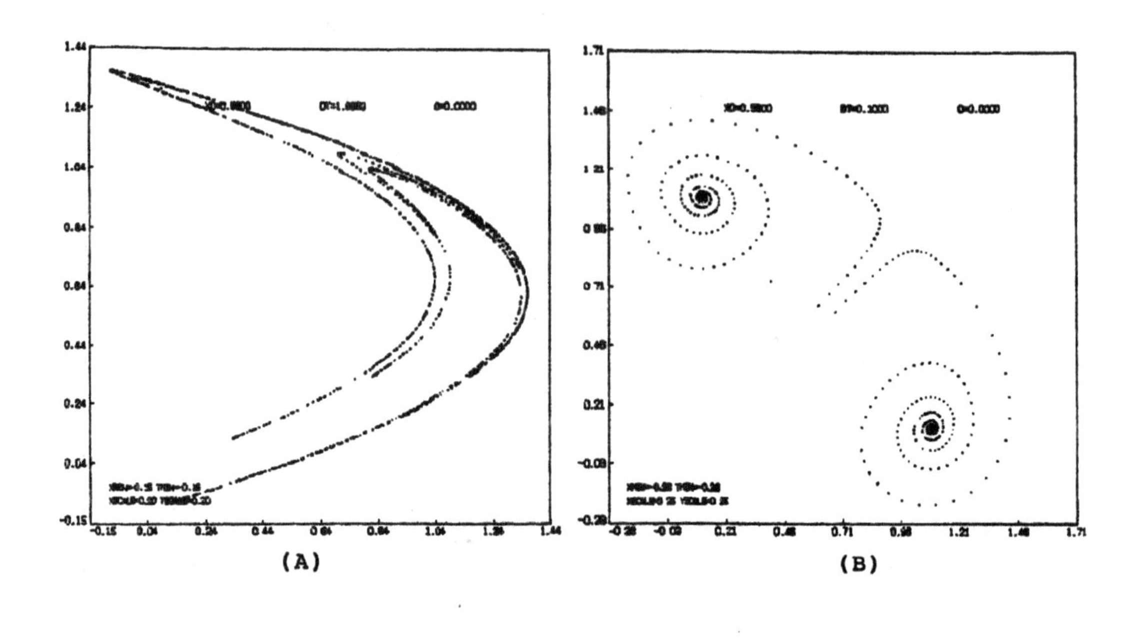

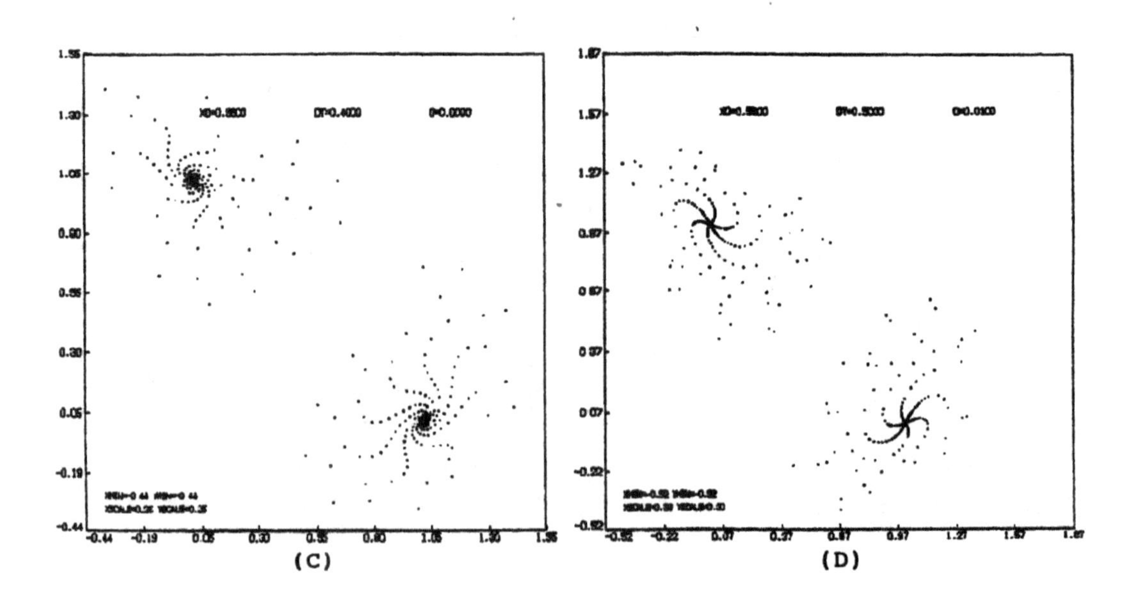

Fig. 3

References

[1] Feng K. On difference schemes and symplectic geometry. In Feng K, ed. Proc 1984 Beijing Symp Diff Geometry and Diff Equations. Beijing: Science Press, 1985: 42-58

[2] Feng K. Canonical difference schemes for Hamiltonian canonical differential equation. In Xiao S, ed. Inter workshop on applied differential equation. Singapore: World Scientific. 1985: 59-71

[3] Feng K. Symplectic geometry and numerical methods in fluid dynamics. In Zhuang F G, Zhu Y L, eds. Proc 10th Inter Conf on Numerical Methods in Fluid Dynamics. Lecture Notes in Physics 264. Berlin: Springer Verlag. 1986: 1-7

[4] Feng K. Difference schemes for Hamiltonian formalism and symplectic geometry. J Comp Math. 1986, 4(3): 279-289

[5] Feng K, Wu H M, Qin M Z. Symplectic difference schemes for linear Hamiltonian canonical systems. J Comp Math. 1990, 8(4): 371-380

[6] Feng K, Wu H M, Qin M Z, Wang D L. Construction of canonical difference schemes for Hamiltonian formalism via generating functions. J Comp Math. 1989, 7(1): 71-96

[7] Ge Z, Feng K. On the approximation of Hamiltonian systems. J Comp Math. 1988, 6(1): 88-97

[8] Li C W, Qin M Z. Symplectic difference schemes for infinite dimensional Hamiltonian systems. J Comp Math. 1988, 6(2): 164-174

[9] Qin M Z. A symplectic difference scheme for the Hamiltonian equation. J Comp Math. 1987, 5: 203-209

[10] Qin M Z. Leap-frog schemes of two kinds of Hamiltonian systems for wave equation. Mathematical Numerical Sinica. 1988, 7(3): 272-281

[11] Ge Z. Generating functions for the Poisson map: [Preprint]. Beijing: CAS. Computing Center, 1987

[12] Arnold V I. Mathematical Methods of Classical Mechanics. New York: Springer. 1978

[13] Siegel C L. Symplectic geometry. Amer J Math. 1943, 65: 1-86

[14] Hua L K. On the theory of automorphic functions of a matrix variable I, II. Amer J Math. 1944, 66: 470-488, 531-563

[15] Li C W. Numerical calculation for the Hamiltonian System: [Master dissertation]. Beijing: CAS. Computing Center, 1986

[16] Lee T D. Discrete Mechanics. CU-TP-267. A Series of Four Lectures Given at the International School of Subnucleax Physics, Erice, August 1983.

[17] Vainberg M. Variational method in nonlinear operator analysis. San Francisco: Holden Day. 1964

[18] Abraham R, Marsden J. Foundation of Mechanics. Mass: Addison-Wesley, Reading. 1978

[19] Chernoff P R, Marsden J E. Properties of infinite dimensional Hamiltonian. Lecture Notes in Mathematics 425. Berlin: Springer.

[20] Miller J H. On the Location of zeros of certain classes of polynomials with application to numerical analysis. J Inst Math Appl. 1971, 8: 397-406

[21] Hènon M, Heiles C. The applicability of the third integral of motion: some numerical experiments. Astron J. 1964, 69: 73-79

[22] Hènon M. A two-dimensional mapping with a strange attractor. Comm Math Phys. 1976, 50: 69-77

[23] Sanz-Serna J M. Studies in numerical nonlinear instability I. SIAM J Sci Stat Comp. 1985, 6(4): 923-938

[24] Yamaguti M, Ushiki S. Chaos in numerical analysis of ordinary differential equations. Phys 3D. 1981: 618-626

[25] Huang M Y. Hamiltonian approximation for the nonlinear wave equation. In this Proceedings.

7 On the Approximation of Linear Hamiltonian Systems①②

论对线性哈密尔顿系统的逼近

Abstract

When we study the oscillation of a physical system near its equilibrium and ignore dissipative effects, we may assume it is a linear Hamiltonian system (H-system), which possesses a special symplectic structure. Thus there arises a question: how to take this structure into account in the approximation of the H-system? This question was first answered by Feng Kang for finite dimensional H-systems[1−4]

We will in this paper discuss the symplectic difference schemes preserving the symplectic structure and its related properties, with emphasis on the infinite dimensional H-systems.

In the first section we propose the notion of symmetry of a difference scheme, and obtain the equivalence between symmetries and the conservation of first integrals. In the second section we discuss hyperbolic equations with constant coefficients in one space variable. This kind of H-system possesses not only a symplectic structure but also a unitary structure. Our result is that a difference scheme is symplectic iff its amplification factors are of modulus one. In the third section we discuss symmetric hyperbolic equations with constant coefficients in several space variables. Although the antisymmetric operator of the symplectic structure is not invertible in that case, we obtain a similar conclusion. In the fourth section, we propose the notion of multiplelevel symplectic difference schemes. Finally, we derive the generating function for K-symplectic transformation, and construct an SDS for a hyperbolic equation with variable coefficients using the generating functions.

§1 Symmetries of Difference Schemes

a) Consider a linear Hamiltonian system with quadratic Hamiltonian $H(z)=\frac{1}{2}z^TAz$:

$$\frac{dz}{dt}=J^{-1}Az, \tag{0}$$

where $J=\begin{bmatrix} 0 & I_n \\ -I_n & O \end{bmatrix}$ and A is a $2n\times 2n$ symmetric matrix, and a difference scheme

$$z^{m+1}=\phi\left(J^{-1}A\right)z^m. \tag{1}$$

---

① Joint with Ge Z

② J Comp Math. 1988, 6(1): 88-97

Definition We say (1) is a symplectic difference scheme if the matrix $\phi = \phi(J^{-1}A)$ is a symplectic matrix.

Now we perform a canonical coordinate transformation $z \to w : z = Pw$, and the H-system written in the new coordinate w is

$$\frac{dw}{dt} = J^{-1}P^TAPw \tag{2}$$

and the scheme (1) is

$$w^{m+1} = P^{-1}\phi\left(J^{-1}A\right)Pw^m. \tag{3}$$

Now we construct the difference scheme for the H-system (2)

$$w^{m+1} = \phi\left(J^{-1}P^TAP\right)w^m. \tag{4}$$

Then there arises a problem: Is scheme (3) equivalent to scheme (4)? The answer leads to the notion of symmetry.

Definition We say a scheme (1) is invariant under a group G of linear symplectic transformations if

$$P^{-1}\phi\left(J^{-1}A\right)P = \phi\left(J^{-1}P^TAP\right)$$

for all A symmetric and all $P \in G$.

Another question of interest is: If a quadratic form $f(z) = \frac{1}{2}z^TBz$ is a first integral of the H-system, is it conserved by scheme (1), i.e.,

$$f\left(z^{m+1}\right) = f\left(z^m\right)$$

for all $z^{m+1} = \phi\left(J^{-1}A\right)z^m$?

Example 1. Euler's mid-point scheme

$$z^{m+1} - z^m = \tau J^{-1}A\left(\left(z^{m+1} + z^m\right)/2\right) \tag{5}$$

where τ is the time step, is symplectic and is invariant under the full symplectic group $Sp(2n)$. All the quadratic first integrals are conserved by the scheme (5)[2].

Theorem 1 *Let the quadratic form $f(z) = \frac{1}{2}z^TBz$ be a first integral of the H-system (0), $G^t = exp\left(tJ^{-1}B\right)$ is the phase flow of the H-system with Hamiltonian $f(z)$, the scheme (1) is symplectic. Then f is conserved by (1) iff (1) is invariant under the phase flow G^t.*

Proof. By definition,

$$\left(G^t\right)^{-1}\phi\left(J^{-1}A\right)G^t = \phi\left(\left(G^t\right)^{-1}J^{-1}AG^t\right) = \phi\left(J^{-1}\left(G^t\right)^TAG^t\right). \tag{6}$$

f is a first integral of the H-System. By the Noether theorem, $(G^t)^T AG^t = A$. Thus it follows from (6) that

$$(G^t)^{-1} \phi(J^{-1}A) G^t = \phi(J^{-1}A), \quad \text{or} \quad \phi(J^{-1}A) G^t = G^t \phi(J^{-1}A).$$

Taking derivative at $t = 0$ and making use of $\left.\frac{d}{dt}\right|_{t=0} G^t = J^{-1}B$, we obtain

$$(\phi(J^{-1}A))^T B\phi(J^{-1}A) = (\phi(J^{-1}A))^T J\phi(J^{-1}A) J^{-1}B = JJ^{-1}B = B.$$

Thus the symmetries imply conservations.

Now we assume conservations, i.e, $(\phi(J^{-1}A))^T B\phi(J^{-1}A) = B$. Then

$$J^{-1}B(\phi(J^{-1}A)) = \phi(J^{-1}A) J^{-1}B,$$

and

$$\phi(J^{-1}A) \exp(J^{-1}Bt) = \exp(J^{-1}Bt) \phi(J^{-1}A).$$

Example 2. The generalized Euler scheme $z^{m+1} = \phi(\tau J^{-1}A) z^m$, where $\phi_p(\lambda)$ is the p-th diagonal Pade approximant of $\exp \lambda$, is invariant under $Sp(2n)$. Hence all the first integrals of quadratic form are conserved by the difference scheme[1,2,4,].

b) We will generalize the above motion to the nonlinear Hamiltonian system:

$$\frac{dz}{dt} = J^{-1}H_z \tag{0$'$}$$

Suppose we have a symplectic difference scheme derived according to a certain rule:

$$z^{m+1} = \phi_{J^{-1}H_z}(z^m) \tag{1$'$}$$

where $\phi_{J^{-1}H_z}$ is a nonlinear symplectic transformation (locally). Now we perform a nonlinear symplectic transformation $z = S(w)$. Then the H-system $(0)'$ written in the coordinate w is

$$\frac{dw}{dt} = J^{-1}\tilde{H}_w$$

where $\tilde{H}(w) = H(S(w))$.

Now the scheme $(1)'$ written in the new coordinate is

$$w^{m+1} = S^{-1} \circ \phi \circ S(w^m).$$

Definition We say the scheme $(1)'$ is invariant under a group G of symplectic transformations, if

$$S^{-1} \circ \phi_{J^{-1}H_z} \circ S = \phi_{J^{-1}\tilde{H}_w}, \quad \forall S \in G.$$

Theorem 2 *Suppose f is a first integral of the H-system $(0)'$, and G^t is the phase flow of the Hamiltonian function f. Then f is conserved up to a constant by the scheme $(1)'$*

$$f \circ \phi_{J^{-1}H_z}(z) = f(z) + c, \quad c \text{ is a constant,}$$

iff the scheme $(1)'$ is invariant under G^t.

The proof will be given in subsection c).

Remark If the scheme has a fixed point, i.e. a point z, such that

$$\phi_{J^{-1}H_z}(z) = z,$$

then the constant $c = 0$. This is often the case in practice.

Example 3. Euler's mid-point scheme

$$z^{n+1} - z^n = \tau J^{-1} H_z \left(\frac{z^n + z^{n+1}}{2} \right)$$

is invariant under any linear symplectic transformation. Thus every first integral of quadratic form is conserved by this scheme.

Example 4. The staggered explicit scheme for separable Hamiltonian function $H(p, q) = U(p) + V(q)$(cf. [1])

$$\begin{aligned} &\frac{1}{\tau}\left(p^{k+1} - p^k\right) = -V_q\left(q^{k+1/2}\right), \\ &\frac{1}{\tau}\left(q^{k+1+1/2} - q^{k+1/2}\right) = U_p\left(p^{k+1}\right) \end{aligned}$$

is invariant under a canonical transformation of the form

$$\begin{bmatrix} A^{-T} & 0 \\ 0 & A \end{bmatrix}.$$

Thus the linear momentum and angular momentum are conserved by the scheme.

c) The proof of Theorem 2.

We will work in the general context of a phase manifold P (cf. [5]) with a symplectic 2-form w. Recall that a function f on P determines a Hamiltonian vector field $J^{-1}df$ such that

$$\langle df, x \rangle = \omega\left(x, J^{-1}df\right), \quad \forall x \in TP.$$

First we assume the symmetry, changing the notation $\phi_{J^{-1}H_z} = \phi_{J^{-1}dH}$ for convenience,

$$\left(G^t\right)^{-1} \circ \varphi_{J^{-1}dH} \circ G^t = \varphi_{J^{-1}(G^t)*dH}.$$

Since f is a first integral of the system $(0)'$,

$$\left(G^t\right)^* dH = d\left(H \circ G^t\right) = dH.$$

Thus

$$G_f^t \circ \phi_{J^{-1}dH} = \phi_{J^{-1}dH} \circ G^t.$$

Taking derivative at $t = 0, a \in P$, we have

$$J^{-1}df \circ \phi_{J^{-1}dH}(a) = (\phi_{J^{-1}dH})_* J^{-1}df(a).$$

Then for $x \in T_{\phi(a)}P, \phi = \phi_{J^{-1}dH}$,

$$\begin{aligned}
\omega\left(J^{-1}df(a), \left(\phi^{-1}\right)_* x\right) &= \phi^*\omega\left(J^{-1}df(a), \left(\phi^{-1}\right)_* x\right) \\
&= \omega\left((\phi)_* J^{-1}df(a), x\right) = \omega\left(J^{-1}df \circ \phi(a), x\right), \\
\left\langle \left(\phi^{-1}\right)^* df(a), x \right\rangle &= \left\langle df(a), \left(\phi^{-1}\right)_* x\right\rangle = \omega\left(\left(\phi^{-1}\right)_* x, J^{-1}df(a)\right) \\
&= \omega\left(x, J^{-1}df \circ \phi(a)\right) = \langle df \circ \phi(a), x\rangle.
\end{aligned}$$

Hence

$$\left(\left(\phi^{-1}\right)^* df - df\right)(\phi(a)) = 0,$$

or,

$$\begin{aligned}
&d(f \circ \phi) - df = 0, \\
&f \circ \phi = f + c \quad (c \text{ is constant}).
\end{aligned}$$

Now we assume the conservation

$$f \circ \phi_{J^{-1}dH} = f + c.$$

We can prove that

$$(\phi_{J^{-1}dH})_* J^{-1}df = J^{-1}df \circ \phi_{J^{-1}dH}.$$

The phase flows of the vector fields $J^{-1}df$ and $(\phi_{J^{-1}dH}) J^{-1}df \circ \phi_{J^{-1}dH}^{-1}$ are $G^t, \phi_{J^{-1}dH} \circ G^t \circ \phi_{J^{-1}dH}^{-1}$ respectively. Therefore

$$G^t = \phi_{J^{-1}dH} \circ G^t \circ \phi_{J^{-1}dH}^{-1}.$$

§2 Symplectic Difference Schemes for Hyperbolic Equations

Because the H system treated here possesses a unitary structure, we will make use of the language of complex number.

a) Consider the phase space $\mathbb{R}^{2n}$ with a canonical coordinates p_i, q_i. We can regard it as an n-dimensional complex space $\mathbb{C}^n$, with the coordinates $z_j = p_j + iq_j$.

Recall that an $n \times n$ complex matrix can be realized as a $2n \times 2n$ real matrix A. On the other hand, a $2n \times 2n$ real matrix A can be complexified into an $n \times n$ complex matrix $C(A)$ such that

$$c(Az) = C(A)c(z), \quad \forall z \in \mathbb{R}^{2n}$$

iff A is of the form

$$A = \begin{bmatrix} A_1 & -B_1 \\ B_1 & A_1 \end{bmatrix}, \quad C(A) = A_1 + iB_1, z = \begin{bmatrix} p \\ q \end{bmatrix}, c(z) = p + iq.$$

In particular, $C\left(J_{2n}^{-1}\right) = I_n i$.

A symplectic matrix can be complexified iff it is an orthogonal matrix at the same time, and $C(A)$ is a unitary matrix. On the other hand, a complexifiable matrix is symplectic iff it is unitary

$$U(n) = Sp(2n) \bigcap O(2n) = Sp(2n) \bigcap GL(n, C).$$

b) Consider a hyperbolic equation with periodic boundary condition:

$$\frac{\partial u}{\partial t} = \frac{\partial u}{\partial x}. \tag{7}$$

Here the antisymmetric operator is $D = \dfrac{\partial}{\partial x}$, the symplectic form is $\omega(w, v) = \displaystyle\int_0^{2\pi} wDvdx$ and the Hamiltonian functional is $H(u) = \dfrac{1}{2}\displaystyle\int_0^{2\pi} u^2 dx$ with functional derivative $H_u = u$. So (7) is an infinite dimensional H-system (cf. [6])

$$\frac{\partial u}{\partial t} = DH_u.$$

Using complex Fourier expansion for the real periodic $u(x, t)$, we have

$$u(x, t) = \sum_{-\infty}^{\infty} c_k(t) e^{ikx}, \quad c_{-k} = \bar{c}_k, \quad \frac{dc_k}{dt} = \sum_{-\infty}^{\infty} ikc_k.$$

Now consider a difference scheme approximating (7):

$$P_1(E)u^{n+1} = P_2(E)u^n \tag{8}$$

where $P_1(\lambda)$ and $P_2(\lambda)$ are functions of the form

$$\sum a_k \lambda^k, \quad k \text{ are integers}$$

and E is the shift operator: $(Eu)(x) = u(x + h)$; h is the space step.

Expand the functions u^n, u^{n+1} in Fourier series

$$u^n = \sum c_k^n e^{ikx}, \quad u^{n+1} = \sum c_k^{n+1} e^{ikx},$$

and substitute them into (8). Then

$$P_1\left(e^{ikh}\right) c_k^{n+1} = P_2\left(e^{ikh}\right) c_k^n.$$

Hence the transformation $u^n \rightarrow u^{n+1}$can be complexified. From the discussion in subsection a, we have

Theorem 3 *The difference scheme* (8) *is symplectic iff the amplification factors are of modulus one*

$$\left|P_2\left(e^{ikh}\right)/P_1\left(e^{ikh}\right)\right| = 1.$$

c) Consider the following hyperbolic system with periodic boundary condition

$$\frac{\partial U}{\partial t} = \frac{\partial}{\partial x}(AU), \tag{9}$$

where $U = (u_1, \cdots, u_m)^T$, A is a constant real symmetric matrix, the antisymmetric operator is $D = I_m \frac{\partial}{\partial x}$, the H functional is $\frac{1}{2}\int U^T AU dx$, and the symplectic two-form is

$$\omega(U, V) = \int U^T DV dx.$$

We have similar Fourier expansions as in subsection b), but with complex vector coefficients C_k instead. Now consider a difference scheme

$$P_1(E)U^{n+1} = P_2(E)U^n \tag{10}$$

where $P_1(\lambda), P_2(\lambda)$ are $n \times n$ matrices, their elements are functions of the form

$$\sum a_k \lambda^k.$$

Let C_k^n, C_k^{n+1} be the Fourier coefficients of U^n, U^{n+1} respectively, then

$$P_1\left(e^{ikh}\right) C_k^{n+1} = P_2\left(e^{ikh}\right) C_k^n.$$

Theorem 4 *The scheme* (10) *is symplectic iff the amplification matrices*

$$P_1^{-1}\left(e^{ikh}\right) P_2\left(e^{ikh}\right)$$

are unitary.

§3 The Case for Hyperbolic Equations in Several Space Variables

a) Consider an equation

$$\frac{\partial u}{\partial t}=\frac{\partial u}{\partial x}+2\frac{\partial u}{\partial y}.$$

Here the H operator is $D=\frac{\partial}{\partial x}+2\frac{\partial}{\partial y}$, which is not invertible. Therefore the notion of canonical transformation needs reconsideration.

b) Consider a scalar hyperbolic equation in several periodic space variables

$$\frac{\partial U}{\partial t}=\left(A_1\frac{\partial}{\partial x_1}+A_2\frac{\partial}{\partial x_2}+\cdots+A_n\frac{\partial}{\partial x_n}\right)U.$$

Let antisymmetric operator $D=A_1\frac{\partial}{\partial x_1}+\cdots+A_n\frac{\partial}{\partial x_n}$, and the action of D on $e^{iK\cdot X}$ is $D\cdot e^{iK\cdot X}=iA\cdot Ke^{iK\cdot X}$, with the notations

$$A=(A_1,\cdots,A_n),\quad K=(k_1,\cdots,k_n),\quad A\cdot K=A_1k_1+\cdots+A_nk_n.$$

We say an index K is resonant if $A\cdot K=0$; otherwise it is non-resonant.

Now the given equation is a H-system with Hamiltonian functional $H(U)=\frac{1}{2}\int U^2dX$ and symplectic form $\omega(U,V)=\int UDVdX$.

Definition A linear bounded invertible transformation T is a symplectic transformation if

(1) $\omega(TU,TV)=\omega(U,V)$ holds for all U,V;

(2) if $DU=0$, then $TU=U$.

c) Consider a difference scheme

$$P_1(E)U^{m+1}=P_2(E)U^m \tag{11}$$

where $P_1(\lambda),P_2(\lambda)$ are functions of the form

$$\sum a_K\lambda_1^{k_1}\cdots\lambda_n^{k_n}=\sum a_K\lambda^K$$

with the notations $E=(E_1,\cdots,E_n),E_iU(X)=U(X+h_ie_i);h=(h_1,\cdots,h_n)$. Then

$$P_1(E)e^{iK\cdot X}=P_1\left(e^{iK\cdot h}\right)e^{iK\cdot X}.$$

Now assume that $U^{m+1}=\sum C_K^{m+1}e^{iK\cdot X},U^m=\sum C_K^me^{iK\cdot X}$. Substituting them into (11), we have

$$P_1\left(e^{iK\cdot h}\right)C_K^{m+1}=P_2\left(e^{iK\cdot h}\right)C_K^m.$$

Theorem 5 *The scheme* (11) *is symplectic iff* $\left|\left(P_1\left(e^{iK\cdot h}\right)\right)^{-1} P_2\left(e^{iK\cdot h}\right)\right| = 1$ *if* K *is nonresonant and* $P_1\left(e^{iK\cdot h}\right) = P_2\left(e^{iK\cdot h}\right)$ *otherwise.*

Proof. Suppose the condition of the theorem is satisfied. The matrix $\left(P_1\left(e^{iK\cdot h}\right)\right)^{-1} \cdot P_2\left(e^{iK\cdot h}\right)$ is denoted by F_K,

$$U^m = \sum B_K e^{iK\cdot X}, \quad U^{m+1} = \sum F_K B_K e^{iK\cdot X},$$

$$V^m = \sum C_K e^{iK\cdot X}, \quad V^{m+1} = \sum F_K C_K e^{iK\cdot X},$$

$$\omega\left(U^m, V^m\right) = \sum_{k_1 \geqslant 0} Im\left(B_K \bar{C}_K\right)(A\cdot K)(2\pi)^n,$$

$$\omega\left(U^{m+1}, V^{m+1}\right) = \sum_{k_1 \geqslant 0} Im\left(B_K F_K \bar{F}_K \bar{C}_K\right)(A\cdot K)(2\pi)^n.$$

If K is non-resonant, $F_K \bar{F}_K = 1$; otherwise $A\cdot K = 0$. Therefore

$$\omega\left(U^m, V^m\right) = \omega\left(U^{m+1}, V^{m+1}\right).$$

If $DV^m = 0$, then $V^m = \sum_{K\in R} C_K e^{iK\cdot X}$, where R is the set of K satisfying $A\cdot K = 0$, $V^{m+1} = \sum_{K\in R} F_K C_K e^{iK\cdot X} = V^m$.

The converse is easy to prove.

§4 Multiple-level Symplectic Difference Schemes

a) We first consider a H-system

$$\frac{\partial z}{\partial t} = J^{-1} A z.$$

Let $v(t) = z(t-\tau)$; then

$$\frac{d}{dt}\begin{bmatrix} z \\ v \end{bmatrix} = \begin{pmatrix} 0 & J^{-1} \\ J^{-1} & 0 \end{pmatrix}\begin{pmatrix} 0 & A \\ A & 0 \end{pmatrix}\begin{pmatrix} z \\ v \end{pmatrix},$$

where the antisymmetric operator is $\begin{bmatrix} 0 & J^{-1} \\ J^{-1} & 0 \end{bmatrix}$.

Consider a 3-level difference scheme of the form

$$z^{n+1} - \phi_1 z^n - \phi_2 z^{n-1} = 0. \tag{12}$$

By introducing a new variable $v^n = z^{n-1}$, (12) can be written in the form

$$\begin{bmatrix} z^{k+1} \\ v^{k+1} \end{bmatrix} = \begin{bmatrix} \phi_1 & \phi_2 \\ I_n & 0 \end{bmatrix}\begin{bmatrix} z^k \\ v^k \end{bmatrix}.$$

Definition We say a difference scheme (12) is symplectic if

$$\begin{bmatrix} \phi_1 & \phi_2 \\ I_n & 0 \end{bmatrix}^T \begin{bmatrix} 0 & J^{-1} \\ J^{-1} & 0 \end{bmatrix} \begin{bmatrix} \phi_1 & \phi_2 \\ I_n & 0 \end{bmatrix} = \begin{bmatrix} 0 & J^{-1} \\ J^{-1} & 0 \end{bmatrix}.$$

After a short computation, we can prove

Lemma 1 *Scheme* (12) *is symplectic iff* ϕ_1 *is an infinitesimal symplectic matrix and* $\phi_2 = I$.

For example, the following schemes are symplectic,

$$z^{n+1} - \phi_m\left(\tau J^{-1} A\right) z^n - z^{n-1} = 0$$

where

$$\phi_0(\lambda) = 2\lambda,$$

$$\phi_m(\lambda) = 2\sum_{k=0}^{m} \frac{\lambda^{2k+1}}{(2k+1)!}.$$

b) A real vector $z \in \mathbb{R}^{2n}$ can also be regarded as a complex vector $c(z)$. Then from

$$z^{n+1} - \phi z^n - z^{n-1} = 0$$

we have

$$c\left(z^{n+1}\right) - c\left(\phi z^n\right) - c\left(z^{n-1}\right) = 0.$$

If matrix ϕ can be complexified as $C(\phi)$, the above can be written as

$$c\left(z^{n+1}\right) - C(\phi) c\left(z^n\right) - c\left(z^{n-1}\right) = 0.$$

Lemma 2 *The necessary and sufficient condition that an infinitesimal symplectic matrix can be complexified is that it is an infinitesimal orthogonal matrix, and the complexified matrix is an infinitesimal unitary matrix.*

Lemma 3 *A complex matrix is an infinitesimal symplectic matrix iff it is an infinitesimal unitary matrix.*

The proofs of these lemmas are easy.

c) Consider a three-level difference scheme for $\dfrac{\partial u}{\partial t} = \dfrac{\partial u}{\partial x}$:

$$u^{n+1} - 2\tau P_1(E) u^n - P_2(E) u^{n-1} = 0.$$

It is easy to prove that it is symplectic iff $P(E)$ is a centered differencing $P_1\left(E^{-1}\right) = -P(E), P_2(E) = I$.

Note that the well known leap-frog scheme is symplectic. Another one with accuracy $O(\tau^4+h^4)$ is

$$\begin{aligned}&\frac{u^{n+1}(x)-u^{n-1}(x)}{2\tau}+\frac{1}{12h}\left(u^n(x+2h)-8u^n(x+h)+8u^n(x-h)-u^n(x-2h)\right)\\=&\frac{\tau^2}{36h^3}\left(u^n(x+2h)-2u^n(x+h)+2u^n(x-h)-u^n(x-2h)\right).\end{aligned}$$

d) The above notion can be generalized to the non-linear H-system

$$\frac{dz}{dt}=J^{-1}H_z.$$

Consider a three-level difference scheme

$$z^{n+1}-\phi_1\left(z^n\right)-\phi_1\left(z^{n-1}\right)=0. \tag{13}$$

Introduce a new variable $v^n=z^{n-1}$, then

$$\begin{aligned}z^{n+1}&=\phi_1\left(z^n\right)+\phi_2\left(v^n\right),\\v^{n+1}&=z^n.\end{aligned}$$

We say (13) is symplectic if the Jacobian $Q=\dfrac{\partial\left(z^{n+1},v^{n+1}\right)}{\partial\left(z^n,v^n\right)}$ satisfies

$$Q^T\begin{bmatrix}0&J^{-1}\\J^{-1}&0\end{bmatrix}Q=\begin{bmatrix}0&J^{-1}\\J^{-1}&0\end{bmatrix}.$$

Theorem 6 *Scheme (D) is symplectic iff* $\dfrac{\partial\phi_1}{\partial z^n}$ *is an infinitesimal symplectic matrix,* $\dfrac{\partial\phi_2}{\partial z^{n-1}}=I$.

For example, the leap-frog scheme

$$z^{n+1}-2\tau J^{-1}H_z\left(z^n\right)-z^{n-1}=0$$

is symplectic.

§5 Symplectic Schemes Based on Generating Functions

We will extend the construction of symplectic difference schemes based on the generating functions in [2].

Suppose K is an invertible skew-symmetric matrix. We say a transformation T is a K-symplectic transformation if its Jacobian matrix is a K-symplectic matrix.

Lemma 4 *A K-symplectic transformation, not too far from the identity, can be given by a generating function ϕ of Euler type*

$$\tilde{w}=T(w), \quad K(\tilde{w}-w)=\phi_w\left(\frac{w+\tilde{w}}{2}\right).$$

Proof. There exists a non-singular matrix P s.t. $P^TKP=J$. Let $z=Pw, \tilde{z}=P\tilde{w}$. Then the transformation $z\to\tilde{z}=P^{-1}\circ T\circ P(z)$ is a J-symplectic transformation, which can be given by a generating function $\tilde{\phi}$:

$$J(\tilde{z}-z)=\tilde{\phi}_z\left(\frac{\tilde{z}+z}{2}\right).$$

Let $\phi(w)=\tilde{\phi}(Pw)$. Then from above,

$$K(\tilde{w}-w)=\phi_w\left(\frac{\tilde{w}+w}{2}\right).$$

Now we are going to construct a symplectic scheme for the following hyperbolic equation with variable coefficient $a(x)$

$$\frac{\partial u}{\partial t}=\frac{\partial}{\partial x}(au).$$

We first discretize in the time direction (cf. [2]),

$$\left(12-6\tau D+\tau^2D^2\right)u^{n+1}=\left(12+6\tau D+\tau^2D^2\right)u^n, \quad D=\frac{\partial}{\partial x},$$

which can be written in terms of a generating functional ϕ

$$\left(u^{n+1}-u^n\right)=\frac{\partial}{\partial x}\phi_u\left(\frac{u^{n+1}+u^n}{2}\right),$$

where $\phi_u=\dfrac{\delta\phi}{\delta u}$ is the variational derivative of

$$\phi(u)=6\tau\int ua\left(12-6\tau D+\tau^2D^2\right)^{-1}udx.$$

We approximate $\dfrac{\partial}{\partial x}$ by $P_1(E)=-\dfrac{1}{12}\left(E-8E+8E^{-1}-E^{-2}\right.$, $\phi(u)$ by $\phi_n(u)=6\tau\int ua(12-6\tau\left(P_1(E)a\right)+\tau^2\left(P_2(E)a\right)^2)^{-1}udx$, $P_2(E)=\dfrac{1}{h^2}\left(E-2I+E^{-1}\right)$. Then $\phi_n(u)$ gives rise to a symplectic transformation $u^n\to u^{n+1}$ by

$$u^{n+1}-u^n=P_1(E)\frac{\delta\phi_n}{\delta u}\left(\frac{u^{n+1}+u^n}{2}\right)$$

or,

$$\begin{aligned}&\left(12-6\tau P_1(E)a+\tau^2\left(P_2(E)a\right)^2\right)u^{n+1}\\=&(12+6\tau P_1(E)a+\tau^2\left(P_2(E)a\right)^2u^n.\end{aligned}$$

This is symplectic and of accuracy $O\left(\tau^4+h^4\right)$.

References

[1] Feng K. On difference schemes and symplectic geometry. In Feng K, ed. Proc 1984 Beijing Symp Diff Geometry and Diff Equations. Beijing: Science Press, 1985: 42–58

[2] Feng K. Difference schemes for Hamiltonian formalism and symplectic geometry. J Comp Math. 1986, 4(3): 279–289

[3] Feng K. Symplectic geometry and numerical methods in fluid dynamics. In Zhuang F G, Zhu Y L, eds. Proc 10th Inter Conf on Numerical Methods in Fluid Dynamics. Lecture Notes in Physics 264. Berlin: Springer Verlag. 1986: 1–7

[4] Feng K, Wu H M, Qin M Z. Symplectic difference schemes for linear Hamiltonian canonical systems. J Comp Math. 1990, 8(4): 371–380

[5] Arnold V I. Mathematical Methods of Classical Mechanics. New York: Springer. 1978

[6] Benjamin, Brooke T. Impulse, flow forces and variational principles. IMA J Applied Math. 1984, 32: 3–68

8 Construction of Canonical Difference Schemes for Hamiltonian Formalism via Generating Functions[①②]

构造哈密尔顿形式的典则差分格式的生成函数法

Abstract

This paper discusses the relationship between canonical maps and generating functions and gives the general Hamilton-Jacobi theory for time-independent Hamiltonian systems. Based on this theory, the general method—*the generating function method*—of the construction of difference schemes for Hamiltonian systems is considered. The transition of such difference schemes from one time-step to the next is canonical. So they are called the canonical difference schemes. The well known Euler centered scheme is a canonical difference scheme. Its higher order canonical generalizations and other families of canonical difference schemes are given. The construction method proposed in the paper is also applicable to time-dependent Hamiltonian systems.

Contents

§0 Introduction

As is well known, Hamiltonian systems have many intrinsic properties: the preservation of phase areas of even dimension and the phase volume, the conservation laws

① Joint with Wu H M, Qin M Z and Wang D L

② J Comp Math. 1989, 7(1): 71–96

of energy and momenta and other symmetries. The *canonicity* of the phase flow for time-independent Hamiltonian systems is the most important property. It ensures the preservation of phase areas and the phase volume. Thus we can hope that preserving the canonicity of transition of difference schemes from one time step to the next is also important in numerical solution of Hamiltonian systems. The first author in [1] has proposed a notion—*canonical difference schemes.* Just as its name implies, the transition of such difference schemes is canonical. In this paper, we give a general method —*generating function method*—for the construction of canonical difference schemes via generating functions. We first establish the relationship between canonical maps and generating functions and then give the general Hamilton-Jacobi theory for time-independent Hamiltonian systems. Given a matrix $\alpha \in \mathbf{CSp}\left(\tilde{J}_{4n}, J_{4n}\right)$, then canonical maps and generating functions can be determined each other under so called transversality conditions. Moreover, to the phase flow of the system with Hamiltonian H there corresponds a time-dependent generating function which satisfies the Hamilton-Jacobi equation related to the given α and H. If the Hamiltonian function is analytic, then the generating function can be expressed as a power series in t, and the series can be determined recursively (Theorem 20). So truncating or approximating it in some way, we can get certain canonical map which approximates the phase flow of the Hamiltonian system. Fixing t as the time step, we obtain difference schemes. In general, such difference scheme is implicit.

In Sec.1, we review some notions and facts about symplectic geometry which can be found in the standard texts, e.g., [2], [3], [4]. Sec.2 concerns linear fractional transformations. This theory is important for next sections. In Sec.3, we discuss the relationship between linear canonical maps and generating functions. It shows the outline of our idea. Sec.4 is the continuation and deepening of section 3. It gives the relationship between nonlinear canonical maps and generating functions. In Sec.5, we give the general Hamilton Jacobi theory. With the aid of the theory, generating functions can be represented as power series in t. It makes the preparation for constructing canonical difference schemes. In Sec.6 it shows the general method for the construction of canonical difference schemes. Many canonical difference schemes, such as Euler centered scheme, 4-th order centered scheme, staggered explicit scheme and others are presented.

We shall limit ourselves to the local case throughout the paper. Moreover, in this paper we use the older terminologies such as canonicity, canonical maps, etc in stead of the modern ones such as symplecticity, symplectic maps, etc. So the *canonical* difference schemes can also be called synonymously or even more preferably *symplectic* difference schemes.

§1 Preliminary Facts about Symplectic Geometry

We now review some notions and facts of symplectic geometry [2–4].

Let $\mathbf{R}^{2n}$ be a $2n$-dim real linear space. The elements of $\mathbf{R}^{2n}$ are $2n$-dim column vectors $z = (z_1, \cdots, z_n, z_{n+1}, \cdots, z_{2n})^T = (p_1, \cdots, p_n, q_1, \cdots, q_n)^T$. The superscript T represents the matrix transpose.

A symplectic form ω_K is a bilinear form defined by an anti-symmetric matrix $K \in \mathbf{GL}(2n)$ as

$$\omega_K(z_1, z_2) = z_1^T K z_2, \qquad \text{for all } z_1, z_2 \in \mathbf{R}^{2n}. \tag{1}$$

The symplectic form ω_J,

$$\omega_J(z_1, z_2) = z_1^T J z_2, \quad J = \begin{bmatrix} 0_n & I_n \\ -I_n & 0_n \end{bmatrix}, \qquad \text{for all } z_1, z_2 \in \mathbf{R}^{2n}, \tag{2}$$

where I_n and 0_n represent the $n \times n$ unit and zero matrix respectively, is called the standard symplectic form of $\mathbf{R}^{2n}$, briefly denoted by ω.

Every $2n \times n$ matrix

$$A = \begin{bmatrix} A_1 \\ A_2 \end{bmatrix} \in \mathbf{M}(2n, n) \quad \text{of rank } n, \quad A_1, A_2 \in \mathbf{M}(n),$$

defines an n-dim subspace $\{A\}$, spanned by its n column vectors. Evidently, $\{A\} = \{B\}$ iff $\exists P \in \mathbf{GL}(n)$ such that

$$AP = B, \quad i.e., \quad \begin{bmatrix} A_1 P \\ A_2 P \end{bmatrix} = \begin{bmatrix} B_1 \\ B_2 \end{bmatrix}.$$

An n-dim subspace $\{A\}$ is called K-Lagrangian if

$$\omega_K(z_1, z_2) = z_1^T K z_2 = 0, \qquad \text{for all } z_1, z_2 \in \{A\}.$$

Evidently, $\{A\}$ is K-Lagrangian iff

$$A^T K A = 0.$$

Define

$$\begin{aligned}
\mathbf{Sp}(K_1, K_2; 2n) &= \left\{M \in \mathbf{GL}(2n) \mid M^T K_2 M = K_1\right\}; \\
\mathbf{CSp}(K_1, K_2; 2n) &= \left\{M \in \mathbf{GL}(2n) \mid M^T K_2 M = \mu K_1, \text{ for some } \mu = \mu(M) \neq 0\right\}; \\
\mathbf{Sp}(K; 2n) = \mathbf{Sp}(K, K; 2n) &= \left\{M \in \mathbf{GL}(2n) \mid M^T K M = K\right\}; \\
\mathbf{CSp}(K; 2n) &= \mathbf{CSp}(K, K; 2n) \\
&= \left\{M \in \mathbf{GL}(2n) \mid M^T K M = \mu K, \text{ for some } \mu = \mu(M) \neq 0\right\}.
\end{aligned}$$

Sp $(K;2n)$ and **CSp** $(K;2n)$ are groups and called the K-symplectic group and the conformal K-symplectic group respectively. $\mathbf{Sp}(2n) \equiv \mathbf{Sp}(J;2n)$ and $\mathbf{CSp}(2n) \equiv \mathbf{CSP}(J;2n)$. They are usually called the symplectic group and the conformal symplectic group respectively.

Proposition 1

$$M \in \mathbf{Sp}\,(K_1,K_2;2n) \textit{ iff } M^{-1} \in \mathbf{Sp}\,(K_2,K_1;2n)\,.$$
$$M \in \mathbf{CSp}\,(K_1,K_2;2n) \textit{ iff } M^{-1} \in \mathbf{CSp}\,(K_2,K_1;2n)\,.$$

Proposition 2 *Let* $M_0 \in \mathbf{Sp}\,(K_1,K_2;2n)$. *Then*

$$\mathbf{Sp}\,(K_1,K_2;2n) = \mathbf{Sp}\,(K_2;2n)\cdot M_0 = M_0\cdot \mathbf{Sp}\,(K_1;2n)\,, \tag{3}$$
$$\mathbf{CSp}\,(K_1,K_2;2n) = \mathbf{CSp}\,(K_{2;}2n)\cdot M_0 = M_0\cdot \mathbf{CSp}\,(K_1;2n)\,, \tag{4}$$

where

$$\mathbf{Sp}\,(K_2;2n)\cdot M_0 = \{M\cdot M_0 | M \in \mathbf{Sp}\,(K_2;2n)\}$$

and others are similar.

§2 Linear Fractional Transformations

Definition 3 Let $\alpha = \begin{bmatrix} A_\alpha & B_\alpha \\ C_\alpha & D_\alpha \end{bmatrix} \in \mathbf{GL}(2m)$. Define a linear fractional transformation

$$\begin{aligned} &\sigma_\alpha : \mathbf{M}(m) \to \mathbf{M}(m), \\ &\qquad M \longrightarrow N = \sigma_\alpha(M) = (A_\alpha M + B_\alpha)\,(C_\alpha M + D_\alpha)^{-1} \end{aligned} \tag{5}$$

under the transversality condition

$$|C_\alpha M + D_\alpha| \neq 0. \tag{6}$$

Proposition 4 *Let* $\alpha \in \mathbf{GL}(2m)$. *Denote* $\alpha^{-1} = \begin{bmatrix} A^\alpha & B^\alpha \\ C^\alpha & D^\alpha \end{bmatrix}$. *Then*

$$|C_\alpha M + D_\alpha| \neq 0 \quad \textit{iff} \quad |MC^\alpha - A^\alpha| \neq 0, \tag{7}$$
$$|A_\alpha M + B_\alpha| \neq 0 \quad \textit{iff} \quad |B^\alpha - MD^\alpha| \neq 0. \tag{8}$$

Proof. From the equation

$$\begin{bmatrix} A_\alpha & B_\alpha \\ C_\alpha & D_\alpha \end{bmatrix}\begin{bmatrix} A^\alpha & B^\alpha \\ C^\alpha & D^\alpha \end{bmatrix} = \begin{bmatrix} A^\alpha & B^\alpha \\ C^\alpha & D^\alpha \end{bmatrix}\begin{bmatrix} A_\alpha & B_\alpha \\ C_\alpha & D_\alpha \end{bmatrix} = I_{2m} \tag{9}$$

i.e.,

$$
\begin{aligned}
&A_\alpha A^\alpha + B_\alpha C^\alpha = A^\alpha A_\alpha + B^\alpha C_\alpha = I_m, \\
&C_\alpha B^\alpha + D_\alpha D^\alpha = C^\alpha B_\alpha + D^\alpha D_\alpha = I_m, \\
&A_\alpha B^\alpha + B_\alpha D^\alpha = A^\alpha B_\alpha + B^\alpha D_\alpha = 0, \\
&C_\alpha A^\alpha + D_\alpha C^\alpha = C^\alpha A_\alpha + D^\alpha C_\alpha = 0,
\end{aligned}
\tag{10}
$$

we obtain the identities

$$
\begin{bmatrix} I & -M \\ C_\alpha & D_\alpha \end{bmatrix}\begin{bmatrix} A^\alpha & B^\alpha \\ C^\alpha & D^\alpha \end{bmatrix} = \begin{bmatrix} A^\alpha - MC^\alpha & B^\alpha - MD^\alpha \\ 0 & I \end{bmatrix},
$$
$$
\begin{bmatrix} I & -M \\ A_\alpha & B_\alpha \end{bmatrix}\begin{bmatrix} A^\alpha & B^\alpha \\ C^\alpha & D^\alpha \end{bmatrix} = \begin{bmatrix} A^\alpha - MC^\alpha & B^\alpha - MD^\alpha \\ I & 0 \end{bmatrix}.
$$

In addition,

$$
\begin{bmatrix} I & -M \\ C_\alpha & D_\alpha \end{bmatrix} = \begin{bmatrix} I & 0 \\ C_\alpha & I \end{bmatrix}\begin{bmatrix} I & -M \\ 0 & C_\alpha M + D_\alpha \end{bmatrix},
$$
$$
\begin{bmatrix} I & -M \\ A_\alpha & B_\alpha \end{bmatrix} = \begin{bmatrix} I & 0 \\ A_\alpha & I \end{bmatrix}\begin{bmatrix} I & -M \\ 0 & A_\alpha M + B_\alpha \end{bmatrix}.
$$

Taking their determinant, we get

$$
\begin{aligned}
&|C_\alpha M + D_\alpha|\,|\alpha|^{-1} = |A^\alpha - MC^\alpha|, \\
&|A_\alpha M + B_\alpha|\,|\alpha|^{-1} = (-1)^m\,|B^\alpha - MD^\alpha|.
\end{aligned}
$$

Q.E.D.

Proposition 5 *The linear fractional transformation* σ_α *in* (5) *can be represented as*

$$
\sigma_\alpha(M) = (MC^\alpha - A^\alpha)^{-1}(B^\alpha - MD^\alpha). \tag{11}
$$

Proof. By (7), (11) is well defined. Therefore we need only to verify the identity

$$
(MC^\alpha - A^\alpha)^{-1}(B^\alpha - MD^\alpha) = (A_\alpha M + B_\alpha)(C_\alpha M + D_\alpha)^{-1},
$$

i.e.,

$$
(B^\alpha - MD^\alpha)(C_\alpha M + D_\alpha) = (MC^\alpha - A^\alpha)(A_\alpha M + B_\alpha). \tag{12}
$$

Expanding it and using the conditions (10), we know that (12) holds. Q.E.D.

Proposition 6

$$
(C^\alpha N + D^\alpha)(C_\alpha M + D_\alpha) = I, \tag{13}
$$

hence

$$|C^{\alpha}N + D^{\alpha}| \neq 0 \quad \textit{iff} \quad |C_{\alpha}M + D_{\alpha}| \neq 0, \tag{14}$$

where $N = \sigma_{\alpha}(M)$. *So under the transversality condition* (6) σ_{α} *has an inverse transformation* $\sigma_{\alpha}^{-1} = \sigma_{\alpha^{-1}}$.

$$M = \sigma_{\alpha^{-1}}(N) = (A^{\alpha}N + B^{\alpha})(C^{\alpha}N + D^{\alpha})^{-1} \tag{15}$$

$$= (NC_{\alpha} - A_{\alpha})^{-1}(B_{\alpha} - ND_{\alpha}). \tag{16}$$

Proof.

$$\begin{aligned}
&(C^{\alpha}N + D^{\alpha})(C_{\alpha}M + D_{\alpha})\\
=&\left(C^{\alpha}(A_{\alpha}M + B_{\alpha})(C_{\alpha}M + D_{\alpha})^{-1} + D^{\alpha}\right)(C_{\alpha}M + D_{\alpha})\\
=&(C^{\alpha}A_{\alpha} + D^{\alpha}C_{\alpha})M + C^{\alpha}B_{\alpha} + D^{\alpha}D_{\alpha}\\
=&I \quad (by (10)).
\end{aligned}$$

This is (13). In Prop. 4, substituting α^{-1} for α and N for M, we get

$$|C^{\alpha}N + D^{\alpha}| \neq 0 \quad \text{iff} \quad |NC_{\alpha} - A_{\alpha}| \neq 0.$$

Hence the equation

$$N = (A_{\alpha}M + B_{\alpha})(C_{\alpha}M + D_{\alpha})^{-1}$$

is solvable for M and the solution is

$$M = (NC_{\alpha} - A_{\alpha})^{-1}(B_{\alpha} - ND_{\alpha}).$$

This is (16). (15) can be got from (11). Q.E.D.

Putting together the above stated, the following four transversality conditions are equivalent mutually:

$$|C_{\alpha}M + D_{\alpha}| \neq 0 \tag{17.1}$$

$$|MC^{\alpha} - A^{\alpha}| \neq 0 \tag{17.2}$$

$$|C^{\alpha}N + D^{\alpha}| \neq 0 \tag{17.3}$$

$$|NC_{\alpha} - A_{\alpha}| \neq 0, \tag{17.4}$$

where

$$N = \sigma_{\alpha}(M) = (A_{\alpha}M + B_{\alpha})(C_{\alpha}M + D_{\alpha})^{-1}$$

$$M = \sigma_{\alpha^{-1}}(N) = (A^{\alpha}N + B^{\alpha})(C^{\alpha}N + D^{\alpha})^{-1}.$$

§3 Linear Canonical and Linear Gradient Maps and Generating Functions

In this section we study linear fractional transformation relating the symplectic group $\mathbf{Sp}(2n)$ and the space $\mathbf{Sm}(2n)$ of symmetric matrices of order $2n$.

Consider the two different symplectic forms on $\mathbf{R}^{4n}$, the natural symplectic form $J_{4n} = \begin{bmatrix} 0 & I_{2n} \\ -I_{2n} & 0 \end{bmatrix}$ and the natural product symplectic form $\tilde{J}_{4n} = \begin{bmatrix} J_{2n} & 0 \\ 0 & -J_{2n} \end{bmatrix}$. Denote $\mathbf{R}^{4n} = (\mathbf{R}^{4n}, J_{4n})$, $\tilde{\mathbf{R}}^{4n} = \left(\mathbf{R}^{4n}, \tilde{J}_{4n}\right)$.

A $2n$-dim subspace $\{X\} = \begin{Bmatrix} X_1 \\ X_2 \end{Bmatrix}$ of $\mathbf{R}^{4n}$, $X_1, X_2 \in \mathbf{M}(2n)$, is J_{4n}-Lagrangian if

$$X^T J_{4n} X = 0,$$

i.e.,

$$X_1^T X_2 - X_2^T X_1 = 0 \quad \text{or} \quad X_1^T X_2 \in \mathbf{Sm}(2n).$$

We call such $4n \times 2n$ matrix $X = \begin{bmatrix} X_1 \\ X_2 \end{bmatrix}$, according to Siegel [8], a symmetric pair. If, moreover, $|X_2| \neq 0$, then $X_1 X_2^{-1} = N \in \mathbf{Sm}(2n)$ and $\begin{Bmatrix} X_1 \\ X_2 \end{Bmatrix} = \begin{Bmatrix} N \\ I \end{Bmatrix}$. Similarly, a $2n$-dim subspace $\{Y\} = \begin{Bmatrix} Y_1 \\ Y_2 \end{Bmatrix}$ is $\tilde{J}_{4n}$-Lagrangian if

$$Y^T \tilde{J}_{4n} Y = 0,$$

i.e.,

$$Y_1^T J_{2n} Y_1 = Y_2^T J_{2n} Y_2.$$

The $4n \times 2n$ matrix $Y = \begin{bmatrix} Y_1 \\ Y_2 \end{bmatrix}$ is called a symplectic pair. $|Y_2| \neq 0$ implies $Y_1 Y_2^{-1} = M \in \mathbf{Sp}(2n)$ and $\begin{Bmatrix} Y_1 \\ Y_2 \end{Bmatrix} = \begin{Bmatrix} M \\ I \end{Bmatrix}$.

Theorem 7 $\boldsymbol{\alpha} = \begin{bmatrix} A_\alpha & B_\alpha \\ C_\alpha & D_\alpha \end{bmatrix} \in \mathbf{GL}(4n)$ *carries every $\tilde{J}_{4n}$-Lagrangian subspace into a J_{4n}-Lagrangian subspace if and only if $\alpha \in \mathbf{CSp}\left(\tilde{J}_{4n}, J_{4n}\right)$, i.e.,*

$$\alpha^T J_{4n} \alpha = \mu \tilde{J}_{4n}, \quad \textit{for some } \mu = \mu(\alpha) \neq 0. \tag{18}$$

Proof. "if" part is obvious, we need only to prove "only if" part.

Taking $\alpha_0 \in \mathbf{Sp}\left(\tilde{J}_{4n}, J_{4n}\right)$(it always exists), by Prop.2 we have

$$\mathbf{CSp}\left(\tilde{J}_{4n}, J_{4n}\right) = \mathbf{CSp}(4n) \cdot \alpha_0.$$

Therefore it suffices to show that if α carries every J_{4n}-Lagrangian subspace into J_{4n}-Lagrangian subspace then $\alpha \in \mathbf{CSp}(4n)$, i.e.,

$$\alpha^T J_{4n} \alpha = \mu J_{4n}, \qquad \text{for some } \mu \neq 0.$$

1° Take the symmetric pair $X = \begin{bmatrix} I_{2n} \\ 0_{2n} \end{bmatrix}$. By assumption,

$$\alpha X = \begin{bmatrix} A_\alpha & B_\alpha \\ C_\alpha & D_\alpha \end{bmatrix} \begin{bmatrix} I \\ 0 \end{bmatrix} = \begin{bmatrix} A_\alpha \\ C_\alpha \end{bmatrix}$$

is also a symmetric pair, i.e., $A_\alpha^T C_\alpha - C_\alpha^T A_\alpha = 0$. Similarly, $B_\alpha^T D_\alpha - D_\alpha^T B_\alpha = 0$.

2° Take the symmetric pairs $X = \begin{bmatrix} S \\ I \end{bmatrix}$, $S \in \mathbf{Sm}(2n)$. Then

$$\alpha X = \begin{bmatrix} A_\alpha & B_\alpha \\ C_\alpha & D_\alpha \end{bmatrix} \begin{bmatrix} S \\ I \end{bmatrix} = \begin{bmatrix} A_\alpha S + B_\alpha \\ C_\alpha S + D_\alpha \end{bmatrix}$$

are also symmetric pairs, i.e.,

$$\begin{aligned} 0 &= (\alpha X)^T J_{4n} (\alpha X) \\ &= \left(S^T A_\alpha^T + B_\alpha^T, S^T C_\alpha^T + D_\alpha^T\right) \begin{bmatrix} 0 & I \\ -I & 0 \end{bmatrix} \begin{bmatrix} A_\alpha S + B_\alpha \\ C_\alpha S + D_\alpha \end{bmatrix} \\ &= S\left(A_\alpha^T C_\alpha - C_\alpha^T A_\alpha\right) S + S\left(A_\alpha^T D_\alpha - C_\alpha^T B_\alpha\right) \\ &\quad - \left(D_\alpha^T A_\alpha - B_\alpha^T C_\alpha\right) S + B_\alpha^T D_\alpha - D_\alpha^T B_\alpha \\ &= S\left(A_\alpha^T D_\alpha - C_\alpha^T B_\alpha\right) - \left(A_\alpha^T D_\alpha - C_\alpha^T B_\alpha\right)^T S, \quad \forall S \in \mathbf{Sm}(2n). \end{aligned}$$

Set $P = A_\alpha^T D_\alpha - C_\alpha^T B_\alpha$. Then the above equation becomes

$$SP = P^T S, \qquad \forall S \in \mathbf{Sm}(2n).$$

It follows that $P = \mu I$, i.e.,

$$A_\alpha^T D_\alpha - C_\alpha^T B_\alpha = \mu I.$$

So

$$\alpha^T J_{4n}\alpha = \begin{bmatrix} A_\alpha & B_\alpha \\ C_\alpha & D_\alpha \end{bmatrix}^T \begin{bmatrix} 0 & I \\ -I & 0 \end{bmatrix} \begin{bmatrix} A_\alpha & B_\alpha \\ C_\alpha & D_\alpha \end{bmatrix}$$
$$= \begin{bmatrix} A_\alpha^T C_\alpha - C_\alpha^T A_\alpha & A_\alpha^T D_\alpha - C_\alpha^T B_\alpha \\ B_\alpha^T C_\alpha - D_\alpha^T A_\alpha & B_\alpha^T D_\alpha - D_\alpha^T B_\alpha \end{bmatrix}$$
$$= \mu \begin{bmatrix} 0 & I \\ -I & 0 \end{bmatrix} = \mu J_{4n}.$$

$\alpha \in \mathbf{GL}(4n)$ implies $\mu \neq 0$. Q.E.D.

Remark 8 The inverse matrix of α is denoted by $\alpha^{-1} = \begin{bmatrix} A^\alpha\ B^\alpha \\ C^\alpha\ D^\alpha \end{bmatrix}$. Then by (18), we have

$$\begin{aligned} A^\alpha &= \mu^{-1} J C_\alpha^T, & B^\alpha &= -\mu^{-1} J A_\alpha^T, \\ C^\alpha &= -\mu^{-1} J D_\alpha^T, & D^\alpha &= \mu^{-1} J B_\alpha^T. \end{aligned} \tag{19}$$

Definition 9 A linear map $z \to \widehat{z} = g(z) = Mz$ resp. $w \to \widehat{w} = f(z) = Nw$: $\mathbf{R}^{2n} \to \mathbf{R}^{2n}$ is called a canonical resp. gradient map if $M \in \mathbf{Sp}(2n)$ resp. $N \in \mathbf{Sm}(2n)$.

For a linear map $z \longrightarrow \widehat{z} = g(z) = Mz : \mathbf{R}^m \to \mathbf{R}^m$, the graph of g,

$$\Gamma_g = \left\{ \begin{bmatrix} \widehat{z} \\ z \end{bmatrix} \in \mathbf{R}^{2m} | \widehat{z} = g(z) = Mz \right\}$$

is an m-dim subspace of $\mathbf{R}^{2m}$ and $\Gamma = \begin{Bmatrix} M \\ I \end{Bmatrix}$, i.e., it is spanned by the column vectors of $\begin{bmatrix} M \\ I \end{bmatrix}$.

Proposition 10 *The graph Γ_g of a linear canonical map g(z) is a $\tilde{J}_{4n}$-Lagrangian subspace. The graph Γ_f of a linear gradient map f is a J_{4n}-Lagrangian subspace.*

Let $\alpha = \begin{bmatrix} A_\alpha\ B_\alpha \\ C_\alpha\ D_\alpha \end{bmatrix} \in \mathbf{CSp}\left(\tilde{J}_{4n}, J_{4n}\right)$. It defines linear transformations

$$\begin{bmatrix} \widehat{w} \\ w \end{bmatrix} = \alpha \begin{bmatrix} \widehat{z} \\ z \end{bmatrix}, \quad \begin{bmatrix} \widehat{z} \\ z \end{bmatrix} = \alpha^{-1} \begin{bmatrix} \widehat{w} \\ w \end{bmatrix}, \tag{20}$$

i.e.,

$$\begin{aligned} \widehat{w} &= A_\alpha \widehat{z} + B_\alpha z, & \widehat{z} &= A^\alpha \widehat{w} + B^\alpha w, \\ w &= C_\alpha \widehat{z} + D_\alpha z, & z &= C^\alpha \widehat{w} + D^\alpha w. \end{aligned} \tag{21}$$

Theorem 11 *Let* $\alpha = \begin{bmatrix} A_\alpha & B_\alpha \\ C_\alpha & D_\alpha \end{bmatrix} \in \mathbf{CSp}\left(\tilde{J}_{4n}, J_{4n}\right)$. *Then* $M \in \mathbf{Sp}(2n)$ *and* M *satisfies* (17.1) *iff* $N = \sigma_\alpha(M) \in \mathbf{Sm}(2n)$ *and* N *satisfies* (17.3). *That is, the linear fractional transformation* $\sigma_\alpha : \{M \in \mathbf{Sp}(2n) \mid |C_\alpha M + D_\alpha| \neq 0\} \longrightarrow \{N \in \mathbf{Sm}(2n) \mid |C^\alpha N + D^\alpha| \neq 0\}$ *is one to one and onto.*

Proof. By (17), we know $|C_\alpha M + D_\alpha| \neq 0$ iff $|C^\alpha N + D^\alpha| \neq 0$.

"only if" part. The map $z \rightarrow \widehat{z} = g(z) = Mz : \mathbf{R}^{2n} \rightarrow \mathbf{R}^{2n}$ is a linear canonical map. So its graph Γ_g is a $\tilde{J}_{4n}$-Lagrangian subspace. Since $\alpha \in \mathbf{CSp}\left(\tilde{J}_{4n}, J_{4n}\right)$, by Theorem 7, $\alpha(\Gamma_g)$ is J_{4n}-Lagrangian. Notice that $\Gamma_g = \begin{Bmatrix} M \\ I \end{Bmatrix}$.

$$\alpha(\Gamma_g) = \left\{\begin{bmatrix} A_\alpha & B_\alpha \\ C_\alpha & D_\alpha \end{bmatrix}\begin{bmatrix} M \\ I \end{bmatrix}\right\} = \begin{Bmatrix} A_\alpha M + B_\alpha \\ C_\alpha M + D_\alpha \end{Bmatrix}.$$

By assumption, $|C_\alpha M + D_\alpha| \neq 0$. Therefore

$$\alpha(\Gamma_g) = \begin{Bmatrix} (A_\alpha M + B_\alpha)(C_\alpha M + D_\alpha)^{-1} \\ I \end{Bmatrix} = \begin{Bmatrix} \sigma_\alpha(M) \\ I \end{Bmatrix}$$

and it is J_{4n}-Lagrangian. That is, $N = \sigma_\alpha(M)$ is symmetric.

Substituting α^{-1} for α and noting $\alpha^{-1} \in \mathbf{CSp}\left(J_{4n}, \tilde{J}_{4n}\right)$, we can get similarly the "if" part. Q.E.D.

Theorem 12 *Let* $\alpha \in \mathbf{CSp}\left(\tilde{J}_{4n}, J_{4n}\right)$. *Let* $z \rightarrow \widehat{z} = g(z) = Mz$ *be a linear canonical map and* M *satisfy* (17.1). *Then there exists a linear gradient map* $w \rightarrow \widehat{w} = f(w) = Nw$ *and a quadratic function—generating function—*$\phi(w) = \frac{1}{2} w^T N w$ *(depending on* α *and* g*) such that*

$$1. \quad f(w) = \nabla\phi(w); \tag{22}$$

$$\begin{aligned} 2. \quad A_\alpha g(z) + B_\alpha z &= f\left(C_\alpha g(z) + D_\alpha z\right) \\ &= \nabla\phi\left(C_\alpha g(z) + D_\alpha z\right), \quad \textit{identically in } z; \end{aligned} \tag{23}$$

$$\begin{aligned} 3. \quad N &= \sigma_\alpha(M) = (A_\alpha M + B_\alpha)(C_\alpha M + D_\alpha)^{-1}; \\ M &= \sigma_{\alpha^{-1}}(N) = (A^\alpha N + B^\alpha)(C^\alpha N + D^\alpha)^{-1}; \end{aligned} \tag{24}$$

$$4. \quad \Gamma_f = \alpha(\Gamma_g), \Gamma_g = \alpha^{-1}(\Gamma_f), \tag{25}$$

where $\nabla\phi = (\phi_{w_1}, \cdots, \phi_{w_{2n}})^T, \phi_w = (\phi_{w_1}, \cdots, \phi_{w_{2n}})$, *so* $\nabla\phi = (\phi_w)^T$.

Proof. The image of Γ_g under α is

$$\alpha(\Gamma_g) - \left\{\begin{bmatrix} \widehat{w} \\ w \end{bmatrix} \in \mathbf{R}^{4n} \mid \widehat{w} = A_\alpha M z + B_\alpha z, w = C_\alpha M z + D_\alpha z\right\}.$$

By assumption, $|C_\alpha M + D_\alpha| \neq 0$, so $w = (C_\alpha M + D_\alpha) z$ is invertible and its inverse is $z = (C_\alpha M + D_\alpha)^{-1} w$. Set $\widehat{w} = (A_\alpha M + B_\alpha)(C_\alpha M + D_\alpha)^{-1} w = Nw = f(w)$. By Theorem 11, $f(w)$ is a linear gradient map. Obviously, $\phi(w) = \frac{1}{2} w^T N w$ satisfies (22).

From the equation

$$N = (A_\alpha M + B_\alpha)(C_\alpha M + D_\alpha)^{-1},$$

it follows that

$$A_\alpha M + B_\alpha = N (C_\alpha M + D_\alpha).$$

So

$$A_\alpha M z + B_\alpha z = N (C_\alpha M z + D_\alpha z), \quad \forall z \in \mathbf{R}^{2n},$$

i.e.,

$$A_\alpha g(z) + B_\alpha z = f (C_\alpha g(z) + D_\alpha z), \qquad \text{identically in } z.$$

Since $\Gamma_g = \begin{Bmatrix} M \\ I \end{Bmatrix}$,

$$\alpha(\Gamma_g) = \left\{ \begin{bmatrix} A_\alpha & B_\alpha \\ C_\alpha & D_\alpha \end{bmatrix} \begin{bmatrix} M \\ I \end{bmatrix} \right\} = \begin{Bmatrix} A_\alpha M + B_\alpha \\ C_\alpha M + D_\alpha \end{Bmatrix}.$$

$|C_\alpha M + D_\alpha| \neq 0$ implies $\alpha(\Gamma_g) = \begin{Bmatrix} N \\ I \end{Bmatrix} = \Gamma_f$. Q.E.D.

Theorem 13 *Let* $\alpha \in \mathbf{CSp}\left(\tilde{J}_{4n}, J_{4n}\right)$. *Let* $\phi(w) = \frac{1}{2} w^T N w$, $N \in \mathbf{Sm}(2n)$ *be a quadratic function,* $f(w) = \nabla \phi = Nw$ *its induced linear gradient map and* N *satisfy* (17.3). *Then there exists a linear canonical map* $z \to \widehat{z} = g(z) = Mz$ *such that*

1. $A^\alpha f(w) + B^\alpha w = g (C^\alpha f(w) + D^\alpha w)$ *identically in* w
2. $M = \sigma_{\alpha^{-1}}(N) = (A^\alpha N + B^\alpha)(C^\alpha N + D^\alpha)^{-1}$

   $N = \sigma_\alpha(M) = (A_\alpha M + B_\alpha)(C_\alpha M + D_\alpha)^{-1}$
3. $\Gamma_g = \alpha^{-1}(\Gamma_f), \quad \Gamma_f = \alpha(\Gamma_g).$

The proof is similar to the one of Theorem 12 and omitted here.

§4 General Canonical and Gradient Maps and Generating Functions

Definition 14 A map $z \to \widehat{z} = g(z) : \mathbf{R}^{2n} \to \mathbf{R}^{2n}$ is called a canonical map if its Jacobian $M(z) = g_z(z) \in \mathbf{Sp}(2n)$ everywhere. A map $w \to \widehat{w} = f(w) : \mathbf{R}^{2n} \to \mathbf{R}^{2n}$ is called a gradient map if its Jacobian $N(w) = f_w(w) \in \mathbf{Sm}(2n)$ everywhere.

Definition 15 An m-dim submanifold U of $(\mathbf{R}^{2m}, K)$ is a K-Lagrangian submanifold if its tangent plane $T_z U$ at z is a, K-Lagrangian subspace of the tangent space $T_z\mathbf{R}^{2m}$ to $\mathbf{R}^{2m}$ at z for any $z \in U$.

Proposition 16 *The graph Γ_g of a canonical map $z \to \widehat{z} = g(z)$ is a $\tilde{J}_{4n}$-Lagrangian submanifold of $\tilde{\mathbf{R}}^{4n}$. The graph Γ_f of a gradient map $w \to \widehat{w} = f(w)$ is a J_{4n}-Lagrangian submanifold of $\mathbf{R}^{4n}$.*

Theorem 17 *Let $\alpha \in \mathbf{CSp}\left(\tilde{J}_{4n}, J_{4n}\right)$. Let $z \to \widehat{z} = g(z) : \mathbf{R}^{2n} \to \mathbf{R}^{2n}$ be a canonical map with Jacobian $M(z) = g_z(z) \in \mathbf{Sp}(2n)$ satisfying (17.1) in (some neighborhood of) $\mathbf{R}^{2n}$. Then there exists a gradient map $w \to \hat{w} = f(w)$ in (some neighborhood of) $\mathbf{R}^{2n}$ with Jacobian $N(w) = f_w(w) \in \mathbf{Sm}(2n)$ satisfying (17.3) and a scalar function —generating function—$\phi(w)$ (depending on α and g) such that*

1. $$f(w) = \nabla\phi(w); \tag{26}$$

2. $$\begin{aligned} A_\alpha g(z) + B_\alpha z &= f\left(C_\alpha g(z) + D_\alpha z\right) \\ &= \nabla\phi\left(C_\alpha g(z) + D_\alpha z\right), \quad \textit{identically in } z; \end{aligned} \tag{27}$$

3. $$\begin{aligned} N &= \sigma_\alpha(M) = \left(A_\alpha M + B_\alpha\right)\left(C_\alpha M + D_\alpha\right)^{-1}; \\ M &= \sigma_{\alpha^{-1}}(N) = \left(A^\alpha N + B^\alpha\right)\left(C^\alpha N + D^\alpha\right)^{-1}; \end{aligned} \tag{28}$$

4. $$\Gamma_f = \alpha\left(\Gamma_g\right), \quad \Gamma_g = \alpha^{-1}\left(\Gamma_f\right). \tag{29}$$

Proof. Under the linear transformation α, the image of Γ_g is

$$\alpha\left(\Gamma_g\right) = \left\{ \begin{bmatrix} \widehat{w} \\ w \end{bmatrix} \in \mathbf{R}^{4n} | \widehat{w} = A_\alpha g(z) + B_\alpha z, \quad w = C_\alpha g(z) + D_\alpha z \right\}.$$

Since Γ_g is a $\tilde{J}_{4n}$-Lagrangian submanifold and $\alpha \in \mathbf{CSp}\left(\tilde{J}_{4n}, J_{4n}\right)$, the tangent plane of $\alpha\left(\Gamma_g\right)$

$$\left\{ \begin{matrix} A_\alpha M(z) + B_\alpha \\ C_\alpha M(z) + D_\alpha \end{matrix} \right\}$$

is a J_{4n}-Lagrangian subspace. So $\alpha\left(\Gamma_g\right)$ is a J_{4n}-Lagrangian submanifold. By assumption, $|C_\alpha M + D_\alpha| \neq 0$, so, by the Implicit Function Theorem,

$$w = C_\alpha g(z) + D_\alpha z \tag{30}$$

is invertible and its inverse is denoted by $z = z(w)$. Set

$$\widehat{w} = f(w) = \left(A_\alpha g(z) + B_\alpha z\right)\big|_{z=z(w)} = A_\alpha g(z(w)) + B_\alpha z(w).$$

Obviously, such $f(w)$ satisfies the identity

$$A_\alpha g(z) + B_\alpha z \equiv f\left(C_\alpha g(z) + D_\alpha z\right).$$

The Jacobian of f is

$$N(w)=f_w(w)=\frac{\partial\widehat{w}}{\partial w}=\frac{\partial\widehat{w}}{\partial z}\frac{\partial z}{\partial w}=\frac{\partial\widehat{w}}{\partial z}\left(\frac{\partial w}{\partial z}\right)^{-1}.$$
$$=\left(A_\alpha M(z)+B_\alpha\right)\left(C_\alpha M(z)+D_\alpha\right)^{-1}=\sigma_\alpha(M(z)).$$

By Theorem 11, it is symmetric. So $f(w)$ is a gradient map. By Poincaré Lemma, there exists a scalar function $\phi(w)$, such that

$$f(w)=\nabla\phi(w).$$

In addition,

$$\Gamma_f=\left\{\begin{bmatrix}\widehat{w}\\ w\end{bmatrix}\in\mathbf{R}^{4n}|\widehat{w}=f(w)=A_\alpha g(z(w))+B_\alpha z(w)\right\}=\alpha\left(\Gamma_g\right).$$

Q.E.D.

Theorem 18 *Let* $\alpha\in\mathbf{CSp}\left(\tilde{J}_{4n},J_{4n}\right)$. *Let* $\phi(w)$ *be a scalar function and* $w\to\widehat{w}=f(w)=\nabla\phi(w)$ *be its induced gradient map and* $N(w)=f_w(w)=\phi_{ww}(w)$, *the Hessian matrix of* $\phi(w)$, *satisfy* (17.3) *in (some neighborhood of)* $\mathbf{R}^{2n}$. *Then there exists a canonical map* $z\to\widehat{z}=g(z)$ *with Jacobian* $M(z)=g_z(z)$ *satisfying* (17.1) *such that*

1. $A^\alpha f(w)+B^\alpha w=g\left(C^\alpha f(w)+D^\alpha w\right),$ *identically in* w;
2. $M=\sigma_{\alpha^{-1}}(N)=\left(A^\alpha N+B^\alpha\right)\left(C^\alpha N+D^\alpha\right)^{-1};$
   $N=\sigma_\alpha(M)=\left(A_\alpha M+B_\alpha\right)\left(C_\alpha M+D_\alpha\right)^{-1};$
3. $\Gamma_g=\alpha^{-1}\left(\Gamma_f\right),\quad\Gamma_f=\alpha\left(\Gamma_g\right).$

§5 Generating Functions for the Phase Flow of Hamiltonian Systems

Consider the Hamiltonian system

$$\frac{dz}{dt}=J_{2n}^{-1}\nabla H(z),\quad z\in\mathbf{R}^{2n},\tag{31}$$

where $H(z)$ is a Hamiltonian function. Its phase flow is denoted as $g^t(z)=g(z,t)=g_H(z,t)$, being a one-parameter group of canonical maps [2,3], i.e.,

$$g^0=\text{identity},\qquad g^{t_1+t_2}=g^{t_1}\circ g^{t_2}$$

and if z_0 is taken as an initial condition, then $z(t)=g^t\left(z_0\right)$ is the solution of (31) with the initial value z_0.

Theorem 19 *Let* $\alpha \in \mathbf{CSp}\left(\tilde{J}_{4n}, J_{4n}\right)$. *Let* $z \to \widehat{z} = g(z,t)$ *be the phase flow of the Hamiltonian system* (31) *and* $M_0 \in \mathbf{Sp}(2n)$. *Set* $G(z,t) = g(M_0 z, t)$ *with Jacobian* $M(z,t) = G_z(z,t)$. *It is a time-dependent canonical map. If* M_0 *satisfies the transversality condition* (17.1), *i.e.,*

$$|C_\alpha M_0 + D_\alpha| \neq 0, \tag{32}$$

then there exists, for sufficiently small $|t|$ *and in (some neighborhood of)* $\mathbf{R}^{2n}$, *a time-dependent gradient map* $w \to \widehat{w} = f(w,t)$ *with Jacobian* $N(w,t) = f_w(w,t) \in \mathbf{Sm}(2n)$ *satisfying the transversality condition* (17.3) *and a time-dependent generating function* $\phi_{\alpha,H}(w,t) = \phi(w,t)$ *such that*

1. $$f(w,t) = \nabla\phi(w,t); \tag{33}$$
2. $$\frac{\partial}{\partial t}\phi(w,t) = -\mu H\left(A^\alpha \nabla\phi(w,t) + B^\alpha w\right); \tag{34}$$
3. $$A_\alpha G(z,t) + B_\alpha z \equiv f\left(C_\alpha G(z,t) + D_\alpha z, t\right) \equiv \nabla\phi\left(C_\alpha G(z,t) + D_\alpha z, t\right); \tag{35}$$
4. $$N = \sigma_\alpha(M) = \left(A_\alpha M + B_\alpha\right)\left(C_\alpha M + D_\alpha\right)^{-1}, \tag{36}$$
$$M = \sigma_{\alpha^{-1}}(N) = \left(A^\alpha N + B^\alpha\right)\left(C^\alpha N + D^\alpha\right)^{-1}.$$

(34) is the most general Hamilton-Jacobi equation for the Hamiltonian system (31) and the linear transformation α.

Proof. Since $g(z,t)$ is differentiable with respect to z and t, so is $G(z,t)$. Condition (32) implies that for sufficiently small $|t|$ and in some neighborhood of $\mathbf{R}^{2n}$,

$$|C_\alpha M(z,t) + D_\alpha| \neq 0 \tag{37}$$

Thus by Theorem 17 there exists a time-dependent gradient map $\widehat{w} = f(w,t)$ such that it satisfies (35) and (36).

Set

$$\begin{aligned}\bar{H}(w,t) &= -\mu H(\widehat{z})|_{\widehat{z}=A^\alpha \widehat{w}(w,t)+B^\alpha w} \\ &= -\mu H\left(A^\alpha \widehat{w}(w,t) + B^\alpha w\right).\end{aligned} \tag{38}$$

Consider the differential 1-form

$$\omega^1 = \sum_{i=1}^{2n} \widehat{w}_i dw_i + \bar{H}(w,t)dt. \tag{39}$$

$$\begin{aligned}d\omega^1 &= \sum_{i,j=1}^{2n} \frac{\partial \widehat{w}_i}{\partial w_j} dw_j \wedge dw_i + \sum_{i=1}^{2n} \frac{\partial \widehat{w}_i}{\partial t} dt \wedge dw_i + \sum_{i=1}^{2n} \frac{\partial \bar{H}}{\partial w_i} dw_i \wedge dt \\ &= \sum_{i<j} \left(\frac{\partial \widehat{w}_i}{\partial w_j} - \frac{\partial \widehat{w}_j}{\partial w_i}\right) dw_j \wedge dw_i + \sum_{i=1}^{2n} \left(\frac{\partial \widehat{w}_i}{\partial t} - \frac{\partial \bar{H}}{\partial w_i}\right) dt \wedge dw_i.\end{aligned} \tag{40}$$

Since $N(w,t)=f_w(w,t)=\partial\widehat{w}/\partial w$ is symmetric, the first term of (40) is zero.

Notice that $\widehat{z}=G(z,t)=g\left(M_0 z,t\right)$,

$$\frac{d\widehat{z}}{dt}=\frac{dg\left(M_0 z,t\right)}{dt}=J^{-1}\nabla H(G(z,t)) \tag{41}$$

So $G(z,t)$ is the solution of the following initial-value problem

$$\begin{cases}\dfrac{d\widehat{z}}{dt}=J^{-1}\nabla H(\widehat{z}),\\ \widehat{z}(0)=M_0 z.\end{cases}$$

Therefore from the equations

$$\widehat{w}=A_\alpha G(z,t)+B_\alpha z,\quad w=C_\alpha G(z,t)+D_\alpha z,$$

it follows that

$$\frac{d\widehat{w}}{dt}=A_\alpha J^{-1}\nabla H(\widehat{z}),\quad \frac{dw}{dt}=C_\alpha J^{-1}\nabla H(\widehat{z}).$$

Since

$$\begin{aligned}\frac{d\widehat{w}}{dt}&=\frac{\partial\widehat{w}}{\partial w}\frac{dw}{dt}+\frac{\partial\widehat{w}}{\partial t},\\ \frac{\partial\widehat{w}}{\partial t}&=\left(A_\alpha-\frac{\partial\widehat{w}}{\partial w}C_\alpha\right)J^{-1}\nabla H(\widehat{z}).\end{aligned}$$

On the other hand,

$$\begin{aligned}\nabla_w\bar{H}(w,t)&=\left(\bar{H}_w(w,t)\right)^T=\mu\left(-H_{\hat{z}}\cdot\left(A^\alpha\frac{\partial\widehat{w}}{\partial w}+B^\alpha\right)\right)^T\\ &=-\mu\left(B^{\alpha T}+\left(\frac{\partial\widehat{w}}{\partial w}\right)^T A^{\alpha T}\right)\nabla H(\bar{z})\\ &=\left(A_\alpha J^{-1}-\frac{\partial\widehat{w}}{\partial w}C_\alpha J^{-1}\right)\nabla H(\widehat{z})\quad(\text{ by (19) and } N\in\mathbf{Sm}(2n))\\ &=\frac{\partial\widehat{w}}{\partial t}.\end{aligned}$$

So $d\omega^1=0$. By Poincaré Lemma there exists, in some neighborhood of $\mathbf{R}^{2n+1}$, a scalar function $\phi(w,t)$, such that

$$\omega^1=\widehat{w}dw+\bar{H}dt=d\phi(w,t),$$

i.e.,

$$\begin{aligned}&f(w,t)=\nabla_w\phi(w,t),\\ &\frac{\partial}{\partial t}\phi(w,t)=-\mu H\left(A^\alpha\nabla_w\phi_{\alpha,H}(w,t)+B^\alpha w\right).\quad \text{Q.E.D.}\end{aligned}$$

Examples of generating functions:

$$\text{(I)}\quad \alpha=\begin{bmatrix}0&0&-I_n&0\\ I_n&0&0&0\\ 0&0&0&I_n\\ 0&I_n&0&0\end{bmatrix},\quad \mu=1,\quad M_0=J,\quad |C_\alpha M_0+D_\alpha|\neq 0;$$

$$w=\begin{bmatrix}q\\ \widehat{q}\end{bmatrix},\quad \phi=\phi(q,\widehat{q},t);$$

$$\widehat{w}=\begin{bmatrix}-p\\ \widehat{p}\end{bmatrix}=\begin{bmatrix}\phi_q\\ \phi_{\widehat{q}}\end{bmatrix},\quad \phi_t=-H\left(\phi_{\widehat{q}},\widehat{q}\right).$$

This is the generating function and H.J. equation of the first kind [3],

$$\text{(II)}\quad \alpha=\begin{bmatrix}0&0&-I_n&0\\ 0&-I_n&0&0\\ 0&0&0&I_n\\ I_n&0&0&0\end{bmatrix},\quad \mu=1,\quad M_0=I,\quad |C_\alpha M_0+D_\alpha|\neq 0;$$

$$w=\begin{bmatrix}q\\ \widehat{p}\end{bmatrix},\quad \phi=\phi(q,\widehat{p},t);$$

$$\widehat{w}=-\begin{bmatrix}p\\ \widehat{q}\end{bmatrix}=\begin{bmatrix}\phi_q\\ \phi_{\widehat{p}}\end{bmatrix},\quad \phi_t=-H\left(\widehat{p},-\phi_{\widehat{p}}\right).$$

This is the generating function and H.J. equation of the second kind [3].

$$\text{(III)}\quad \alpha=\begin{bmatrix}-J_{2n}&J_{2n}\\ \frac{1}{2}I_{2n}&\frac{1}{2}I_{2n}\end{bmatrix},\mu=1,M_0=I,\quad |C_\alpha M_0+D_\alpha|\neq 0;$$

$$w=\frac{1}{2}(z+\widehat{z}),\quad \phi=\phi(w,t);$$

$$\widehat{w}=J(z-\widehat{z})=\nabla\phi,\quad \phi_t=-H\left(w-\frac{1}{2}J^{-1}\nabla\phi\right).$$

This is the Poincaré's generating function and H.J. equation.

If the Hamiltonian function $H(z)$ depends analytically on z then we can give through recursions the explicit expression of the corresponding generating functions.

Theorem 20 *Let $H(z)$ depend analytically on z. Then $\phi_{\alpha,H}(w,t)$ is expressible as a convergent power series in t for sufficiently small $|t|$, with recursively determined* *coefficients:*

$$\phi(w,t)=\sum_{k=0}^{\infty}\phi^{(k)}(w)t^k, \tag{42}$$

$$\phi^{(0)}(w)=\frac{1}{2}w^T N_0 w,\quad N_0=(A_\alpha M_0+B_\alpha)(C_\alpha M_0+D_\alpha)^{-1}, \tag{43}$$

$$\phi^{(1)}(w)=-\mu(\alpha)H(E_0 w),\quad E_0=A^\alpha N_0+B^\alpha=M_0(C_\alpha M_0+D_\alpha)^{-1}, \tag{44}$$

$$k\geqslant 1,\phi^{(k+1)}(w)=-\frac{\mu(\alpha)}{k+1}\sum_{m=1}^{k}\frac{1}{m!}\sum_{i_1,\cdots,i_m=1}^{2,l}\sum_{\substack{j_1+\cdots+j_m=k\\ j_l\geqslant 1}}H_{z_{i_1},\cdots,z_{i_m}}(E_0 w)$$
$$\times\left(A^\alpha\nabla\phi^{(j_1)}\right)_{i_1}\cdots\left(A^\alpha\nabla\phi^{(j_m)}\right)_{i_m}, \tag{45}$$

where $H_{z_{i_1},\cdots,z_{i_m}}(E_0 w)$ *is the m-th partial derivative of H(z) w.r.t.* $z_{i_1},\cdots,z_{i_m}$, *evaluated at* $z=E_0 w$ *and* $\left(A^\alpha\nabla\phi^{(j_l)}(w)\right)_{i_l}$ *is the* i_l*-th component of the column vector* $A^\alpha\nabla\phi^{(j_l)}(w)$.

Proof. Under our assumption, the generating function $\phi_{\alpha,H}(w,t)$ depends analytically on w and t in some neighborhood of $\mathbf{R}^{2n}$ and for small $|t|$. Expand it as a power series as follows

$$\phi(w,t)=\sum_{k=0}^{\infty}\phi^{(k)}(w)t^k.$$

Differentiating it with respect to w and t, we get

$$\nabla\phi(w,t)=\sum_{k=0}^{\infty}\nabla\phi^{(k)}(w)t^k, \tag{46}$$

$$\frac{\partial}{\partial t}\phi(w,t)=\sum_{k=0}^{\infty}(k+1)t^k\phi^{(k+1)}(w). \tag{47}$$

By (33),

$$\nabla\phi^{(0)}(w)=\nabla\phi(w,0)=f(w,0)=N_0 w.$$

So we can take $\phi^{(0)}(w)=\frac{1}{2}w^T N_0 w$. Denote $E_0=A^\alpha N_0+B^\alpha$. Then

$$A^\alpha\nabla\phi(w,t)+B^\alpha w=E_0 w+\sum_{k=1}^{\infty}A^\alpha\nabla\phi^{(k)}(w)t^k.$$

Expanding $H(z)$ at $z=E_0w$, we get

$$
\begin{aligned}
&H\left(A^\alpha\nabla\phi(w,t)+B^\alpha w\right)\\
=&H\left(E_0w+\sum_{k=1}^{\infty}A^\alpha\nabla\phi^{(k)}(w)t^k\right)\\
=&H\left(E_0w\right)+\sum_{m=1}^{\infty}\frac{1}{m!}\sum_{i_1,\cdots,i_m=1}^{2n}\sum_{j_1,\cdots,j_m=1}^{\infty}t^{j_1+\cdots+j_m}H_{z_{i_1},\cdots,z_{i_m}}\left(E_0w\right)\\
&\times\left(A^\alpha\nabla\phi^{(j_1)}(w)\right)_{i_1}\cdots\left(A^\alpha\nabla\phi\left(j_m\right)(w)\right)_{i_m}\\
=&H\left(E_0w\right)+\sum_{m=1}^{\infty}\frac{1}{m!}\sum_{i_1,\cdots,i_m=1}^{2n}\sum_{k\geqslant m}t^k\sum_{\substack{j_1+\cdots+j_m=k\\ j_l\geqslant 1}}H_{z_{i_1},\cdots,z_{i_m}}\left(E_0w\right)\\
&\times\left(A^\alpha\nabla\phi^{(j_1)}(w)\right)_{i_1}\cdots\left(A^\alpha\nabla\phi^{(j_m)}(w)\right)_{i_m}\\
=&H\left(E_0w\right)+\sum_{k=1}^{\infty}t^k\sum_{m=1}^{k}\frac{1}{m!}\sum_{i_1,\cdots,i_m=1}^{2n}\sum_{\substack{j_1+\cdots+j_m=k\\ j_l\geqslant 1}}H_{z_{i_1},\cdots,z_{i_m}}\left(E_0w\right)\\
&\times\left(A^\alpha\nabla\phi^{(j_1)}\right)_{i_1}\cdots\left(A^\alpha\nabla\phi^{(j_m)}\right)_{i_m}.
\end{aligned}
$$

Substituting this into the R.H.S. of (34) and (47) into the L.H.S. of (34), then comparing the coefficients of t^k on both sides, we get the recursions (44) and (45). Q.E.D.

In the next section when we use generating functions $\phi_{\alpha,H}$ to construct difference schemes we always assume $M_0=I$. For the sake of convenience, we restate Theorem 19 and Theorem 20 as follows.

Theorem 21 *Let* $\alpha\in\mathbf{CSp}\left(\tilde{J}_{4n},J_{4n}\right)$. *Let* $\boldsymbol{z}\to\widehat{\boldsymbol{z}}=g(z,t)$ *be the phase flow of the Hamiltonian system* (31) *with Jacobian* $M(z,t)=g_z(z,t)$. *If*

$$|C_\alpha+D_\alpha|\neq 0,$$

then there exists, for sufficiently small $|t|$ *and in (some neighborhood of)* R^{2n}, *a time-dependent gradient map* $w\to\widehat{w}=f(w,t)$ *with Jacobian* $N(w,t)=f_w(w,t)\in\mathbf{Sm}(2n)$ *satisfying the transversality condition* (17.3) *and a time-dependent generating function* $\phi_{\alpha,H}(w,t)=\phi(w,t)$ *such that*

1. $f(w,t)=\nabla\phi(w,t)$;
2. $\dfrac{\partial\phi}{\partial t}=-\mu H\left(A^\alpha\nabla\phi(w,t)+B^\alpha w\right)$;
3. $A_\alpha g(z,t)+B_\alpha z\equiv f\left(C_\alpha g(z,t)+D_\alpha z,t\right)\equiv\nabla\phi\left(C_\alpha g(z,t)+D_\alpha z,t\right)$;
4. $N=\sigma_\alpha(M)=\left(A_\alpha M+B_\alpha\right)\left(C_\alpha M+D_\alpha\right)^{-1}$,
   $M=\sigma_{\alpha^{-1}}(N)=\left(A^\alpha N+B^\alpha\right)\left(C^\alpha N+D^\alpha\right)^{-1}$.

Theorem 22 *Let $H(z)$ depend analytically on z. Then $\phi_{\alpha,H}(w,t)$ is expressible as a convergent power series in t for sufficiently small $|t|$, with the recursively determined coefficients:*

$$\begin{aligned}
\phi(w,t) &= \sum_{k=0}^{\infty}\phi^{(k)}(w)t^k,\\
\phi^{(0)}(w) &= \frac{1}{2}w^T N_0 w, \quad N_0=(A_\alpha+B_\alpha)(C_\alpha+D_\alpha)^{-1},\\
\phi^{(1)}(w) &= -\mu(\alpha)H(E_0 w), \quad E_0=(C_\alpha+D_\alpha)^{-1},\\
k\geqslant 1, \phi^{(k+1)}(w) &= -\frac{\mu(\alpha)}{k+1}\sum_{m=1}^{k}\frac{1}{m!}\sum_{i_1,\cdots,i_m=1}^{2n}\sum_{\substack{j_1+\cdots+j_m=k\\ j_l\geqslant 1}} H_{z_{i_1},\cdots,z_{i_m}}(E_0 w)\\
&\quad\times\left(A^\alpha\nabla\phi^{(j_1)}\right)_{i_1}\cdots\left(A^\alpha\nabla\phi^{(j_m)}\right)_{i_m}.
\end{aligned}$$

§6 Construction of Canonical Difference Schemes

In this section we consider the construction of canonical difference schemes for the Hamiltonian system (31). By Theorem 18, for a given time-dependent scalar function $\psi(w,t):\mathbf{R}^{2n}\times\mathbf{R}\to\mathbf{R}$, we can get a time-dependent canonical map $\tilde{g}(z,t)$. If $\psi(w,t)$ approximates some generating function $\phi_{\alpha,H}(w,t)$ of the Hamiltonian system (31) then $\tilde{g}(z,t)$ approximates the phase flow $g(z,t)$. Then fixing t as a time step, we can get a difference scheme—the canonical difference scheme—whose transition from one time-step to the next is canonical. By Theorem 22, generating functions $\phi(w,t)$ can be expressed as a power series. So a natural way to approximate $\phi(w,t)$ is to take the truncation of the series. More precisely, we have

Theorem 23 *Using Theorems 21 and 22, for sufficiently small $\tau>0$ as the time-step, define*

$$\psi^{(m)}(w,\tau)=\sum_{i=0}^{m}\phi^{(i)}(w)\tau^i, \quad m=1,2,\cdots. \tag{48}$$

Then the gradient map

$$w\to\widehat{w}=\tilde{f}(w,\tau)=\nabla\psi^{(m)}(w,\tau) \tag{49}$$

defines an implicit canonical difference scheme $z=z^k\to z^{k+1}=\widehat{z}$,

$$A_\alpha z^{k+1}+B_\alpha z^k=\nabla\psi^{(m)}\left(C_\alpha z^{k+1}+D_\alpha z^k,\tau\right) \tag{50}$$

of m-th order of accuracy.

Proof. Since $\psi^{(m)}(w,0) = \phi(w,0)$, $\psi^{(m)}_{ww}(w,0) = \phi_{ww}(w,0) = f_w(w,0) = N(w,0)$ satisfies the transversality condition (17.3), i.e., $|C^\alpha N(w,0)+D^\alpha| \neq 0$. Thus for sufficiently small τ and in some neighborhood of $\mathbf{R}^{2n}$, $N^{(m)}(w,\tau) = \psi^{(m)}_{ww}(w,\tau)$ satisfies the transversality condition (17.3), i.e., $\left|C^\alpha N^{(m)}(w,\tau)+D^\alpha\right| \neq 0$. By Theorem 18, the gradient map $w \to \widehat{w} = \tilde{f}(w,\tau) = \nabla\psi^{(m)}(w,\tau)$ defines implicitly a time-dependent canonical map $z \to \widehat{z} = \tilde{g}(z,\tau)$ by the equation

$$A_\alpha \widehat{z} + B_\alpha z = \nabla\psi^{(m)}\left(C_\alpha\widehat{z} + D_\alpha z, \tau\right).$$

That is, the equation

$$A_\alpha z^{k+1} + B_\alpha z^k = \nabla\psi^{(m)}\left(C_\alpha z^{k+1} + D_\alpha z^k, \tau\right)$$

is an implicit canonical difference scheme.

Since $\psi^{(m)}(w,\tau)$ is the m-th approximant to $\phi(w,\tau)$, so is $\tilde{f}(w,\tau) = \nabla\psi^{(m)}(w,\tau)$ to $f(w,\tau)$. It follows that the canonical difference scheme given by (50) is of m-th order of accuracy. Q.E.D.

It is not difficult to show that the generating function $\phi(w,t)$ of type (III) is odd in t. Hence Theorem 23 leads to a family of canonical difference schemes of arbitrary even order accuracy.

Theorem 24 *Let* $\alpha = \begin{bmatrix} -J_{2n} & J_{2n} \\ \frac{1}{2}I_{2n} & \frac{1}{2}I_{2n} \end{bmatrix}$. *For sufficiently small $\tau > 0$ as the time-step, define*

$$\psi^{(2m)}(w,\tau) = \sum_{i=1}^{m} \phi^{(2i-1)}(w)\tau^{2i-1}, \quad m = 1,2,\cdots \tag{51}$$

Then the gradient map

$$w \to \widehat{w} = \tilde{f}(w,\tau) = \nabla\psi^{(2m)}(w,\tau)$$

defines implicitly canonical difference schemes $z = z^k \to z^{k+1} = \widehat{z}$,

$$z^{k+1} = z^k - J^{-1}\nabla\psi^{(2m)}\left(\frac{1}{2}\left(z^{k+1}+z^k\right), \tau\right) \tag{52}$$

of 2m-th order of accuracy. The case $m = 1$ is the Euler centered scheme.

For the linear Hamiltonian system (31) with the quadratic Hamiltonian $H(z) = \frac{1}{2}z^T S z$, $S \in \mathbf{Sm}(2n)$, the generating function of type (III) is the quadratic form

$$\begin{aligned}\phi(w,\tau) &= -\frac{1}{2}w^T\left(2J\tanh\frac{\tau}{2}L\right)w \\ &= -\sum_{i=1}^{\infty} a_{2i-1} w^T J\left(\frac{\tau}{2}L\right)^{2i-1} w, \quad L = J^{-1}S,\end{aligned} \tag{53}$$

where

$$\tanh\lambda = \lambda - \frac{1}{3}\lambda^3 + \frac{2}{15}\lambda^5 - \frac{17}{312}\lambda^7 + \cdots = \sum_{i=1}^{\infty} a_{2i-1}\lambda^{2i-1},$$
$$a_{2i-1} = 2^{2i}\left(2^{2i}-1\right)B_{2i}/(2i)!, \quad B_{2i}\text{— Bernoulli numbers,}$$
$$J\tanh\frac{\tau}{2}L \in \mathbf{Sm}(2n).$$

We can easily get (53) by a simple way.

We know, the phase flow of the linear Hamiltonian system

$$\frac{dz}{dt} = J^{-1}Sz = Lz, \quad L = J^{-1}S,$$

is e^{tL}. Set $z = z(0), \hat{z} = e^{\tau L}z(0) = e^{\tau L}z$. Then

$$\widehat{w} = J(z-\widehat{z}) = J\left(I - e^{\tau L}\right)z, \quad w = \frac{1}{2}(z+\widehat{z}) = \frac{1}{2}\left(I + e^{\tau L}\right)z.$$

Hence

$$\widehat{w} = 2J\left(I - e^{\tau L}\right)\left(I + e^{\tau L}\right)^{-1} w = 2J\frac{e^{-\frac{\tau}{2}L} - e^{\frac{\tau}{2}L}}{e^{-\frac{\tau}{2}L} + e^{\frac{\tau}{2}L}}w$$
$$= -2J\tanh\left(\frac{\tau}{2}L\right)w.$$

Taking the truncation of (53),

$$\psi^{(2m)}(w,\tau) = -\sum_{i=1}^{m} a_{2i-1}w^T J\left(\frac{\tau}{2}L\right)^{2i-1} w,$$

we get the gradient map

$$\widehat{w} = \nabla_w \psi^{(2m)} = -2J\sum_{i=1}^{m} a_{2i-1}\left(\frac{\tau}{2}L\right)^{2i-1} w.$$

Noting that $\widehat{w} = J(z-\widehat{z}), w = \frac{1}{2}(z+\widehat{z})$, we have

$$J(z-\widehat{z}) = -J\sum_{2i-1}^{m} a_{2i-1}\left(\frac{\tau}{2}L\right)^{2i-1}(z+\widehat{z}).$$

So the 2m-th order difference scheme is

$$z^{k+\mathrm{i}} - z^k = \sum_{i=1}^{m} a_{2i-1}\left(\frac{\tau}{2}L\right)^{2i-1}\left(z^{k+1}+z^k\right).$$

Set $m=1$, we get the Euler centered scheme.

If we take the diagonal Pade approximants to $\tanh\lambda$,

$$\frac{R_m(\lambda)}{Q_m(\lambda)} - \tanh\lambda = 0\left(|\lambda|^{2m+1}\right),$$

where

$$
\begin{aligned}
&R_0(\lambda) = 0, \qquad R_1(\lambda) = \lambda, \\
&R_m(\lambda) = (2m-1)R_{m-1}(\lambda) + \lambda^2 R_{m-2}(\lambda), \quad m = 2, 3, \cdots, \\
&Q_0(\lambda) = 1, \qquad Q_1(\lambda) = 1, \\
&Q_m(\lambda) = (2m-1)Q_{m-1}(\lambda) + \lambda^2 Q_{m-2}(\lambda), \quad m = 2, 3, \cdots,
\end{aligned}
$$

then we get another type symplectic difference schemes

$$
z^{k+1} - z^k = \frac{R_m\left(\dfrac{\tau}{2}L\right)}{Q_m\left(\dfrac{\tau}{2}L\right)} \left(z^{k+1} + z^k\right),
$$

i.e.,

$$
z^{k+1} = \frac{Q_m\left(\dfrac{\tau}{2}L\right) + R_m\left(\dfrac{\tau}{2}L\right)}{Q_m\left(\dfrac{\tau}{2}L\right) - R_m\left(\dfrac{\tau}{2}L\right)} z^k. \tag{54}
$$

Suppose that $P_m(\lambda)/P_m(-\lambda)$ is the diagonal Pade approximant to e^λ where

$$
\begin{aligned}
&P_0(\lambda) = 1, \\
&P_1(\lambda) = 2 + \lambda, \\
&P_m(\lambda) = 2(2m-1)P_{m-1}(\lambda) + \lambda^2 P_{m-2}(\lambda), \quad m = 2, 3, \cdots.
\end{aligned}
$$

Then because

$$
\tanh \lambda = \frac{e^\lambda - e^{-\lambda}}{e^\lambda + e^{-\lambda}} = \frac{-1 + e^{2\lambda}}{1 + e^{2\lambda}},
$$

the diagonal Pade approximant $R_m(\lambda)/Q_m(\lambda)$ to $\tanh \lambda$ is

$$
\frac{R_m(\lambda)}{Q_m(\lambda)} = \frac{-1 + P_m(2\lambda)/P_m(-2\lambda)}{1 + P_m(2\lambda)/P_m(-2\lambda)} = \frac{P_m(2\lambda) - P_m(-2\lambda)}{P_m(2\lambda) + P_m(-2\lambda)}.
$$

Hence

$$
R_m(\lambda) = c_m \left(P_m(2\lambda) - P_m(-2\lambda)\right), \quad Q_m(\lambda) = c_m \left(P_m(2\lambda) + P_m(-2\lambda)\right),
$$

where $c_m = const \neq 0$. It follows that

$$
P_m(2\lambda) = \frac{1}{2c_m}\left(Q_m(\lambda) + R_m(\lambda)\right), \quad P_m(-2\lambda) = \frac{1}{2c_m}\left(Q_m(\lambda) - R_m(\lambda)\right).
$$

In fact $c_m = 2^{-m}$. So (54) becomes

$$
z^{k+1} = \frac{P_m(\tau L)}{P_m(-\tau L)} z^k.
$$

Examples of canonical difference schemes.

By Theorem 22, as $\mu=1$,

$$\begin{aligned}
\phi^{(0)}(w) &= \frac{1}{2}w^T N_0 w, \qquad N_0=(A_\alpha+B_\alpha)(C_\alpha+D_\alpha)^{-1},\\
\phi^{(1)}(w) &= -H(E_0 w), \qquad E_0=(C_\alpha+D_\alpha)^{-1},\\
\phi^{(2)}(w) &= \frac{1}{2}(\nabla H)^T A^\alpha E_0^T(\nabla H)(E_0 w),\\
\phi^{(3)}(w) &= -\frac{1}{3}(\nabla H)^T A^\alpha \nabla_w \phi^{(2)} - \frac{1}{6}\left(A^\alpha \nabla \phi^{(1)}\right)^T H_{zz}\left(A^\alpha \nabla \phi^{(1)}\right)\\
&= -\frac{1}{6}(\nabla H)^T A^\alpha\left(E_0^T H_{zz} A^\alpha E_0^T \nabla H + E_0^T H_{zz} E_0 A^{\alpha T}\nabla H\right)\\
&\quad -\frac{1}{6}(\nabla H)^T E_0 A^{\alpha T} H_{zz} A^\alpha E_0^T \nabla H\\
&= -\frac{1}{6}\left\{(\nabla H)^T A^\alpha E_0^T H_{zz}\left(A^\alpha E_0^T + E_0 A^{\alpha T}\right)\nabla H + (\nabla H)^T E_0 A^{\alpha T} H_{zz} A^\alpha E_0^T \nabla H\right\}.
\end{aligned}$$

Here we use, instead of the component notation in Theorem 22, the matrix notation, H_{zz} denotes the Hessian matrix of H; all derivatives of H are evaluated at $z=E_0 w$.

Type (II).

$$\alpha=\begin{bmatrix}0&0&-I_n&0\\0&-I_n&0&0\\0&0&0&I_n\\I_n&0&0&0\end{bmatrix},\quad \alpha^T=\alpha^{-1}=\begin{bmatrix}0&0&0&I_n\\0&-I_n&0&0\\-I_n&0&0&0\\0&0&I_n&0\end{bmatrix}.$$

$$\begin{aligned}
w &= \begin{bmatrix}q\\ \hat{p}\end{bmatrix},\quad \widehat{w}=-\begin{bmatrix}p\\ \widehat{q}\end{bmatrix}.\\
N_0 &= -\begin{bmatrix}0&I\\I&0\end{bmatrix},\quad E_0=\begin{bmatrix}0&I\\I&0\end{bmatrix},\quad A^\alpha E_0^T=-\begin{bmatrix}0&0\\I&0\end{bmatrix},\\
\phi^{(1)}(w) &= -H(\widehat{p},q),\\
\phi^{(2)}(w) &= -\frac{1}{2}\sum_{i=1}^{n}\left(H_{q_i}H_{p_i}\right)(\widehat{p},q),\\
\phi^{(3)}(w) &= -\frac{1}{6}\sum_{i,j=1}^{n}\left(H_{p_ip_j}H_{q_i}H_{q_j}+H_{q_iq_j}H_{p_i}H_{p_j}+H_{q_ip_j}H_{p_i}H_{q_j}\right)
\end{aligned}$$

where $H(z)=H(p_1,\cdots,p_n,q_1,\cdots,q_n), H_{z_i}=\partial H/\partial z_i$.

a. First order scheme.

$$\psi^{(1)}(w,\tau)=\phi^{(0)}(w)+\tau\phi^{(1)}(w).$$

The equation $\widehat{w}=\nabla\psi^{(1)}(w,\tau)$ defines a first order canonical difference scheme

$$\begin{cases} p_i^{k+1}=p_i^k-\tau H_{q_i}\left(p^{k+1},q^k\right),\\ q_i^{k+1}=q_i^k+\tau H_{p_i}\left(p^{k+1},q^k\right).\end{cases}\quad i=1,\cdots,n. \tag{55}$$

When H is separable, $H = U(p) + V(q), H_{q_i}\left(p^{k+1}, q^k\right) = V_{q_i}\left(q^k\right), H_{p_i}\left(p^{k+1}, q^k\right) = U_{p_i}\left(p^{k+1}\right)$. At this time, (55) becomes

$$\left\{\begin{array}{l} p_i^{k+1} = p_i^k - \tau V_{q_i}\left(q^k\right), \\ q_i^{k+1} = q_i^k + \tau U_{p_i}\left(p^{k+1}\right). \end{array}\right. \quad i = 1, \cdots, n. \tag{56}$$

Evidently, (56) is an explicit difference scheme of 1-st order of accuracy. If we set $q's$ at half-integer times $t = \left(k + \frac{1}{2}\right)\tau$, then (56) becomes

$$\left\{\begin{array}{l} p_i^{k+1} = p_i^k - \tau V_{q_i}\left(q^{k+\frac{1}{2}}\right), \\ q_i^{k+\frac{1}{2}+1} = q_i^{k+\frac{1}{2}} + \tau U_{p_i}\left(p^{k+1}\right). \end{array}\right. \quad i = 1, \cdots, n. \tag{57}$$

(57) is a staggered explicit scheme of 2-nd order of accuracy.

b. Second order scheme.

$$\psi^{(2)}(w, \tau) = \psi^{(1)}(w) + \tau^2 \phi^{(2)}(w).$$

The induced gradient map is

$$\widehat{\omega} = \nabla_w \psi^{(2)} = -\begin{bmatrix} \hat{p} \\ q \end{bmatrix} - \tau \begin{bmatrix} \nabla_q H \\ \nabla_p H \end{bmatrix} - \frac{\tau^2}{2} \begin{bmatrix} \nabla_q \left(\sum\limits_{i=1}^n H_{q_i} H_{p_i}\right) \\ \nabla_p \left(\sum\limits_{i=1}^n H_{q_i} H_{p_i}\right) \end{bmatrix}.$$

So the second order scheme is

$$\left\{\begin{array}{l} p_i^{k+1} = p_i^k - \tau H_{q_i}\left(p^{k+1}, q^k\right) - \dfrac{\tau^2}{2}\left(\sum\limits_{j=1}^n H_{q_j} H_{p_j}\right)_{q_i}\left(p^{k+1}, q^k\right) \\ q_i^{k+1} = q_i^k + \tau H_{p_i}\left(p^{k+1}, q^k\right) + \dfrac{\tau^2}{2}\left(\sum\limits_{j=1}^n H_{q_j} H_{p_j}\right)_{p_i}\left(p^{k+1}, q^k\right) \end{array}\right. \quad i = 1, \cdots, n.$$

This scheme is already implicit even when $H(z)$ is separable.

c. The third order scheme is, for $i = 1, \cdots, n$,

$$\begin{aligned} p_i^{k+1} = {} & p_i^k - \tau H_{q_i}\left(p^{k+1}, q^k\right) - \frac{\tau^2}{2} \sum_{j=1}^n \left(H_{q_i} H_{p_j}\right)_{q_i}\left(p^{k+1}, q^k\right) \\ & - \frac{\tau^3}{6} \sum_{l,j=1}^n \left(H_{p_l p_j} H_{q_l} H_{q_j} + H_{q_l q_j} H_{p_l} H_{p_j} + H_{p_l q_j} H_{q_l} H_{p_j}\right)_{q_i}\left(p^{k+1}, q^k\right), \\ q_i^{k+1} = {} & q_i^k + \tau H_{p_i}\left(p^{k+1}, q^k\right) + \frac{\tau^2}{2} \sum_{j=1}^n \left(H_{qj} H_{p_j}\right)_{p_i}\left(p^{k+1}, q^k\right) \\ & + \frac{\tau^3}{6} \sum_{l,j=1}^n \left(H_{p_l p_j} H_{q_l} H_{q_j} + H_{q_l q_j} H_{p_l} H_{p_j} + H_{p_l q_j} H_{q_l} H_{p_j}\right)_{p_i}\left(p^{k+1}, q^k\right). \end{aligned}$$

Type (III).

$$\alpha=\begin{bmatrix}-J_{2n} & J_{2n}\\ \frac{1}{2}I_{2n} & \frac{1}{2}I_{2n}\end{bmatrix},\quad \alpha^{-1}=\begin{bmatrix}\frac{1}{2}J_{2n} & I_{2n}\\ -\frac{1}{2}J_{2n} & I_{2n}\end{bmatrix}.$$

$$w=\frac{1}{2}(\hat{z}+z),\quad \widehat{w}=J(z-\widehat{z}).$$

$$N_0=0,\quad E_0=I,\quad A^{\alpha}E_0^T+E_0A^{\alpha T}=0.$$

$$\phi^{(0)}=\phi^{(2)}=\phi^{(4)}=0,$$

$$\phi^{(1)}(w)=-H\left(\frac{1}{2}(\hat{z}+z)\right),$$

$$\phi^{(3)}(w)=\frac{1}{24}(\nabla H)^TJH_{zz}J\nabla H,$$

$$\psi^{(2)}(w,\tau)=-\tau H,$$

$$\psi^{(4)}(w,\tau)=-\tau H+\frac{\tau^3}{24}(\nabla H)^TJH_{zz}J\nabla H.$$

By Theorem 24 the second order scheme is

$$J(z-\hat{z})=\hat{w}=\nabla_w\psi^{(2)}(w,t)=-\tau\nabla H\left(\frac{1}{2}(z+\hat{z})\right),$$

i.e.,

$$z^{k+1}=z^k+\tau J^{-1}\nabla H\left(\frac{1}{2}\left(z^{k+1}+z^k\right)\right).$$

The 4-th order scheme is

$$J(z-\hat{z})=\hat{w}=\nabla_w\psi^{(4)}(w,t)=-\tau\nabla H\left(\frac{1}{2}(z+\hat{z})\right)+\frac{r^2}{24}\nabla_z\left((\nabla H)^TJH_{zz}J\nabla H\right)$$

i.e.,

$$z^{k+1}=z^k+\tau J^{-1}\nabla H\left(\frac{1}{2}\left(z^{k+1}+z^k\right)\right)-\frac{\tau^3}{24}J^{-1}\nabla_z\left((\nabla H)^TJH_{zz}J\nabla H\right)\left(\frac{1}{2}\left(z^{k+1}+z^k\right)\right).$$

References

[1] Feng K. On difference schemes and symplectic geometry. In Feng K, ed. Proc 1984 Beijing Symp Diff Geometry and Diff Equations. Beijing: Science Press, 1985. 42-58

[2] Abraham R., Marsden J. Foundations of Mechanics. Mass: Addison-Wesley, Reading. 1978

[3] Arnold V I. Mathematical Methods of Classical Mechanics. New York: Springer. 1978

[4] Artin E. Geometrical Algebra, New York. 1957

[5] Baker G A, Graves-Morris P. Pade Approximants, Part I: Basic Theory. London: AddisonWeslay. 1981

[6] Feng K, Wu H M, Qin M Z. Symplectic difference schemes for linear Hamiltonian canonical systems. J Comp Math. 1990, 8(4): 371-380

[7] Hua L K. On the theory of automorphic functions of a matrix variable I, II. Amer Jour Math. 1944, 66: 470-488, 531-563

[8] Siegel C K. Symplectic geometry. Amer Jour Math. 1943, 65: 1-86

9 Symplectic Difference Schemes for Linear Hamiltonian Canonical Systems①②

线性哈密尔顿系统的辛差分格式

Abstract

In this paper, we present some results of a study, specifically within the framework of symplectic geometry, of difference schemes for numerical solution of the linear Hamiltonian systems. We generalize the Cayley transform with which we can get different types of symplectic schemes. These schemes are various generalizations of the Euler centered scheme. They preserve all the invariant first integrals of the linear Hamiltonian systems.

§1 Introduction

Recently, it becomes evident that the Hamiltonian formalism plays a fundamental role in mathematical physics. One needs only to recall a few examples: classical mechanics, quantum mechanics, hydrodynamics of a perfect fluid, plasma physics, and accelerator physics.

The evolution of Hamiltonian systems has the important property of being symplectic, i.e.,the sum of the areas of the canonical variable pairs, projected on any two-dimensional surface in a phase space, is time invariant. In numerically solving these equations it is necessary to replace them with finite difference equations which preserve this symplectic evolution property. In [1] the first author proposed a systematic study of symplectic difference schemes for Hamiltonian systems from the viewpoint of symplectic geometry. We present here some developments for linear Hamiltonian systems.

An outline of this paper is as follows: Section 2 is devoted to a review of well known facts concerning symplectic structures and Hamiltonian mechanics. In Section 3 we review some properties of the symplectic matrix and the infinitesimal symplectic matrix. In Section 4 we review some linear symplectic difference schemes. Constructions of linear symplectic schemes based on the Padé approximation are described in §5. Generalized Cayley transform and its corresponding symplectic schemes and conservation laws are presented in §6.

---

① Joint with Wu H M, Qin M Z

② J Comput Math. 1990, 8(4): 371-380

§2 Some Facts from Hamiltonian Mechanics and Symplectic Geometry

In this section we will review some facts from Hamiltonian mechanics and symplectic geometry which are fundamental to what follows. Consider the following system of differential equations on R^{2n}

$$\begin{aligned} \frac{dp_i}{dt} &= -\frac{\partial H}{\partial q_i}, \\ \frac{dq_i}{dt} &= \frac{\partial H}{\partial p_i}, \quad i=1,2,\cdots,n \end{aligned} \tag{2.1}$$

where $H(p,q)$ is some real valued smooth function on R^{2n}. We call (2.1) a canonical system of differential equations with Hamiltonian H. We denote $p_i = z_i, q_i = z_{i+n}, z = (z_1,\cdots,z_{2n})'$, and $\frac{\partial H}{\partial z} = \left(\frac{\partial H}{\partial z_1},\cdots,\frac{\partial H}{\partial z_{2n}}\right)' \in R^{2n}$, Then (2.1) becomes

$$\frac{dz}{dt} = J^{-1}\frac{\partial H}{\partial z} \tag{2.2}$$

with

$$J = \begin{pmatrix} 0 & I_n \\ -I_n & 0 \end{pmatrix}, \quad J' = -J = J^{-1} \tag{2.3}$$

where I_n is the identity matrix. The phase space R^{2n} is equipped with a standard symplectic structure defined by the "fundamental" differential 2-form

$$\omega = \sum_1^n dp_i \wedge dq_i.$$

Let g be a diffeomorphism of R^{2n}:

$$z = \begin{pmatrix} p \\ q \end{pmatrix} \to g(z) = \begin{bmatrix} g_1(z) \\ \vdots \\ g_{2n}(z) \end{bmatrix} = \begin{bmatrix} \hat{p}(p,q) \\ \hat{q}(p,q) \end{bmatrix}$$

g is called a symplectic transformation if g preserves the 2-form ω, i.e.,

$$\sum_1^n d\hat{p}_i \wedge d\hat{q}_i = \sum_1^n dp_i \wedge dq_i.$$

This is equivalent to the condition that

$$\left(\frac{\partial g}{\partial z}\right)' J \left(\frac{\partial g}{\partial z}\right) \equiv J$$

i.e. the Jacobian matrix $\dfrac{\partial g}{\partial z}$ is symplectic everywhere,

$$\frac{\partial g}{\partial z}=\begin{bmatrix}\dfrac{\partial \hat{p}}{\partial p} & \dfrac{\partial \hat{p}}{\partial q}\\ \dfrac{\partial \hat{q}}{\partial p} & \dfrac{\partial \hat{q}}{\partial q}\end{bmatrix}.$$

For every pair of smooth functions $H(z)$, $F(z)$ on R^{2n}, the Poisson bracket is defined as

$$\{H,F\}=\left(\frac{\partial H}{\partial z}\right)' J\left(\frac{\partial F}{\partial z}\right)=\left(\frac{\partial H}{\partial q}\right)'\frac{\partial F}{\partial p}-\left(\frac{\partial H}{\partial p}\right)'\frac{\partial F}{\partial q}.$$

The bracket operation is anti-symmetric, bilinear and satisfies the Jacobi identity. A function F on R^{2n} is called an invariant integral of the canonical system (2.2) with Hamiltonian H if

$$F(z(t))=\text{ const.}$$

independent of t for every solution $z(t)$ of (2.2). A necessary and sufficient condition for the above invariant is

$$\{H,F\}=0.$$

The Hamiltonian H itself is an invariant integral of the system (2.2). The fundamental property of the canonical system (2.2) with Hamiltonian function H is that there exists a one parameter group of symplectic transformations $g^t=g_H^t$, called the phase flow of function H, such that the solution $z(t)$ of (2.2) with initial value $z(0)$ is given by

$$z(t)=g_H^t(z(0)).$$

So the time evolution of a Hamiltonian system is always symplectic.

§3 Some Properties of Symplectic Matrices

Let us now briefly sketch some properties of the symplectic matrix.

Definition 1 A matrix S of order $2n$ is called symplectic if it satisfies the relation

$$S'JS=J \tag{3.1}$$

where S' is the transpose of S. All symplectic matrices form a group $Sp(2n)$.

Definition 2 A matrix B of order $2n$ is called infinitesimal symplectic if

$$JB+B'J=0. \tag{3.2}$$

All infinitesimal symplectic matrices form a Lie algebra $sp(2n)$ with commutation operation $[A,B]=AB-BA$, and $sp(2n)$ is the Lie algebra of Lie group $Sp(2n)$.

We have the following well-known propositions:

P1. det $S=1$, if $S \in Sp(2n)$

P2. $S^{-1}=-JS'J=J^{-1}S'J$, if $S \in Sp(2n)$

P3. $SJS'=J$, if $S \in Sp(2n)$

P4. Let

$$S=\begin{pmatrix} A & B \\ C & D \end{pmatrix},$$

where A,B,C,D are $n\times n$ matrices. Then $S \in Sp(2n)$ iff

$$\begin{aligned} &AB'-BA'=0, \quad CD'-DC'=0, \quad AD'-BC'=I, \\ &A'C-C'A=0, \quad B'D-D'B=0, \quad A'D-C'B=I. \end{aligned}$$

P5. Matrices

$$\begin{pmatrix} I & B \\ 0 & I \end{pmatrix}, \quad \begin{pmatrix} I & 0 \\ D & I \end{pmatrix}$$

are symplectic if $B'=B, D'=D$.

P6. Matrix

$$\begin{pmatrix} A & 0 \\ 0 & D \end{pmatrix} \in Sp(2n) \quad \text{if} \quad A=(D')^{-1}.$$

P7. Matrix $S=M^{-1}N \in Sp(2n)$ iff $M'JM=N'JN$.

P8. Matrix

$$\begin{pmatrix} Q & I-Q \\ -(I-Q) & Q \end{pmatrix} \in Sp(2n) \text{ iff } Q^2=Q, Q'=Q.$$

P9. If $B \in sp(2n)$, then $\exp(B) \in Sp(2n)$.

P10. If $B \in sp(2n)$, and $|I+B| \neq 0$, then $F=(I+B)^{-1}(I-B) \in Sp(2n)$, the Cayley transform of B.

P11. If $B \in sp(2n)$, then $(B^{2m})'J=J(B^{2m})$.

P12. If $B \in sp(2n)$, then $(B^{2m+1})'J=-J(B^{2m+1})$.

P13. If $f(x)$ is an even polynomial and $B \in sp(2n)$, then $f(B')J=Jf(B)$.

P14. If $g(x)$ is an odd polynomial and $B \in sp(2n)$, then $g(B) \in sp(2n)$, i.e., $g(B')J+Jg(B)=0$.

§4 Some Symplectic Schemes for Linear Hamiltonian Systems

A Hamiltonian system (2.1) is called linear if the Hamiltonian is a quadratic form of z

$$H(z)=\frac{1}{2}z'Cz, \quad C'=C,$$

and J is a standard antisymmetric matrix

$$J=\begin{bmatrix} 0 & I_n \\ -I_n & 0 \end{bmatrix}, \quad J'=-J=J^{-1}, \quad \det J=1.$$

Then the canonical system (2.1) becomes

$$\frac{dz}{dt}=Bz, \quad B=J^{-1}C, \quad C'=C, \tag{4.1}$$

where $B=J^{-1}C$ is infinitesimal symplectic. The solution of (4.1) is

$$z(t)=g^t z(0), \quad g^t=\exp(tB) \tag{4.2}$$

where g^t as the exponential transform of infinitesimal symplectic tB, is symplectic (P. 14).

Consider now a quadratic form $F(z)=\frac{1}{2}z'Az$. The Poisson bracket of two quadratic forms H, F is also a quadratic form

$$\{H,F\}=\frac{1}{2}z'(AJC-CJA)z.$$

The condition for the quadratic form F to be an invariant integral of the linear Hamiltonian system $\frac{dz}{dt}=J^{-1}Cz$ can be expressed in any one of the following equivalent ways:

$$F\left(\left(\exp\left(tJ^{-1}C\right)\right)z\right)\equiv F(z), \tag{4.3a}$$

$$\{H,F\}=0, \tag{4.3b}$$

$$\left(\exp\left(tJ^{-1}C\right)\right)'A\left(\exp\left(tJ^{-1}C\right)\right)=A, \tag{4.3c}$$

$$AJC=CJA. \tag{4.3d}$$

In [1] some types of the symplectic scheme are proposed. The first is called the time-centered Euler scheme

$$\frac{z^{n+1}-z^n}{\tau}=B\frac{z^{n+1}+z^n}{2}. \tag{4.4}$$

The transition $z^n\to z^{n+1}$ is given by the following

$$z^{n+1}=F_\tau z^n, \quad F_\tau=\phi\left(-\frac{\tau}{2}B\right), \quad \phi(\lambda)=\frac{1-\lambda}{1+\lambda}, \tag{4.5}$$

where F_τ, as the Cayley transform of infinitesimal symplectic $\left(-\frac{\tau}{2}B\right)$, is symplectic (**P10**).

The second is the staggered explicit scheme for separable Hamiltonian. For a separable Hamiltonian

$$H(p,q) = \frac{1}{2}(p',q')\,S\begin{pmatrix} p \\ q \end{pmatrix} = \frac{1}{2}p'Up + \frac{1}{2}q'Vq \tag{4.6}$$

where

$$S = \begin{bmatrix} U & 0 \\ 0 & V \end{bmatrix},$$

$U' = U$ positive definite and $V' = V$, the canonical equation (4.1) becomes

$$\begin{aligned} \frac{dp}{dt} &= -Vq, \\ \frac{dq}{dt} &= Up. \end{aligned} \tag{4.7}$$

The staggered explicit scheme is

$$\begin{aligned} &\frac{1}{\tau}\left(p^{n+1} - p^n\right) = -Vq^{n+\frac{1}{2}}, \\ &\frac{1}{\tau}\left(q^{n+\frac{1}{2}+1} - q^{n+\frac{1}{2}}\right) = Up^{n+1}. \end{aligned} \tag{4.8}$$

The $p's$ are set at integer times $t = n\tau$, and $q's$ at half-integer times $t = \left(n + \frac{1}{2}\right)\tau$. The transition

$$w^n = \begin{bmatrix} p^n \\ q^{n+\frac{1}{2}} \end{bmatrix} \longrightarrow \begin{bmatrix} p^{n+1} \\ q^{n+\frac{1}{2}+1} \end{bmatrix} = w^{n+1}$$

is given by the following

$$w^{n+1} = F_\tau w^n,$$

where

$$F_\tau = \begin{bmatrix} I & 0 \\ -U & I \end{bmatrix}^{-1} \cdot \begin{bmatrix} I & -V \\ 0 & I \end{bmatrix} \tag{4.9}$$

as the product of two symplectic matrices is symplectic (**P5**), and has second order of accuracy.

§5 Construction of Symplectic Schemes Based on Padé Approximation

We know that the trajectory $z(t) = g^t z_0$ is the solution satisfying the initial condition $z(0) = z_0$. In a linear system g^t coincides with its own Jacobian. One might ask how the approximations of $\exp(tB)$ are. This is most simply described in terms of Padé rational approximations. Here we consider the rational approximation to exp(x) defined by

$$\exp(x) \sim \frac{n_{lm}(x)}{d_{lm}(x)} = g_{lm}(x) \tag{5.1}$$

where

$$n_{lm}(x) = \sum_{k=0}^{m} \frac{(l+m-k)!m!}{(l+m)!k!(m-k)!}(x)^k, \tag{5.2}$$

$$d_{lm}(x) = \sum_{k=0}^{l} \frac{(l+m-k)!l!}{(l+m)!k!(l-k)!}(-x)^k. \tag{5.3}$$

For each pair of nonnegative integers l and m, the Taylor's series expansion of $\frac{n_{lm}(x)}{d_{lm}(x)}$ about the origin shows that

$$\exp(x) - \frac{n_{lm}(x)}{d_{lm}} = o\left(|x|^{m+l+1}\right), \quad |x| \longrightarrow 0. \tag{5.4}$$

The resulting $(l+m)$-th order Padé approximation of exp(x) is denoted by g_{lm}.

Theorem 1 *Let B be an infinitesimal symplectic matrix, then for sufficiently small $|t|$, $g_{lm}(tB)$ is symplectic iff $l = m$, i.e. $g_{ll}(x)$ is the (l,l) diagonal Padé approximant to exp(x).*

Proof. "If" part. Let $n_{ll}(x) = f(x) + g(x), d_{ll}(x) = f(x) - g(x)$, where $f(x)$ is an even polynomial and $g(x)$ is an odd polynomial. In order to prove $g_{ll}(tB) \in Sp(2n)$, we need only to verify (**P7**)

$$(f(tB) + g(tB))'J(f(tB) + g(tB)) = (f(tB) - g(tB))'J(f(tB) - g(tB)). \tag{5.5}$$

By (P.13,14), the L.H.S of (5.5) turns into

$$(f(tB') + g(tB'))\,J(f(tB) + g(tB)) = J(f(tB) - g(tB))(f(tB) + g(tB)). \tag{5.6}$$

Similarly for the R.H.S., we have

$$(f(tB') - g(tB'))\,J(f(tB) - g(tB)) = J(f(tB) + g(tB))(f(tB) - g(tB)). \tag{5.7}$$

Comparing (5.6) and (5.7) completes the proof of "If" part of the theorem.

"Only If" part. Without loss of generality, we may take $l > m$. We need only to notice that, in Proposition 7, the order of the polynomial on the right hand is higher than that on the left hand.

From Theorem 1, we can obtain a sequence of symplectic difference schemes based on the diagonal (k,k) approximant Padé table. For the (1.1) approximant, we have the Euler centered scheme (4.4).

$$z^{n+1}=z^n+\frac{\tau B}{2}\left(z^n+z^{n+1}\right),\quad F_\tau^{(1.1)}=\phi^{(1,1)}(\tau B),\quad \phi^{(1,1)}(\lambda)=\frac{1+\dfrac{\lambda}{2}}{1-\dfrac{\lambda}{2}}.$$

For the (2.2) approximant, we have

$$z^{n+1}=z^n+\frac{\tau B\left(z^n+z^{n+1}\right)}{2}+\frac{\tau^2B^2}{12}\left(z^n-z^{n+1}\right), \tag{5.8}$$

its transition is

$$F_\tau^{(2.2)}=\phi^{(2,2)}(\tau B),\quad \phi^{(2.2)}(\lambda)=\frac{1+\dfrac{\lambda}{2}+\dfrac{\lambda^2}{12}}{1-\dfrac{\lambda}{2}+\dfrac{\lambda^2}{12}}. \tag{5.9}$$

Evidently this scheme has fourth-order accuracy. For (3.3), we get

$$z^{n+1}=z^n+\frac{\tau B}{2}\left(z^n+z^{n+1}\right)+\frac{\tau^2B^2}{10}\left(z^n-z^{n+1}\right)+\frac{\tau^3B^3}{120}\left(z^n+z^{n+1}\right), \tag{5.10}$$

$$F_\tau^{(3.3)}=\phi^{(3,3)}(\tau B),\quad \phi^{(3.3)}(\lambda)=1+\frac{\lambda}{2}+\frac{\lambda^2}{10}+\frac{\lambda^3}{120}\Big/\left(1-\frac{\lambda}{2}+\frac{\lambda^2}{10}-\frac{\lambda^3}{120}\right). \tag{5.11}$$

This scheme has sixth order accuracy. For (4.4), we obtain

$$\begin{aligned}z^{n+1}&=z^n+\frac{\tau B}{2}\left(z^n+z^{n+1}\right)+\frac{3\tau^2B^2}{28}\left(z^n-z^{n+1}\right)\\&\quad+\frac{\tau^3B^3}{84}\left(z^n+z^{n+1}\right)+\frac{\tau^4B^4}{1680}\left(z^n-z^{n+1}\right),\end{aligned} \tag{5.12}$$

$$F_\tau^{(4.4)}=\phi^{(4,4)}(\tau B),$$

$$\phi^{(4.4)}(\lambda)=1+\frac{\lambda}{2}+\frac{3\lambda^2}{28}+\frac{\lambda^3}{84}+\frac{\lambda^4}{1680}\Big/\left(1-\frac{\lambda}{2}+\frac{3\lambda^2}{28}-\frac{\lambda^3}{84}+\frac{\lambda^4}{1680}\right). \tag{5.13}$$

It has eighth order accuracy.

Theorem 2 *The difference schemes*

$$z^{k+1}=g_{ll}(\tau B)z^k,\quad l=1,2,3,\cdots$$

for the system (4.2) *are symplectic of 2l-th order accuracy.*

§6 Generalized Cayley Transformation and its Application

A matrix B is called non-exceptional if

$$\det(I+B)\neq 0. \tag{6.1}$$

We introduce a matrix S by

$$I+S=2(I+B)^{-1} \tag{6.2}$$

with the inversion

$$I+B=2(I+S)^{-1}. \tag{6.3}$$

S is likewise non-exceptional, and we have the Cayley transform[2]

$$S=(I-B)(I+B)^{-1}=(I+B)^{-1}(I-B), \tag{6.4}$$

$$B=(I-S)(I+S)^{-1}=(I+S)^{-1}(I-S). \tag{6.5}$$

Let A be an arbitrary matrix. The equation

$$S'AS=A \tag{6.6}$$

expresses the condition that the substitution of S into both variables z, w leaves invariant the bilinear form $z'Aw$.

Lemma 1 *If the non-exceptional matrices B and S are connected by* (6.4) *and* (6.5), *and A is any matrix, then*

$$S'AS=A \tag{6.7}$$

iff

$$B'A+AB=0. \tag{6.8}$$

Proof. Take the transpose of (6.5) one gets

$$I-S'=B'(I+S').$$

Multiplying on the right by AS and noting (6.6), one finds

$$A(S-I)=B'A(S+I)$$

and hence, multiplying $(S+I)^{-1}$ on the right one gets

$$-AB=B'A.$$

Conversely, if assuming (6.8) and multiply the transposed equation

$$S'(I+B')=I-B'$$

of (6.4) on the right by A, we have

$$S'A(I-B)=A(I+B)$$

which yields (6.7) on post-multiplication by $(I+B)^{-1}$.

Let $\phi(\lambda)=(1-\lambda)/(1+\lambda)$, then the Cayley transform of B is denoted by $\phi(B)=(I+B)^{-1}(I-B)$. By taking successively $A=J$ and $A=A'$ in Lemma 1, we get

Theorem 3 *The Cayley transform of a non-exceptional infinitesimal symplectic (symplectic) matrix is a non-exceptional symplectic (infinitesimal symplectic) matrix. Let* $B=J^{-1}C, C'=C, B\in sp(2n)$, *det* $(I+\tau B)\neq 0, A'=A$. *Then*

$$(\phi(\tau B))'A(\phi(\tau B))=A \tag{6.9}$$

iff

$$B'A+AB=0. \tag{6.10}$$

In other words, a quadratic form $F(z)=\frac{1}{2}z'Az$ *is invariant under the symplectic transformation* $\phi(\tau B)$ *iff* $F(z)$ *is an invariant integral of the Hamiltonian system* (4.1).

Theorem 4 *Let* $\psi(\lambda)$ *be a function of complex variable* λ *satisfying*

1° $\psi(\lambda)$ *is analytic with real coefficients in a neighborhood* D *of* $\lambda=0$.

2° $\psi(\lambda)\psi(-\lambda)=1$ *in* D.

3° $\psi_\lambda(0)\neq 0$. *Let* A *be a matrix of order 2n. Then*

$$(\psi(\tau B))'A(\psi(\tau B))=A \tag{6.11}$$

for all τ *with sufficiently small* $|\tau|$ *iff*

$$B'A+AB=0. \tag{6.12}$$

Proof. The condition 2° implies $\psi^2(0)\neq 0$. So $\psi(0)\neq 0$. If

$$(\psi(\tau B))'A(\psi(\tau B))=A$$

for all τ with sufficiently small $|\tau|$, then differentiating both sides of the above equation with respect to τ, we get

$$B'\left(\psi_\lambda(\tau B)\right)'A\psi(\tau B)+(\psi(\tau B))'AB\psi_\lambda(\tau B)=0.$$

Set $\tau=0$, it becomes

$$(B'A+AB)\,\psi(0)\psi_\lambda(0)=0,$$

i.e.

$$B'A+AB=0.$$

Conversely, if $B'A+AB=0$, then it is not difficult to verify that the equations

$$\psi_\lambda\left(\tau B'\right)A=A\psi_\lambda(-\tau B),\quad \psi\left(\tau B'\right)A=A\psi(-\tau B)$$

hold for any analytic function ψ. From condition 2° it follows that

$$\psi_\lambda(\lambda)\psi(-\lambda)-\psi(\lambda)\psi_\lambda(-\lambda)=0,$$

so

$$\begin{aligned}\frac{d}{d\tau}\left(\psi(\tau B)'A\psi(\tau B)\right)&=\frac{d}{d\tau}\left(\psi\left(\tau B'\right)A\psi(\tau B)\right)\\&=B'\psi_\lambda\left(\tau B'\right)A\psi(\tau B)+\psi\left(\tau B'\right)AB\psi_\lambda(\tau B)\\&=B'A\psi_\lambda(-\tau B)\psi(\tau B)+AB\psi\left(-\tau B'\right)\psi_\lambda(\tau B)\\&=\left(B'A+AB\right)\psi_\lambda(-\tau B)\psi(\tau B)=0\end{aligned}$$

i.e. $\psi\left(\tau B'\right)A\psi(\tau B)=\psi(0)A\psi(0)=A\psi^2(0)=A$. The proof is completed.

By taking successively $A=J$ and $A'=A$ in Theorem 4 and using (4.3) we get

Theorem 5 *Take $|\tau|$ sufficiently small so that τB has no eigenvalue at the pole of the function $\psi(\lambda)$ in Lemma 2. Then $\psi(\tau B)\in Sp(2n)$ iff $B\in sp(2n)$. Let $B=J^{-1}C, C'=C, A'=A$. Then*

$$\left(\psi\left(\tau J^{-1}C\right)\right)'A\left(\psi\left(\tau J^{-1}C\right)\right)=A \tag{6.13}$$

iff

$$AJC=CJA. \tag{6.14}$$

In other words, a quadratic form $F(z)=\frac{1}{2}z'Az$ is invariant under the symplectic transformation $\psi(\tau B)$ iff $F(z)$ is an invariant integral of the system (4.1).

The transform $\psi(\tau B)$ based on Theorem 4 includes exponential transform $\exp(\tau B)$, Cayley transform $\phi\left(-\frac{\tau}{2}B\right)$ and diagonal Padé transform as special cases. Taking $\psi(\lambda)$ in Theorem 4 as a rational function, then necessarily $\psi(\lambda)=\frac{P(\lambda)}{P(-\lambda)}$, $P(\lambda)$ is a polynomial, and is often normalized by setting $P(0)=1, P'(0)\neq 0$.

Theorem 6 *Let $P(\lambda)$ be a polynomial, $P(0)=1, P'(0)\neq 0$, and*

$$\exp(\lambda)-\frac{P(\lambda)}{P(-\lambda)}=O\left(|\lambda|^{m+1}\right). \tag{6.15}$$

Then

$$P(-\tau B)z^{m+1}=P(\tau B)z^m,$$

i.e.

$$z^{m+1}=\frac{P(\tau B)}{P(-\tau B)}z^m \tag{6.16}$$

is a symplectic scheme of order m for the linear system (4.1). *This difference scheme and the original system* (4.1) *have the same set of quadratic invariants.*

In order to find rational approximants $\dfrac{P(x)}{P(-x)}$ to $\exp(x)$, we may express $\exp(x)$ in various ways. For example,

$$(1)\quad \exp(x) \sim \frac{n_{ll}(x)}{n_{ll}(-x)} = \frac{d_{ll}(-x)}{d_{ll}(x)}, \qquad (2)\quad \exp(x) = 1 + \tanh\frac{x}{2} \Big/ \left(1 - \tanh\frac{x}{2}\right),$$

$$(3)\quad \exp(x) = \frac{e^{\frac{x}{2}}}{e^{-\frac{x}{2}}}, \qquad (4)\quad \exp(x) = \frac{1}{2}(1+e^{x}) \Big/ \left(\frac{1}{2}(1+e^{-x})\right).$$

Each denominator and numerator in the above expressions can be expanded about the origin in Taylor's series. The first term of the approximation gives the function $\psi(x) = \left(1+\dfrac{x}{2}\right) / \left(1-\dfrac{x}{2}\right)$ which yields the Euler centered scheme. Keeping $m(>1)$ terms in the expansions for both the denominator and numerator we will get function $\psi(x)$ which will extend the Euler centered schemes. The schemes obtained in this way are all symplectic schemes, however the order of accuracy of the first two kinds of schemes is higher than that of the last two kinds. For example, if in the formula (4) the first three terms of the expansions of the denominator and numerator are kept, then the 4-th order symplectic scheme is obtained. However, the same kind of truncation gives 6-th order schemes from (1) and (2).

References

[1] Feng K. On difference schemes and symplectic geometry. In Feng K, ed. Proc 1984 Beijing Symp Diff Geometry and Diff Equations. Beijing: Science Press, 1985: 42-58

[2] Weyl H. Classical Groups. Princeton. 1946

10 Hamiltonian Algorithms for Hamiltonian Dynamical Systems[①②]

哈密尔顿动力系统的哈密尔顿算法

§1 Background and Motivation

It is well known that two great contributions in natural science are given by Newton in the 17th century, one being the fundamental law of motion $F = Ma$ in physics, the other being the invention of differential calculus in mathematics. For the motion with n-degree freedom, let

$$\text{vector of position } q = (q_1, \ldots, q_n),$$

$$\text{vector of velocity } \dot{q} = \frac{dq}{dt}(q_1, \ldots, q_n),$$

$$\text{vector of acceleration } \ddot{q} = \frac{d^2}{dt^2}(q_1, \ldots, q_n),$$

force field in conservative mechanical systems be represented by

$$\text{force (vector) } F = -\frac{\partial V}{\partial q},$$

which is the negative gradient of a potential function $V = V(q)$. Then,

$$M\frac{d^2 q}{dt^2} = -\frac{\partial V}{\partial q}. \tag{1.1}$$

This is a standard form of equation of motion and is a system of second order differential equations in the configuration space R^n, usually called the standard form (or Newton's form) in classical mechanics.

In the 18th century Euler and Lagrange introduced the action function

$$L = T - V = \text{ kinetic energy } - \text{ potential energy}$$

to be the difference between the kinetic and potential energies, i.e.

$$L(q, \dot{q}) = T(\dot{q}) - V(q) = \frac{1}{2}(\dot{q}, M\dot{q}) - V(q).$$

① Joint with Qin M Z

② Progress in Natural Science, 1991 (1:2): 105-106

They show that the actual trajectory is an extremal of the functional on the space of trajectories joining the initial position $q(t_0)$ and final position $q(t_1)$ i.e.

$$\delta \int_{t_0}^{t_1} L(q, \dot{q}) dt = 0. \tag{1.2}$$

In fact, this extremal equation of variational principle (Euler-Lagrange equation)

$$\frac{d}{dt}\frac{\partial L}{\partial \dot{q}} - \frac{\partial L}{\partial q} = 0 \tag{1.2$'$}$$

is equivalent to Newton's equation (1.1). Eq. (1.2) is called Lagrange variational form for classical mechanics.

In the 19th century Hamilton proposed another form by using momentum $p = Mq$ instead of velocity. Then, $T = \frac{1}{2}(p, M^{-1}q)$, total energy $H = T + V$ is represented by a function of p and q, namely,

$$H(p, q) = T(p) + V(q) = \frac{1}{2}(p, M^{-1}p) + V(q),$$

and furthermore,

$$\begin{cases} \dot{p} = -\dfrac{\partial H}{\partial q}, \\ \dot{q} = \dfrac{\partial H}{\partial p}, \end{cases} \tag{1.3}$$

which is called the Hamiltonian system in canonical form. It is seen that (1.3) is a first order differential equation system with variable $(p_1, \cdots, p_n, q_1, \cdots, q_n)$ in $2n$-dimensional phase space or symplectic space and it is in a form with remarkable simplicity and symmetry property. Outline of these three basic forms of equations for classical mechanics can be found in almost all textbooks of theoretical physics and theoretical mechanics. These different mathematical forms describe the same physical law. They are equivalent, but look very different and therefore give different technical approaches in practical "problem solving". So equivalent mathematical formalisms do not imply equal effectiveness in practice. Concerning this we have our understanding throughout our own experience.

In the early 1960s the first author of this paper developed the finite element method—a systematical algorithm for solving stationary problems. This kind of physical problem has two equivalent mathematical formulations: one is Newtonian form, i.e. solving elliptic equation of second order; the other is variational form, i.e. solving extremal of energy functional. The key to success of the finite element method is that variational forms are used as its foundation. After working out the finite element method and its mathematical foundation, the author had tried to apply the idea of this method to dynamical problems of continuum mechanics and failed.

Thus, Lagrange formalism seems not to be the best choice of basis for the numerical methods on nonstationary problem, let alone the Newtonian formalism. So, a more judicious choice might be the Hamiltonian formalism. This inspired our interest towards the Hamiltonian formalism. We also considered the historical inception to Hamiltonian system. Hamilton first originated a mathematical formalism for the foundation of geometrical optics in the 1820's and several years later he applied it to a quite different field, classical mechanics, as an alternative and mathematical equivalent of the then well-founded Newtonian and Lagrangian formalisms. Hamilton had moderate expectations about his own contributions for still wider applications and for a possible formation of a new discipline[1]. However, the acceptance of the Hamiltonian formalism had been slow and sceptical; it was considered generally "beautiful but useless"[2]. Klein, a famous mathematician, while giving a high appreciation to the mathematical elegance of the theory, but suspected its applicability, and said: "Physicists can make use of these theories only very little, and engineers nothing at all"[3]. This "non-Kleinian" remark of Klein was refuted definitively, at least so far as physicists are concerned, by the founding of quantum mechanics, whose formalism is Hamiltonian. Schrödinger, one of the founders of quantum mechanics, has said, "Hamiltonian principle has become the cornerstone of modern physics, $\cdots$. If you wish to apply the modern theory to any particular problem, you must start with putting in Hamiltonian form."[4]

§2 The Significance of Hamiltonian Systems

It is by now almost beyond dispute that, all real physical processes with negligible dissipation—no matter whether it is classical, quantum, or relativistic, and of finite or infinite degrees of freedom—can always be cast in suitable Hamiltonian form.

Finite degrees of freedom. Such as celestial and satellite mechanics, rigid bodies and rigid multi-bodies—including robotic motions, geometrical optics, plasma confinement, accelerator design, optimal control, WKB and ray asymptotics, etc.

Infinite degrees of freedom. Such as ideal fluid dynamics, elasticity, electrodynamics, nonlinear waves, solitons, quantum mechanics and quantum field theory, relativity, etc.

From above, we can see that Hamiltonian system is ubiquitous and it can express different laws of physics in a unified mathematical form. So, a systematic research and development of numerical methodology for Hamiltonian systems is well motivated. If it could turn out to be successful, it would imply wide-ranging applications.

Then is started a search for the existing literature on the numerical methods for Hamiltonian systems. Hamiltonian systems, in finite or infinite dimensions are special forms of ordinary or partial differential equations. From the 18th century up to date, there has existed cumulative wealth of literature on numerical methods for differential equations, no matter whether the methods are in general use or for special problems. However, we were surprised by the puzzling fact that, the pertinent works related specifically to Hamiltonian differential equations are virtually, if not absolutely, void. This gives us added incentive to make an inquiry into this grossly neglected but fertile and promising field.

§3 The Approach Based on Symplectic Geometry

We have illustrated the background, the motivation and the significance of the research. In the following, we shall discuss the scientific approach adopted.

A basic but unwritten rule for the research on numerical methods is that the properties of the original problem should be preserved as much as possible under discretization. The best way to achieve this aim is to work out the analog within the same framework of the original. For example, finite element method is such a method that discrete problems are kept in the same Sobolev space consisting of the solutions of the original problem such that some essential properties such as symmetry, positive definite property, conservation law are preserved, guaranteeing the effectiveness and the reliability in practical performance. Some numerical methods which are not so successful provide us with contrary examples to support this rule.

According to the above guideline, numerical methods for Hamiltonian systems which preserve the essential properties of the system are proposed. We called them Hamiltonian algorithms, which can be worked out in the same framework of Hamiltonian systems. In the following, we shall show that symplectic geometry is the mathematical framework of Hamiltonian system. So, Hamiltonian algorithm should be originated from within the framework of symplectic geometry. This is our basic approach for research.

We illustrate concisely symplectic geometry by analogy with Euclidean geometry. An Euclidean structure in Euclidean space R^n is determined by a bilinear, symmetric and nondegenerate inner product

$$\langle x, y\rangle = \langle x, Iy\rangle, \quad I = I_n \text{ is identity.} \tag{3.1}$$

When $x \neq 0, \langle x, x\rangle$ is positive, and so, we can define the length $\|x\| = \sqrt{\langle x, x\rangle} > 0$. Linear operators which keep inner product (i.e. length) or satisfy $A'A = I$ form a group $O(N)$, i.e. orthogonal group. It is a classical Lie group and its Lie algebra $o(n)$ consists of antisymmetric operators, i.e. $A' + A = A'I + IA = 0$, or of infinitesimal orthogonal operators.

Symplectic geometry is the geometry of phase space; a symplectic structure in symplectic space, i.e. phase space R^{2n} is determined by a bilinear, anti-symmetric and nondegenerate inner product-symplectic inner product:

$$[x, y] = \langle x, Jy\rangle, J = J_{2n} = \begin{pmatrix} 0 & +I_n \\ -I_n & 0 \end{pmatrix}. \tag{3.2}$$

When $n = 1$,

$$[x, y] = \begin{vmatrix} x_1 & y_1 \\ x_2 & y_2 \end{vmatrix}, \tag{3.3}$$

which is the area of the parallelogram formed by vectors x, y. In general, symplectic inner product is area measure. Since symplectic inner product is anti-symmetric, $[x, x] = 0$ for any x, and so it is impossible to define length by symplectic structure. This distinguishes symplectic geometry from Euclidean geometry. Linear operator which keep symplectic inner product, i.e. $A'JA = J$, form a group $Sp(2n)$, i.e. symplectic group. It is also a classical Lie group, and its Lie algebra consists of infinitesimal symplectic operators B, satisfying $B'J + JB = 0$, which is denoted by $sp(2n)$. In odd dimensions anti-symmetric matrix always degenerates, and thus, the dimensions of symplectic space should be even, so should the dimensions of phase space.

In symplectic geometry, a symplectic operator (or, in classical language, canonical transformation) is a diffeomorphism g of R^{2n} onto itself with Jacobi matrix g_z everywhere symplectic. Naturally, this transformation plays an important role in symplectic geometry. Let us come back to the Hamiltonian dynamics. If a pair of n-dimensional vectors are presented as a $2n$-dimensional vector $z = (p, q)$, then canonical Hamiltonian Eq. (1.3) can be written as

$$\frac{dz}{dt} = J^{-1}\nabla H(z), \tag{3.4}$$

and symplectic transformation keeps the canonical form Hamiltonian Eq. (3.4). The fundamental theorem on Hamiltonian formalism says that the solution $z(t)$ of the canonical system (3.4) of energy H with initial value $z(t_0)$ can be generated by a one-parameter group $g^{t,t_0} = g_H^{t,t_0}$, called the phase flow of H, of symplectic operators of R^{2n} (locally in t) such that

$$z(t) = g_H^{t,t_0} z(0),$$

i.e. g_H^{t,t_0} transforms the state at time t_0 to the state at time t. Therefore the evolutions of Hamiltonian dynamical systems are exclusively canonical transformations. This is the fundamental property of Hamiltonian formalism. When the system H is time-independent, then $g_H^{t,t_0} = g_H^{t-t_0,0}$, i.e. the phase flow depends only on the difference of parameters $t - t_0$. For simplicity, we denote it by $g^t = g_H^{t,0}$.

§4 Main Results

Next, we give a brief survey of results of the authors and their colleagues[5,15].

Every difference scheme, explicit or implicit can be regarded as a mapping from time t^n to time t^{n+1}. If the mapping is symplectic we call the difference scheme symplectic scheme.

For example, we consider the linear Hamiltonian system i.e. $H(z) = \frac{1}{2}z'Sz, S' = S$. Then Eq. (3.4) becomes

$$\frac{dz}{dt} = J^{-1}Sz = Lz, \quad L = J^{-1}S.$$

It is easy to see that the weighted Euler scheme

$$\frac{1}{2}\left(z^{k+1} - z^k\right) = L\left(\alpha z^{k+1} + (1-\alpha)z^k\right) \tag{4.1}$$

for the linear system is symplectic iff $\alpha = \frac{1}{2}$ i.e. it is a time-centered Euler scheme with transition matrix F_τ:

$$z^{k+1} = F_\tau z^k, \quad F_\tau = \varphi(\tau L), \quad \varphi(\lambda) = \frac{1+\lambda/2}{1-\lambda/2}, \tag{4.2}$$

where F_τ, as the Cayley transform of infinitesimally symplectic τL, is symplectic.

For the separable Hamiltonian

$$H(z) = H(p,q) = \frac{1}{2}p'Up + \frac{1}{2}q'Vq \quad (U, V \text{ are symmetric matrices }), \tag{4.3}$$

we have Hamiltonian system

$$\begin{aligned} \frac{dp}{dt} &= -Vq, \\ \frac{dq}{dt} &= Up. \end{aligned} \tag{4.4}$$

An explicit scheme

$$\begin{aligned} p^{k+1} &= p^k - \tau V q^k, \\ q^{k+1} &= q^k + \tau U p^{k+1}, \end{aligned} \quad k \geqslant 0 \tag{4.5}$$

can be rewritten in the form

$$\begin{pmatrix} q^{k+1} \\ p^{k+1} \end{pmatrix} = \begin{pmatrix} I & -\tau V \\ \tau U & I - \tau^2 UV \end{pmatrix} \begin{pmatrix} q^k \\ q^k \end{pmatrix} = g^k \begin{pmatrix} p^k \\ q^k \end{pmatrix}, \tag{4.6}$$

and

$$g^k = \begin{pmatrix} I & -\tau V \\ \tau U & I - \tau^2 UV \end{pmatrix} = \begin{pmatrix} I & 0 \\ -\tau U & I \end{pmatrix} \begin{pmatrix} I & -\tau V \\ 0 & I \end{pmatrix}. \tag{4.7}$$

g^k is symplectic matrix because the right side in Eq. (4.7) is product of two symplectic matrices.

We consider the diagonal Padê approximants $P_m(\lambda)/P_{-m}(\lambda)$ to the exponential function e^λ, where

$$\begin{aligned}
&P_0(\lambda) = 1, \\
&P_1(\lambda) = 2 + \lambda, \\
&P_2(\lambda) = 12 + 6\lambda + \lambda^2, \\
&\quad \cdots\cdots \\
&P_m(\lambda) = 2(2m-1)P_{m-1}(\lambda) + \lambda^2 P_{m-2}(\lambda).
\end{aligned}$$

We can prove by using generalized Cayley transformation that the difference scheme,

$$z^{k+1} = \frac{P_m(\tau L)}{P_m(-\tau L)} z^k, \quad m = 1, 2, \cdots \tag{4.8}$$

is symplectic, A-stable, of $2m$th order of accuracy, and of the same set of quadratic invariant integrals including $H(z)$ as that of system (4.9)

$$\frac{dz}{dt} = Lz. \tag{4.9}$$

For the general non-linear Hamiltonian system (3.4), the time-centered Euler scheme

$$z^{k+1} = z^k + \tau J^{-1} H_z\left(\frac{z^{k+1} + z^k}{2}\right) \tag{4.10}$$

is symplectic too, but the trapezoid scheme

$$z^{k+1} = z^k + \tau J^{-1}\left(H_z\left(z^{k+1}\right) + H_z\left(z^k\right)\right)/2 \tag{4.11}$$

is not symplectic.

For the nonlinear case of the separable Hamiltonian $H(p,q) = U(p) + V(q)$, like scheme (4.5), the scheme

$$\begin{aligned}
p^{k+1} &= p^k - \tau V_q\left(q^k\right), \\
q^{k+1} &= q^k + \tau U_p\left(p^{k+1}\right)
\end{aligned} \tag{4.12}$$

is symplectic. In order to increase the order of accuracy we propose time staggered explicit difference scheme

$$\begin{aligned}
p^{k+1} &= p^k - \tau V_q\left(q^{k+\frac{1}{2}}\right), \\
q^{k+1+\frac{1}{2}} &= q^{k+\frac{1}{2}} + \tau U_p\left(p^{k+1}\right).
\end{aligned} \tag{4.13}$$

Next we consider construction of canonical difference schemes via generating functions. Let

$$\alpha = \begin{bmatrix} A_\alpha & B_\alpha \\ C_\alpha & D_\alpha \end{bmatrix}, \quad \alpha^{-1} = \begin{bmatrix} A^\alpha & B^\alpha \\ C^\alpha & D^\alpha \end{bmatrix}, \tag{4.14}$$

and α be a non-singular real matrix of order $4n$ satisfying

$$\alpha^T \begin{bmatrix} 0 & I_{2n} \\ I_{2n} & 0 \end{bmatrix} \alpha = \mu \begin{bmatrix} -I_{2n} & 0 \\ 0 & I_{2n} \end{bmatrix}, \quad \text{for some } \mu \neq 0. \tag{4.15}$$

α defines a linear Darboux transformation in product space $R^{2n} \times R^{2n}$ by

$$\begin{bmatrix} \hat{w} \\ w \end{bmatrix} = \alpha \begin{bmatrix} \hat{z} \\ z \end{bmatrix}, \quad \begin{bmatrix} \hat{z} \\ z \end{bmatrix} = \alpha^{-1} \begin{bmatrix} \hat{w} \\ w \end{bmatrix}, \tag{4.16}$$

namely,

$$\begin{aligned} \hat{w} &= A_\alpha \hat{z} + B_\alpha z, \\ w &= C_\alpha \hat{z} + D_\alpha z, \end{aligned} \qquad \hat{z}, z, \hat{w}, w \in R^{2n}. \tag{4.17}$$

Let $z \to \hat{z} = g(z,t) = G(z,t)M_0, G(z,t)$ be a phase flow of Hamiltonian system. It is a time-dependent canonical transformation.

The Jacobian

$$M(z,t) = \frac{\partial g(z,t)}{\partial z} \in Sp(2n), \quad M(z,0) = M_0.$$

If

$$\det |C_\alpha M_0 + D_\alpha| \neq 0 \tag{4.18}$$

holds, then there exists, for sufficiently small $|t|$ and in (some neighborhood of) R^{2n}, a time-dependent gradient transformation $w \to \hat{w} = f(w,t)$ with Jacobian $\dfrac{\partial f(w,t)}{\partial w} = N(w,t) \in Sm(2n)$ (i.e. everywhere symmetric), and a time-dependent generating function $\Phi_{\alpha,H}(w,t)$ with gradient

$$\begin{gathered} f(w,t) = \nabla \Phi_{\alpha,H}, \\ [w - \nabla \Phi_{\alpha,H}(w,t)]|_{\hat{w}=A_\alpha \hat{z}+B_\alpha z, w=C_\alpha \hat{z}+D_\alpha z} = 0 \end{gathered}$$

is an implicit representation of the canonical transformation $\hat{z} = g(z,t)$. The generating function $\Phi(w,t)$ satisfies the Hamilton-Jacobi equation

$$\Phi_t(w,t) = -\mu H\left(C^\alpha \Phi_w(w,t) + D^\alpha w\right), \quad w = C_\alpha \hat{z} + D_\alpha z. \tag{4.19}$$

On the other hand, for a given time-dependent scalar function $\psi(w,t): R^{2n}\times R\to R$, we can get a time-dependent canonical map $\tilde{g}(z,t)$. If $\psi(w,t)$ approximates the generating function $\Phi_{\alpha,H}(w,t)$ of the Hamiltonian system (3.1), then $\tilde{g}(z,t)$ approximates the phase flow $g(z,t)$.

For sufficiently small $\tau>0$ as the time-step, define

$$\psi^{(m)}(w,\tau)=\sum_{k=0}^{m}\Phi^{(k)}(w)\tau^{k}, \tag{4.20}$$

where

$$\Phi^{(0)}(w)=\frac{1}{2}w^{T}N_0w, N_0=\left(A_\alpha M_0+B_\alpha\right)\left(C_\alpha M_0+D_\alpha\right)^{-1},$$
$$\Phi^{(1)}(w)=-\mu H\left(E_0w\right), E_0=\left(C_\alpha M_0+D_\alpha\right)^{-1},$$

and as $k\geqslant 1$,

$$\Phi^{(k+1)}(w)=-\frac{\mu}{k+1}\sum_{m=1}^{k}\frac{1}{m!}D_z^m H\left(E_0w\right)\sum_{k_1+\cdots+k_m=k}\left(A^\alpha\nabla\Phi^{k_1}(w),\ldots,A^\alpha\nabla\Phi^{k_m}(w)\right).$$

Then $\psi^{(m)}(w,\tau)$ is the mth approximant of $\Phi(w,\tau)$, and the gradient map

$$w\to\hat{w}=f(w,\tau)=\nabla\psi^{(m)}(w,\tau) \tag{4.21}$$

defines an implicit canonical map $z\to\hat{z}=g(w,t)$ by the equation

$$\left[\hat{w}-\nabla\psi_{\alpha,H}(w,t)\right]|_{\hat{w}=A_\alpha\hat{z}+B_\alpha z,w=C_\alpha\hat{z}+D_\alpha z}=0, \tag{4.22}$$

i.e.

$$A_\alpha\hat{z}+B_\alpha z=\nabla\psi^{(m)}\left(C_\alpha\hat{z}^{k+1}+D_\alpha z^{k},\tau\right). \tag{4.23}$$

Choose

$$\alpha=\begin{bmatrix}-I_n & 0 & 0 & 0\\ 0 & 0 & I_n & 0\\ 0 & I_n & 0 & 0\\ 0 & 0 & 0 & I_n\end{bmatrix},\quad \mu=1, M_0=J_{2n}.$$

We get generating function of the 1st type for the case $\left|\frac{\partial\hat{q}}{\partial p}\right|\neq 0$:

$$\Phi=\Phi(\hat{q},q,t,),\quad -\hat{p}=\Phi_{\hat{q}},\quad p=\Phi_q.$$

Its Hamiltonian-Jacobi equation is

$$\Phi_t+H\left(\Phi_q,q\right)=0.$$

Choose

$$\alpha = \begin{bmatrix} -I_n & 0 & 0 & 0 \\ 0 & 0 & I_n & 0 \\ 0 & I_n & 0 & 0 \\ 0 & 0 & 0 & I_n \end{bmatrix}, \quad \mu = 1, \quad M_0 = J_{2n}.$$

We get generating function of the 2nd type for the case $\left|\dfrac{\partial \hat{q}}{\partial p}\right| \neq 0$:

$$\Phi = \Phi(p, \hat{q}, t,), \quad -\hat{p} = \Phi_{\hat{q}}, \quad -q = \Phi_p, \quad \Phi_t + H\left(p, -\Phi_p\right) = 0.$$

Choose Poincare generating transformation

$$\boldsymbol{\alpha} = \begin{bmatrix} -J_{2n} & J_{2n} \\ \dfrac{1}{2}J_{2n} & \dfrac{1}{2}J_{2n} \end{bmatrix}, \quad \mu = -1, \quad M_0 = I_{2n}.$$

We obtain Poincare generating function

$$\Phi = \Phi(w, t), \quad w = \frac{1}{2}(\hat{z} + z), \quad \hat{w} = J(z - \hat{z}) = \Phi_w,$$

$$\Phi_t - H\left(w - \frac{1}{2}J\Phi_w\right) = 0.$$

For the linear Darboux transformation α satisfying conditions $A_\alpha + B_\alpha = 0, C_\alpha + D_\alpha = I_{2n}$, it can be represented in the following general form

$$\alpha = \begin{vmatrix} -J & J \\ \dfrac{1}{2}(I + V) & \dfrac{1}{2}(I - V) \end{vmatrix}, \tag{4.24}$$

where $V^T J + JV = 0$. Therefore, we can construct an infinite variety of symplectic difference schemes via this method.

We now take

$$V = \begin{vmatrix} (1 - 2\theta)I_n & 0 \\ 0 & -(1 - 2\theta)I_n \end{vmatrix}. \tag{4.25}$$

It can be easily verified that $V^T J + JV = 0$. When $\theta = 1/2$, the scheme is centered scheme ($V = 0$), which preserves all quadratic first integrals of the Hamiltonian system (1.3). When $\theta \neq 1/2$, the symplectic difference schemes constructed by (4.25) preserve only the quadratic first integrals with the form $\dfrac{1}{2}z^T \begin{bmatrix} 0 & B \\ B^T & 0 \end{bmatrix} z = p^T Bq$ of the Hamiltonian system[17].

A generalization of the above theory and method to the infinite dimensional Hamiltonian equations is also given. Suppose B is a reflexive Banach space and B^* its dual. E^n is an Euclidean space and n is dimension. The generalized coordinate in the Banach space is function $q(r,t)$: $E^n \times R \to R$. We have $q(r,t) \in B$, which corresponds to the configuration space. We introduce $p(r,t)$, generalized momentum, where $r \in E^n, t \in R$. For $\forall t \in T$, $p(r,t) \in B^*$, which corresponds to momentum space, and $B \times B^*$ is the phase space.

Let H be a functional in Hamiltonian mechanics. We have the Hamiltonian equation in $B \times B^*$

$$\begin{aligned} \frac{dp}{dt} &= -\frac{\delta H(p,q,t)}{\delta q}, \\ \frac{dq}{dt} &= +\frac{\delta H(p,q,t)}{\delta p}, \end{aligned} \tag{4.26}$$

where $\delta/\delta q, \delta/\delta p$ are variational derivatives. We denote $z = (p,q) \in B \times B^*$. Then we get formally the form of (4.26)

$$\frac{dz}{dt} = J^{-1}\left(\frac{\delta H}{\delta z}\right). \tag{4.27}$$

We cite references on further results of Hamiltonian algorithms, including symmetry and conservation law[9,15,25,27], symplectic Runge-Kutta method[24], multi-stage symplectic schemes[19], leap-frog scheme[21,15], schemes which preserve energy, explicit symplectic schemes and its applications in the Arnold diffusion, chaos computation by symplectic schemes[13], Poisson transformation and its generating function theory and its application[16], applications in quantum mechanics[25], etc.

§5 Conclusion

Our experiences and achievements show that:

1. The numerical methods on Hamiltonian systems is a fertile and promising field and it is worthy of study.

2. Hamiltonian (symplectic) algorithm for Hamiltonian systems can be worked out and would be of great significance. Apart from some rare exceptions, almost all the conventional schemes for ODE's are non-symplectic when applied to Hamiltonian systems. Unavoidably they introduce, above all, artificial dissipation and other parasitic effects which are alien to Hamiltonian dynamics, leading eventually to grave, qualitative distortions. Many nonconventional schemes are proposed, and these schemes have the advantage of preserving the properties of the systems, especially in the aspects of global, structural properties, stability and long-term tracking capabilities, as confirmed by extensive numerical experimentation[9,13].

3. This judiciously chosen symplectic approach is quite natural and justified in view of the innate relationships between Hamiltonian dynamics and symplectic geometry, and the approach turns out to be fruitful and successful. The algorithms are constructed and analysed within the framework of symplectic geometry. For constructing algorithms, we developed the generating function theory in the context of Darboux transformation, which itself is a matter of independent significance.

4. We have also begun to study the Hamiltonian algorithms for quantum mechanics[25]. Hamiltonian algorithms for infinite dimensional Hamiltonian systems-including quantum mechanics and other continuum medium physics-are of promising in applications.

5. Hamiltonian algorithms are also available to dissipative systems. By doubling the variable, we extended dissipative systems as Hamiltonian systems and obtained Hamiltonian algorithms for them[20,28]. On the other hand, Hamiltonian algorithms can be directly and naturally adapted to Hamiltonian systems with dissipative perturbation given by Rayleigh function. Studies in this field are just beginning and which are also of extensive promise.

References

[1] Hamilton W R. General methods in dynamics, The *MathematicalPapers*, Vo 1, 2. Cambridge Univ. Press. 1940

[2] SYNGE, J. L. The life and early work of Sir William Rowan Hamilton. Scripta Mathematica. 1945: 13-24.

[3] Klein F. Vorlesungen über die Entwicklung der Mathematik in 19 Jahrhundert. Teubner. 1926

[4] Schrödinger E. Scripta Mathematica. 1944, 10: 92-94

[5] Feng K. On difference schemes and symplectic geometry. In Feng K, ed. Proc 1984 Beijing Symp Diff Geometry and Diff Equations. Beijing: Science Press, 1985: 42-58

[6] Feng K. Difference schemes for Hamiltonian formalism and symplectic geometry. J Comp Math. 1986, 4(3): 279-289

[7] Feng K. Symplectic geometry and numerical methods in fluid dynamics. In Zhuang F G, Zhu Y L, eds. Proc 10'th Inter Conf on Numerical Methods on Fluid Dynamics. Beijing. 1986. Lecture Notes in Physics. 1986, 264: 1-7. Berlin: Springer Verlag

[8] Feng K. Canonical difference schemes for Hamiltonian canonical differential equations. In Proc Inter Workshop on Applied Diff Equations. Singapore: World Scientific. 1986: 59-73

[9] Feng K, Qin M Z. The symplectic methods for computation of Hamiltonian equations. In Zhu Y L, Guo Ben-yu, ed. Proc Conf on Numerical Methods for PDE's. Berlin: Springer, 1987: 1-37. Lect Notes in Math 1297

[10] Feng K, Wu H M, Qin M Z, Wang D L. Construction of canonical difference schemes for Hamiltonian formalism via generating functions. J Comp Math. 1989, 7(1): 71-96

[11] Feng K, Wang D L. Symplectic difference schemes for Hamiltonian systems in general symplectic structures. J Comp Math. 1991, 9(1): 86-96

[12] Feng K, Wang D L. A note on conservation laws of symplectic difference schemes for Hamiltonian systems. J Comp Math. 1991, 9(3): 229-237

[13] Feng K. The Hamiltonian way for computing Hamiltonian dynamics. In Spigler R, ed. Applied and Industrial Mathematics. Netherlands: Kluwer, 1991: 17-35

[14] Feng K, Wu H M, Qin M Z. Symplectic difference schemes for the linear Hamiltonian canonical systems. J Comp Math. 1990, 8(4): 371-380

[15] Ge Z, Feng K. On the approximation of Hamiltonian systems. J Comp Math. 1988, 6(1): 88-97

[16] Ge Zhong, Generating function for the Poisson map, preprint, Computing Center, Chinese Academy of Sciences, 1988

[17] Li C W, Qin M Z. A symplectic difference scheme for the infinite dimensional Hamiltonian system. J Comp Math. 1988, 6(2): 164-174

[18] Qin M Z. A symplectic difference scheme for the Hamiltonian equation, ibid. 1987, 5: 203-209

[19] Qin M Z, Wang D L, Zhang M Q. Explicit symplectic difference schemes for separable Hamiltonian systems. J Comp Math. 1991, 9(3): 211-221

[20] Wu Y H. The generating function for the solution of ODE and its discrete methods. Computer Math Applic. 1988, 15(12): 1041-1050

[21] Qin M Z. Leap-frog schemes of two kinds of Hamiltonian equation. Math Numer Sinica. 1988, 7(3): 272-281

[22] Qin M Z. Canonical difference scheme for the Hamiltonian equation. Math Methods in Appl Science. 1989, 11(3): 543-557

[23] Qin M Z, Zhang M Q. Multi-stage symplectic schemes of two kinds of Hamiltonian system for wave equation. Computer Math Applic. 1990, 19(10): 51-62

[24] Qin M Z, Zhang M Q. Symplectic Runge-Kutta schemes for Hamiltonian systems:[Preprint]. Beijing: Computing Center, Chinese Academy of Sciences. 1989

[25] Qin M Z, Zhemg M Q. Explicit Runge-Kutte-like schemes to solve certain quantum operator equations of motion. J Statistical Physics. 1990, 65(5/6): 839-844

[26] Qin M Z, Zhang M Q. High-order accurate difference schemes for KdV equation and its conservation law: [Preprint]. Beijing: Computing Center, Chinese Academy of Sciences. 1989

[27] 汪道柳. Hamilton 系统的辛型数值方法. 博士论文. 北京: 中国科学院计算中心. 1988

[28] 葛忠. 数值分析中的辛几何及其应用. 博士论文. 北京: 中国科学院计算中心. 1988

11 The Hamiltonian Way for Computing Hamiltonian Dynamics①

计算哈密尔顿动力学的哈密尔顿方法

Abstract

We present a survey of a recent comprehensive study on the numerical methods for Hamiltonian systems based on symplectic geometry, together the motivations for the research, justification for the symplectic approach adopted, some of the main results, their ramifications and their implications[1−9].

§1 Backgrounds and Motivations

Hamilton originated a mathematical formalism for the foundation of geometrical optics in 1820's and several years later he applied it to a quite different field, i.e., classical mechanics, as an alternative and mathematically equivalent formalism to the then well-founded Newtonian and Lagrangian ones. Hamilton had moderate expectations about his own contributions for still wider applications and for a possible formation of a new discipline [10]. However, the acceptance of Hamiltonian formalism had been slow and sceptical; it was considered generally as "beautiful but useless" [11]. This attitude was best summarized by Klein's recognition of the mathematical elegance of the theory but with an additional remark of caution: "Physicists can make use of these theories only very little, and engineers nothing at all" [12].

This "non-Kleinian" remark of Klein was refuted definitively, at least so far as physicists are concerned, by the founding of quantum mechanics, whose formalism is Hamiltonian. It is by now almost beyond dispute that, all real physical processes with negligible dissipation can always be cast in suitable Hamiltonian form; and, as Schroedinger has put it, "Hamiltonian principle has become the cornerstone of modem physics, $\cdots$. If you wish to apply the modern theory to any particular problem, you must start with putting in Hamiltonian form"[13].

At present, Hamiltonian formalism lay at the basis of many diverse fields of physical and engineering sciences such as classical mechanics, satellite orbits, rigid bodies and

① ln Spigler R, ed. Applied and Industrial Mathematics, the Netherlands: Kluwer Academic Publisher. 1991: 17-35

robotic motions, geometrical optics, plasma confinement, accelerator design, optimal control, WKB and ray asymptotics, fluid dynamics, elasticity, electrodynamics, nonlinear waves, solitons, quantum mechanics, relativity, etc. So, a systematic research and development of numerical methodology for Hamiltonian systems is well motivated. If it turns out to be successful, it would imply wide-ranging applications and give additional evidences for the refutation of Klein's remark by showing that Hamiltonian formalism is not only beautiful but also useful, and, not only useful for theoretical analysis but also useful for practical computations.

After a search into the cumulated wealth of existing literature on numerical methods for differential equations, we were surprised by the puzzling fact that, the pertinent works related specifically to Hamiltonian differential equations are virtually, if not absolutely, void. This gives us added incentive to make an inquiry into this grossly neglected but fertile and promising field.

§2 The Symplectic Approach

After the initiation of Hamilton, developed further by Jacobi, the most significant progress of the formalism was its geometrization, started since Poincaré, resulted in socalled *symplectic geometry*, a new discipline as anticipated by Hamilton, which has become the natural language for Hamiltonian dynamics. Symplectic geometry has the merit to make the deep-lying and intricate properties of dynamics explicit, transparent and easy to grasp. Hamiltonian systems are simply canonical systems on symplectic manifold, i.e., phase space endowed with symplectic structures, the dynamical evolutions are simply symplectic transformations. see e.g., [14]. The desirable algorithms should preserve as much as possible the relevant symplectic properties of the original system. The best way to achieve this aim is to work out the analog within the same framework of the original. So we conceive, construct, analyze and assess numerical algorithms exclusively from within the framework of symplectic geometry. This judiciously chosen approach, inspired by the author's previous experience with the development of the finite element method for solving elliptic equations, is quite natural and justified in view of the innate relationships between Hamiltonian dynamics and symplectic geometry. The approach turns out to be fruitful and successful, and leads to the effective construction as well as the theoretical understanding of an abundance of what we call *symplectic difference schemes*, or symplectic algorithms, or simply *Hamiltonian algorithms*, since they present the *proper* way, i.e., *the Hamiltonian way* for computing *Hamiltonian dynamics*.

Symplectic geometry was developed, due to historical circumstance, mostly for the purposes of theoretical analysis; it lacks the computational component; for example, there is no such theory as that of symplectic approximations for symplectic operators. So, our program might be considered also as an attempt to fill this blank. Generating functions play a key role in the classical transformation theory in symplectic geometry, but they are rather of limited scope in the present setting. So we worked out an extended theory of generating functions in the context of linear Darboux transformations [2,4,6]. This theory leads to our general methodology for the construction and analysis of symplectic schemes. It seems to have its own independent interest in symplectic geometry too.

§3 Some Preliminaries

In real linear space R^{2n}, the vectors will be represented by column matrices $z = (z_1, \dots z_{2n})'$, the prime $'$ denotes matrix transpose. The *standard symplectic form* in R^{2n} is given by the anti-symmetric, non-degenerate bilinear form

$$\begin{gathered} \omega(x,y) = x'Jy, \quad \forall x,y \in R^{2n}, \\ J = J_{2n} = \begin{pmatrix} 0 & -I \\ I & 0 \end{pmatrix}, \quad J' = -J = J^{-1}. \end{gathered} \tag{1}$$

The symplectic group $Sp(2n)$ consists of the real *symplectic* matrices S preserving ω i.e., satisfying $S'JS = J$. The corresponding Lie algebra $sp(2n)$ consists of real *infinitesimally symplectic* matrices L satisfying $L'J+JL = 0$, or equivalently $L = JA$, for some symmetric matrix A. The *symplectic product* for two symmetric matrices A, B is defined as $\{A, B\} = AJB - BJA$. Then the space of real *symmetric* matrices $sm(2n)$, under the symplectic product, forms a Lie algebra which is isomorphic to $sp(2n)$ under the usual commutator bracket.

A diffeomorphism g of R^{2n} onto itself is called a *symplectic* operator (or, in classical language, *canonical* transformation) if its Jacobian matrix g_z is everywhere symplectic. A differential map of R^{2n} into itself is called an *infinitesimally symplectic* operator (or, transformation) if its Jacobian matrix is everywhere infinitesimally symplectic. A differential map f of R^{2n} into itself is called a *symmetric* operator (or, transformation) if its Jacobian matrix is everywhere symmetric; in this case, there exists, at least locally, a real function ϕ of R^{2n}, unique up to an additive constant, such that $f = \nabla\phi$, where $\nabla = (\partial/\partial z_1, \dots, \partial/\partial z_{2n})'$. The space of smooth real functions forms a Lie algebra (of infinite dimensions) $C^\infty(R^{2n})$ under the *Poisson bracket* defined by

$$\{\phi, \psi\} = (\nabla\phi)'J(\nabla\psi). \tag{2}$$

An operator S is symplectic iff it preserves the Poisson bracket, i.e., $\{\phi \circ S, \psi \circ S\} =$

$\{\phi, \psi\} \circ S$. The totality of quadratic forms $\phi_A(z) = \left(\frac{1}{2}\right) z'Az, A' = A$ forms a Lie subalgebra of $C^\infty(R^{2n})$ under Poisson bracket, which is isomorphic to the Lie algebra $sm(2n)$ under the symplectic product.

Choose a smooth function $H \in C^\infty(R^{2n})$, it defines a *Hamiltonian canonical system*

$$\frac{dz}{dt} = J\nabla H(z) \tag{3}$$

on the symplectic space, i.e., phase space R^{2n}, endowed with the symplectic form $\omega(x, y) = x'Jy$. H is called the *Hamiltonian* or *energy* function defining the system. It is customary to split the $z \in R^{2n}$ into two parts $z = (p, q)$, $p_k = z_k, q_k = z_{n+k}, k = 1, \cdots, n$. In classical language p, q are called position and momentum vectors conjugate to each other. Then $H(z) = H(p, q)$, and equation (3) can be written as

$$\begin{aligned} \frac{dp}{dt} &= -H_q(p, q) \\ \frac{dq}{dt} &= H_p(p, q) \end{aligned} \tag{4}$$

The *fundamental theorem on Hamiltonian formalism* says that the solution $z(t)$ of the canonical system (3) of energy H with initial value $z(0)$ can be generated by a *one-parameter group* $g^t = g_H^t$, called the *phase flow* of H, of symplectic operators of R^{2n} (locally in t) such that

$$z(t) = g_H^t z(0)$$

The system (3) can be written as

$$\frac{d}{dt}\left(g_H^t z\right) = J(\nabla H) \circ g_H^t z, \quad \forall z \in R^{2n}$$

The symplecticity of the phase flow implies the class of *conservation laws of phase area* of even dimensions $2m$, $m = 1, \cdots, n$, for the Hamiltonian system (3), the case $m = n$ is the Liouville's conservation law.

Moreover, the Hamiltonian system possesses another class of *conservation laws* related to the *energy* $H(z)$. Consider any smooth function $F(z)$ in R^{2n}, then the composite function $F \circ g_H^t(z)$ satisfies the equation

$$\frac{d}{dt} F \circ \left(g_H^t z\right) = \{F, H\} \circ g_H^t z$$

$F(z)$ is called an *invariant* function under the system (3) with Hamiltonian H if $F = F \circ g_H^t$ for all t, this is equivalent to the condition $\{F, H\} = 0$. The energy H itself is always an variant function.

§4 Simple Symplectic Difference Schemes

Consider now the difference schemes for Hamiltonian system (3), restricted mainly to the case of single step (i.e., 2-level) schemes. Time t is discretized into $t = 0$, $\pm\tau, \pm 2\tau, \cdots$; τ is the step size, $z(k\tau) \approx z^k$. Each 2-level-scheme is characterized by a *transition* operator relating the old and new states by

$$\hat{z} = G^\tau z, \quad z = z^k, \quad \hat{z} = z^{k+1} \tag{5}$$

$G^\tau = G_H^\tau$ depends on τ, H and the mode of discretization. From the standard-point of symplectic geometry, it is natural and even mandatory to require G_H^τ to be symplectic. A difference scheme for Hamiltonian system (3) is called *symplectic* (or, *canonical*) if its transition operator G_H^τ is symplectic for all Hamiltonian functions H and step sizes τ. We shall also consider difference schemes non-symplectic in this general sense, but are symplectic for special classes of Hamiltonians. The true solution relating the old and new states is

$$z((k+1)\tau) = g_H^t z(k\tau).$$

So we are confronted with the problem of symplectic approximation to the symplectic phase flow g_H^τ. For the latter there is a Lie expansion in τ

$$(g_H^\tau z)_i = z_i + \tau \{z_i, H\} + \frac{\tau^2}{2!} \{\{z_i, H\}, H\} + \cdots, \quad i = 1, \cdots, 2n. \tag{6}$$

Although the truncations give "legitimate" approximations, but they are non-symplectic in general, so undesirable.

Consider the simplest one-legged weighted Euler scheme for (3)

$$\hat{z} = z + \tau J(\nabla H)(c\hat{z} + (1-c)z) \tag{7}$$

where c is a real constant. It is easy to show that it is symplectic iff $c = \frac{1}{2}$. The only symplectic case corresponds to the *time-centered Euler scheme*

$$\hat{z} = z + \tau J(\nabla H)\left(\frac{\hat{z}+z}{2}\right) \tag{8}$$

This simple proposition illustrate a general situation: apart from some very rare exceptions, the vast majority of conventional schemes are non-symplectic. However, if we allow c in (7) to be a *real matrix of order* $2n$, we get a far-reaching generalization: (7) *is symplectic* iff

$$c = \frac{1}{2}(I_{2n} + J_{2n}B), \quad B' = B \tag{9}$$

The symplectic case corresponds to

$$\hat{z} = z + \tau J(\nabla H)\left(\frac{1}{2}(I+JB)\hat{z} + \frac{1}{2}(I-JB)z\right), \quad B' = B. \tag{10}$$

They are all implicit, (10) is solvable for new $\hat{z}$ in terms of old z whenever $|\tau|$ is small enough. (10) enjoys many properties in common with (3), they may be considered as *Hamiltonian difference equations.*

Let the transition in (10) to be

$$z^{k+1} = G^{\tau}_{H,B} z^k, \tag{11}$$

for $B = 0$, the accuracy order is 2; for $B \neq 0$, the order is 1. However, the alternating composite schemes given by

$$z^{k+1} = G^{\tau}_{H,\pm B} z^k, \tag{12}$$

where

$$\begin{aligned} G^{\tau}_{H,\pm B} &= G^{\frac{\tau}{2}}_{H,B} \circ G^{\frac{\tau}{2}}_{H,-B} \\ G^{\tau}_{H,\mp B} &= G^{\frac{\tau}{2}}_{H,-B} \circ G^{\frac{\tau}{2}}_{H,B} \end{aligned} \tag{13}$$

have accuracy order 2.

So we get a great variety of simple symplectic schemes of order 1 or 2, classified according to type matrices $B \in sm(2n)$, which is a linear space of dimension $2n^2 + n$.

The conservation properties of the above symplectic schemes are well-understood. In fact we have proved:

Let $\phi_A(z) = \frac{1}{2} z' A z$, $A' = A$ be a quadratic form and A commutes symplectically with the type matrix B, i.e., $AJB = BJA$. Then ϕ_A is invariant under the system with Hamiltonian H iff ϕ_A is invariant under symplectic difference schemes (10) of types $G^{\tau}_{H,B}$, $G^{\tau}_{H,-B}$.

We list some of the most important types B together with the corresponding form of symplectic matrices A of the conserved quadratic invariants ϕ_A:

$B = 0$, A arbitrary.

$B = \pm\begin{pmatrix} 0 & I_n \\ I_n & 0 \end{pmatrix} = \pm E_{2n}$, $A = \begin{pmatrix} 0 & b \\ b' & 0 \end{pmatrix}$, b arbitrary; angular momentum type.

$B = \pm I_{2n}$, $A = \begin{pmatrix} a & b \\ -b & a \end{pmatrix}$, $a' = a$, $b' = -b$; Hermitian type.

$B = \pm\begin{pmatrix} I_n & 0 \\ 0 & -I_n \end{pmatrix} = \pm I_{n,n}$, $A = \begin{pmatrix} a & b \\ -b & -a \end{pmatrix}$, $a' = a$, $b' = -b$.

§5 Explicit Symplectic Schemes

The symplecticity of Hamiltonian dynamical evolutions implies some kind of symmetry between the past and the future. So symplectic schemes are implicit by nature. However, for special classes of Hamiltonians of practical importance, and for certain type matrices B (see 4), the corresponding simple symplectic schemes will be practically explicit, leading to simple and fast algorithms. For example, consider the "separable" Hamiltonians of the form

$$H(p,q)=\phi(p)+\psi(q), \tag{14}$$

where ϕ,ψ are functions of n variables, then

$$H_p(p,q)=\phi_p(p),\quad H_q(p,q)=\psi_q(q).$$

Most of the energy functions in classical mechanics are separable in the above sense. Apply symplectic schemes (10) of type E_{2n}, and we get explicit schemes

$$\begin{aligned}&\text{Type}\quad E_{2n};\quad \hat{p}=p-\tau\psi_q(q),\quad \hat{q}=q+\tau\phi_p(\hat{p}).\\&\text{Type}\quad -E_{2n};\quad \hat{p}=p-\tau\psi_q(\hat{q}),\quad \hat{q}=q+\tau\phi_p(p).\end{aligned} \tag{15}$$

The accuracy of this scheme can be raised without additional labor by putting p (or q) at integer times and q (or p) at half-integer times, we get *time-staggered symplectic* scheme of order 2, explicit for separable Hamiltons:

$$p^{k+1}=p^k-\tau\psi_q\left(q^{k+\frac{1}{2}}\right),\quad q^{k+1+\frac{1}{2}}=q^{k+\frac{1}{2}}+\tau\phi_p\left(p^{k+1}\right). \tag{16}$$

This explicit symplectic scheme was used and explained in detail by Feynman in his Lectures for computing the motions of an oscillating spring and of a planet around the sun [11]. The staggering (in time and/or in space) is a useful numerical technique for improving accuracy and symmetry.

The oldest and simplest among all difference schemes is the explicit Euler method

$$\hat{z}=z+\tau J\nabla H(z)=(I+\tau J\nabla H)(z)=G_H^\tau(z). \tag{17}$$

It would be interesting to ask: under what conditions on H the operator G_H^τ will be symplectic. We have proved: If, in R^{2n}

$$H(p,q)=\phi(Ap+Bq),\quad AB'=BA' \tag{18}$$

where ϕ is a function of n variable, A,B are matrices of order n; then G_H^τ is symplectic and exact, i.e., $G_H^\tau=g_H^\tau$.

If, moreover

$$H(p,q)=\sum_{i=1}^{m} H_i(p,q), \quad H_i(p,q)=\phi_i\left(A_i p+B_i q\right), \quad A_i B_i'=B_i A_i' \tag{19}$$

Where ϕ_i are functions of n variables, A_i, B_i are matrices of order n. Let $[s_1, s_2, \cdots, s_m]$ be an arbitrary permutation of $[1,2,\cdots,m]$; then the following scheme is symplectic and of order 1:

$$\begin{aligned} z \rightarrow \hat{z}=\left(G^\tau\left[s_1, s_2, \cdots, s_m\right]\right) z, \quad & G^\tau\left[s_1, s_2, \cdots, s_m\right] \\ & =\left(G^\tau\left[s_1\right]\right) \circ\left(G^\tau\left[s_2\right]\right) \circ \cdots \circ\left(G^\tau\left[s_m\right]\right) \end{aligned} \tag{20}$$

where $G^\tau[i]=G_{H_i}^\tau$. Moreover, the following alternating composite is symplectic and of order 2:

$$z \rightarrow \hat{z}=\left(G^{\frac{\tau}{2}}\left(\left[s_1, s_2, \cdots, s_m\right]\right) \circ\left(G^{\frac{\tau}{2}}\left[s_m, \cdots, s_2, s_1\right]\right) z\right. \tag{21}$$

Note that the explicit symplectic schemes of order 1 and 2 of type $\pm E_{2n}$ for separable Hamiltonian in the sense of (14) are special cases of the above situation. We have computed with success using (19-21) the systems with Hamiltonians [18,19]

$$H_k(p,q)=\sum_{j=1}^{k} \cos\left(p \cos\frac{2\pi j}{k}+q \sin\frac{2\pi j}{k}\right) \tag{22}$$

with k-fold rotational symmetry in phase plane. They are not separable in the sense of (14) when $k \neq 1,2,4$.

For the general characterization for the symplecticity of explicit Euler schemes, we have proved: $G_H^\tau=I+\tau J\nabla H$ in R^{2n} is symplectic iff

$$H_{zz} J H_{zz}=0, \quad \forall z \in R^{2n} \tag{23}$$

where H_{zz} is the Hessian matrix of H.

Furthermore we have an as yet incomplete but interesting characterization of (23) as follows:

A necessary and sufficient condition for $H_{zz} J H_{zz} \equiv 0$ in R^{2n} is that

$H(z)=H(p,q)$ can be expressed, under suitable *linear symplectic and orthogonal transformation* $z=Sw, w=(u,v)$, as

$$H(S(u,v))=\phi(u)+\psi(v), \quad \psi \text{ is linear}, \quad S \in Sp(2n) \cap O(2n) \tag{24}$$

i.e., separable in sense of (14) with additional linearity in ψ.

The "sufficiency" part follows easily from the propositions stated above. However, the "necessity" part is not yet fully established. We have proved for the case that H is a quadratic form. Moreover for $n=1$, (23) is equivalent to the Monge-Ampere equation $|H_{zz}|=0$, so H defines a global surface of zero Gaussian curvature in R^3, the assertion is a consequence of a global theorem on developable surface in R^3 [16,17]. Since the global strong degeneracy condition (23) should imply drastic simplification of the profile of function H, so we have reasons to suggest the conjecture (23) $\Rightarrow$ (24).

§6 Symplectic Schemes for Linear Hamiltonian Systems

By a linear Hamiltonian system here we mean the system with a quadratic form $H(z)=\phi_A(z)=(z'Az)/2, \quad A'=A$ as the Hamiltonian. Then the system becomes

$$\frac{dz}{dt}=J\nabla\phi_A(z)=JAz \tag{25}$$

with phase flow

$$g_H^t=\exp(tJA)=I+\sum_{k=1}^{\infty}\frac{t^k}{k!}(JA)^k \tag{26}$$

as linear symplectic operators. All the constructions of symplectic schemes in Section 4 and in Sections 7 and 8 below can be applied, but here we present the approach of symplectic approximations to the exponential [1, 2, 9]. The Taylor series in (27) can be arbitrarily truncated, but apart from some rare exceptions, symplectic approximations can not be provided. The natural way is looking for rational and especially the Padé approximations of the exponential function $\exp(\lambda)$.

$$\exp(\lambda)-\frac{P_{m,l}(\lambda)}{Q_{m,l}(\lambda)}=O\left(|\lambda|^{m+l+1}\right), \quad |\lambda|\sim 0 \tag{27}$$

$P_{m,l}$ and $Q_{m,l}$ is of degree m, l respectively. It can be proved that, the matrix Padé transform $P_{m,l}(\tau JA)/Q_{m,l}(\tau JA)$ is symplectic iff the Padé approximant is diagonal, i.e., $m=l$; then $P_{m,m}(\lambda)=P_m(\lambda)$, $Q_{m,m}(\lambda)=P_m(-\lambda)$, and

$$\begin{aligned}
&P_0(\lambda)=1,\\
&P_1(\lambda)=2+\lambda,\\
&P_2(\lambda)=12+6\lambda+\lambda^2,\\
&P_m(\lambda)=2(2m-1)P_{m-1}(\lambda)+\lambda^2P_{m-2}(\lambda)
\end{aligned} \tag{28}$$

We get the *Padé symplectic* scheme

$$P_m(-\tau JA)\hat{z}=P_m(\tau JA)z, \quad \hat{z}=G^\tau z=\frac{P_m(\tau JA)}{P_m(-\tau JA)}z. \tag{29}$$

which is of order $2m$. For $m=1$ we get the time-centered Euler scheme (8).

It is known that every linear system (26) in R^{2n} with $H = \phi_A$ is completely integrable in the sense that there exists n invariant quadratic forms, $\phi_{A_1} = \phi_A, \cdots, \phi_{A_n}$ which are functionally independent and mutually commuting $\{\phi_{A_i}, \phi_{A_j}\} = 0$, $i, j = 1, \cdots, n$. The *Padé symplectic schemes* have the remarkable property that they possess *the same set of invariant quadratic forms* as that of the original linear Hamiltonian system. If the original linear phase flow is not only symplectic but also unitary, then the Padé schemes are also unitary, so they are useful for quantum mechanical applications.

In general, the linear symplectic schemes for linear Hamiltonian system, even when they do not preserve the quadratic invariants (including the energy) of the original system, we still can construct for any linear symplectic transition operator G_H^τ (in fact even for any $G \in Sp(2n)$) a set of n independent, mutually commuting quadratic forms ϕ_{B_i} which are invariant under G. This set is near to the corresponding set of the original when τ is small, so they behave well in the aspects of orbital structure and stability.

§7 Symplectic Leap-frog Schemes

The concept of symplecticity for multi-level schemes for Hamiltonian system is more complicated than the 2-level case. We discuss only the 3-level schemes. For system (3) with Hamiltonian H and phase flow operator g_H^t we have the following exact relations between the solution at 3 successive moments, we fix $\tau > 0$ and t:

$$z(t) \to z(t+\tau) = g_H^\tau z(t)$$
$$z(t) \to z(t-\tau) = g_H^{-\tau} z(t)$$

Then we have

$$z(t+\tau) - z(t-\tau) = \delta_H^\tau z(t), \quad \delta_H^\tau = g_H^\tau - g_H^{-\tau} \tag{30}$$

where δ_H^τ is the exact difference operator for the Hamiltonian system, and can be shown to be infinitesimally symplectic, i.e., its Jacobian matrix belongs to $sp(2n)$ everywhere.

We take the general form of 3-level leap-frog schemes as

$$z^{k+1} - z^{k-1} = \Delta_H^\tau z^k \tag{31}$$

here the coefficients chosen before z^{k+1} and z^{k-1} are justified on grounds of symmetry. The *leap-frog* scheme (31) is called *symplectic* if the operator Δ_H^τ is *infinitesimally symplectic*. So our problem is to seek for approximations Δ_H^τ to δ_H^τ preserving infinitesimal symplecticity.

For linear Hamiltonian system (25, 26) we have

$$\delta_H^\tau = \exp(\tau JA) - \exp(-\tau JA) = 2\sinh(\tau JA) = 2\sum_{j=1}^{\infty} \frac{(-1)^j}{(2j-1)!}(\tau JA)^{2j-1} \tag{32}$$

here all the truncations are infinitesimally symplectic, so we get symplectic leap-frog schemes

$$\begin{aligned} \text{order } 2: &\quad z^{k+1} - z^{k-1} = 2\tau JAz^k \\ \text{order } 2m: &\quad z^{k+1} - z^{k-1} = 2\left(\sum_{j=1}^{m} \frac{(-1)^j}{(2j-1)!}(\tau JA)^{2j-1}\right) z^k. \end{aligned}$$

Similarly, for non-linear Hamiltonian system, from the Lie series (6) of phase flow we get

$$\left(\delta_H^\tau z\right)_i = 2\sum_{j=1}^{\infty} \frac{\tau^{2j-1}}{(2j-1)!} \left\{\left\{\cdots\left\{\left\{z_i^k, H\right\}, H\right\}\cdots, H\right\}, H\right\} \tag{33}$$

whose truncations provide *symplectic leap-frog* schemes ①

$$\text{order } 2: \quad z^{k+1} - z^{k-1} = 2\tau J\nabla H\left(z^k\right) \tag{34}$$

$$\text{order } 2m: \quad z_i^{k+1} - z_i^{k-1} = \delta_{H,m}^\tau z^k \tag{35}$$

where $\delta_{H,m}^\tau z$ is the m-term truncation of (33).

For these schemes we can establish the conservation property as: If $F(z) = z'Bz$ is a quadratic invariant under g_H^t, then follows the "cross invariance" under the scheme (34)

$$\left(z^{k+1}\right)' Bz^k = \left(z^k\right)' Bz^{k-1}.$$

All the symplectic leap-frog schemes are explicit. The simplest 2nd order scheme (34) is one of the rare cases of existing methods which are symplectic and seems to be the oldest one. It is interesting to note that, according to linearized analysis, the stability domain for (34) is only a segment on the imaginary axis in the complex plane, it is absolutely unstable when applied to asymptotically stable systems. However, it works well with Hamiltonian systems, which correspond to the critical case with pure imaginary characteristic exponents and are never asymptotically stable owing to the symplecticity of dynamics.

① This is incorrect. They are non-symplectic. The author noticed it and gave the note on his reprint.

§8 Fractional Transforms and Generating Functions

A matrix α of order $4n$ is called a *Darboux matrix* if

$$\alpha' J_{4n}\alpha = \tilde{J}_{4n} \quad J_{4n} = \begin{pmatrix} 0 & -I_{2n} \\ I_{2n} & 0 \end{pmatrix}, \quad \tilde{J}_{4n} = \begin{pmatrix} J_{2n} & 0 \\ 0 & -J_{2n} \end{pmatrix},$$
$$\alpha = \begin{pmatrix} a & b \\ c & d \end{pmatrix}, \quad \alpha^{-1} = \begin{pmatrix} a_1 & b_1 \\ c_1 & d_1 \end{pmatrix}. \tag{36}$$

Each Darboux matrix induces a (linear) *fractional transform* between symplectic and symmetric matrices

$$\begin{aligned} &\sigma_\alpha: \quad Sp(2n) \to sm(2n) \\ &\sigma_\alpha(S) = (aS+b)(cS+d)^{-1} = A, \quad \text{for} \quad |cS+d| \neq 0 \end{aligned} \tag{37}$$

with inverse transform $\sigma_\alpha^{-1} = \sigma_{\alpha^{-1}}$

$$\begin{aligned} &\sigma_\alpha^{-1}: \quad sm(2n) \to Sp(2n), \\ &\sigma_\alpha^{-1}(A) = (a_1A+b_1)(c_1A+d_1)^{-1} = S, \quad \text{for} \quad |c_1A+d_1| \neq 0. \end{aligned} \tag{38}$$

The above machinery can be extended to generally non-lineax operators in R^{2n}. Let $Symp(2n)$ denote the totality of symplectic operators, and $symm(2n)$ denote the totality of symmetric operators. Each $f \in symm(2n)$ corresponds, at least locally a real function ϕ (up to a constant) such that $f(w) = \nabla\phi(w)$. Then we have

$$\begin{aligned} &\sigma_\alpha: \quad Symp(2n) \to symm(2n), \\ &\sigma_\alpha(g) = (ag+b)(cg+d)^{-1} = \nabla\phi, \quad \text{for} \quad |cg_z+d| \neq 0, \end{aligned} \tag{39}$$

or alternatively

$$ag(z) + bz = (\nabla\phi)(cg(z)+dz), \tag{40}$$

where ϕ is called the *generating function* of Darboux type α for the symplectic operator g. Then

$$\begin{aligned} &\sigma_\alpha^{-1}: \quad symm(2n) \to Symp(2n), \\ &\sigma_\alpha^{-1}(\nabla\phi) = (a_1\nabla\phi+b_1)(c_1\nabla\phi+d_1)^{-1} = g, \quad \text{for} \quad |c_1\phi_{ww}+d_1| \neq 0 \end{aligned} \tag{41}$$

or alternatively

$$a_1\nabla\phi(w) + b_1(w) = g(c_1\nabla\phi(w)+d_1w), \tag{42}$$

where g is called the symplectic operator of Darboux type α for the generating function ϕ.

For the study of symplectic difference algorithms we may narrow down the class of Darboux matrices to the subclass of *normal Darboux matrices*, i.e., those satisfying $a+b=0$, $c+d=I_{2n}$. The normal Darboux matrices α can be characterized as

$$\alpha=\begin{pmatrix} a & b \\ c & d \end{pmatrix}=\begin{pmatrix} J & -J \\ \frac{1}{2}(I+JB) & \frac{1}{2}(I-JB) \end{pmatrix}, \quad B'=B$$

$$\alpha^{-1}=\begin{pmatrix} a_1 & b_1 \\ c_1 & d_1 \end{pmatrix}=\begin{pmatrix} \frac{1}{2}(JBJ-J) & I \\ \frac{1}{2}(JBJ+J) & I \end{pmatrix}. \tag{43}$$

The fractional transform induced by normal Darboux matrix establishes a 1-1 correspondence between *symplectic operators near identity* and *symmetric operators near nullity*. Then the determinantal conditions could be taken for granted. Those B listed in 4 correspond to the most important normal Darboux matrices.

For each Hamiltonian H with its phase flow g_H^t and for each normal Darboux matrix α, we get the *generating function* $\phi=\phi_H^t=\phi_{H,\alpha}^t$ of *normal Darboux type* α for the *phase flow* of H by

$$\nabla\phi_{H,\alpha}^t=\left(ag_H^t+b\right)\left(cg_H^t+d\right)^{-1}, \qquad \text{for small } |t|. \tag{44}$$

$\phi_{H,\alpha}^t$ satisfies the *Hamilton-Jacobi* equation

$$\frac{\partial}{\partial t}\phi=-H\left(a_1\nabla\phi(w)+b_1w\right)=-H\left(c_1\nabla\phi(w)+d_1w\right) \tag{45}$$

and can be expressed by Taylor series in t

$$\phi(w,t)=\sum_{k=1}^{\infty}\phi^{(k)}(w)t^k, \quad |t| \text{ small}. \tag{46}$$

The coefficients can be determined recursively

$$\phi^{(1)}(w)=-H(w), \qquad \text{and for} \quad k\geqslant 0, \quad a_1=\frac{1}{2}(JBJ-J):$$

$$\phi^{(k+1)}(w)=\frac{-1}{k+1}\sum_{m=1}^{k}\frac{1}{m!}\sum_{i_1,\cdots,i_n=1}^{2n}\sum_{\substack{j_1+\cdots+j_m=k\\ j_l\geqslant 1}}H_{z_{i_1}\cdots z_{i_m}}(w)$$
$$\times\left(a_1\nabla\phi^{(j_1)}(w)\right)_{i_1}\cdots\left(a_1\nabla\phi^{(j_m)}(w)\right)_{i_m}. \tag{47}$$

Let ψ^τ be a truncation of $\phi_{H,\alpha}^\tau$ up to certain power t^m, say. Using inverse transform σ_α^{-1} we get the symplectic operator

$$G^\tau=\sigma_\alpha^{-1}\left(\nabla\psi^\tau\right), \quad |\tau| \text{ small}$$

which depends on τ, H, α (or equivalently B) and the mode of truncation. It is a symplectic approximation to the phase flow g_H^τ and can serve as the transition operator of a symplectic difference scheme (for the Hamiltonian system (3))

$$\hat{z} = G^\tau z \tag{48}$$

which, by (39,40,41) can be written as

$$J\hat{z} - Jz = \nabla\psi^\tau(c\hat{z} + (I - c)z), \quad c = \frac{1}{2}(I + JB). \tag{49}$$

Thus, using the machinery of phase flow generating functions we have constructed, for each H and each normal Daxboux matrix, an hierarchy of symplectic schemes by truncation. The simple symplectic schemes in 4 corresponds to the lowest truncation. The conservation properties of all these higher order schemes are the same as stated in 4.

§9 Numerical Experiments and Discussions

Numerical experimentation shows strong evidences of superior performance of symplectic schemes over the non-symplectic ones. Most of the conventional schemes are non-symplectic when applied to Hamiltonian systems. Unavoidably they introduce, above all, artificial dissipation and other parasitic effects which are alien to Hamiltonian dynamics, leading eventually to grave, qualitative distortions.

On the other hand, symplectic schemes tend to purify (sometimes even simplify) the algorithms, inhibiting artificial dissipations and all other non-Hamiltonian distortions. We present a comparative study for the harmonic oscillator, Fig 1, 2 and for a nonlinear oscillator, Fig 3, 4. For some other result, see [4]. The non-symplectic Runge-Kutta scheme yields persistent spiraling of phase trajectories with artificial creation of attractors (impossible for Hamiltonian system), irrespective of the higher order of the scheme and the smallness of step sizes. The symplectic scheme always give clear-cut invariant elliptical trajectories even for low order of schemes and large step sizes. The striking contrast is, at least, in the aspects of global and structural preservation and long-term tracking capabilities.

The most important characteristic property of Hamiltonian dynamics is the intricate *coexistence* of *regular* and *chaotic* motions, as implied by the celebrated KAM theorem [14]. The advantage of symplectic schemes consists primarily in the essential fidelity to the original in this aspect. The behavior of symplectic schemes should be and could be understood and analyzed in the context of KAM theory on preservation and break-down of invariant tori. There are evidences showing that an analog of KAM theorem should hold for symplectic difference schemes, we have a related result for a simple special case.

Symplectic schemes for Hamiltonian systems do not imply non-Hamiltonian perturbations, but they do inevitably imply Hamiltonian perturbations. An example is the break-down of invariant tori when the scheme is applied to integrable or near integrable systems. The effect is indiscernible for small step size, but becomes pronounced with large step sizes, see Fig 5. Another example is the *artificial chaos* in the form of Arnold diffusion. We computed by explicit symplectic schemes (21) of order 2, the Hamiltonian system H_k (22) with k-fold rotational symmetry in 1 degree of freedom, suggested by Sagdeev, Zaslavsky *et al* [18] in the context of chaotic stream lines of Lagrangian turbulence, see also [19-21]. Fig 6 gives some results for $k = 4$, when initial point is chosen on the original separatrix, the diffusion shows a remarkable combination of *regularity* and *chaoticity*, almost rectilinear motion along the "highways" and random turning at the "crossroads". Fig 7, 8 give some results for $k = 3, 5$. For $k = 5$ the orbit is reminiscent of Alhambra fresco decoration or Penrose tiling or quasi-crystal with 10-fold symmetry. It is worth to note the connections of symplectic schemes with these widely different topics.

Acknowledgements This research program is supported by the National Natural Science Foundation of China. The author expresses his gratitude to his collaborators M.Z. Qin, H.M. Wu, Z. Ge, D.L. Wang, Y.H. Wu, M.Q. Zhang for close and fruitful cooperation, and in particular to M.Z. Qin for his involvement in the work and supplying his unpublished numerical results and to M.Q. Zhang for discussion and assistance in preparing the manuscript. Thanks are also due to Prof. R. Spigler for the invitation and support for the author's participation of Venice-1 Symposium 1989 to present this paper.

Fig.1. Harmonic oscillator $H = (p^2 + a^2q^2)/2, \quad a = 2.$

Runge-Kutta method, order 4, non-symplectic.

stepsize$=\tau$, total number of steps $= N$, number of steps per plot $= M$.

Single orbit is computed in all cases with the same initial point.

Upper-left: $\tau = 0.5, N = 2000, M = 1$. Upper-right: $\tau = 0.1, N = 2000, M = 1$.

Lower-left: $\tau = 0.3, N = 2000, M = 1$. Lower-right: $\tau = 0.1, N = 200000, M = 100$.

For large stepsize 0.5 (upper-left), the orbit spirials quickly around and towards the origin with a wrong formation of an artificial attractor there. The dissipation effect diminishes with smaller stepsize 0.3 (lower-left), but still pronounced. At very small stepsize 0.1 (upper-right), the orbit appears to be well-behaved as a closed ellipse when total number of steps is limited to 2000 as before. However, the artificial dissipation eventually becomes pronounced again when computed to twenty thousands steps (lower-right).

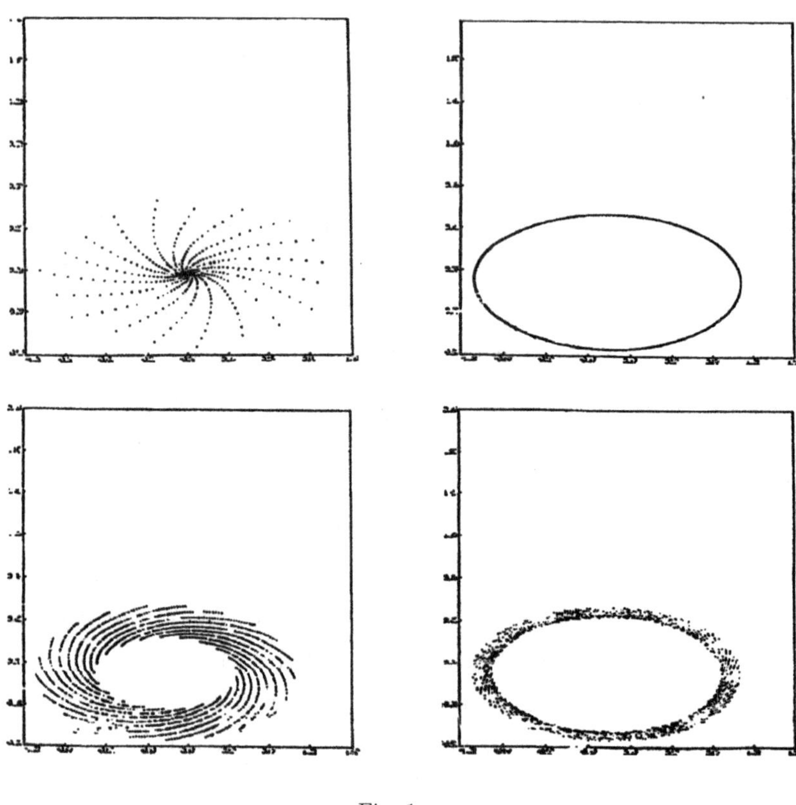

Fig. 1

Fig.2. Harmonic oscillator $H = (p^2 + a^2 q^2)/2, \quad a = 2$.

Explicit symplectic composite scheme, order 2

stepsize $= \tau$, total number of steps $= N$, number of steps per plot $= M$.

Single orbit is computed in all cases with the same initial point.

Upper-left: $\tau = 0.5, N = 2000, M = 1$. Upper-right: $\tau = 0.1, N = 2000, M = 1$.

Lower-left: $\tau = 0.3, N = 2000, M = 1$. Lower-right: $\tau = 0.1, N = 1000000, M = 500$.

The arrangement of calculation is just the same as in Fig. 1 for comparison. In all cases the symplectic computing gives always clear-cut invariant ellipses even for the very long run of a million steps (upper-right). The artificial dissipation is fully inhibited. The computational stability and the ability for long time tracking and the preservation of the structure of the phase portrait is truely remarkable. We note that the specific symplectic scheme used here does not preserve the quadratic energy exactly but it does possess an invariant quadratic form whose level lines are ellipses very close to and practically indistinguishable from the original ellipses.

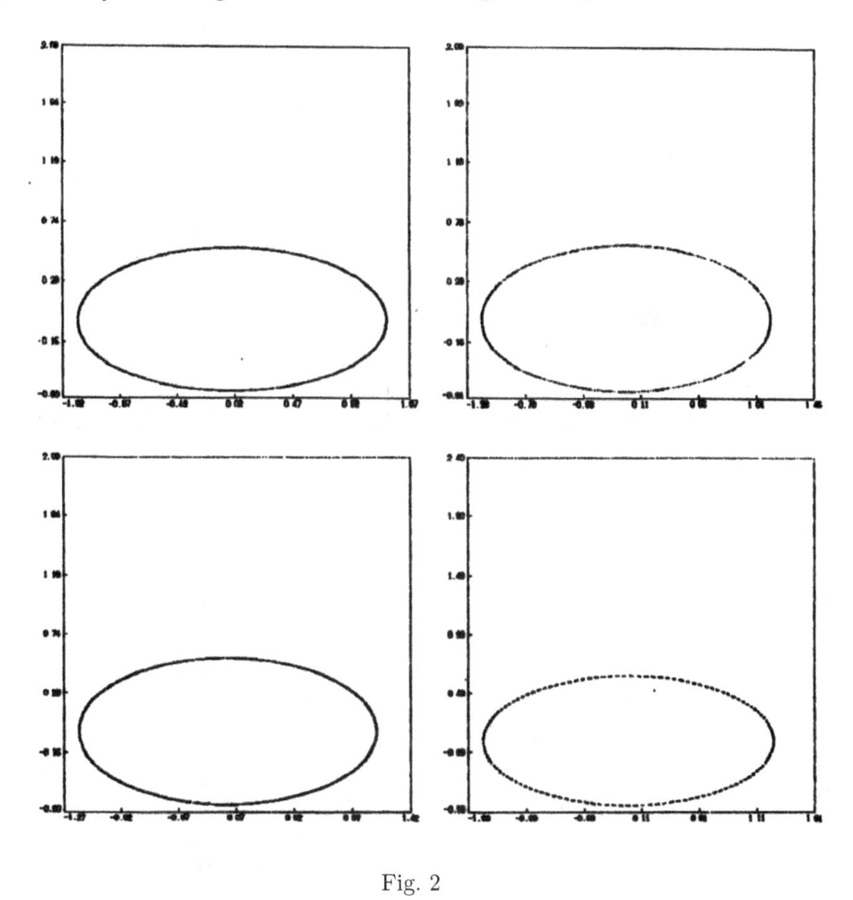

Fig. 2

Fig.3. Nonlinear oscillator $H = \left(p^2 + a^2q^2 + a^4q^4/12\right)/2, \quad a = 2.$

Runge-Kutta method, order 4, non-symplectic.

stepsize $= \tau$, total number of steps $= N$, number of steps per plot $= M$.

Single orbit is computed in all cases with the same initial point.

Upper-left: $\tau = 0.5, N = 2000, M = 1$. Upper-right: $\tau = 0.1, N = 2000, M = 1$.

Lower-left: $\tau = 0.3, N = 2000, M = 1$. Lower-right: $\tau = 0.1, N = 200000, M = 100$.

This nonlinear oscillator is the harmonic one plus a nonlinear perturbation. At low energy values the invariant orbits are only nearly elliptical in form. For larger stepsizes 0.5 (upper-left), and 0.3 (lower-left), the spiraling structures of the orbits are much more complicated than the corresponding linear cases in the left hand side of Fig.1. At very small stepsize 0.1 (upper-right and lower-right), the situation is again similar to the linear case with the eventual cumulative artificial dissipation noticeable in the long run.

Artificial dissipation is inevitable for both linear and nonlinear cases in non-symplectic schemes.

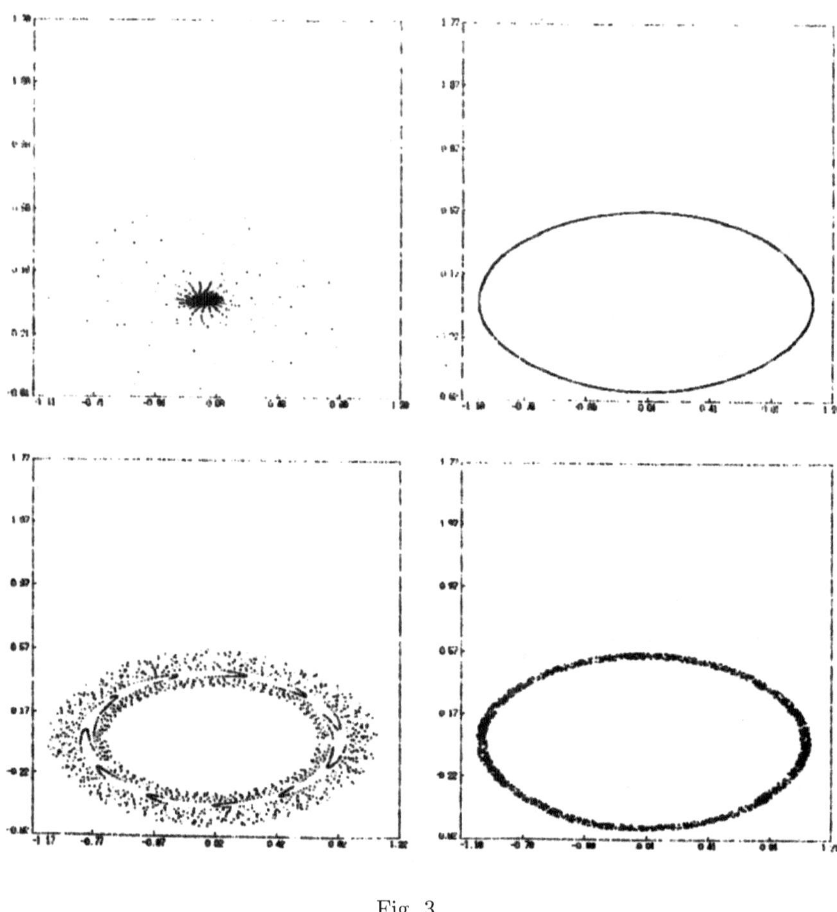

Fig. 3

Fig.4. Nonlinear oscillator $H = \left(p^2 + a^2q^2 + a^4q^4/12\right)/2, \quad a = 2.$

Explicit symplectic composite scheme, order 2.

stepsize $= \tau$, total number of steps $= N$, number of steps per plot $= M$.

Single orbit is computed in all cases with the same initial point.

Upper-leift: $\tau = 0.5, N = 2000, M = 1$. Upper-right: $\tau = 0.1, N = 2000, M = 1$.

Lower-left: $\tau = 0.3, N = 2000, M = 1$. Lower-right: $\tau = 0.1, N = 1000000, M = 500$.

Here again the symplectic computing gives always clear-cut invariant closed orbits including the run of a million steps without any dissipation. The computed orbits are only nearly elliptical but hardly discernible from the ellipses in Fig.2. The apparent identity of all the orbits in Fig. 1 and 2 shows the stability and robustness of the method. From Fig. 1-4 one sees the sharp contrast between the performances of the two methods.

The symplectic method is far superior at least in the aspects of computational stability, global and structural preservation and long-term simulation. Incidentally the specific 2nd order symplectic method adopted is 4 times faster than the 4th order Runge-Kutta method.

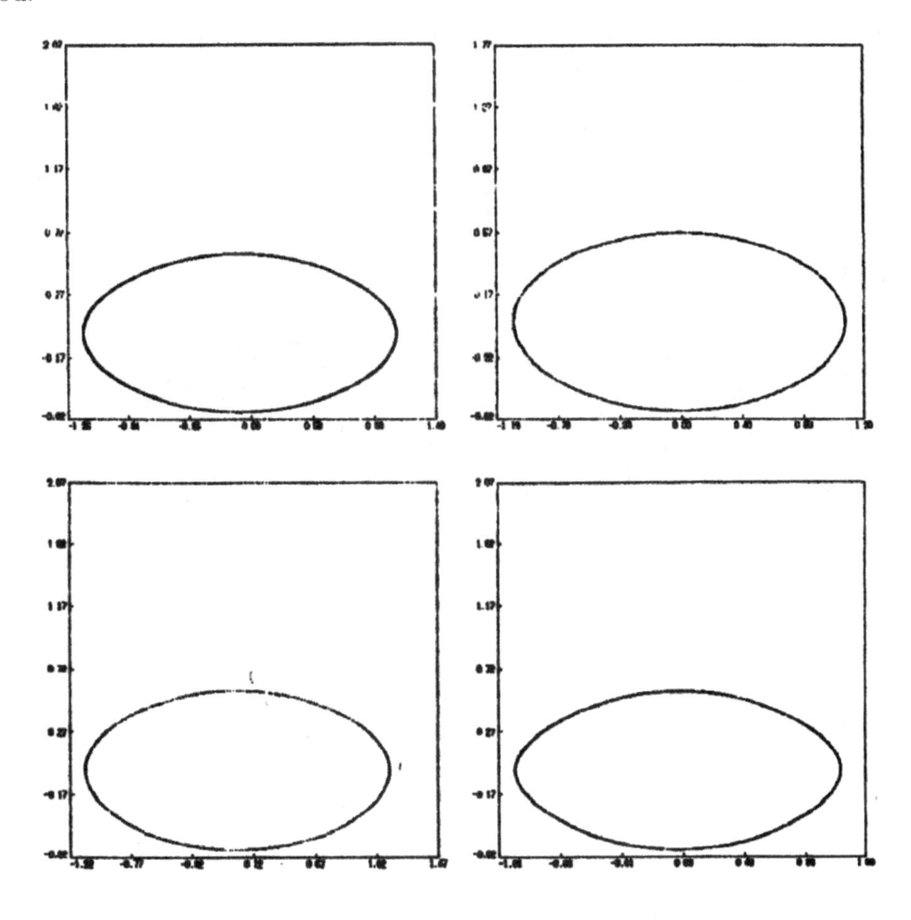

Fig. 4

Fig.5. Preservation and breakdown of invariant tori for nonlinear oscillator $H = (p^2 + a^2q^2 + a^4q^4/12)/2, \quad a = 2$.

Explicit symplectic composite scheme, order 2.

stepsize$=\tau$, total number of steps$=N$, number of steps per plot $= M$.

Five orbits are computed in all cases with the same set of five initial points.

Upper-left: $\tau = 0.5, N = 2000, M = 1$. Upper-right: $\tau = 0.1, N = 2000, M = 1$.

Lower-left: $\tau = 0.3, N = 2000, M = 1$. Lower-right: $\tau = 0.1, N = 1000000, M = 500$.

The five initial points are (0.5, 0.25), (0.65, 0.535), (0.375, 0.55), (0.685, 0.65),

(0.495, 0.375). For smaller stepsize 0.3671 (upper-right), all the invariant tori are preserved practically, the effect of segmentation is almost imperceptible. However, for larger stepsizes (upper-left, lower-left, lower-right), while some invariant tori are preserved, the breakdown becomes pronounced. Some invariant tori are severely segmented and some are broken down into islands. This phenomenon is typical in the perturbation theory of Hamiltonian dynamics.

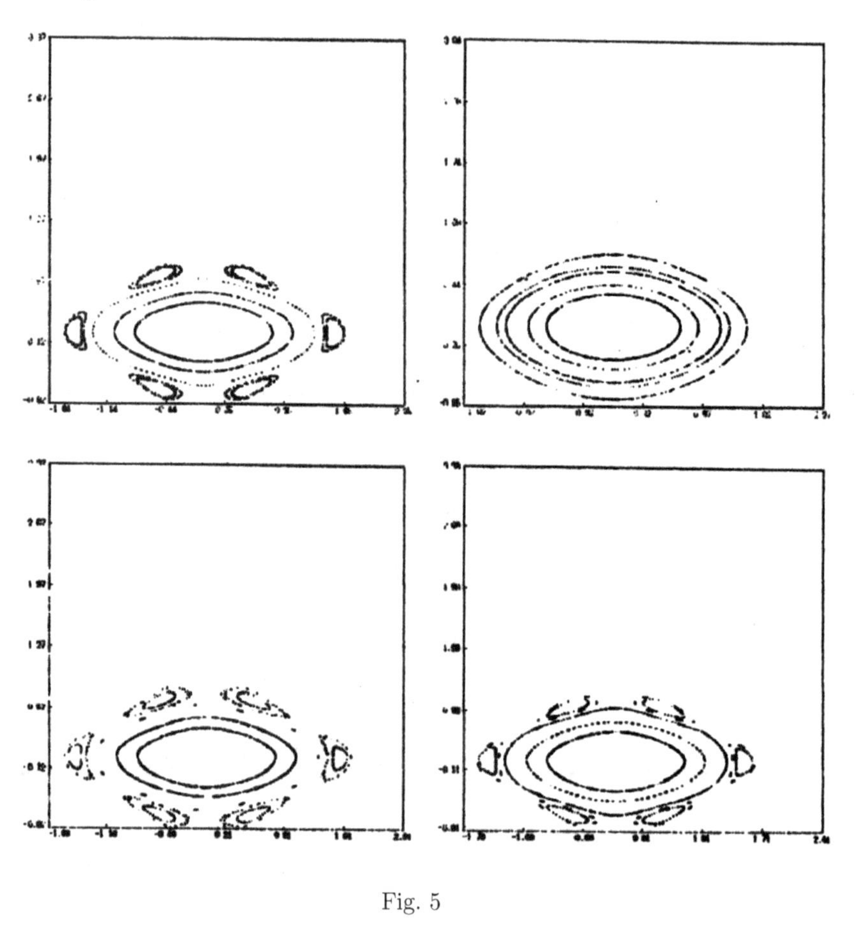

Fig. 5

Fig.6. Hamiltonian with 4-fold rotational symmetry, using H_k in (22), $k = 4$. Explicit symplectic composite scheme, order 2 (21).

stepsize=τ, total number of steps=N, number of steps per plot = M.

Upper-part: $\tau = 1.3, N = 10000, M = 1$.

Lower-left: $\tau = 0.20000002, N = 40000, M = 1$.

Lower-right: $\tau = 0.2, N = 40000, M = 1$.

In the symplectic scheme for system with Hamiltonian H_k, the initial point, chosen on or very near to a separatrix, moves within an Arnold web which is a network of canals with the separatrices as its skeleton. The canals become thinner with smaller τ. The separatrix network of H_4 is a square lattice. For larger $\tau = 1.3$ (upper part) the diffusion pattern is more wide-spread and coarse. For small τ the motion is almost rectilinear along the lattice plus random turnings at the crossings, see the cases of $\tau = 0.2$ (lower-right) and $\tau = 0.20000002$ (lower-left), which, in addition, show the extreme sensitivity of the diffusion pattern to the stepsize τ.

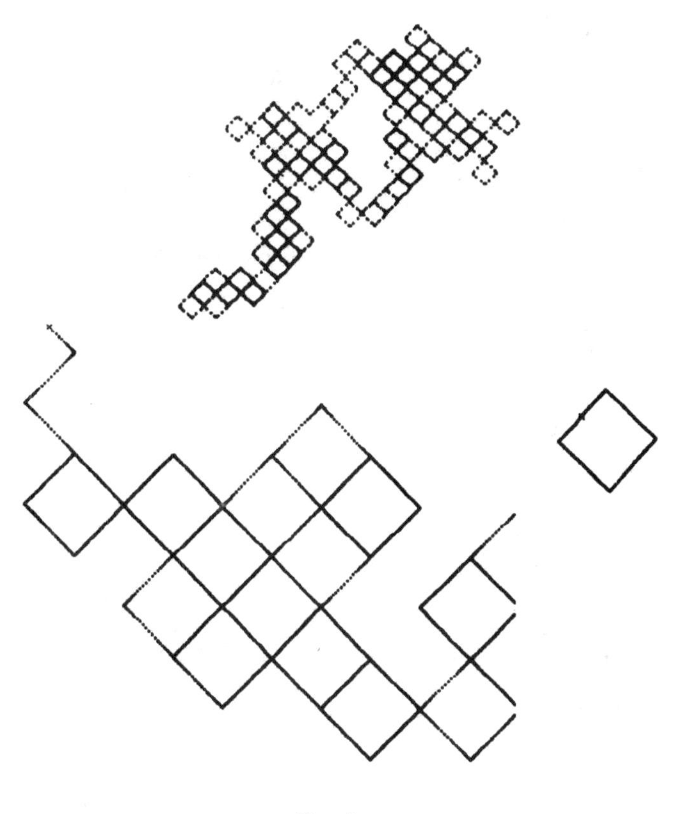

Fig. 6

Fig.7. Hamiltonian with 3-fold rotational symmetry, using H_k in (22), $k = 3$.
Explicit symplectic composite scheme, order 2 (21).
stepsize$=\tau$, total number of steps$=N$, number of steps per plot $= M$.
Single orbit is computed in all cases.
Upper-left: $\tau = 1.4, N = 4000, M = 2$. Upper-right: $\tau = 1.04, N = 10000, M = 1$.
Lower-left: $\tau=0.7002, N=10000, M = 2$. Lower-right: $\tau = 0.7, N=25000, M=5$.

The separatrix network of H_3 (which is proportional to H_6) is a periodic lattice composed of triangles and hexagons. For large $\tau = 1.4$ (upper-left) we get a very coarse web; then at $\tau = 1.04$ (upper-right) the diffusion path is still not fully straightened. For smaller $\tau = 0.7002$ (lower-left) the diffusion is quite uniform and wide-spread. However, with slightly different $\tau = 0.7$ (lower-right), the diffusion behavior becomes grossly different. In an early stage up to 14000 steps the point is confined on an initial triangle, then it moves out to a far away hexagon and stays there again for a long time and then in certain way repeats such oscillating visits between these two entities.

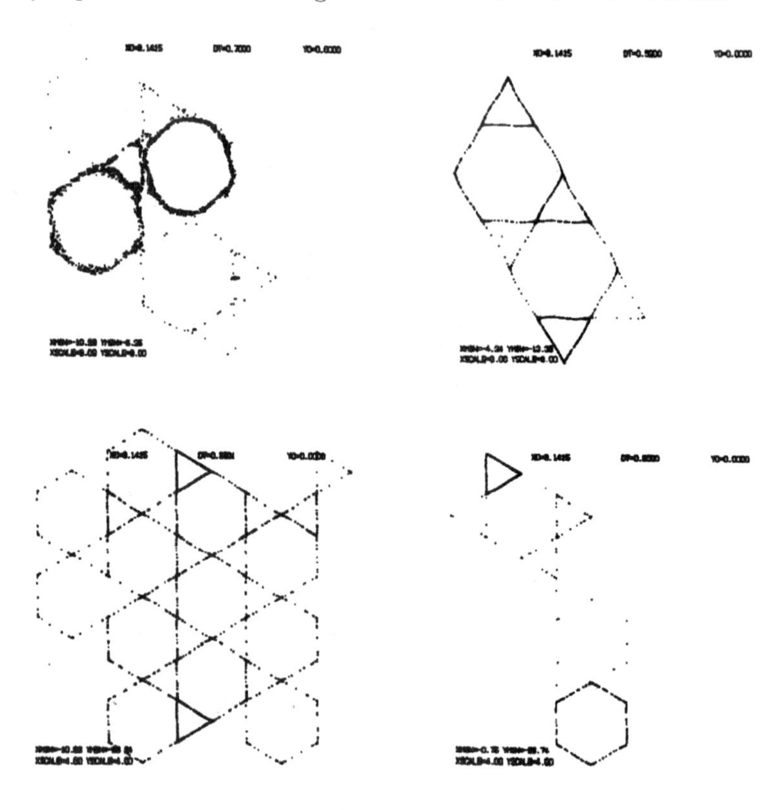

Fig. 7

Fig.8. Hamiltonian with 5-fold rotational symmetry, using H_k in (22), $k = 5$.

Explicit symplectic composite scheme, order 2 (21).

stepsize$=\tau$, total number of steps$=N$, number of steps per plot $= M$.

Single orbit is computed in all cases.

Upper-left: $\tau = 0.5, N = 100000, M = 10$. Upper-right: $\tau=0.56, N=50000, M=5$.

Lower-left: $\tau = 0.701, N = 20000, M = 1$. Lower-right: $\tau = 0.701, N = 100000$, $M = 10$.

The separatrix network of H_5 (proportional to H_{10}) is not periodic but quasiperiodic and not connected, looks like a Penrose tiling with 10-fold symmetry. For small values $\tau = 0.5$ (upper-left) and $\tau = 0.56$ (upper-right) we get trapped diffusion, resulting in rosette patterns with 10-fold symmetry. For larger $\tau = 0.701$ (lower left and right, for evolution) we get the spread-out pattern (see also [21]) which looks like certain Moslem decoration and the electron diffraction pattern of a 2D decagonal quasicrystal, cf. [22].

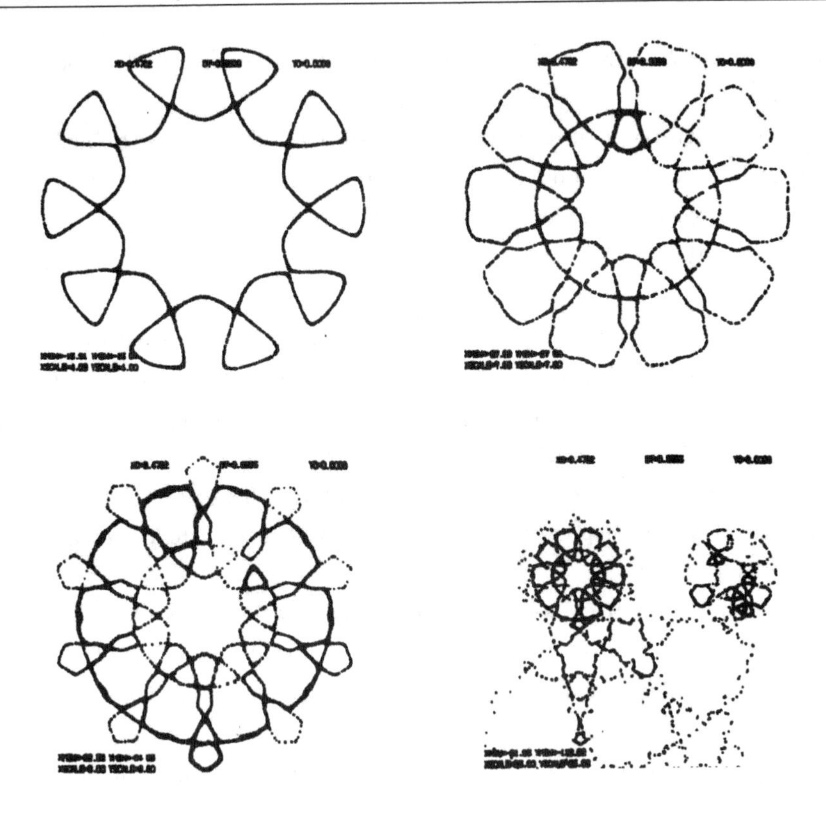

Fig. 8

References

[1] Feng K. In Feng K, ed. On difference schemes and symplectic geometry. Proc 1984 Beijing Symp Diff Geometry and Diff Equations. Beijing: Science Press, 1985: 42-58

[2] Feng K. Difference schemes for Hamiltonian formalism and symplectic geometry. J Comp Math. 1986 4(3): 279-289

[3] Feng K. In Zhunag F G, ed. Symplectic geometry and numerical methods in fluid dynamics. Proc 10th Inter Conf Numer Methods in Fluid Dynamics. Lect. Notes in Phys 264. Berlin: Springer. 1986: 1-7

[4] Feng K, Qin M Z. In Zhu Y L, Guo Ben-yu, ed. The symplectic methods for computation of Hamiltonian equations. Proc Conf on Numerical Methods for PDE's. Lect Notes in Math 1297. Berlin: Springer, 1987: 1-37

[5] Ge Z, Feng K. On the approximation of Hamiltonian systems. J Comp Math. 1988 6(1): 88-97

[6] Feng K, Wu H M, Qin M Z, Wang D L. Construction of canonical difference schemes for Hamiltonian formalism via generating functions. J Comp Math. 1989 7(1): 71-96

[7] Li C W, Qin M Z. A symplectic difference scheme for the infinite dimensional Hamiltonian system. J Comp Math. 1988 6(2): 164-174

[8] Wu Y H. The generating function for the solution of ODE and its discrete methods. Computer Math & Appl. 1989 15(12): 1041-1050

[9] Qin M Z. Canonical difference scheme for the Hamiltonian equation. Math Meth in Appl Sci. 1989 11: 543-557

[10] Hamilton W R. General methods in dynamics. Mathematical Papers, V 2. Cambridge, 1940

[11] Synge J L. The life and early work of Sir William Rowan Hamilton. Scripta Math. 1944 10: 13-24

[12] Klein F. Entwickelung der Mathematik im 19 Jahrhundert. Teubner. 1928

[13] Schröedinger E. What Is Life? The Physical Aspect of the Living Cell. Cambridge University Press, 1944

[14] Arnold V I. Mathematical Methods of Classical Mechanics. Moscow: Nauka, 1974

[15] Feynman R. Lectures on Physics, Vol 1. Reading: Addison-Wesley. 1963: 148-160

[16] Pogorelov A V. On spherical image maps whose Jacobians do not change sign. Dokl Akad Nauk. SSSR. 1956 111: 757-759, 945-947

[17] Hartman P, Nirenberg L. On continuous mappings of bounded variation; An extension of Gauss' theorem on the spherical representation of surfaces of bounded exterior curvature. Amer Jour Math. 1959 81: 901-920

[18] Chernikov A A, Sagdeev R Z, Zaslavsky G M. Stochastic webs. Physica D. 1988 33: 65-76

[19] Arnold V I. Remarks on quasicrystallic symmetries. Physica D. 1988 33: 21-25

[20] Arnol'd V. I. Instability of dynamical systems with many degrees of freedom. Dokl.akad.nauk Sssr, 1964, 5(1): 9-12

[21] Beloshapkin V V, Sagdeev R Z, Zaslavsky G M, et al. Chaotic streamlines in preturbulent states. Nature. 1989 337: 543-557

[22] Fung K K , Yang C Y , Zhou Y Q , et al. Icosahedrally related decagonal quasicrystal in rapidly cooled Al-14-at.%-Fe alloy. Physical Review Letters, 1986, 56(19): 2060-2063.

12 Hamiltonian Algorithms and a Comparative Numerical Study①②

哈密尔顿算法及其数值对比研究

Abstract

We discuss some of the contributions, made by the authors and their research group, on the numerical methods for Hamiltonian systems. Our main concern will be the Hamiltonian algorithms, presenting the proper way for computing Hamiltonian dynamics. These algorithms are conceived developed and analysed systematically within the framework of symplectic geometry. This approach is natural since the dynamical evolution of Hamiltonian systems are exclusively symplectic transformations. The vast majority of conventional methods are non-symplectic; they inevitably imply artificial dissipation and other parasitic artifacts of non-Hamiltonian distortions. The Hamiltonian algorithms are clean algorithms, free from all kinds of non-Hamiltonian pollutions. They actually give outstanding performance, far superior than the conventional non-symplectic methods, especially in the aspects of global, structural properties and long-term tracking capabilities. We give in detail some comparative numerical experimentation; in many cases the contrast is quite striking.

§1 Introduction

We study the numerical methods, or rather, the "proper" numerical methods, for solving dynamical problems expressed in the form of a *canonical system* of differential equations,

$$\frac{dp_i}{dt} = -\frac{\partial H}{\partial q_i}, \quad \frac{dq_i}{dt} = \frac{\partial H}{\partial p_i}, \quad i = 1, \cdots, n, \tag{1.1}$$

with given *Hamiltonian* or *energy* function $H(p_1, \cdots, p_n, q_1, \cdots, q_n)$.

The canonical system (1.1) with remarkable elegance and symmetry was introduced by Hamilton. The approach, together with the analytic works of Jacobi, developed further into a well-established mathematical formalism for classical mechanics, which is an alternative of, and equivalent to, the Newtonian and Lagrangian formalisms. The geometrization of the Hamiltonian formalism was undertaken by Poincare. This gave

① Joint with Qin M Z

② Computer Physics Communications. 1991 65: 173 -187

rise to a new discipline, called *symplectic geometry*, which serves as the mathematical foundation of the Hamiltonian formalism and has the merit of making the underlying intricate properties and symmetries of Hamiltonian systems more explicit, simple and easy to grasp. By now it is almost certain that all real physical processes with negligible dissipation can be cast, in some way or other, into the Hamiltonian formalism, so the latter is becoming one of the most useful tools in the mathematical arsenal of physical and engineering sciences. It covers classical mechanics, dynamics of continuous media, quantum mechanics, relatively, etc. In this view, a systematic study of numerical methods of Hamiltonian systems is motivated and might eventually lead to wide applications. We conceive, develop, analyse and evaluate difference schemes and algorithms systematically from within the framework of symplectic geometry. The approach proves to be very effective and successful. We actually derive in this way an infinite variety of "unconventional" symplectic schemes with outstanding performance.

In the following, vectors are always represented by column matrices, matrix transpose is denoted by prime $'$. Let $z=(z_1,\cdots,z_n,z_{n+1},\cdots,z_{2n})'=(p_1,\cdots,p_n,q_1,\cdots,q_n)'$,

$$\nabla H = H_z = \left(\frac{\partial H}{\partial p_1},\cdots,\frac{\partial H}{\partial p_n},\frac{\partial H}{\partial q_1},\cdots,\frac{\partial H}{\partial q_n}\right)',$$

$$J = J_{2n} = \begin{pmatrix} 0 & -I \\ I & 0 \end{pmatrix}, \quad J' = -J = J^{-1}.$$

Equation (1.1) can be written as

$$\frac{dz}{dt} = J\nabla H(z) \tag{1.2}$$

defined in phase space $\mathbb{R}^{2n}$ with a standard symplectic structure given by the non-singular anti-symmetric closed differential 2-form

$$\omega = \sum_{i=1}^{n} dz_i \wedge dz_{n+i}. \tag{1.3}$$

So, at each point $z \in \mathbb{R}^{2n}$, we have the non-degenerate anti-symmetric bilinear form

$$\omega(\xi,\eta) = \xi' J_{2n}\eta \tag{1.4}$$

for tangent vectors ξ,η. According to the *Darboux Theorem*, the symplectic structure given by any non-singular anti-symmetric closed differential 2-form can be brought to the above standard form, at least locally, by a suitable change of coordinates.

The linear operator (matrix) S in $\mathbb{R}^{2n}$ is called symplectic if it preserves the symplectic form (1.4), i.e., satisfies

$$S'JS = J. \tag{1.5}$$

Symplectic matrices form a Lie group — *symplectic group* Sp$(2n)$ — whose Lie algebra sp$(2n)$ consists of *infinitesimal symplectic* matrices L satisfying

$$L'J + JL = 0, \tag{1.6}$$

or equivalently

$$L = JA, \quad \text{where} \quad A' = A. \tag{1.7}$$

So sp$(2n)$ is closely related to the space sm$(2n)$ of *symmetric* matrices. We define the *symplectic product* of every pair A_1, A_2 of symmetric matrices by

$$\{A_1, A_2\} = A_1JA_2 - A_2JA_1. \tag{1.8}$$

Then sm$(2n)$ becomes a Lie algebra which is isomorphic to the Lie algebra sp$(2n)$ under the usual commutator bracket $[L_1, L_2] = L_1L_2 - L_2L_1$.

An operator $g : \mathbb{R}^{2n} \to \mathbb{R}^{2n}$ is called symplectic (in classical language, *canonical transformation*) if it is a diffeomorphism with Jacobian matrix g_* everywhere symplectic. The totality of symplectic operators is denoted by symp$(2n)$. An operator $f : \mathbb{R}^{2n} \to \mathbb{R}^{2n}$ is called *symmetric* if its Jacobian matrix f_* is everywhere symmetric. The totality of symmetric operators is denoted by symm$(2n)$. By the Poincare lemma, for each symmetric operator there is, at least locally, a scalar function $\phi(z)$, unique up to a constant, such that $f = \nabla\phi$.

The *Fundamental Theorem on Hamiltonian Formalism* says that the solution $z(t)$ of the canonical system with energy H can be generated by a *one-parameter group* $g^t = g_H^t$ of symplectic operators of $\mathbb{R}^{2n}$. g_H^t is called the phase flow of the given system.

The symplecticity of g_H^t implies the preservation of the 2-form ω, 4-form $\omega \wedge \omega, \cdots$, $2n$-form $\omega \wedge \omega \wedge \cdots \wedge \omega$. They constitute the class of *conservation laws of phase area of* even dimensions for the Hamiltonian system (1.2), including the Liouville theorem.

Moreover, the Hamiltonian system possesses another class of conservation laws related to the *energy* $H(z)$. A function $\phi(z)$ is said to be an *invariant integral* of the system if it is invariant under phase flow g_H^t,

$$\phi(z) = \phi\left(g_H^t z\right), \quad \forall z \in \mathbb{R}^{2n},$$

which is equivalent to

$$\{\phi, H\} = 0,$$

where the Poisson bracket for two functions $\phi(z), \psi(z)$ are defined as

$$\{\phi, \psi\} = \phi_z' J \psi_z.$$

H itself is always an invariant integral, see, e.g., ref. [23]. The above situations in Hamiltonian systems suggest the following *guidelines* for the numerical study of dynamical problems: The problem should be expressed in some suitable *Hamiltonian formalism.* The numerical schemes should preserve as much as possible the characteristic properties and inner symmetries of the original system. The transition $z \to \hat{z}$ from the kth time step $z^k = z$ to the next time step $z^{k+1} = \hat{z}$ should be symplectic. Such schemes are called *symplectic difference schemes*, or *symplectic algorithms*, or more generally, *Hamiltonian algorithms.*

The terminology Hamiltonian algorithms will be understood in a broader context. Since there are many different categories of Hamiltonian systems and symplectic structures, the Hamiltonian evolutions for each category may have their own specific transformation properties. For example, for classical mechanics, for which (1.2) is the standard form, the evolutions are symplectic, so, for this case, the Hamiltonian algorithms are symplectic algorithms. For quantum mechanics, we have complex Hilbert space of ∞ dimensions carrying unitary structure. The evolutions are not only symplectic but also unitary (symplectic plus orthogonal), so the Hamiltonian algorithms will be unitary algorithms. In the following we discuss only symplectic Hamiltonian algorithms.

§2 Symplectic Difference Schemes and Generating Functions

We study the 2-level difference schemes for solving Hamiltonian system (1.2). Time t is discretized into $t = 0, \pm s, \pm 2s, \cdots$, with step size s, $z^k \approx z(ks), k = 0, \pm 1, \pm 2, \cdots$. Each 2-level-scheme is represented by a transition

$$\hat{z} = G^s z, \quad z = z^k, \quad \hat{z} = z^{k+1}. \tag{2.1}$$

$G^s = G^s_H$ depends on s, H and the mode of discretization. The difference scheme (2.1) for Hamiltonian system (1.2) is called symplectic if its transition operator G^s_H is symplectic for all H and s.

It is easy to see that the weighted Euler scheme

$$\hat{z} = z + sJ(\nabla H)(c\hat{z} + (1-c)z), \tag{2.2}$$

where c is a real constant, is symplectic iff $c = \dfrac{1}{2}$. The unique symplectic case is the time-centered Euler scheme

$$\hat{z} = z + sJ(\nabla H)\left(\frac{\hat{z}+z}{2}\right). \tag{2.3}$$

This illustrates the fact that, apart from some rare exceptions, the vast majority of conventional schemes are non-symplectic. However, we have a far-reaching generalization when c is allowed to be a $2n \times 2n$ matrix: (2.2) is symplectic iff

$$c = \frac{1}{2}(I_{2n} + J_{2n}B), \quad B' = B. \tag{2.4}$$

Such symplectic schemes are all implicit, (2.2) is solvable for new $\hat{z}$ in terms of old z when $|s|$ is small. The transition of (2.2) and (2.4) will be written as

$$\hat{z} = G^s_{H,B} z. \tag{2.5}$$

For $B = 0$, we get (2.3) with order 2, for $B \neq 0$, the order is 1, but the composite schemes

$$G^{s/2}_{H,B} \circ G^{s/2}_{H,-B}, \qquad G^{s/2}_{H,-B} \circ G^{s/2}_{H,B} \tag{2.6}$$

have order 2. The above situation illustrates the fact, that an abundance of unconventional symplectic schemes for Hamiltonian systems can be constructed.

The above construction of symplectic algorithms presents only a special case of a general methodology based on *generating functions* as explained below.

A matrix α of order $4n$ is called a *Darboux matrix* if

$$\begin{aligned} &\alpha' J_{4n} \alpha = \tilde{J}_{4n}, \quad J_{4n} = \begin{pmatrix} 0 & -I_{2n} \\ I_{2n} & 0 \end{pmatrix}, \quad \tilde{J}_{4n} = \begin{pmatrix} J_{2n} & 0 \\ 0 & -J_{2n} \end{pmatrix}, \\ &\alpha = \begin{pmatrix} a & b \\ c & d \end{pmatrix} \quad \alpha^{-1} = \begin{pmatrix} a_1 & b_1 \\ c_1 & d_1 \end{pmatrix}. \end{aligned} \tag{2.7}$$

For the study of symplectic algorithms we may impose two more conditions $a + b = 0$, $c + d = I_{2n}$. Then it is called a *normal Darboux matrix*, which can be characterized as

$$\begin{aligned} &\alpha = \alpha_B = \begin{pmatrix} a & b \\ c & d \end{pmatrix} = \begin{pmatrix} J & -J \\ \frac{1}{2}(I + JB) & \frac{1}{2}(I - JB) \end{pmatrix}, \quad B' = B \\ &\alpha^{-1} = \alpha_B^{-1} = \begin{pmatrix} a_1 & b_1 \\ c_1 & d_1 \end{pmatrix} = \begin{pmatrix} \frac{1}{2}(JBJ - J) & I \\ \frac{1}{2}(JBJ + J) & I \end{pmatrix}. \end{aligned} \tag{2.8}$$

Note that the same form of matrix appeared in (2.4).

Each Darboux matrix induces a (linear) *fractional transform* between *symplectic matrices near identity* and *symmetric matrices near nullity*:

$$\begin{aligned} &\sigma_\alpha : \qquad \mathrm{Sp}(2n) \to \mathrm{sm}(2n) \\ &\sigma_\alpha(S) = (aS + b)(cS + d)^{-1} = A, \qquad \text{for} \quad |cS + d| \neq 0, \end{aligned} \tag{2.9}$$

with inverse transform $\sigma_\alpha^{-1} = \sigma_{\alpha^{-1}}$

$$\begin{aligned}\sigma_\alpha^{-1}: &\quad \mathrm{sm}(2n) \to \mathrm{Sp}(2n),\\ \sigma_\alpha^{-1}(A) &= (a_1 A + b_1)(c_1 A + d_1)^{-1} = S, \quad \text{for} \quad |c_1 A + d_1| \neq 0.\end{aligned} \tag{2.10}$$

The above machinery can be extended to generally non-linear operators in $\mathbb{R}^{2n}$ to establish a local 1-1 correspondence between symplectic operators near identity and symmetric operators near nullity. We write each symmetric operator f as the gradient $\nabla\phi$ of a scalar function ϕ.

$$\begin{aligned}\sigma_\alpha: &\quad \mathrm{Symp}(2n) \to \mathrm{symm}(2n),\\ \sigma_\alpha(g) &= (ag + b)(cg + d)^{-1} = \nabla\phi, \quad \text{for} \quad |cg_z + d| \neq 0,\end{aligned} \tag{2.11}$$

or alternatively in the implicit form

$$ag(z) + bz = (\nabla\phi)(cg(z) + dz), \tag{2.12}$$

where $\phi = \phi_g = \phi_{g,\alpha}$ is called the generating function of Darbonx type α for the symplectic operator g. Then

$$\begin{aligned}\sigma_\alpha^{-1}: &\quad \mathrm{symm}(2n) \to \mathrm{Symp}(2n),\\ \sigma_\alpha^{-1}(\nabla\phi) &= (a_1\nabla\phi + b_1)(c_1\nabla\phi + d_1)^{-1} = g, \quad \text{for} \quad |c_1\phi_{ww} + d_1| \neq 0,\end{aligned} \tag{2.13}$$

or alternatively in the implicit form

$$a_1\nabla\phi(w) + b_1(w) = g\left(c_1\nabla\phi(w) + d_1 w\right), \tag{2.14}$$

where $g = g_\phi = g_{\phi,\alpha}$ is called the symplectic operator for the generating function ϕ of Darboux type α.

For each Hamiltonian H with its phase flow g_H^t and for each normal Darboux matrix α, we get the generating function $\phi = \phi_{g_H^t}$, or for short $\phi = \phi_H^t = \phi_{H,\alpha}^t$, of normal Darboux type α for the phase flow of g_H^t by

$$\nabla\phi_{H,\alpha}^t = \left(ag_H^t + b\right)\left(cg_H^t + d\right)^{-1}, \quad \text{for small } |t|. \tag{2.15}$$

$\phi_{H,\alpha}^t$ satisfies the *Hamilton-Jacobi* equation

$$\frac{\partial}{\partial t}\phi = -H\left(a_1\nabla\phi(w) + w\right) = -H\left(c_1\nabla\phi(w) + w\right) \tag{2.16}$$

and can be expressed by Taylor series in t

$$\phi = \phi^t(w) = \phi(w,t) = \sum_{k=1}^{\infty}\phi^{(k)}(w)t^k, \quad \text{for small } |t|. \tag{2.17}$$

The coefficients can be determined recursively

$$\phi^{(1)}(w) = -H(w), \quad \text{and for} \quad k \geqslant 0, \quad a_1 = \frac{1}{2}(JBJ - J),$$

$$\phi^{(k+1)}(w) = \sum_{m=1}^{k} \frac{1}{(k+1)m!} \sum_{i_1,\cdots,i_m=1}^{2n} \sum_{\substack{j_1+\cdots+j_m=k \\ j_l \geqslant 1}} \phi^{(1)}_{w_{i_1}\cdots w_{i_m}}(w)$$

$$\times \left(a_1 \nabla \phi^{(j_1)}_w(w)\right)_{i_1} \cdots \left(a_1 \nabla \phi^{(j_m)}_w(w)\right)_{i_m}. \tag{2.18}$$

Let ψ^s be a truncation of $\phi^s_{H,\alpha}$ up to certain power s^m, say. Using inverse transform σ_α^{-1} we get the symplectic operator

$$G^s = \sigma_\alpha^{-1}(\nabla \psi^s), \quad |s| \text{ small},$$

which depends on s, H, α (or equivalently B) and the mode of truncation. It is a symplectic approximation to the phase flow g^s_H and can serve as the transition operator of a symplectic difference scheme (for the Hamiltonian system (1.2))

$$\hat{z} = G^s z \tag{2.19}$$

which can be written implicitly as

$$J\hat{z} - Jz = \nabla \psi^s(c\hat{z} + (I - c)z), \quad c = \frac{1}{2}(I + JB). \tag{2.20}$$

So we have constructed, for each normal Daxboux type α_B, an hierarchy of symplectic schemes of arbitrary high order. The scheme (2.2), (2.4) corresponds to the lowest truncation $\psi^s(w) = -sH(w)$. For the general theory of generating functions in the context of Darboux matrices, see refs. [3, 6, 7].

For the Hamiltonian of the special form

$$H(p,q) = \sum_{i=1}^{m} \phi_i(A_i p + B_i q), \quad A_i B_i' = B_i A_i', \quad \phi_i \text{ are functions in } n \text{ variables}, \tag{2.21}$$

explicit symplectic schemes of composite type can be constructed [7]. The Hamiltonian of the form $H(p,q) = T(p) + V(q)$ is a special case of (2.21). The corresponding symplectic schemes of order 1 of Darboux type α_B with

$$B = \pm E_{2n} := \pm \begin{pmatrix} 0 & I_n \\ I_n & 0 \end{pmatrix}$$

become explicit:

$$B = -E_{2n}, \quad \hat{p} = p - sV_q(q), \quad \hat{q} = q + sT_p(\hat{p}), \tag{2.22}$$

$$B = E_{2n}, \quad \hat{q} = q + sT_p(p), \quad \hat{p} = p - sV_q(\hat{q}). \tag{2.23}$$

The order of accuracy is raised to 2 by time-staggering, i.e., by putting q (or p) at integer times and p (or q) at half-integer times. This is the first case of symplectic schemes ever to be used and rediscovered many times [1-5,17,24-26] and might even be traced back as early to Von Zeipel. For high-order composite schemes based on (2.22), (2.23) see refs. [1- 9,12,13,17]. For symplectic Runge-Kutta schemes, see refs. [19,20,27-30]. For symplectic leap-frog schemes, see refs. [5, 7, 9, 11]. For Hamiltonian algorithms in ∞ dimensions, see refs. [1,4,5,9,10,18,20,25]. For Hamiltonian algorithms on Poisson manifolds, see refs. [16,22]. For applications of generating functions to difference schemes of general ODE's, see ref. [14]. For symplectic schemes based on generating functions of the 1st and 3rd kinds, see refs. [31]. For convergence of symplectic schemes, see ref. [15]. For other numerical experiments and discussions, see refs. [7,28,31,37].

For the conservation properties of symplectic schemes (2.20), we know [1-9,21]:

Let the quadratic form $\phi_A(z) = \frac{1}{2}z'Az, A' = A$, be an invariant integral of system (1.2). Then, ϕ_A remains to be invariant under the symplectic schemes of Darboux type α_B if A commutes symplectically with B, i.e., $AJB = BJA$.

We list some of the most important α_B with the corresponding form of symplectic matrices A of the conserved quadratic invariants ϕ_A:

$$B = 0, \qquad A \quad \text{arbitrary.}$$

$$B = \pm\begin{pmatrix} 0 & I_n \\ I_n & 0 \end{pmatrix} := \pm E_{2n}, \quad A = \begin{pmatrix} 0 & b \\ b' & 0 \end{pmatrix}, \quad b \text{ arbitrary; angular momentum type.}$$

$$B = \pm I_{2n}, \qquad A = \begin{pmatrix} a & b \\ -b & a \end{pmatrix}, \quad a' = a, \quad b' = -b; \text{ Hermitian type.}$$

$$B = \pm\begin{pmatrix} I_n & 0 \\ 0 & -I_n \end{pmatrix} := \pm I_{n,n}, \; A = \begin{pmatrix} a & b \\ -b & -a \end{pmatrix}, \quad a' = a, \quad b' = -b.$$

The conservation property is best with type $B = 0$, this type of generating functions was first introduced by Poincaré. This classical generating functions of 2nd and 3rd kinds correspond to $B = -E$ and $B = +E$.

For non-linear Hamiltonian systems the energy H is in general not preserved under symplectic schemes [9]. However, analysis from KAM theorems [32, 33] and experimental evidence from symplectic computing strongly suggest that, for any symplectic difference schemes G^s for canonical system with Hamiltonian H, there should exist, for small $|s|$, in a suitable region whose Lebesgue measure tends to fill locally the phase space as $s \to 0$, a function $h^s(z) = h(s,z)$ which is preserved by G^s; moreover, $h(s,z) = H(z) + O(|s|^m)$ and G^s is of mth order of accuracy.

§3 Numerical Experiments and Discussion

Extensive numerical experiments strongly show the advantages of the symplectic methods over the conventional non-symplectic ones. The stability problem of Hamiltonian systems is always related to the "neutral" case with pure imaginary characteristic exponents, they can never by asymptotically stable owing to the symplecticity of dynamics. The conventional methods were designed mostly for asymptotically stable systems, therefore, for Hamiltonian systems they introduce unavoidably artificial dissipation and other parasitic effects which are alien to Hamiltonian dynamics, leading eventually to serious and even qualitative distortions. On the contrary, symplectic methods tend to purify (sometimes even simplify) the algorithms, inhibiting artificial dissipations and all other non-Hamiltonian distortions. Hamiltonian algorithms are clean, free from non-Hamiltonian pollutions.

We present a comparative study for a non-linear oscillator in Fig. 1. The non-symplectic 4th order Runge-Kutta scheme yields persistent spiraling of phase trajectories with artificial creation of the attractor (impossible for Hamiltonian system), see Fig. 1A. The artificial dissipation diminishes with step-size s within fixed time limit of the integration, see Fig. 1B, C. However, the dissipative effect becomes pronounced eventually for long time simulations, see Fig. 1D for 200000 steps. On the contrary, the symplectic method gives clear-cut invariant closed trajectories even for low order of schemes and large step size, see Fig. 1E, F for a million steps.

The symplectic scheme generally does not preserve energy $H(z)$ however, in general it possesses a function $h^s(z)$ which depends on s and is near to H and invariant under the given scheme. For example, for the simple harmonic oscillator $H = \frac{1}{2}(p^2 + a^2q^2)$ and for the first-order explicit scheme (2.22), we have

$$h^s(p,q) = \frac{1}{2}\left(p^2 + a^2q^2\right) - \frac{1}{2}sa^2pq. \tag{3.1}$$

The contour lines $h^2(p,q,a) =$ constant represent ellipse, parabola, hyperbola when $s < 2/a$, $s = 2/a$, $s > 2/a$ respectively. Fig. 2 show that the points produced by the scheme lies on the level lines of h^s.

We know that the classical 4th order Runge-Kutta method is non-symplectic, but for the Harmonic oscillator $H = \frac{1}{2}(p^2 + a^2q^2)$ it becomes symplectic when $s = 2\sqrt{2}/a$, which is also the critical value of s for the stability of the method. Fig. 3 shows the numerical results for step sizes around that critical value.

In Fig.4, Fig.5 and Fig.6 we present the numerical orbits in the phase plane for a

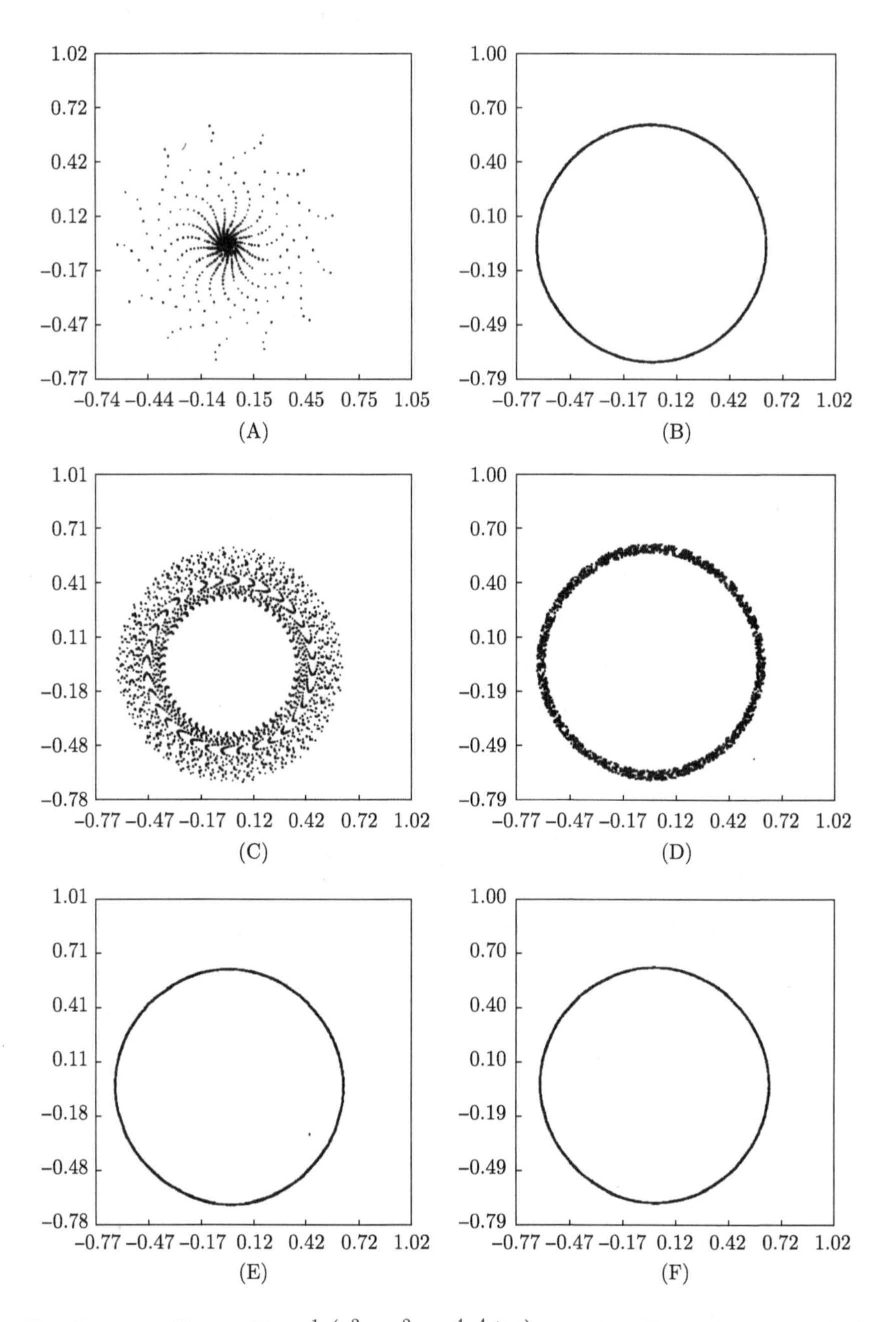

Fig. 1 Non-linear oscillator $H = \frac{1}{2}\left(p^2 + q^2 + a^4 q^4/12\right), a = 1$. Runge-Kutta method, order 4, non-symplectic (A)−(D). Explicit symplectic composite scheme, order 2 (E)−(F). Stepsize = s, total number of steps = N, number of steps per plot = M. (A) $s = 1.0, N = 2000, M = 1$; (B) $s = 0.2, N = 2000, M = 1$; (C) $s = 0.6, N = 2000, M = 1$; (D) $s = 0.2, N = 200000, M = 500$; (E) $s = 0.6, N = 2000, M = 1$; (F) $s = 0.2, N = 1000000, M = 500$.

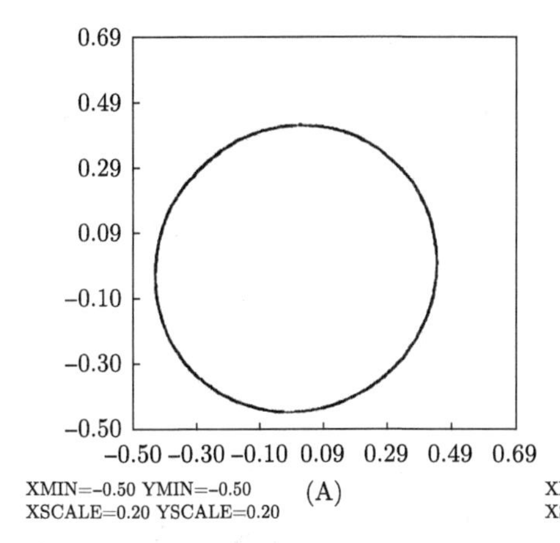

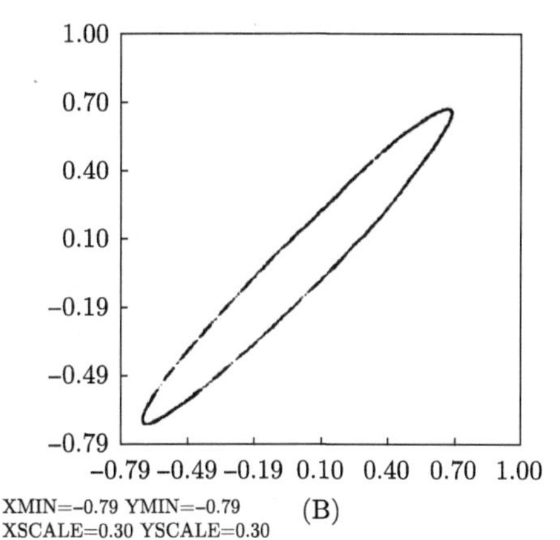

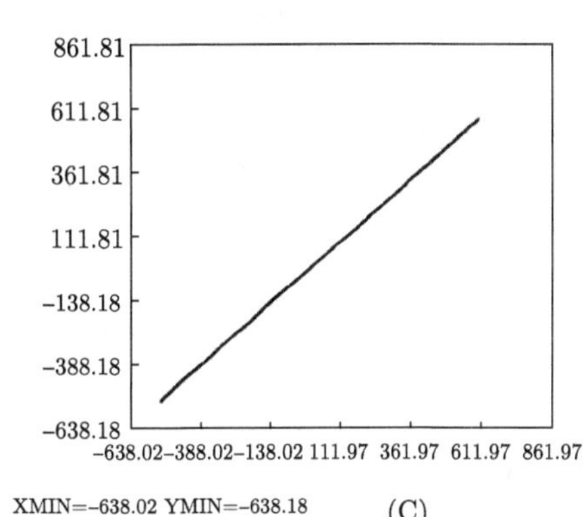

Fig. 2 Harmonic oscillator $H=(p^2+a^2q^2)/2$, $a=1$. Explicit Symplectic scheme, order 1. Stepsize $= s$, total number of steps $= N$, number of steps per plot $= M$. (A) $s = 0.1$, $N = 2000$, $M = 1$; (B) $s = 1.9$, $N = 2000$, $M = 1$; (C) $s = 2.0$, $N = 2000$, $M = 1$. In (A) the orbit appears as an ellipse closed to the circle. In (B) the orbit appears as a tilted oblate ellipse. In (C) the orbit degenerates into a line, $s > 2.0$ corresponds to the unstable case.

Hamiltonian $H_3(p, q)$ of 3-fold rotational symmetry, which is a special case of

$$H_k(p,q) = \sum_{j=1}^{k} \cos\left(p\cos\frac{2\pi j}{k} + q\sin\frac{2\pi j}{k}\right), \tag{3.2}$$

with k-fold rotational symmetry, recently introduced by Sagdeev et al. [34-36]. The separatrices of H_3 are 3 families of parallel lines, forming a periodic lattice composed of hexagons and equilateral triangles. The intersections are saddle points, the centers of basic cells are stable equilibrium points. We use mainly a 2nd-order explicit composite symplectic scheme and occasionally the 4th order Runge-Kutta for comparison. Fig. 4A shows a single symplectic orbit with the initial point located on the separatrix lattice. The motion is almost rectilinear along the "highways" plus almost random turning at the crossings. The resulting Arnold diffusion pattern is a web of finite width (decreasing with step size s) with the original separatrix lattice at its backbone [34]. The theoretical orbits within each cell are concentric closed curves (invariant tori) around the center. The contrasts between a symplectic orbit (up to 1 million steps) and a Runge-Kutta orbit (with much fewer steps) are shown in Fig 4B, C for a hexagonal cell and in Fig

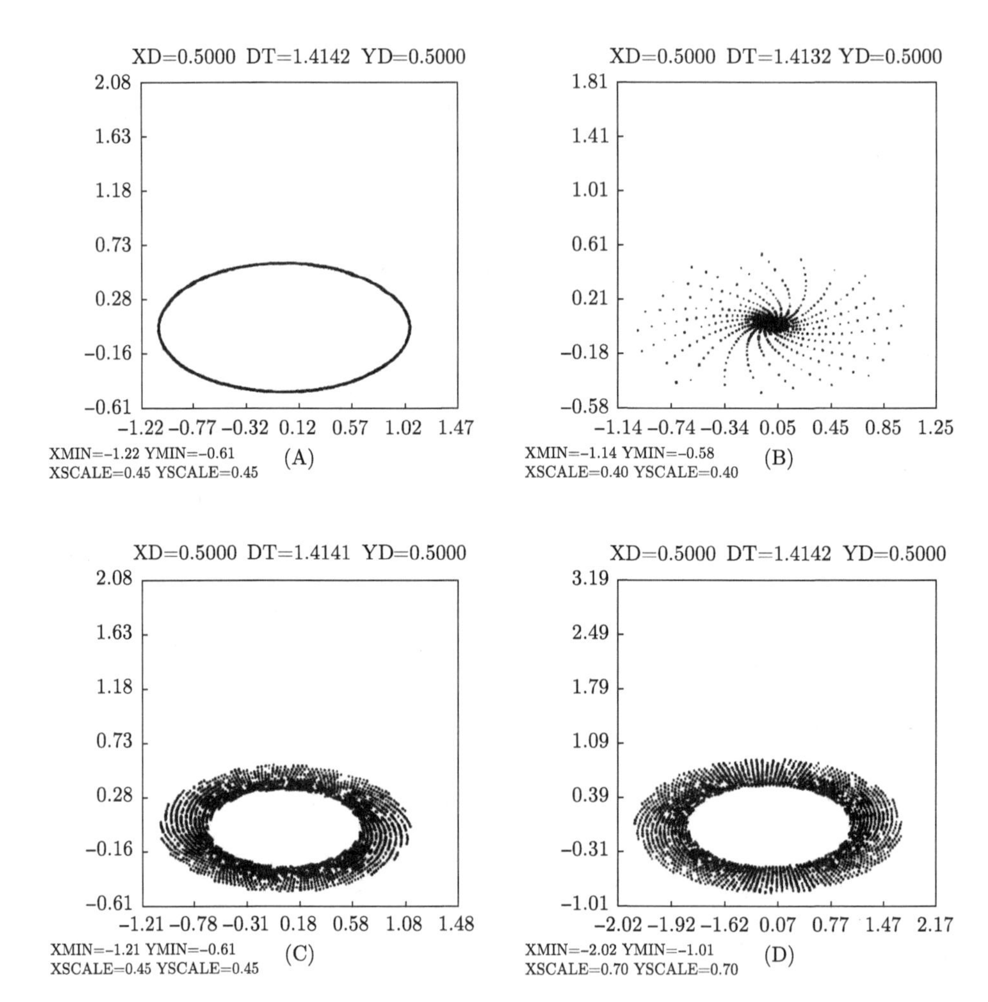

Fig. 3 Harmonic oscillator $H = \left(p^2 + a^2q^2\right)/2, a = 2$. Runge-Kutta method, order 4, nonsymplectic. Stepsize $= s$, total number of steps $= N$, number of steps per plot $= M$. The single orbit is computed in all case with the same initial point. (A) $s = \sqrt{2}, N = 200000, M = 1$; (B) $s = \sqrt{2} - 0.001, N = 2000, M = 1$; (C) $s = \sqrt{2} - 0.00005, N = 2000, M = 1$; (D) $s = \sqrt{2} + 0.00005, N = 2000, M = 1$; Critical $s = \sqrt{2}$ gives a clear-cut invariant ellipse even for the very long run of a 200000 steps. We take $s = \sqrt{2} - \epsilon$; when $\epsilon = 0.001$ the orbit spirals quickly around and towards the origin with a wrong formation of an artificial attractor there. The dissipation effect diminishes with $\epsilon = 0.00005$ but is still pronounced. A negative perturbation $\epsilon = -0.00005$ produces a divergence effect.

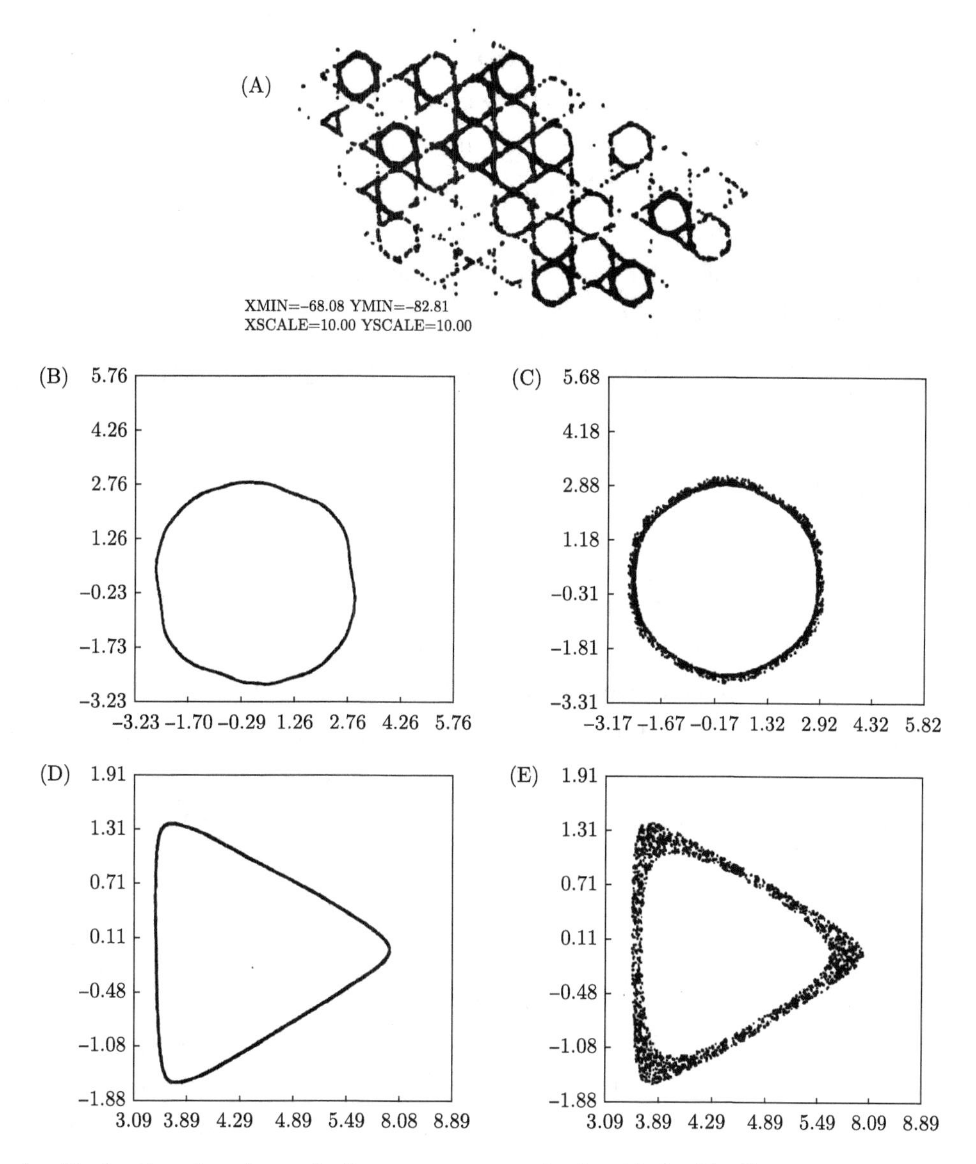

Fig. 4 Stochastic web and symplectic versus non-symplectic orbit for H_3. Stepsize $= s$, total number of steps $= N$, number of steps per plot $= M$. (A) Symplectic, $s = 0.7, N = 50000, M = 5$; $p_0 = 3.14159265, q_0 = 0$; (B) Symplectic, $s = 0.6, N = 1000000, M = 500$; $p_0 = 2.90159265, q_0 = 0$; (C) Runge-Kutta, $s = 0.6, N = 50000, M = 25$; $p_0 = 2.90159265, q_0 = 0$; (D) Symplectic, $s = 0.6, N = 1000000, M = 500; p_0 = 3.17159265, q_0 = 0$; (E) Runge-Kutta, $s = 0.6, N = 20000, M = 10; p_0 = 3.17159265, q_0 = 0$.

4D, E for a triangular cell.

Fig. 5 and Fig. 6 show the symplectic computed phase portraits in the hexagon and the triangle, respectively. At large s, a part of the invariant tori break down into chaotic orbits of periodic archipelagoes and also a part of the invariant tori are preserved, see Fig. 5A, and 6A, C. This is in conformity with the KAM theorem [23]. Preservation of invariant tori at every small step-sizes in the near-center region is also evident from Fig. 5D-5F and 6C-6F. The almost "eternal" confinement to closed orbits in symplectic computation is really striking.

Using symplectic scheme we have carried out an exhaustive numerical analysis for the Toda Hamiltonian, in Fig. 7,

$$H=\frac{1}{2}\left(p_1^2+p_2^2\right)+\frac{1}{24}\left[\exp\left(2q_2+2\sqrt{3}q_1\right)+\exp\left(2q_2-2\sqrt{3}q_1\right)+\exp\left(-4q_2\right)\right]-\frac{1}{8} \tag{3.3}$$

and for the quartic dynamical model with Hamiltonian, in Fig 8,

$$H=\frac{1}{2}\left(p_1^2+p_2^2\right)+q_1^4+q_2^4+2\alpha q_1^2q_2^2. \tag{3.4}$$

For the Toda system we get the result that the trajectories were regular for arbitrary energy $H=E$, with all intersections of a trajectory with the Poincaré section $p_1=0$ falling on smooth invariant curves. For the Toda Hamiltonian we found that, with increasing energy from 1 to 256 the motion remains to be regular. Numerical experiments are in conformity for the integrability of the Toda systems. For the quartic model, $\alpha=-2$ corresponds to the zero-dimensional model of the classical Yang-Mills field,

$$H=\frac{1}{2}\left(p_1^2+p_2^2\right)+q_1^2p_2^2, \tag{3.5}$$

through rotation of $\frac{1}{4}\pi$ in the q-plane. Numerical experiment shows the motion is irregular everywhere, therefore it is in conformity for the non-integrability of Yang-Mills.

In the case $\alpha=2$ numerical experiment shows the existence of irregular motions, which occupy a small fraction of the total area of the surface of the section, the remaining part consists of regular motion. This is also in conformity with the mixed-type integrability property of the system.

The remarkable performances of symplectic algorithms could be understood fully only in the theory of preservation and breakdown of invariant tori in the context of the KAM theorems. From all the numerical and theoretical evidences so far accumulated, one might ascertain that symplectic algorithms provide the solution to the long-standing computational problem of long-time predictions for Hamiltonian dynamics.

Acknowledgements The authors express their gratitude to H. M . Wu, Z. Ge, D. L. Wang, Y. H. Wu and M. Q. Zhang for close and fruitful collaboration, and in particular

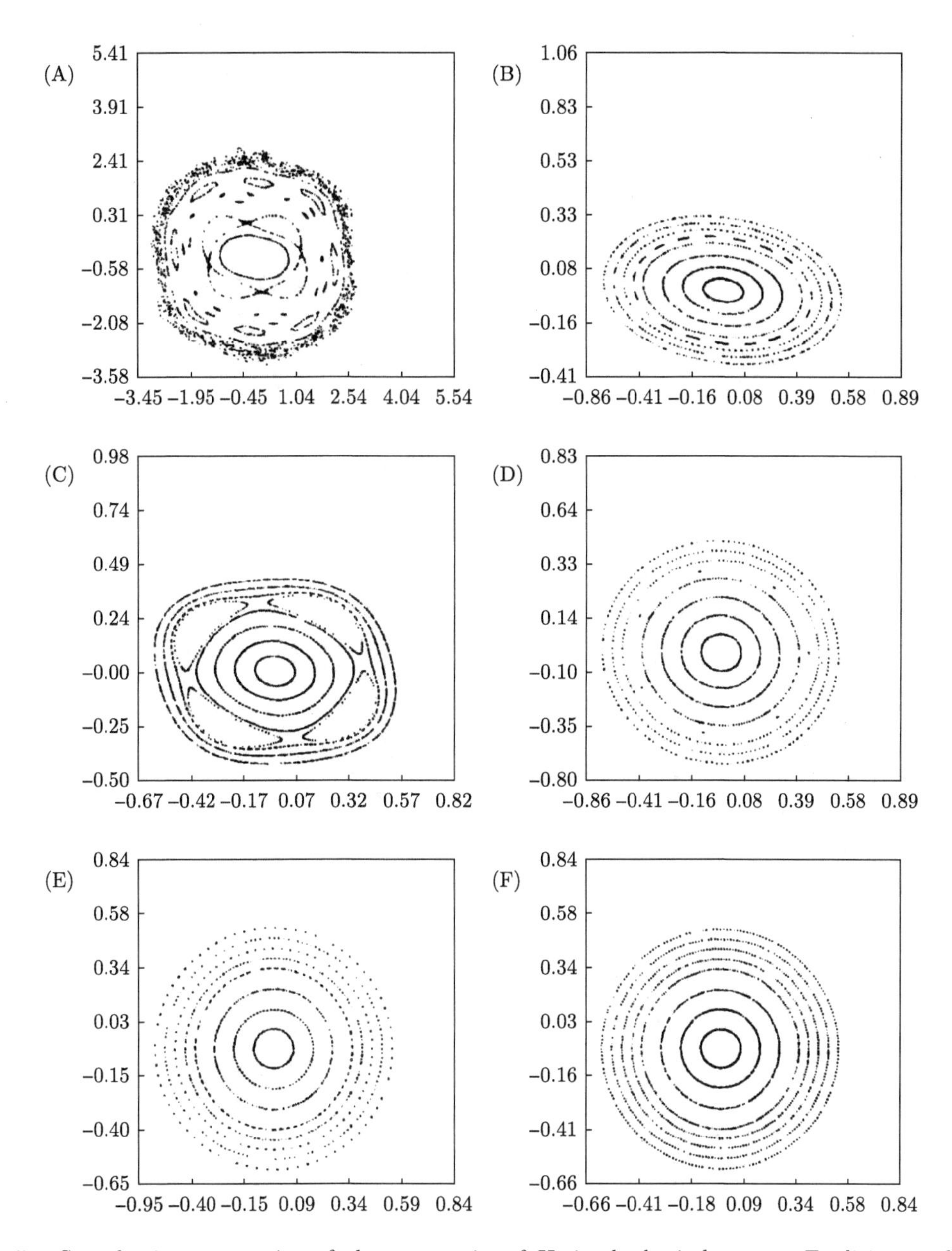

Fig. 5 Symplectic computation of phase portraits of H_3 in the basic hexagon. Explicit symplectic composite scheme, order 2. Stepsize $= s$, total number of steps $= N$, number of steps per plot $= M$. (A) $s = 0.65, N = 250, M = 1$, 8 symplectic orbits near the boundary; (B) $s = 0.6, N = 250, M = 1$; (C) $s = 0.5, N = 400, M = 1$; (D) $s = 0.3, N = 200, M = 1$; (E) $s = 0.03, N = 200, M = 1$; (F) $s = 0.01, N = 220, M = 1$; 8 symplectic orbits near the center; same initial datum are used in (E)-(F).

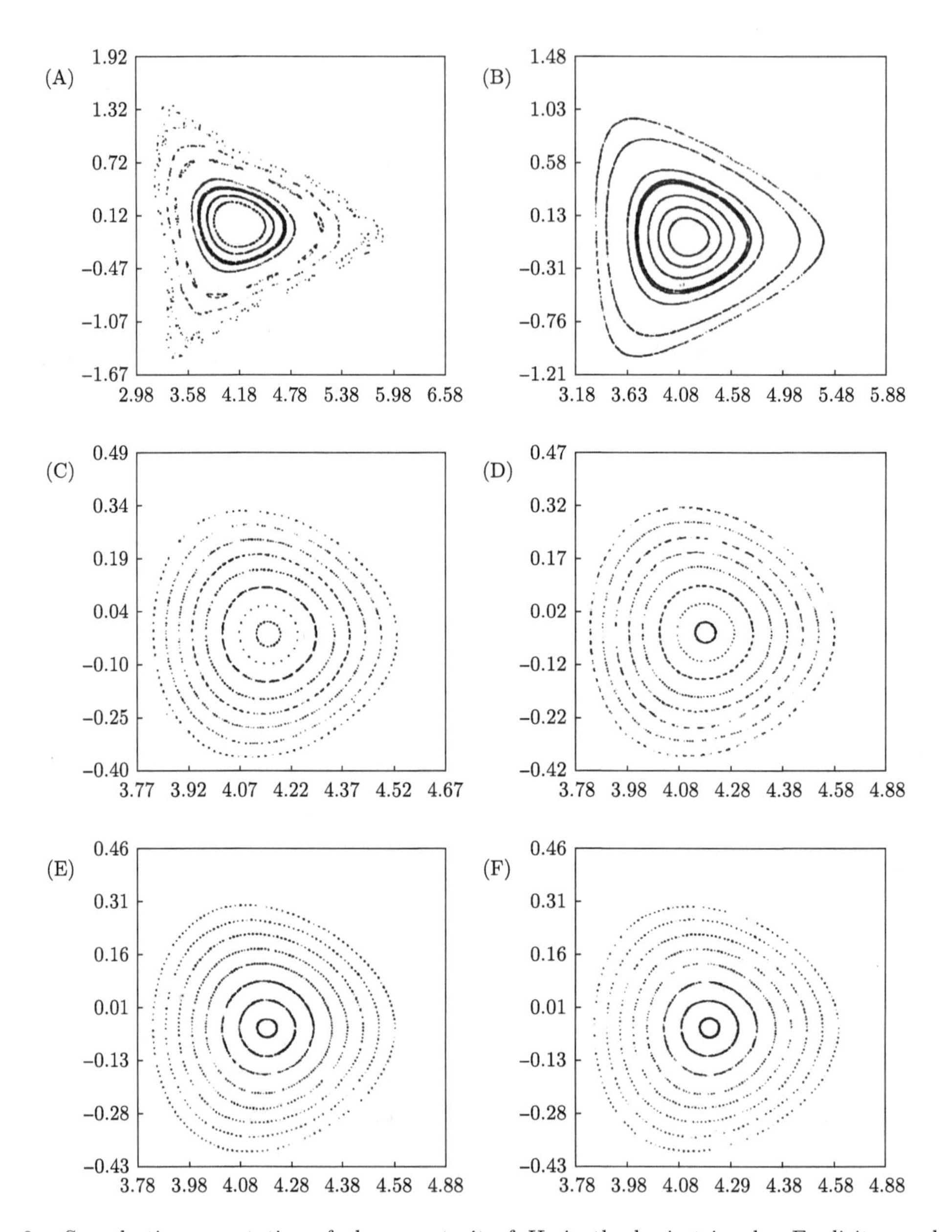

Fig. 6 Symplectic computation of phase portrait of H_3 in the basic triangle. Explicit symplectic composite scheme, order 2. Stepsize = s, total number of steps = N, number of steps per plot = M. (A) $s = 0.6, N = 250, M = 1$; (B) $s = 0.3, N = 250, M = 1$; 8 symplectic orbits near the boundary; same initial data are used. (C) $s = 0.2, N = 200, M = 1$; (D) $s = 0.06, N = 200, M = 1$; (E) $s = 0.03, N = 200, M = 1$; (F) $s = 0.01, N = 200, M = 1$; 8 symplectic orbits near the center; same initial datum are used.

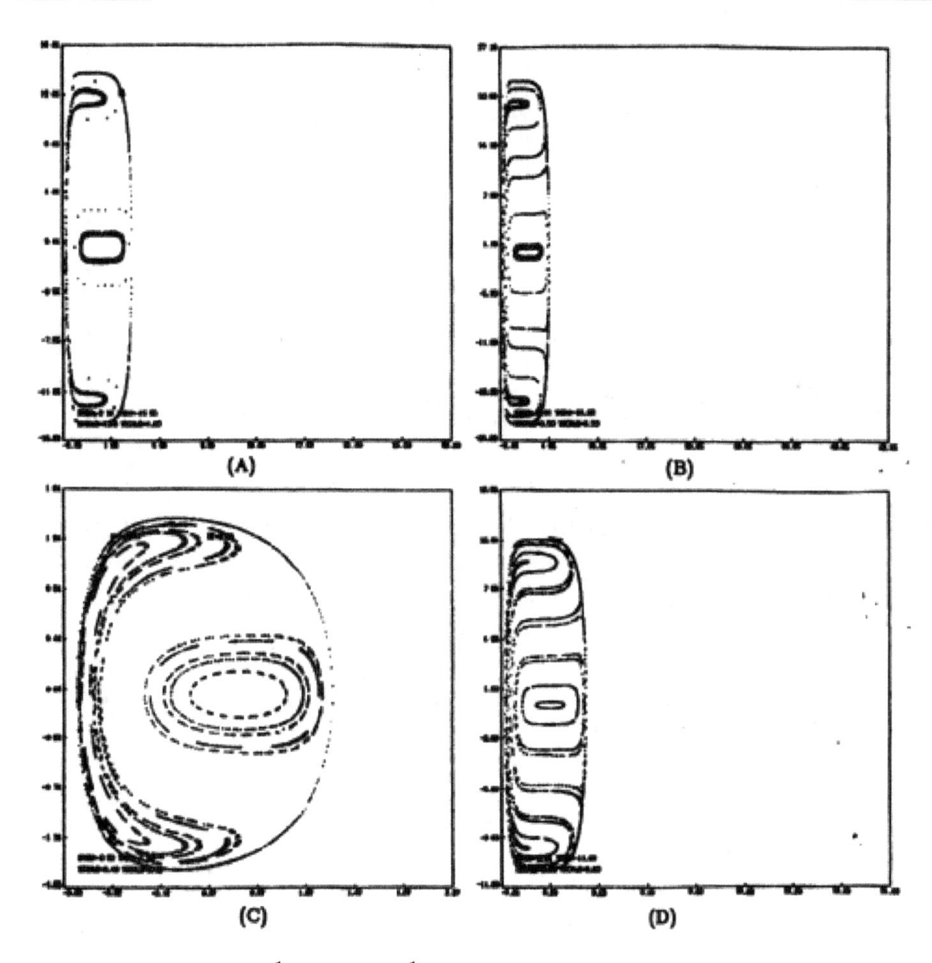

Fig. 7 The Toda lattice $H = \frac{1}{2}(p_1^2+p_2^2)+\frac{1}{24}[\exp(2q_2+2\sqrt{3}q_1)+\exp(2q_2-2\sqrt{3}q_1)+\exp(-4q_2)]-\frac{1}{8}$; Explicit symplectic scheme, order 1. Stepsize $= s$, total number of steps $= N$. (A) $s = 0.005, H = 100, N = 80000$; (B) $s = 0.005, H = 256, N = 80000$; (C) $s = 0.005, H = 1, N = 80000$; (D) $s = 0.005, H = 50, N = 80000$. In all figures we see that all of the surfaces of the section are covered by invariant curves (everywhere regular).

to M. Q. Zhang for discussion and assistance in preparing the manuscript. Thanks are also due to Prof. K. Gustafson and Prof. W. Wyss for the invitation for Feng's participation of the IMACS 1st International Conference on Computational Physics, Boulder, 1990.

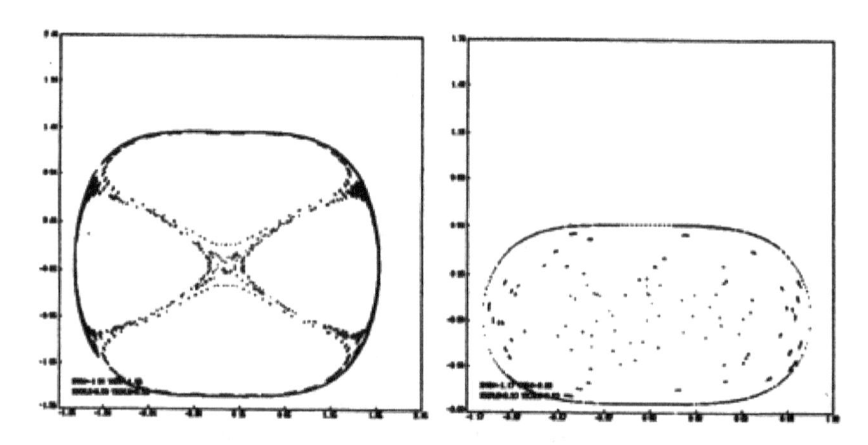

Fig. 8 Quartic dynamical model $H = \frac{1}{2}(p_1^2 + p_2^2) + q_1^4 + q_2^4 + 2\alpha q_1^2 q_2^2$; Explicit symplectic scheme, order 1. Stepsize $= s$, total number of steps $= N$.(A)$s = 0.005$, $N = 80000$, $H = 0.125, \alpha = 2$; (B)$s = 0.005, N = 80000, H = 50, \alpha = -2$; Since the potential part of the Hamiltonian is homogeneous, all energy surfaces $H =$ constant are similar. Therefore, for each α it is sufficient to take only one energy surface. (A) shows the Poincaré section. We see that most of the surface of the section is covered by good invariant curves. On the other hand orbits starting near the unstable axial orbit form a figure "g" stochastic layer. (B) shows the Poincare section. We see that no evidence of regular motion has been found (everywhere chaotic).

References

[1] Feng K. On difference schemes and symplectic geometry. In Feng K, ed. Proc 1984 Beijing Symp Diff Geometry and Diff Equations. Beijing: Science Press, 1985: 42-58

[2] Feng K. Canonical difference schemes for Hamiltonian canonical differential equations. In Proc Intern Workshop on Applied Diff Equations. Singapore: World Scientific, 1986: 59-73

[3] Feng K. Difference schemes for Hamiltonian formalism and symplectic geometry. J Comp Math. 1986 4(3): 279-289

[4] Feng K. Symplectic geometry and numerical methods in fluid dynamics. In Zhuang F G, Zhu Y L, eds. Proc 10,th Intern Conf on Numerical Methods on Fluid Dynamics. Beijing. 1986. Lecture Notes in Physics. 1986 264: 1-7. Berlin: Springer Verlag

[5] Feng K, Qin M Z. The symplectic methods for the computation of Hamiltonian equations. In Zhu Y L, Guo Ben-yu, ed. Proc Conf on Numerical Methods for PDE's. Berlin: Springer, 1987: 1-37. Lect Notes in Math 1297

[6] Feng K, Wu H M, Qin M Z, Wang D L. Construction of canonical difference schemes for Hamiltonian formalism via generating functions. J Comp Math. 1989 7(1): 71-96

[7] Feng K. The Hamiltonian way for computing Hamiltonian dynamics. In Spigler R, ed. Applied and Industrial Mathematics. Netherlands: Kluwer, 1991: 17-35

[8] Feng K, Wu H M, Qin M Z. Symplectic difference schemes for the linear Hamiltonian canonical systems. J Comp Math. 1990 8(4): 371-380

[9] Ge Z, Feng K. On the approximation of Hamiltonian systems. J Comp Math. 1988 6(1): 88-97

[10] Li C W, Qin M Z. Symplectic difference schemes for infinite dimensional Hamiltonian systems. J Comp Math. 1988 6(2): 164-174

[11] Qin M Z. A symplectic difference scheme for the Hamiltonian equation. J Comp Math. 1987 5: 203-209

[12] Qin M Z, Wang D L, Zhang M Q. Explicit symplectic difference schemes for separable Hamiltonian systems. J Comp Math. 1991 9(3): 211-221

[13] Wu Y H. Symplectic transformations and symplectic difference schemes. Chinese J Numer Math Applic. 1990 12(1): 23-31

[14] Wu Y H. The generating function of the solution of ODE and its discrete methods. Computer Math Applic. 1988 15(12): 1041-1050

[15] Zhang M Q, Qin M Z. A note on convergence of symplectic schemes. J Comp Math. 19

[16] Ge Z, Marsden J. Lie-Poisson Hamilton-Jacobi theory and Lie-Poisson integrators. Phys Lett A. 1988 133(3): 137-139

[17] Qin M Z. Canonical difference scheme for the Hamiltonian equation. Math Methods in Appl Science. 1989 11(3): 543-557

[18] Qin M Z, Zhang M Q. Multi-stage symplectic schemes of two kinds of Hamiltonian system for wave equation. Computer Math Applic. 1990 19(10): 51-62

[19] Qin M Z, Zhang M Q. Symplectic Runge-Kutta schemes for Hamiltonian systems. J Comp Math. 1992: 205-215

[20] Qin M Z, Zhang M Q. Explicit Runge-Kutta-like schemers to solve certain quantum operator equations of motion. J Stat Physics. 1990 65(5/6): 839- 844

[21] Feng K, Wang D L. A note on conservation laws of symplectic difference schemes for Hamiltonian systems. J Comp Math. 1991 9(3): 229-237

[22] Wang D L. Poisson difference schemes for Hamiltonian systems on Poisson manifolds. J Comp Math. 1991 9(2): 115-124

[23] Arnold V I. Mathematical Methods of Classical Mechanics. New York: Springer, 1978

[24] Feynman R. Lectures on Physics, Vol 1. Reading: Addison-Wesley. 1963: 148-160

[25] Buneman O. Ideal gas dynamics in Hamiltonian form with benefit for numerical schemes. Phys Fluids. 1980 23: 1716

[26] Ruth R D. A canonical integration technique. IEEE Trans on Nuclear Sciences. 1983 NS-30: 2669-2671

[27] Sanz-Serna J M. Runge-Kutta Schemes for Hamiltonian Systems. BIT. 1988 28: 877-883

[28] Sanz-Serna J M. The numerical integration of Hamiltonian systems. In Proc Conf on Comp Differential Equations. Oxford: Oxford University Press, 1989

[29] Suris Y B. On the preservation of the symplectic structure for numerical integration of Hamiltonian systems. In Fillipov S S, ed. Numerical Solution of ODE's. Moscow: USSR Academy of Sciences, 1988: 148

[30] Lasagni F M. Canonical Runge-Kutta methods. Z Angew Math Phys. 1988 39: 952-953.

[31] Channell P J, Scovel J C. Symplectic Integration of Hamiltonian Systems. Nonlinearity. 1990 3: 213-259

[32] Chierchia L, Gallavotti G. Smooth prime integrals for quasi-integrable Hamiltonian systems. Nuovo Cimento. 1982 B 67: 277-295

[33] Pöschel J. Integrability of Hamiltonian systems on cantor sets. Commun Pure Appl Math. 1982 35: 653-696

[34] Chernikov A A, Sagdeev R Z, Zaslavsky G M. Stochastic webs. Physica D. 1988 33: 65

[35] Arnold V I. Remarks on quasicrystallic symmetries. Physica D. 1988 33: 21-25

[36] Beloshapkin V V, Chernikov A A, Natenzon M Y, Petrovichev B A, Sagdeev R Z, Zaslavsky G M. Chaotic streamlines in preturbulent states. Nature. 1989 337: 133-137

[37] Tang Y F, Feng K. Symplectic computation of geodesic flows on closed surfaces and Kepler motion: [Preprint]. Beijing: Academia Sinica. Computing Center. 1990

13 A Note on Conservation Laws of Symplectic Difference Schemes for Hamiltonian Systems①②

哈密尔顿系统辛差分格式守恒律的注记

Abstract

In this paper we consider the necessary conditions of conservation laws of symplectic difference schemes for Hamiltonian systems and give an example which shows that there does not exist any centered symplectic difference scheme such that it preserves all Hamiltonian energy.

§1 Introduction

It is well known that Hamiltonian systems have many intrinsic properties: preservation of phase areas and phase volume, conservation laws of energy and momenta and etc. In order to remain the first property in numerical solution of Hamiltonian systems, Feng Kang in [1] first introduced a new notion — symplectic difference schemes for Hamiltonian systems and developed, with his colleagues, a systematical method — generating function method — to construct such schemes. This method has been widely deepened and extended [8,10-13]. Meanwhile, symplectic difference schemes constructed in [2,4] preserve a kind of quadratic first integrals of Hamiltonian systems. In particular, any centered symplectic difference scheme preserves all quadratic first integrals of Hamiltonian systems. But generally it can not preserve first integrals other than quadratic form.

In section 2, in order to fit the requirement of the next sections, we review the construction of the symplectic difference schemes for Hamiltonian systems by generating function method developed in [2-4]. In section 3, we give another proof of a theorem in [5] and prove the sufficient condition of the theorem is also necessary for first order symplectic difference schemes. In addition, we give a condition under which a first inte-

---

① Joint with Wang D L

② J Comp Math. 1991 9(3): 229-237

gral of a Hamiltonian system is also a first integral of the centered symplectic difference schemes. In section 4, we give a simple example. It shows that general symplectic difference schemes can not preserve the non-quadratic first integrals, especially, can not preserve the energy of a nonlinear Hamiltonian system.

§2 Review of the Construction of Symplectic Difference Schemes

Let $\mathbf{R}^{2n}$ be a $2n$-dim real space. Its elements are $2n$-dim column vectors $z = (z_1, \cdots, z_n, z_{n+1}, \cdots, z_{2n})^T = (p_1, \cdots, p_n, q_1, \cdots, q_n)^T$. The superscript T stands for the matrix transpose. Let $\mathbf{C}^\infty(\mathbf{R}^{2n})$ be the set of all real smooth functions on $\mathbf{R}^{2n}$. $\forall H \in \mathbf{C}^\infty(\mathbf{R}^{2n}), \nabla H(z) = (H_{z_1}, \cdots, H_{z_{2n}})^T$, the gradient of H. Denote

$$J_{2n} = \begin{bmatrix} 0 & I_n \\ -I_n & 0 \end{bmatrix}, \quad J^{-1} = J^T = -J, \tag{1}$$

where I_n and 0 represent unit and zero matrices respectively. A mapping $z \to \hat{z} = g(z)$ is said to be symplectic if its Jacobian is symplectic, i.e.,

$$g_z^T(z) J g_z(z) = J. \tag{2}$$

Consider the Hamiltonian system

$$\frac{dz}{dt} = J^{-1}\nabla H(z), \qquad z \in \mathbf{R}^{2n}, \tag{3}$$

where $H(z) \in C^\infty(\mathbf{R}^{2n})$ is the Hamiltonian. Its phase flow is denoted by $g^t(z) = g(z,t)$. It is a one-parameter (local) group of symplectic mappings. A function $F(z)$ is the first integral of the Hamiltonian system (3) if and only if their Poisson bracket is equal to zero, i.e.,

$$\{F, H\} = (\nabla F)^T J^{-1} \nabla H = 0. \tag{4}$$

A difference scheme approximating to (3) is called symplectic if its transition from one time-step to the next is a symplectic mapping. [4] has proposed a method, called the generating function method, to construct systematically symplectic difference schemes for the Hamiltonian system (3). We now review the method. The details can be found in [4].

Let

$$\alpha = \begin{bmatrix} -J & J \\ \frac{1}{2}(I+V) & \frac{1}{2}(I-V) \end{bmatrix}, \quad \alpha^{-1} = \begin{bmatrix} \frac{1}{2}J(I+V^T) & I \\ -\frac{1}{2}J(I-V^T) & I \end{bmatrix}, \tag{5}$$

where $V^T J + JV = 0$, i.e., $V \in \mathbf{sp}(2n)$. Then α defines linear transformations

$$\begin{bmatrix} \hat{w} \\ w \end{bmatrix} = \alpha \begin{bmatrix} \hat{z} \\ z \end{bmatrix}, \quad \begin{bmatrix} \hat{z} \\ z \end{bmatrix} = \alpha^{-1} \begin{bmatrix} \hat{w} \\ w \end{bmatrix}, \tag{6}$$

i.e.,

$$\begin{aligned} \hat{w} &= J(z - \hat{z}), & \hat{z} &= w + \frac{1}{2} J\left(I + V^T\right) \hat{w} \\ w &= \frac{1}{2}(\hat{z} + z) + \frac{1}{2} V(\hat{z} - z), & z &= w - \frac{1}{2} J\left(I - V^T\right) \hat{w}. \end{aligned} \tag{7}$$

If $\hat{z} = g(z, t)$ is the phase flow of the Hamiltonian system (3), then the equation

$$w + \frac{1}{2} J(I + V^T)\hat{w} = g(w - \frac{1}{2} J(I - V^T)\hat{w}, t) \tag{8}$$

defines implicitly a time-dependent gradient mapping $w \to \hat{w} = f(w, t)$, i.e., its Jacobian $f_w(w, t) \in \mathbf{Sm}(2n)$ everywhere symplectic. Hence there exists a scalar function, called the generating function, $\phi(w, t)$ such that

$$f(w, t) = \nabla \phi(w, t). \tag{9}$$

This generating function $\phi(w, t)$ satisfies the Hamilton Jacobi equation

$$\frac{\partial}{\partial t} \phi(w, t) = -H(w + A \nabla \phi(w, t)), \tag{10}$$

where $A = \frac{1}{2} J(I + V^T)$. The phase flow $\widehat{z} = g(z, t)$ can be conversely determined by $\phi(w, t)$

$$\widehat{z} - z = -J^{-1} \nabla \phi\left(\frac{1}{2}(\widehat{z} + z) + \frac{1}{2} V(\widehat{z} - z), t\right). \tag{11}$$

Moreover, if $H(z)$ is analytic, then $\phi(w, t)$ can be expanded as a convergent power series in t for sufficiently small $|t|$

$$\phi(w, t) = \sum_{k=1}^{\infty} \phi^{(k)}(w) t^k. \tag{12}$$

Its coefficients $\phi^{(k)}, k \geqslant 1$ can be determined by the following recursive formulas

$$\phi^{(1)}(w) = -H(w), \tag{13}$$

$$k \geqslant 1, \quad \phi^{(k+1)}(w) = -\frac{1}{k+1} \sum_{m=1}^{k} \frac{1}{m!} \sum_{\substack{k_1 + \cdots + k_m = k \\ k_i \geqslant 1}} D^m H(w)\left(A \nabla \phi^{(k_1)}, \cdots, A \nabla \phi^{(k_m)}\right). \tag{14}$$

Therefore in this case, the phase flow $\hat{z} = g(z,t)$ is the solution of the implicit equation

$$\widehat{z} - z = -\sum_{k=1}^{\infty} t^k J^{-1} \nabla \phi^{(k)} \left(\frac{1}{2}(\widehat{z} + z) + \frac{1}{2} V(\widehat{z} - z) \right). \tag{15}$$

Taking its m-th approximant, we then get a symplectic difference scheme with m-th order of accuracy

$$\begin{aligned} z^{k+1} - z^k &= J\nabla\psi^{(m)} \left(\frac{1}{2}\left(z^{k+1} + z^k\right) + \frac{1}{2} V\left(z^{k+1} - z^k\right), \tau \right) \\ &= \sum_{i=1}^{m} \tau^i J \nabla \phi^{(i)} \left(\frac{1}{2}\left(z^{k+1} + z^k\right) + \frac{1}{2} V\left(z^{k+1} - z^k\right) \right), \end{aligned} \tag{16}$$

where $\tau > 0$ is the time-step. When $V = 0, \phi(w,t)$ is odd in t. Hence the symplectic difference scheme (16) is of even order $(m = 2l)$

$$z^{k+1} - z^k = \sum_{i=1}^{l} \tau^{2i} J \nabla \phi^{(2i)} \left(\frac{1}{2}\left(z^{k+1} + z^k\right) \right). \tag{17}$$

§3 On Conservation Laws

Theorem 1 ([5]) *If* $F(z) = \dfrac{1}{2} z^T S z, S \in \mathbf{Sm}(2n)$ *is a quadratic first integral of the Hamiltonian system* (3) *and*

$$V^T S + SV = 0, \tag{18}$$

then F(z) is also the invariant of the symplectic difference scheme (16), *i.e.,*

$$F\left(z^{k+1}\right) = F\left(z^k\right), \quad k \geqslant 0. \tag{19}$$

For $V = 0$, i.e., the case of centered symplectic difference schemes, (18) is always valid. So all centered symplectic difference schemes preserve all quadratic first integrals of the Hamiltonian system (3). This result is first noticed by Ge Zhong, Wu Yu-hua and Wang Dao-liu. For the general case, it is obtained in [5]. Here we give another proof.

Proof of Theorem 1. Since $F(z)$ is the first integral of the system (3),

$$\frac{1}{2} \widehat{z}^T S \hat{z} = \frac{1}{2} z^T S z.$$

It can be rewritten as

$$\frac{1}{2}(\widehat{z} + z)^T S (\widehat{z} - z) = 0. \tag{20}$$

From (18) it follows that

$$\begin{aligned}&\frac{1}{2}(V(\hat{z}-z))^TS(\hat{z}-z)=\frac{1}{2}(\hat{z}-z)^TV^TS(\hat{z}-z)\\=&\frac{1}{4}(\hat{z}-z)^T\left(V^TS+SV\right)(\hat{z}-z)=0,\qquad\forall\hat{z},z\in\mathbf{R}^{2n}.\end{aligned}$$

Combining it with (20), we have

$$\left(\frac{1}{2}(\widehat{z}+z)+\frac{1}{2}V(\widehat{z}-z)\right)^TS(\widehat{z}-z)=0.$$

Using (15), it becomes

$$\left(\frac{1}{2}(\hat{z}+z)+\frac{1}{2}V(\hat{z}-z)\right)^TSJ\sum_{j=1}^{\infty}t^j\nabla\phi^{(j)}\left(\frac{1}{2}(\hat{z}+z)+\frac{1}{2}V(\hat{z}-z)\right)=0.$$

From this we get

$$w^TSJ\nabla\phi^{(j)}(w)=0,\quad\forall j\geqslant1,\forall w\in\mathbf{R}^{2n}.$$

Then when we take $w=\frac{1}{2}\left(z^{k+1}+z^k\right)+\frac{1}{2}V\left(z^{k+1}-z^k\right)$, we have

$$w^TS\left(z^{k+1}-z^k\right)=\sum_{j=1}^{m}\tau^jw^TSJ\nabla\phi^{(j)}(w)=0.$$

Since

$$\begin{aligned}w^TS\left(z^{k+1}-z^k\right)&=\left(\frac{1}{2}\left(z^{k+1}+z^k\right)+\frac{1}{2}V\left(z^{k+1}-z^k\right)\right)^TS\left(z^{k+1}-z^k\right)\\&=\frac{1}{2}\left(z^{k+1}\right)^TSz^{k+1}-\frac{1}{2}\left(z^k\right)^TSz^k,\end{aligned}$$

it leads to that $F(z)$ is the quadratic invariant of the symplectic difference scheme (16).

We now take

$$V=\begin{bmatrix}(1-2\theta)I_n&0\\0&-(1-2\theta)I_n\end{bmatrix}.\tag{21}$$

It can be easily verified that $V^TJ+JV=0$, i.e., $V\in\mathbf{sp}(2n)$. Let $S\in\mathbf{Sm}(2n)$. Denote $S=\begin{bmatrix}A&B\\B^T&D\end{bmatrix}$, $A^T=A,D^T=D$. Then

$$V^TS+SV=2\begin{bmatrix}(1-2\theta)A&0\\0&-(1-2\theta)D\end{bmatrix}.$$

Hence (18) is equivalent to

$$(1-2\theta)A=0, \quad (1-2\theta)D=0.$$

It means that either $\theta=1/2$ or $A=D=0$. When $\theta=1/2$, the scheme (16) is centered. It preserves all quadratic first integrals of the Hamiltonian system (3). When $\theta \neq 1/2$, the symplectic difference scheme (16) only preserves the quadratic first integrals with the form $p^T Bq$ of the Hamiltonian system (3).

Theorem 1 for the first order symplectic difference scheme is invertible. Precisely speaking, we have the following theorem.

Theorem 2 *Let* $F(z)=\dfrac{1}{2}z^T Sz$, $S \in \mathbf{Sm}(2n)$ *be a quadratic first integral of the Hamiltonian system* (3) *and in some neighborhood of* $\mathbf{R}^{2n}$ *the Hessian of* H, H_{zz}, *non-degenerate. If* $F(z)$ *is also an invariant of the first order symplectic difference scheme*

$$z^{k+1}-z^k=\tau J^{-1}\nabla H\left(\frac{1}{2}\left(z^{k+1}+z^k\right)+\frac{1}{2}V\left(z^{k+1}-z^k\right)\right), \tag{22}$$

i.e., $F\left(z^{k+1}\right)=F\left(z^k\right)$, $k \geqslant 0$, *then*

$$V^T S+SV=0. \tag{23}$$

Proof. By assumption, F is the first integral of (3), then

$$\{F, H\}=z^T SJ^{-1}\nabla H(z)=0, \quad \forall z \in \mathbf{R}^{2n}.$$

Hence

$$\left(\frac{1}{2}\left(z^{k+1}+z^k\right)+\frac{1}{2}V\left(z^{k+1}-z^k\right)\right)^T S\left(z^{k+1}-z^k\right)=\tau w^T SJ^{-1}\nabla H(w)=0, \tag{24}$$

where $w=\frac{1}{2}\left(z^{k+1}+z^k\right)+\frac{1}{2}V\left(z^{k+1}-z^k\right)$. By hypothesis, $F(z)$ is also the invariant of (22). It means

$$\frac{1}{2}\left(z^{k+1}\right)^T Sz^{k+1}=\frac{1}{2}\left(z^k\right)^T Sz^k,$$

i.e.,

$$\frac{1}{2}\left(z^{k+1}+z^k\right)^T S\left(z^{k+1}-z^k\right)=0. \tag{25}$$

Combining (24) and (25), we get

$$\frac{1}{2}\left(z^{k+1}-z^k\right)^T V^T S\left(z^{k+1}-z^k\right)=\frac{1}{4}\left(z^{k+1}-z^k\right)^T\left(V^T S+SV\right)\left(z^{k+1}-z^k\right)=0$$

Because H_{zz} is non-degenerate in some neighborhood of $\mathbf{R}^{2n}$, it leads to $z^{k+1} \neq z^k$ for sufficiently small τ. Hence we get the conclusion (23).

Lemma 3 *Let $V \in \mathbf{sp}(2n)$. Then*

$$V^T S + SV = 0, \quad \forall S \in \mathbf{Sm}(2n) \cap \mathbf{GL}(2n)$$

if and only if V=0.

Theorem 4 *There exists none of the first order symplectic difference scheme with the form* (22) *which preserves all Hamiltonian energy.*

It follows from Theorem 2, Lemma 3 and Theorem 6 in the next section.

We now consider the conservation properties of centered symplectic difference schemes (17). In this case, $V = 0$. Thus (7) becomes

$$\widehat{z} = w + \frac{1}{2}J\widehat{w}, \quad z = w - \frac{1}{2}J\widehat{w}.$$

Set

$$\begin{aligned} u_k =& \frac{1}{2}J\nabla\phi^{(k)}(w), \quad k \geqslant 1, \\ u =& \frac{1}{2}J\nabla\phi(w,t) = \sum_{k=1}^{\infty}\frac{1}{2}J\nabla\phi^{(k)}(w)t^k = \sum_{k=1}^{\infty}u_k t^k. \\ \tilde{u} =& \frac{1}{2}J\nabla\psi(w,\tau) = \sum_{k=1}^{m}\frac{1}{2}J\nabla\phi^{(k)}(w)\tau^k = \sum_{k=1}^{m}u_k t^k. \end{aligned}$$

Since the generating function $\phi(w,t)$ is odd in $t, u_{2k} = 0$.

Lemma 5 *F(z) is the first integral of the Hamiltonian system* (3) *if and only if*

$$\sum_{j=1,odd}^{k}\frac{1}{j!}\sum_{\substack{k_1+\cdots+k_j=k\\ k_i\geqslant 1,odd}} D^jF(w)\left(u_{k_1},\cdots,u_{k_j}\right) = 0, \quad \forall \text{ odd } k. \tag{26}$$

F(z) is preserved by the centered symplectic difference scheme (17) *with m-th order of accuracy if and only if*

$$\sum_{j=1,odd}^{k}\frac{1}{j!}\sum_{\substack{k_1+\cdots+k_j=k\\ 1\leqslant k_i\leqslant m,odd}} D^jF(w)\left(u_{k_1},\cdots,u_{k_j}\right) = 0, \quad \forall \text{ odd } k. \tag{27}$$

Proof. Suppose that $F(z)$ is a first integral of the Hamiltonian system (3). Then

$$F(\hat{z}) = F(z),$$

i.e.,

$$F(w+u) = F(w-u). \tag{28}$$

Expanding the left and right hand sides of the equation above, we get

$$F(w+u)=F(w)+\sum_{k=1}^{\infty}t^k\sum_{j=1}^{k}\frac{1}{j!}\sum_{\substack{k_1+\cdots+k_j=k\\ k_i\geqslant 1,odd}}D^jF(w)(u_{k_1},\cdots,u_{k_j}),$$

$$F(w-u)=F(w)+\sum_{k=1}^{\infty}t^k\sum_{j=1}^{k}\frac{(-1)^j}{j!}\sum_{\substack{k_1+\cdots+k_j=k\\ k_i\geqslant 1,odd}}D^jF(w)(u_{k_1},\cdots,u_{k_j}).$$

Hence (28) is equivalent to

$$\sum_{j=1,odd}^{k}\frac{1}{j!}\sum_{\substack{k_1+\cdots+k_j=k\\ k_i\geqslant 1,odd}}D^jF(w)\left(u_{k_1},\cdots,u_{k_j}\right)=0,\quad \forall k\geqslant 1. \tag{29}$$

Similarly, the symplectic difference scheme (17) preserves $F(z)$ if and only if

$$F(w+\tilde{u})=F(w-\tilde{u}). \tag{30}$$

It is equivalent to

$$\sum_{j=1,\text{ odd}}^{k}\frac{1}{j!}\sum_{\substack{k_1+\cdots+k_j=k\\ 1\leqslant k_i\leqslant m,odd}}D^jF(w)\left(u_{k_1},\cdots,u_{k_j}\right)=0,\quad \forall k\geqslant 1. \tag{31}$$

As k is even, the second summation of (29) or (31) is empty. So for even k, (29) and (31) are always valid.

When $F(z)$ is a quadratic form, $D^mF=0, m\geqslant 3$. In this case, (26) and (27) become respectively

$$DF\left(u_k\right)=0,\quad \forall k\geqslant 1, \tag{32}$$

and

$$DF\left(u_k\right)=0,\quad \forall m\geqslant k\geqslant 1. \tag{33}$$

Of course, (32) implies (33). We thus obtain the conclusion again: all centered symplectic difference schemes preserve all quadratic first integrals of the Hamiltonian system (3).

§4 An Example

We now give an example. It shows that in general, the first integral other than quadratic form of Hamiltonian systems can not be preserved by symplectic difference schemes.

Theorem 6 *Let $n=1, H(p,q)=p^2q$ be the Hamiltonian. Then H can not be preserved by any centered symplectic difference scheme* (17).

Lemma 7 *Let H be as above. Then $\phi^{(2k-1)}(w), k \geqslant 1$, determined by* (13) *and* (14), *have the expression*

$$\phi^{(2k-1)}(w)=C_k p^{2k} q, \quad k \geqslant 1, \tag{34}$$

where $C_k=(-1)^k\left|C_k\right|, k \geqslant 1$ are determined by the following recursive formula

$$C_1=-1, \quad C_k=\frac{(-1)^k}{4} \sum_{j=1}^{k-1}\left|C_j\right|\left|C_{k-j}\right|, \quad k \geqslant 1. \tag{35}$$

Proof. By induction with respect to k.

For $k=1$, by (13), $\phi^{(1)}(w)=-H(p,q)=-p^2 q$, (34) and (35) are valid. Suppose for $k-1, k-2, \cdots$, (34) and (35) are also valid. Then since it is a polynomial of degree 3, $D^m H=0, m \geqslant 4$. Using the notation above, $u_{2j}=0$,

$$u_{2j-1}=\frac{1}{2} J \nabla \phi^{(2j-1)}=\frac{1}{2} C_j\left[\begin{array}{c}p^{2j} \\ -2jp^{2j-1}q\end{array}\right], \quad j=1, \cdots, k.$$

$$D^2 H\left(u_{2j-1}, u_{2(k-j+1)-1}\right)=\left(u_{2j-1}\right)^T H_{zz} u_{2(k+1-j)-1}=-\frac{1}{2}(2k+1) C_j C_{k-j+1} p^{2(k+1)} q.$$

Hence by (14),

$$\begin{aligned}
\phi^{(2k+1)} & =-\frac{1}{2k+1} \sum_{m=1}^{2k} \frac{1}{m!} \sum_{\substack{k_1+\cdots+k_m=2k \\ k_i \geqslant 1, odd}} D^m H(w)\left(u_{k_1}, \cdots, u_{k_m}\right) \\
& =-\frac{1}{2(2k+1)} \sum_{j+i=k+1} D^2 H\left(u_{2j-1}, u_{2i-1}\right) \\
& =-\frac{1}{2(2k+1)} \sum_{j=1}^k D^2 H\left(u_{2j-1}, u_{2(k+1-j)-1}\right) \\
& =\frac{1}{4} \sum_{j=1}^k C_j C_{k+1-j} p^{2(k+1)} q. \\
C_{k+1} & =\frac{1}{4} \sum_{j=1}^k C_j C_{k+1-j}=\frac{1}{4} \sum_{j=1}^k(-1)^j\left|C_j\right|(-1)^{k+1-j}\left|C_{k+1-j}\right| \\
& =\frac{(-1)^{k+1}}{4} \sum_{j=1}^k\left|C_j\right|\left|C_{k+1-j}\right| .
\end{aligned}$$

Proof of Theorem 6. Since H is the Hamiltonian, it is of course the first integral of the Hamiltonian system (3). By Lemma 5, we have

$$DH\left(u_k\right)+\frac{1}{3!} \sum_{\substack{k_1+k_2+k_3=k \\ k_i \geqslant 1, odd}} D^3 H(w)\left(u_{k_1}, u_{k_2}, u_{k_3}\right)=0, \quad \forall \text { odd } k. \tag{36}$$

But H is preserved by the centered symplectic difference scheme (17) if and only if

$$DH(u_k) + \frac{1}{3!} \sum_{\substack{k_1+k_2+k_3=k \\ 1\leqslant k_i \leqslant m, odd}} D^3H(w)(u_{k_1}, u_{k_2}, u_{k_3}) = 0, \quad 1 \leqslant k \leqslant m, odd, \tag{37}$$

$$\frac{1}{3!} \sum_{\substack{k_1+k_2+k_3=k \\ 1\leqslant k_i \leqslant m, odd}} D^3H(w)(u_{k_1}, u_{k_2}, u_{k_3}) = 0, \quad k \geqslant l, odd. \tag{38}$$

Consider the term of $k = m + 1 = 2l + 1$. By Lemma 7,

$$\begin{aligned} DH(u_{2l+1}) &= \frac{1}{2} C_{l+1} (2pq, p^2) \begin{pmatrix} p^{2l+2} \\ -2(l+1)p^{2l+1}q \end{pmatrix} \\ &= -lC_{l+1} p^{2l+3} q \neq 0. \end{aligned}$$

Hence (38) is not valid for $k = m + 1$. It implies that the centered symplectic difference scheme (17) does not preserve $H(p, q)$.

References

[1] Feng K. On difference schemes and symplectic geometry. In Feng K, ed. Proc 1984 Beijing Symp Diff Geometry and Diff Equations. Beijing: Science Press, 1985. 42-58

[2] Feng K. Difference schemes for Hamiltonian formalism and symplectic geometry. J Comp Math. 1986 4(3): 279-289

[3] Feng K, Qin M Z. The symplectic methods for computation of Hamiltonian equations. In Zhu Y L, Guo Ben-yu, ed. Proc Conf on Numerical Methods for PDE's. Berlin: Springer, 1987. 1-37. Lect Notes in Math 1297

[4] Feng K, Wu H M, Qin M Z, Wang D L. Construction of canonical difference schemes for Hamiltonian formalism via generating functions. J Comp Math. 1989 7(1): 71-96

[5] Feng K, Wang D L, Ge Z, Li C W. Calculus of generating functions: [Preprint]. Beijing: CAS. Computing Center, 1987

[6] Feng K, Wu H M, Qin M Z. Symplectic difference schemes for the linear Hamiltonian canonical systems. J Comp Math. 1990 8(4): 71-96

[7] Ge Z, Feng K. On the approximation of Hamiltonian systems. J Comp Math. 1988 6(1): 88-97

[8] Li C W, Qin M Z. Symplectic difference schemes for infinite dimensional Hamiltonian systems. J Comp Math. 1988 6(2): 164-174

[9] Qin M Z, Wang D L, Zhang M Q. Explicit symplectic difference schemes for separable Hamiltonian systems. J Comp Math. 1991 9(3): 211-221

[10] Wang D L. Poisson difference schemes for Hamiltonian systems on Poisson manifolds. J Comp Math. 1991 9(2): 115-124

[11] Wu Y H. The generating function of the solution of ODE and its discrete methods. Computer Math Applic. 1988 15(12): 1041-1050

14 Symplectic Difference Schemes for Hamiltonian Systems in General Symplectic Structure①②

一般辛结构下的哈密尔顿系统的辛差分格式

Abstract

We consider the construction of phase flow generating functions and symplectic difference schemes for Hamiltonian systems in general symplectic structure with variable coefficients.

§1 Introduction

The standard symplectic structure w on $\mathbf{R}^{2n}$ is of the form

$$w = \sum_{i<j} J_{ij} dz_i \wedge dz_j, \quad J = \begin{bmatrix} 0 & I_n \\ -I_n & 0 \end{bmatrix}. \tag{1}$$

On such symplectic manifold, the Hamiltonian system has the simplest form

$$\frac{dz}{dt} = J^{-1} \Delta H(z) \tag{2}$$

with the Hamiltonian $H(z)$. The phase flow of the Hamiltonian system (2) preserves the symplectic structure (1). Therefore it preserves phase areas and the phase volume of the phase space. Feng Kang et al. in [2-4] developed a generating function method to construct systematically symplectic difference schemes with arbitrary order of accuracy to approximate the system (2). The transition of such difference schemes from one time-step to the next is a symplectic mapping. So they preserve the symplectic structure. It leads to the preservations of phase areas and the phase volume of the phase space.

Generally, a general symplectic structure on $\mathbf{R}^{2n}$ with variable coefficients

$$w = \sum_{i<j} K_{ij}(z) dz_i \wedge dz_j, \tag{3}$$

① Joint with Wang D L

② J Comp Math. 1991 9(1): 86-96

which is a non-degenerate, closed 2-form. In this case, the Hamiltonian system of the form

$$\frac{dz}{dt}=K^{-1}(z)\nabla H(z),\quad K_{ji}=-K_{ij},\tag{4}$$

where $H(z)$ is the Hamiltonian. As above, the phase flow of the system (4) preserves the symplectic structure (3). In numerical simulation for (4), usual discretization can not preserve the symplectic structure (3). By Darboux's theorem, we can, of course, transform (4) into (2) and then use the method in [2–4]. But (i) it is difficult to find out the transformation, and (ii) it is also interesting to discretize (4) directly such that the transition preserves the symplectic structure (3).

In this paper, we try to construct symplectic difference schemes in same way as in [4]. In [4], the key point is to introduce a linear transformation from the symplectic manifold $(\mathbf{R}^{4n},\tilde{J}_{4n})$ into the symplectic manifold $(\mathbf{R}^{4n},J_{4n})$ which transforms the $\tilde{J}_{4n}$-Lagrangian submanifold into J_{4n}-Lagrangian submanifold. Of course the inverse transformation transforms the J_{4n}-Lagrangian submanifolds into the $\tilde{J}_{4n}$-Lagrangian submanifolds. In fact, we can also take nonlinear transformations for the same purpose. This was first noted by Feng Kang and has been used in [5]. For other related developments, see [6–15]. In this paper, we use nonlinear transformations to reach our purpose.

In Section 2, we give the relationship between the $K(z)$-symplectic mappings and the gradient mappings. In Section 3, we consider the generating functions of the phase flow of the Hamiltonian system (4) and the corresponding Hamilton-Jacobi equation. When the Hamiltonian function is analytic, then the generating function can be expanded as a power series in t and its coefficients can be recursively determined. With the aid of such an expression, in Section 4 we give a systematic method to construct $K(z)$-symplectic difference schemes with arbitrary order of accuracy.

We shall only consider the local case throughout the paper.

§2 Generating Functions for $K(z)$-Symplectic Mappings

Let $\mathbf{R}^{2n}$ be a $2n$-dimensional real space. The elements of $\mathbf{R}^{2n}$ are $2n$-dimensional column vectors $z=(z_1,\cdots,z_n,z_{n+1},\cdots,z_{2n})^T$. The superscript T represents the matrix transpose.

A symplectic form w on $\mathbf{R}^{2n}$ is a non-degenerate, closed 2-form, defined by

$$w=\frac{1}{2}\sum_{i,j=1}^{2n}K_{ij}(z)dz_i\wedge dz_j.\tag{5}$$

Denote the entries of $K(z)$ by $K_{ij}(z),i,j=1,\cdots,2n$. Then $K(z)$ is anti-symmetric. The non-degeneracy and closedness of w imply that det $K(z)\neq 0$ and $K_{ij}(z)$ is subject to the condition

$$\frac{\partial K_{ij}(z)}{\partial z_l} + \frac{\partial K_{jl}(z)}{\partial z_i} + \frac{\partial K_{li}(z)}{\partial z_j} = 0, \quad i,j,l = 1,\cdots,2n. \tag{6}$$

From now on, we always identify the symplectic form w with $K(z)$.

Denote

$$J_{4n} = \begin{bmatrix} 0 & I_{2n} \\ -I_{2n} & 0 \end{bmatrix}, \quad \tilde{K}(\hat{z}, z) = \begin{bmatrix} K(\hat{z}) & 0 \\ 0 & -K(z) \end{bmatrix}.$$

Evidently they define two symplectic structures on $\mathbf{R}^{4n}$:

$$\Omega = \sum_{i=1}^{2n} dw_i \wedge dw_{2n+i}, \quad \tilde{\Omega} = \frac{1}{2}\sum_{i,j=1}^{2n} \left(k_{ij}(\hat{z}) d\hat{z}_i \wedge d\hat{z}_j - k_{ij}(z) dz_i \wedge dz_j\right).$$

A $2n$-dimensional submanifold $L \subset \mathbf{R}^{4n}$ is a J_{4n}-and $\tilde{K}(\hat{z}, z)$-Lagrangian submanifold if $i_L^*\Omega = 0$ and $i_L^*\tilde{\Omega} = 0$, where $i_L : L \to \mathbf{R}^{4n}$ is the inclusion. Suppose that in local coordinates L has the expression

$$L = \left\{ \begin{pmatrix} \hat{z} \\ z \end{pmatrix} \in \mathbf{R}^{4n} | z = z(x), \hat{z} = \hat{z}(x), x \in U \subset \mathbf{R}^{2n}, \text{ open set} \right\}.$$

Then L is a J_{4n}-Lagrangian submanifold if and only if

$$(T_x L)^T J_{4n} T_x L = 0,$$

i.e.,

$$\left(\hat{z}_x^T(x), z_x^T(x)\right) \begin{bmatrix} 0 & I_{2n} \\ -I_{2n} & 0 \end{bmatrix} \begin{pmatrix} \hat{z}_x(x) \\ z_x(x) \end{pmatrix} = \hat{z}_x^T(x) z_x(x) - z_x^T(x) \hat{z}_x(x) = 0,$$

where $T_x L$ is the tangent space to L at x. L is a $\tilde{K}(\hat{z}, z)$-Lagrangian submanifold if and only if

$$(T_x L)^T \tilde{K}(\hat{z}(x), z(x)) T_x L = 0,$$

i.e.,

$$\begin{aligned} &\left(\hat{z}_x^T(x), z_x^T(x)\right) \begin{bmatrix} K(\hat{z}(x)) & 0 \\ 0 & -K(z(x)) \end{bmatrix} \begin{pmatrix} \hat{z}_x(x) \\ z_x(x) \end{pmatrix} \\ &= \hat{z}_x^T(x) K(\hat{z}(x)) \hat{z}_x(x) - z_x^T(x) K(z(x)) z_x(x) = 0. \end{aligned}$$

A smooth mapping $g : z \to \hat{z} = g(z)$ from $\mathbf{R}^{2n}$ to itself is called $K(z)$-symplectic if its Jacobian $M(z) = g_z(z)$ satisfies

$$M^T(z) K(g(z)) M(z) = K(z). \tag{7}$$

Therefore its graph

$$\Gamma_g = \left\{ \begin{bmatrix} \hat{z} \\ z \end{bmatrix} \in \mathbf{R}^{4n} | \hat{z} = g(z), z \in \mathbf{R}^{2n} \right\}$$

is a $\tilde{K}(\hat{z}, z)$-Lagrangian submanifold for

$$\begin{aligned}(T_z\Gamma_g)^T \tilde{K}(g(z), z) T_z\Gamma_g &= \left(M^T(z), I\right) \begin{bmatrix} K(g(z)) & 0 \\ 0 & -K(z) \end{bmatrix} \begin{pmatrix} M(z) \\ I \end{pmatrix} \\ &= M^T(z)K(g(z))M(z) - K(z) = 0.\end{aligned}$$

Similarly, let $w \to \hat{w} = f(w)$ be a gradient mapping from $\mathbf{R}^{2n}$ to itself. Then the graph of f

$$\Gamma_f = \left\{ \begin{bmatrix} \hat{w} \\ w \end{bmatrix} \in \mathbf{R}^{4n} | \hat{w} = f(w), \quad w \in \mathbf{R}^{2n} \right\}$$

is a J_{4n}-Lagrangian submanifold[4].

Define nonlinear transformation from $\mathbf{R}^{4n}$ to itself

$$\alpha : \begin{pmatrix} \hat{z} \\ z \end{pmatrix} \to \begin{pmatrix} \hat{w} \\ w \end{pmatrix} = \alpha \begin{pmatrix} \hat{z} \\ z \end{pmatrix}, \quad \alpha^{-1} : \begin{pmatrix} \hat{w} \\ w \end{pmatrix} \to \begin{pmatrix} \hat{z} \\ z \end{pmatrix} = \alpha^{-1} \begin{pmatrix} \hat{w} \\ w \end{pmatrix} \tag{8}$$

i.e.,

$$\begin{aligned} \hat{w} = \alpha_1(\hat{z}, z), \quad \hat{z} = \alpha^1(\hat{w}, w), \\ w = \alpha_2(\hat{z}, z), \quad z = \alpha^2(\hat{w}, w). \end{aligned} \tag{9}$$

Denote

$$\alpha_*(\hat{z}, z) = \begin{bmatrix} \dfrac{\partial \hat{w}}{\partial \hat{z}} & \dfrac{\partial \hat{w}}{\partial z} \\ \dfrac{\partial w}{\partial \hat{z}} & \dfrac{\partial w}{\partial z} \end{bmatrix} = \begin{bmatrix} A_\alpha & B_\alpha \\ C_\alpha & D_\alpha \end{bmatrix},$$

$$\alpha_*^{-1}(\hat{w}, w) = \begin{bmatrix} \dfrac{\partial \hat{z}}{\partial \hat{w}} & \dfrac{\partial \hat{z}}{\partial w} \\ \dfrac{\partial z}{\partial \hat{w}} & \dfrac{\partial z}{\partial w} \end{bmatrix} = \begin{bmatrix} A^\alpha & B^\alpha \\ C^\alpha & D^\alpha \end{bmatrix},$$

where α_* is the Jacobian of α.

Let α be a diffeomorphism from $\mathbf{R}^{4n}$ to itself and satisfy the condition

$$\alpha_*^T J_{4n} \alpha_* = \tilde{K}(\hat{z}, z), \tag{10}$$

i.e.,

$$\begin{bmatrix} A_\alpha & B_\alpha \\ C_\alpha & D_\alpha \end{bmatrix}^T \begin{bmatrix} 0 & I_{2n} \\ -I_{2n} & 0 \end{bmatrix} \begin{bmatrix} A_\alpha & B_\alpha \\ C_\alpha & D_\alpha \end{bmatrix} = \begin{bmatrix} K(\hat{z}) & 0 \\ 0 & -K(z) \end{bmatrix}. \tag{11}$$

It follows from (11) that

$$\begin{aligned} A^\alpha &= -K^{-1}(\hat{z})C_\alpha^T, \quad B^\alpha = K^{-1}(\hat{z})A_\alpha^T, \\ C^\alpha &= K^{-1}(z)D_\alpha^T, \quad D^\alpha = -K^{-1}(z)B_\alpha^T. \end{aligned} \tag{12}$$

By Darboux's theorem, such diffeomorphism exists, at least locally.

Theorem 1 *Let α be a diffeomorphism of $\mathbf{R}^{4n}$ satisfying (10) as above. Then α carries every $\tilde{K}$-Lagrangian submanifold into a J_{4n}-Lagrangian submanifold and conversely α^{-1} carries every J_{4n}-Lagrangian submanifold into a $\tilde{K}$-Lagrangian submanifold.*

Proof. Let L be a $\tilde{K}$-Lagrangian submanifold. The image of L under α is $\alpha(L)$. Its tangent space is

$$T_*(\alpha(L)) = \alpha_* \cdot T_*L.$$

So by (10),

$$T_*(\alpha(L))^T J_{4n} T_*(\alpha(L)) = (T_*L)^T \alpha_*^T J_{4n} \alpha_* T_*L = (T_*L)^T \tilde{K} T_*L = 0.$$

The converse is similar.

We now introduce a lemma in [4].

Lemma 2 *Let* $\begin{bmatrix} A & B \\ C & D \end{bmatrix} \in \mathrm{GL}(4n)$. *Denote* $\begin{bmatrix} A & B \\ C & D \end{bmatrix}^{-1} = \begin{bmatrix} A_1 & B_1 \\ C_1 & D_1 \end{bmatrix}$. *Define a linear fractional transformation as follows.*

$$\sigma: \begin{aligned} &\mathbf{M}(2n) \to \mathbf{M}(2n), \\ &M \to N = \sigma(M) = (AM+B)(CM+D)^{-1} \end{aligned} \tag{13}$$

under the transversality condition

$$|CM+D| \neq 0.$$

Then the following four conditions are equivalent mutually :

$$|CM+D| \neq 0, \tag{14}$$

$$|MC_1 - A_1| \neq 0, \tag{15}$$

$$|C_1N + D_1| \neq 0, \tag{16}$$

$$|NC - A| \neq 0. \tag{17}$$

The linear fractional transformation defined by (13) *can be represented as*

$$N = \sigma(M) = (MC_1 - A_1)^{-1}(B_1 - MD_1). \tag{18}$$

In this case, the inverse transformation must exist and can be represented as

$$M = \sigma^{-1}(N) = \begin{cases} (A_1N+B_1)(C_1N+D_1)^{-1}, & (19) \\ (NC-A)^{-1}(B-ND). & (20) \end{cases}$$

Theorem 3 *Let α be defined as above. Let $z \to \hat{z} = g(z)$ be a $K(z)$-symplectic mapping in (some neighborhood of) $\mathbf{R}^{2n}$ with Jacobian $g_z(z) = M(z)$. If (in some neighborhood of $\mathbf{R}^{2n}$) M satisfies the transversality condition*

$$|C_\alpha(g(z), z)M(z) + D_\alpha(g(z), z)| \neq 0, \tag{21}$$

then there exists uniquely in (some neighborhood of) $\mathbf{R}^{2n}$ a gradient mapping $w \to \tilde{w} = f(w)$ with Jacobian $f_w(w) = N(w)$ and a scalar function-generating function-$\phi(w)$ such that

$$f(w) = \nabla\phi(w), \tag{22}$$

$$\alpha_1(g(z), z) = f(\alpha_2(g(z), z)) = \nabla\phi(\alpha_2(g(z), z)), \quad \textit{identically in } z, \tag{23}$$

$$N = (A_\alpha M + B_\alpha)(C_\alpha M + D_\alpha)^{-1}, \quad M = (A^\alpha N + B^\alpha)(C^\alpha N + D^\alpha)^{-1}. \tag{24}$$

Proof. The image of the graph Γ_g under α is

$$\alpha(\Gamma_g) = \left\{ \begin{pmatrix} \hat{w} \\ w \end{pmatrix} \in \mathbf{R}^{4n} | \hat{w} = \alpha_1(g(z), z), \quad w = \alpha_2(g(z), z) \right\}.$$

By (21),

$$\frac{\partial w}{\partial z} = \frac{\partial \alpha_2}{\partial \hat{z}} \cdot \frac{\partial \hat{z}}{\partial z} + \frac{\partial \alpha_2}{\partial z} = C_\alpha M + D_\alpha$$

is nonsingular. So by the inverse function theorem, $w = \alpha_2(g(z), z)$ is invertible. The inverse function is denoted by $z = z(w)$. Set

$$\hat{w} = f(w) = \alpha_1(g(z), z)|_{z=z(w)}. \tag{25}$$

Then

$$N = \frac{\partial f}{\partial w} = \left(\frac{\partial \alpha_1}{\partial \hat{z}} \frac{\partial g}{\partial z} + \frac{\partial \alpha_1}{\partial z} \right) \left(\frac{\partial \alpha_2}{\partial \hat{z}} \frac{\partial g}{\partial z} + \frac{\partial \alpha_2}{\partial z} \right)^{-1} = (A_\alpha M + B_\alpha)(C_\alpha M + D_\alpha)^{-1}.$$

Notice that the image $\alpha(\Gamma_g)$ is a J_{4n}-Lagrangian submanifold. That is, the tangent space $T_*(\alpha(\Gamma_g))$ is J_{4n}-Lagrangian. It means

$$\left((A_\alpha M + B_\alpha)^T, (C_\alpha M + D_\alpha)^T \right) \begin{pmatrix} 0 & I_{2n} \\ -I_{2n} & 0 \end{pmatrix} \begin{pmatrix} A_\alpha M + B_\alpha \\ C_\alpha M + D_\alpha \end{pmatrix}$$
$$= (A_\alpha M + B_\alpha)^T (C_\alpha M + D_\alpha) - (C_\alpha M + D_\alpha)^T (A_\alpha M + B_\alpha) = 0,$$

i.e.,

$$N = (A_\alpha M + B_\alpha)(C_\alpha M + D_\alpha)^{-1} \text{ symmetric.}$$

It shows that $\hat{w} = f(w)$ is a gradient mapping. By the Poincare lemma, there is a scalar function $\phi(w)$ such that

$$f(w) = \nabla\phi(w).$$

That is (22). (23) follows from the construction of $f(w)$ and $z(w)$. In fact by (25),

$$f(w) = \alpha_1(g(z), z) \circ z(w). \tag{26}$$

Since $z(w) \circ \alpha_2(g(z), z) \equiv z$, substituting $w = \alpha_2(g(z), z)$ into (26), we get (23) at once.

Proposition 4 *$f(w)$ obtained in Theorem 3 is also the solution of the following implicit equation*

$$\alpha^1(f(w), w) = g\left(\alpha^2(f(w), w)\right). \tag{27}$$

Theorem 5 *Let α be as in Theorem 3. Let $w \to \hat{w} = f(w)$ be a gradient mapping in (some neighborhood of) $\mathbf{R}^{2n}$ with Jacobian $f_w(w) = N(w)$. If in some neighborhood of $\mathbf{R}^{2n}$, N satisfies the condition*

$$|C^\alpha(f(w), w)N(w) + D^\alpha(f(w), w)| \neq 0, \tag{28}$$

then there exists uniquely, in some neighborhood of $\mathbf{R}^{2n}$, a $K(z)$-symplectic mapping $z \to \hat{z} = g(z)$ with Jacobian $g_z(z) = M(z)$ such that

$$\alpha^1(f(w), w) = g(\alpha^2(f(w), w)), \tag{29}$$

$$M = (A^\alpha N + B^\alpha)(C^\alpha N + D^\alpha)^{-1}, \quad N = (A_\alpha M + B_\alpha)(C_\alpha M + D_\alpha)^{-1}. \tag{30}$$

Similarly to Proposition 4, $g(z)$ is the solution of the implicit equation

$$\alpha_1(g(z), z) = f\left(\alpha_2(g(z), z)\right). \tag{31}$$

The proof is similar to that of Theorem 3 and is omitted here.

§3 Generating Functions for the Phase Flow of Hamiltonian Systems

We now consider the general Hamiltonian system

$$\frac{dz}{dt} = K^{-1}(z)\nabla H(z), \quad z \in \mathbf{R}^{2n}, \tag{32}$$

with the Hamiltonian $H(z)$. Its phase flow is denoted by $g^t(z) = g(z,t) = g_H(z,t)$. $g(z,t)$ is a one-parameter group of $K(z)$-symplectic mappings, at least local in z and t, i.e.,

$$g^0 = \text{identity}, \quad g^{t_1+t_2} = g^{t_1} \circ g^{t_2};$$

if z_0 is taken as an initial condition, then $z(t)=g^t(z_0)=g(z_0,t)=g_H(z_0,t)$ is the solution of (32) with the initial value z_0; in addition

$$g_z^T(z,t)K(g(z,t))g_z(z,t)=K(z),\quad \forall t\in\mathbf{R} \tag{33}$$

Theorem 6 *Let α be as above and $z\to\hat{z}=g(z,t)$ be the phase flow of the Hamiltonian system (32) with Jacobian $M(z,t)=g_z(z,t)$. If at some point z_0,*

$$|C_\alpha(z_0,z_0)+D_\alpha(z_0,z_0)|\neq 0, \tag{34}$$

then there exists, for sufficiently small $|t|$ and in some neighborhood of $\mathbf{R}^{2n}$, a time-dependent gradient mapping $w\to\hat{w}=f(w,t)$ with Jacobian $N(w,t)=f_w(w,t)$ symmetric and a time-dependent generating function $\phi_{\alpha,H}(w,t)=\phi(w,t)$ such that

$$f(w,t)=\nabla\phi(w,t), \tag{35}$$

$$\frac{\partial\phi(w,t)}{\partial t}=-H(\hat{z}(\nabla\phi(w,t),w))=-\hat{H}(\nabla\phi(w,t),w), \tag{36}$$

$$\alpha_1(g(z,t),z)=\nabla\phi(\alpha_2(g(z,t),z),t)\text{, identically in some neighborhood of } z_0, \tag{37}$$

$$N=(A_\alpha M+B_\alpha)(C_\alpha M+D_\alpha)^{-1},\quad M=(A^\alpha N+B^\alpha)(C^\alpha N+D^\alpha)^{-1}, \tag{38}$$

where $\hat{H}(\hat{w},w)=H(\hat{z})\circ\alpha^1(\hat{w},w)$.

(36) is also called the Hamilton-Jacobi equation for the Hamiltonian system (32) and the nonlinear transformation α.

Theorem 7 *Let $H(z)$ and α be analytic. Then the generating function $\phi_{\alpha,H}(w,t)=\phi(w,t)$ can be expanded as a convergent power series in t for sufficiently small $|t|$*

$$\phi(w,t)=\sum_{k=0}^{\infty}\phi^{(k)}(w)t^k, \tag{39}$$

and $\phi^{(k)}(w), k\geqslant 0$, can be recursively determined by the following equations

$$\phi^{(0)}(w)=\int f(w,0)dw+\text{ const.}, \tag{40}$$

$$\phi^{(1)}(w)=-\hat{H}(f(w,0),w), \tag{41}$$

$$\phi^{(k+1)}(w)=-\frac{1}{k+1}\sum_{m=1}^{k}\frac{1}{m!}\sum_{\substack{k_1+\cdots k_m=k\\ k_i\geqslant 1}}D_{\hat{w}}^m\hat{H}\left(\nabla\phi^{(k_1)}(w),\cdots,\nabla\phi^{(k_m)}(w)\right),\quad k\geqslant 1, \tag{42}$$

where the integral of (40) is taken on the curve connecting w and $w_0=\alpha_2(z_0,z_0)$, $\hat{H}(\hat{w},w)$

$= H \circ \alpha^1(\hat{w}, w)$. *Here we have used the notation of multi-linear forms,* e.g.,

$$
\begin{aligned}
&D_{\hat{w}}^m \hat{H}(f(w,0),w)\left(\nabla\phi^{(k_1)},\cdots,\nabla\phi^{(k_m)}\right)\\
&=\sum_{i_1,\cdots,i_m=1}^{2n}\hat{H}_{\hat{w}_{i_1},\cdots,\hat{w}_{i_m}}(f(w,0),w)\left(\nabla\phi^{(k_1)}\right)_{i_1}\cdots\left(\nabla\phi^{(k_m)}\right)_{i_m}.
\end{aligned}
$$

$\left(\nabla\phi^{(k_l)}\right)_{i_l}$ *is the* i_l*- th component of the column vector* $\nabla\phi^{(k_l)}$.

Proof. Differentiating (39) with respect to w and t, we get

$$
\nabla\phi(w,t)=\sum_{k=0}^{\infty}\nabla\phi^{(k)}(w)t^k, \tag{43}
$$

$$
\frac{\partial\phi(w,t)}{\partial t}=\sum_{k=0}^{\infty}(k+1)t^k\phi^{(k+1)}(w). \tag{44}
$$

By (35),

$$
\nabla\phi^{(0)}(w)=\nabla\phi(w,0)=f(w,0).
$$

So we can take $\phi^{(0)}(w)=\int_{w_0}^{w}f(w,0)dw$. Expanding $\hat{H}(f(w,t),w)$ in $(f(w,0),w)$ and gathering the terms with the same order, we get

$$
\begin{aligned}
\hat{H}(\nabla\phi(w,t),w)=&\hat{H}\left(f(w,0)+\sum_{k=1}^{\infty}t^k\nabla\phi^{(k)}(w),w\right)\\
=\hat{H}(f(w,0),w)+&\sum_{k=1}^{\infty}t^k\sum_{m=1}^{k}\frac{1}{m!}\sum_{\substack{k_1+\cdots+k_m=k\\k_i\geqslant 1}}D_{\hat{w}}^m\hat{H}(f(w,0),w)(\nabla\phi^{(k_1)},\cdots,\nabla\phi^{(k_m)}).
\end{aligned}
$$

Using the Hamilton-Jacobi equation (36) and comparing with (44), we get (41) and (42).

§4 $K(z)$-Symplectic Difference Schemes

In the previous section, we have established the relationship between the phase flow $g(z,t)$ and the generating function $\phi(w,t)$. And when H and α are analytic, the generating function has a power series expansion. With the aid of the expansion, we can systematically construct $K(z)$-symplectic difference schemes, i.e., the transition of such difference schemes from one time-step to the next is $K(z)$-symplectic.

Theorem 8 *Let* α *be given as in Theorem* 7 *and H(z) analytic. For sufficiently small* $\tau>0$ *as the time-step. Take*

$$
\psi^{(m)}(w,\tau)=\sum_{i=0}^{m}\phi^{(i)}(w)\tau^i,\quad m=1,2,\cdots \tag{45}
$$

where $\phi^{(i)}(w)$ *are determined by* (40),(41) *and* (42). *Then* $\psi^{(m)}(w,\tau)$ *defines a K(z)-symplectic difference scheme* $z=z^k\to z^{k+1}=\hat{z}$,

$$\alpha_1\left(z^{k+1},z^k\right)=\nabla_w\psi^{(m)}\left(\alpha_2\left(z^{k+1},z^k\right),\tau\right) \tag{46}$$

of m-th order of accuracy.

Proof. By hypothesis,

$$|C(z_0,z_0)+D(z_0,z_0)|\neq 0.$$

So by Lemma 2, $|C^\alpha N+D^\alpha|\neq 0$ where $N=(A_\alpha+B_\alpha)(C_\alpha+D_\alpha)^{-1}=\phi_{ww}(w_0,0)=\psi^{(m)}_{ww}(w_0,0)$, the Hessians of $\phi(w_0,0)$ and $\psi^{(m)}(w_0,0)$ respectively, and $w_0=\alpha_2(z_0,z_0)$. Thus for sufficiently small τ and in some neighborhood of w_0, $\left[C^\alpha N^{(m)}(w,\tau)+D^\alpha\right]\neq 0$, where $N^{(m)}(w,\tau)=\psi^{(m)}_{ww}(w,\tau)$. By Theorem 5, $\nabla\psi^{(m)}(w,\tau)$ defines a time-dependent $K(z)$-symplectic mapping which is expressed by (31). It means that (46) is a $K(z)$-symplectic difference scheme.

Since $\psi^{(m)}(w,\tau)$ is the m-th approximate to $\phi(w,\tau)$, so is $f^{(m)}(w,\tau)=\nabla\psi^{(m)}(w,\tau)$ to $f(w,t)$. It follows that the difference scheme (46) is of m-th order of accuracy.

Example Take $n=1$ and

$$K=\begin{bmatrix}0 & p\\ -p & 0\end{bmatrix},\quad H(p,q)=\frac{1}{2}\left(p^2-q^2\right). \tag{47}$$

The corresponding Hamiltonian system is

$$\frac{dz}{dt}=K^{-1}\nabla H=\begin{bmatrix}0 & -p^{-1}\\ p^{-1} & 0\end{bmatrix}\begin{bmatrix}p\\ -q\end{bmatrix}=\begin{bmatrix}p^{-1}q\\ 1\end{bmatrix},$$

i.e.,

$$\frac{dp}{dt}=p^{-1}q,\quad \frac{dq}{dt}=1. \tag{48}$$

The phase flow $\hat{p}=g_1(p,q,t),\hat{q}=g_2(p,q,t)$ is

$$\hat{p}^2=p^2+2qt+t^2,\quad \hat{q}=q+t. \tag{49}$$

Take the transformation α as

$$\begin{bmatrix}\hat{z}\\ z\end{bmatrix}\to\begin{bmatrix}\hat{w}\\ w\end{bmatrix}=\alpha\begin{bmatrix}\hat{z}\\ z\end{bmatrix}:\quad \begin{aligned}&\hat{P}=\frac{1}{4}\hat{p}^2+\hat{q}-q,\quad \hat{Q}=\frac{1}{4}\hat{p}^2+\hat{q}+q,\\ &P=-\frac{1}{4}p^2+\hat{q}-q,\quad Q=\frac{1}{4}p^2+\hat{q}+q,\end{aligned} \tag{50}$$

where $\hat{w} = (\hat{P}, \hat{Q})^T, w = (P, Q)^T$. Its inverse transformation α^{-1} is

$$\begin{bmatrix} \hat{w} \\ w \end{bmatrix} \to \begin{bmatrix} \hat{z} \\ z \end{bmatrix} = \alpha^{-1} \begin{bmatrix} \hat{w} \\ w \end{bmatrix} : \quad \begin{array}{l} \hat{p}^2 = 2(\hat{P} + \hat{Q} - P - Q), \hat{q} = \dfrac{1}{2}(P + Q), \\ p^2 = 2(\hat{P} - \hat{Q} - P + Q), \quad q = \dfrac{1}{2}(\hat{Q} - \hat{P}). \end{array} \tag{51}$$

The Jacobian of α is

$$\alpha_* = \begin{bmatrix} \frac{2}{3}\hat{p} & 1 & \frac{1}{2}p & -1 \\ \hat{p} & 1 & 0 & 1 \\ \hat{p} & 1 & 0 & -1 \\ \frac{1}{2}\hat{p} & 1 & \frac{1}{2}p & 1 \end{bmatrix}. \tag{52}$$

By direct computation, we can immediately know that α satisfies (10), α also satisfies (34) as

$$|C_\alpha + D_\alpha| = -\frac{1}{2}p \neq 0 \quad \text{for } p \neq 0.$$

It follows from (50) that

$$\hat{H}(\hat{P}, \hat{Q}, P, Q) = \hat{P} - \hat{Q} - P + Q - \frac{1}{8}(\hat{P} - \hat{Q})^2, \quad \nabla\phi^{(0)}(w) = f(w, 0) = \begin{pmatrix} -P \\ Q \end{pmatrix}.$$

$$\begin{aligned} \text{So } \phi^{(0)}(P, Q) &= \frac{1}{2}\left(Q^2 - P^2\right), \\ \phi^{(1)}(P, Q) &= -\hat{H}\left(\nabla\phi^{(0)}(P, Q), P, Q\right) = 2P + \frac{1}{8}(P + Q)^2, \\ \nabla\phi^{(1)}(P, Q) &= \begin{pmatrix} 2 + \frac{1}{4}(P + Q) \\ \frac{1}{4}(P + Q) \end{pmatrix}. \end{aligned}$$

The first-order scheme is

$$\begin{aligned} \frac{1}{4}\left(p^{k+1}\right)^2 + q^{k+1} - q^k &= \frac{1}{4}\left(p^k\right)^2 - q^{k+1} + q^k + \tau\left(2 + \frac{1}{4}(P + Q)\right), \\ \frac{1}{4}\left(p^{k+1}\right)^2 + q^{k+1} + q^k &= \frac{1}{4}\left(p^k\right)^2 + q^{k+1} + q^k + \frac{\tau}{4}(P + Q), \end{aligned}$$

i.e.,

$$\left(p^{k+1}\right)^2 = \left(p^k\right)^2 + 2\tau q^{k+1}, \quad q^{k+1} = q^k + \tau. \tag{53}$$

Referring to (49), we have seen that (53) is indeed its first order approximation. Its transition matrix is $\begin{bmatrix} \left(p^{k+1}\right)^{-1} p^k & \tau\left(p^{k+1}\right)^{-1} \\ 0 & 1 \end{bmatrix}$. Direct computation gives that it is $K(z)$-symplectic.

Discussion The difficulty of this method is to find out nonlinear transformations α satisfying (10). It in fact is to realize Darboux's theorem in $4n$ dimensions. It seems as difficult as to directly transform $K(z)$ into J_{2n}. But in our case, we have more parameters to select. Hence we hope the problem will be easier than the original one.

References

[1] Feng K. On difference schemes and symplectic geometry. In Feng K, ed. Proc 1984 Beijing Symp Diff Geometry and Diff Equations. Beijing: Science Press, 1985. 42-58

[2] Feng K. Difference schemes for Hamiltonian formalism and symbectic geometry. J Comp Math. 1986 4(3): 279-289

[3] Feng K, Qin M Z. The symplectic methods for computation of Hamiltonian equations. In Zhu Y L, Guo Ben-yu, ed. Proc Conf on Numerical Methods for PDE's. Berlin: Springer, 1987.1-37. Lect Notes in Math 1297

[4] Feng K, Wu H M, Qin M Z, Wang D L. Construction of canonical difference schemes for Hamiltonian formalism via generating functions. J Comp Math. 1989 7(1): 71-96

[5] Feng K, Wang D L, Ge Z, Li C W. Calculus of generating functions: [Preprint]. Beijing: CAS. Computing Center, 1987

[6] Feng K. Canonical difference schemes for Hamiltonian canonical differential equations. In Xiao S, Pu F, ed. Proc Inter Workshop on Applied Differential Equations. Singapore: World Scientific, 1986. 59-73

[7] Feng K. The Hamiltonian way for computing Hamiltonian dynamics. In Spigler R, ed. Applied and Industrial Mathematics. Netherlands: Kluwer, 1991. 17-35

[8] Feng K. Symplectic geometry and numerical methods in fluid dynamics. In Zhang F G, Zhu Y L, ed. Proc 10th Inter Conf on Numer Methods in Fluid Dynamics. Berlin: Springer, 1986. Lect Notes in Physics 264

[9] Feng K, Wu H M, Qin M Z. Symplectic difference schemes for the linear Hamiltonian canonical systems. J Comp Math. 1990 8(4): 71-96

[10] Ge Z, Feng K. On the approximation of Hamiltonian systems. J Comp Math. 1988 6(1): 88-97

[11] Li C W, Qin M Z. Symplectic difference schemes for infinite dimensional Hamiltonian systems. J Comp Math. 1988 6(2): 164-174

[12] Qin M Z. A symplectic difference schemes for the Hamiltonian equation. J Comp Math. 1991 5: 203-209

[13] Wang D L. Poisson difference schemes for Hamiltonian systems on Poisson manifolds. J Comp Math. 1991 9(2): 115-124

[14] Wu H M, Feng K. Symplectic difference schemes for canonical Hamiltonian systems. The 3'rd Annual Meeting of Chinese Society of Computational Mathematics. Beijing, 1985

[15] Wu Y H. The generating function of the solution of ODE and its discrete methods. Computer Math Applic. 1988 15(12): 1041-1050

15 How to Compute Properly Newton's Equation of Motion? ①

如何正确计算牛顿运动方程?

§1 Problems and Motivations

The main theme of modern scientific computing is the numerical solution of various differential equations of mathematical physics bearing the names such as Newton, Euler, Lagrange, Laplace, Navier-Stokes, Maxwell, Boltzmann, Einstein, Schrödinger, Yang-Mills, etc. At the top of the list is the most celebrated Newton's equation of motion. The historical, theoretical and practical importance of Newton's equation hardly needs any comment, so is the importance of the numerical solution of such equations. On the other hand, starting from Euler, right down to the present computer age, a great wealth of scientific literature on numerical methods for differential equations has been accumulated, and a great variety of algorithms, software packages and even expert systems had been developed. Under such circumstances we still feel motivated to raise the following two questions:

Question 1: *Are the existing numerical methods adequate for computing Newton's equations of motion?*

Question 2: *What is the proper way to compute Newton's equations of motion?*

Question 1 seems to have never been seriously raised before, so question 2 seems to have never been studied systematically.

We start with the case of Newton's equations of motion in conservative force field in configuration space $\mathbf{R}^n$, which is physically more basic and mathematically more difficult:

$$M\ddot{q} = -V_q(q) \tag{1}$$

where $q = (q_1, \cdots, q_n)$ is the position variable, M is the inertia matrix, $V(q)$ is the potential energy. It is well known that the conservative Newton's equations has two

① In Ying L A, Cuo B Y, ed. Proc of 2nd Conf on Numerical Methods for Partial Differential Equations. Singapore: World Scientific, 1992. 15-22 alternative mathematically equivalent formalisms: the Lagrangian formalism as a variational principle and the Hamiltonian formalism in which the Newton's system of differential equations of second order in configuration space is reduced to Hamilton's system of canonical equations of first order in phase space with doubled dimension.

The mathematically equivalent formalisms express the same physical laws, but they look very different in form, so they lead to different technical approaches in problem solving, they are not equally effective in practice. So a judicious choice at the start among the alternative mathematically equivalent formalisms is very crucial for the eventual success or failure in problem solving.

We choose the *Hamiltonian* formalism as the basis, i.e. instead of the Newton's system (1) we consider the Hamilton's canonical system

$$\dot{p} = -H_q(p, q), \quad \dot{q} = H_p(p, q)$$

i.e.,

$$\dot{z} = JH_z(z), \quad J = J_{2n} = \begin{pmatrix} 0 & -I_n \\ I_n & 0 \end{pmatrix} \tag{2}$$

in phase space $\mathbf{R}^{2n}$, defined by the Hamiltonian function $H(z) = H(p, q)$ in $2n$ variables $(z_1, \cdots, z_{2n}) = (p_1, \cdots, p_n, q_1, \cdots, q_n)$ representing the total energy. Our motivations are: (1) Hamiltonian formalism is distinctive in its *simplicity, symmetry* and *elegance* in form. The laws of motion become more explicit and transparent in Hamiltonian formalism. (2) Hamiltonian formalism is also distinctive in its *ubiquity* and *universality*. It is now beyond doubt that any real physical process with negligible dissipation of classical or relativistic or quantum nature in finite or infinite number of degrees of freedom can always be cast in suitable Hamiltonian form. So, if the algorithms for classical Hamiltonian systems turn to be successful, they are expected to have wide applications. We started with a search in the vast volume of archives on numerical methodology for differential equations; however, we found that the pertinent study specifically for Hamiltonian equations is virtually void, in paradoxical contradiction with the importance and general applicability of Hamiltonian formalism. This situation also motivate us to think hard and to raise the questions as mentioned above.

§2 Hamiltonian Algorithms for Hamiltonian Systems

Our technical approach is *symplectic geometry* i.e., the geometry of phase space in which an anti-symmetric area metric is basic, in contrast with the Euclidean or Riemannian geometry in which a symmetric distance metric is basic. The fundamental theorem of classical mechanics expressed in symplectic geometry becomes "all dynamical evolutions of Hamiltonian system are *symplectic* (canonical) transformations" , i.e. those preserving the anti-symmetric symplectic metric, or more precisely, smooth 1-1 transformation $g : \mathbf{R}^{2n} \to \mathbf{R}^{2n}$ with Jacobian matrix $g_z(z)$ satisfying

$$(g_z(z))^T J_{2n} g_z(z) \equiv J_{2n}.$$

It is natural and mandatory to require that the marching of the discrete algorithm should be also symplectic. Such algorithms are called *symplectic* or *canonical* or *Hamiltonian*. We conceive, design, analyse and assess the algorithms conciously and exclusively within the framework of symplectic geometry. This approach turns out to be successful and fruitful, giving a great variety of symplectic algorithms with remarkable properties and outstanding performance and with sound theoretical foundation [1,2].

A comparative numerical study for non-symplectic conventional methods and symplectic methods has been carried out on the following set of "benchmarks" : 1. Simple harmonic oscillator 2. Nonlinear Duffing oscillator 3. Huygens oscillator 4. Cassini oscillator 5. Stationary flow on plane periodic and quasi-periodic lattices 6. Double harmonic oscillator and Lissajous figures 7. Geodesic flow on ellipsoidal surface 8. Kepler motion. Extensive computer experimentation affirms indisputably the high quality and superiority of symplectic methods; especially in the global, structural and long-term tracking aspects, all conventional non-symplectic methods irretrievably fail without exception while all symplectic methods decently succeed without exception, possessing remarkable very-long-term tracking capability. The contrast is sharp and overwhelming.

Fig.1 shows the linear harmonic oscillator $H = \frac{1}{2}(p^2 + 4q^2)$ computed by different methods. (A): 4th order Runge Kutta method, step size $s = 0.4$, dissipative, a false attractor is formed at the center. (B): 4th order Adams method, $s = 0.2$, anti-dissipative, diverging outwards. (C): Leap-flog method, i.e., explicit 3-level method which is symplectic for linear Hamiltonian systems but non-symplectic for non-linear Hamiltonian systems, $s = 0.1$. Three consecutive 1000 points after the 1st step, after the 5 million-th step and after the 10 million-th step are plotted on the same figure, their coincidence shows the long-term tracking capability of symplectic scheme.

Fig.2: Non-linear Duffing oscillater $H = \frac{1}{2}(p^2 + q^2) - \frac{2}{3}q^4$. (A) and (B): The same Leap-flog method, non-symplectic now, $s = 0.2$, 10000 steps. The first 20 points (not clearly shown in the plot) are on the right track of a near elliptic orbit like that of (C), but long-time tracking goes wrong. (C): Explicit symplectic method of order 2 with $s = 0.1$ up to 100 million steps. Three consecutive 1000 points after the first step, after the 50 million-th step and after the 100 million-th step are plotted on the same figure, their remarkable coincidence shows the very long-term tracking capability of symplectic method for non-linear system. Here some function $= H + O(s^2)$, instead of H itself, is exactly conserved.

Fig.3: Non-linear Huygens oscillater $H = p^2 - q^2 + q^4$, having two elliptic fixed points. (A) and (B): Runge-Kutta method, 2 artificial attractors at the elliptic fixed points are formed. The same initial point outside the separatrix is attracted randomly either to the left or to the right by minute changes of step sizes from 0.10000005 in (A) to 0.10000004 in (B). (C): Symplectic computation of 4 typical orbits each up to 100000000 steps, showing the striking robustness of the algorithm.

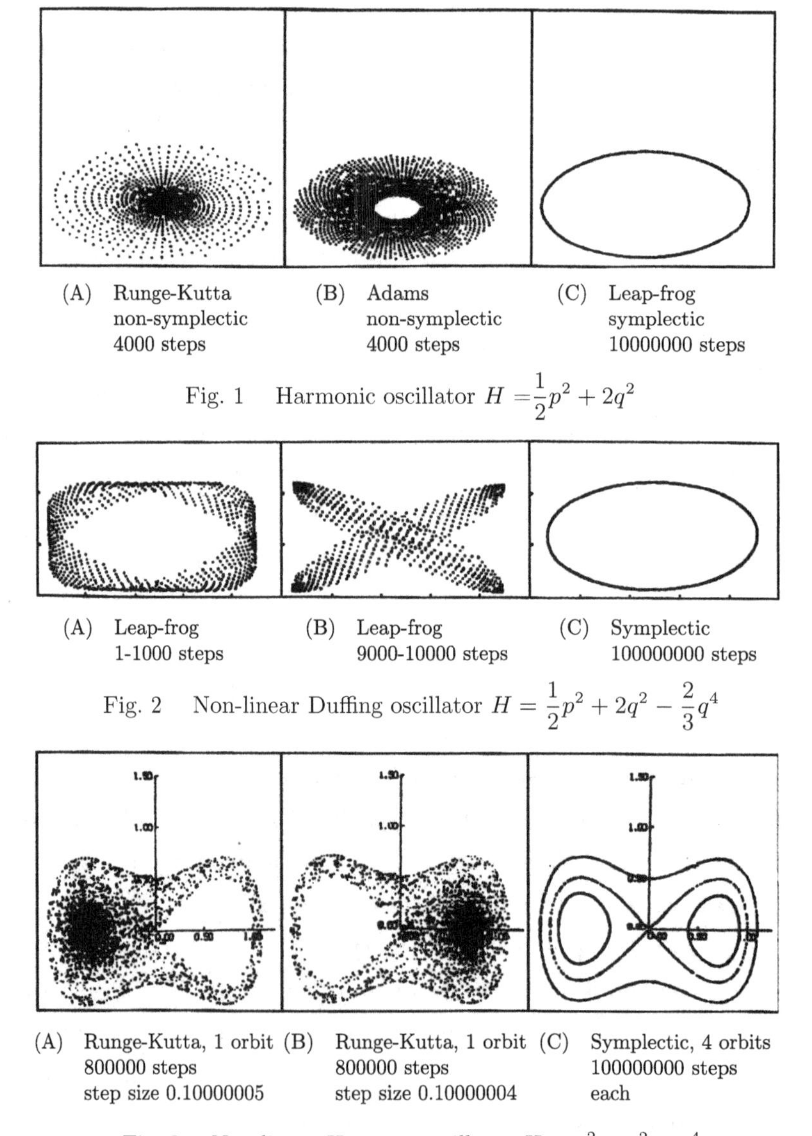

Fig. 1 Harmonic oscillator $H = \frac{1}{2}p^2 + 2q^2$

Fig. 2 Non-linear Duffing oscillator $H = \frac{1}{2}p^2 + 2q^2 - \frac{2}{3}q^4$

Fig. 3 Non-linear Huygens oscillator $H = p^2 - q^2 + q^4$

Fig.4-6 show the geodesic flows on ellipsoidal surface and Kepler motion with Hooke potential. Initial data in both cases correspond to rational or irrational frequency ratio, which give closed or dense orbit. (A): Total failure of Runge-Kutta method in long-term tracking. (B): Success of symplectic method in long-term tracking, exhibiting the closedness or denseness of orbits, [7].

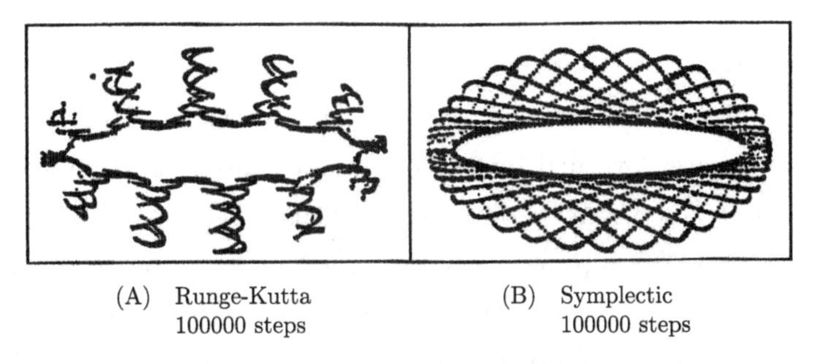

(A) Runge-Kutta 100000 steps (B) Symplectic 100000 steps

Fig. 4 Geodesics on ellipsoid, frequency ratio 11 : 16, closed orbit

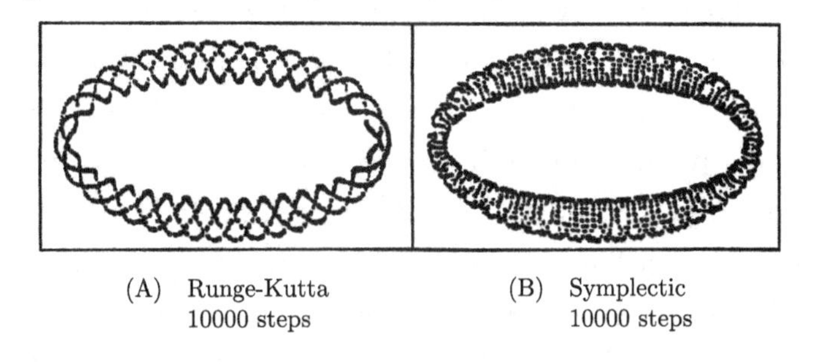

(A) Runge-Kutta 10000 steps (B) Symplectic 10000 steps

Fig. 5 Geodesics on ellipsoid, frequency ratio $\sqrt{5} : 4$, dense orbit.

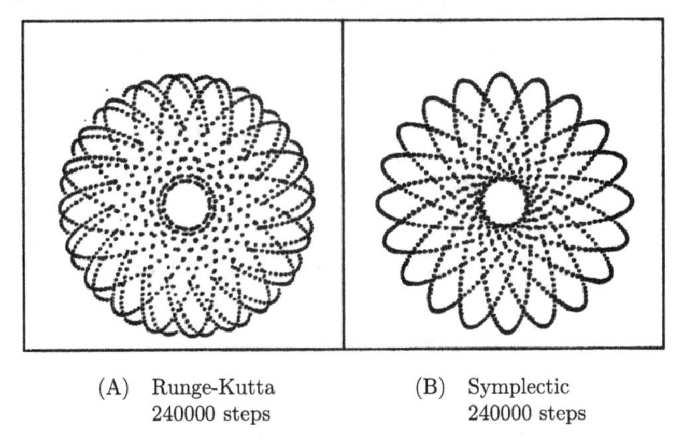

(A) Runge-Kutta 240000 steps (B) Symplectic 240000 steps

Fig. 6 Kepler motion, frequency ratio 11 : 20, closed orbit.

Apart from very rare exceptions, almost all conventional methods are non-symplectic, usually they were designed for asymptotically stable systems, providing dissipative mechanisms to guarantee computational stability. Hamiltonian systems can never be asymptotically stable, so the non-symplectic methods inevitably bring in artificial dissipation, false attractors and/or various parasitic effects which are alien to Hamiltonian dynamics, leading eventually to grave distortions. Although practically applicable for short-term simulation of transcient phenomena, they are impotent and possibly misleading in global and structural study and long-term tracking. Since Newton's equations are mathematically equivalent to Hamilton's so we get at the quite unexpected *negative answer* to Question 1.

Symplectic methods are inherently free from artificial dissipation and all kinds of non-Hamiltonian pollutions. They are "*clean*" algorithms. The dynamical evolution of Hamiltonian systems are governed by two classes of conservation laws. The first class consists in the conservation of even-dimensional phase area, i.e., the Liouville-Poincaré conservation laws, they are automatically preserved under symplectic algorithms. The second class consists in the conservation of energy and other integrals of motion, i.e., Noether conservation laws. The crucial fact is that any symplectic algorithm possesses its own *formal* integrals of motion depending on the step-size parameter and approximating the original integrals of motion (with the same order of accuracy as that of the method itself) and being invariant under the algorithm [4,5]. It can also be proved that the majority of invariant tori of integrable systems are preserved under symplectic algorithms; resulting in a new version of the celebrated K.A.M. theorems and the theoretical infinite-time tracking capability of the symplectic algorithms [6], although practically realizable only approximately on computers with finite number of digits.

The high degree of parallelism and proximity between the structure of the time-discrete Hamiltonian system and that of the original system guarantees and explains in a fundamental way the remarkable high performance of Hamiltonian algorithms. So a *proper* way to compute Newton's equations of motion is *first to Hamiltonize the equations and then to compute them by Hamiltonian algorithm.* This constitutes our answer, by no means trivial, to Question 2.

§3 Recommendations and Conclusions

We list some very simple and useful Hamiltonian algorithms $[1, 2, 8]$, $s =$ step-size (real number).

1. Centered Eulerian scheme, 2nd order, implicit, preserving all quadratic integrals of motion:

$$z^k \to z^{k+1} =: C^s z^k,$$

$$z^{k+1} = z^k + sJH_z\left(\frac{1}{2}\left(z^{k+1} + z^k\right)\right).$$

2. Alternating explicit Eulerian schemes for canonically separable Hamiltonian $H(p, q) = \phi(p) + \psi(q)$, preserving all quadratic integrals of motion of angular momentum

type $p^T Bq$:

$$
\begin{aligned}
z^k &\rightarrow z^{k+1} =: A_1^s z^k, \\
p^{k+1} &= p^k - s\psi_q\left(q^k\right), \\
q^{k+1} &= q^k + s\phi_p\left(p^{k+1}\right),
\end{aligned}
$$

or

$$
\begin{aligned}
z^k &\rightarrow z^{k+1} =: A_2^s z^k, \\
q^{k+1} &= q^k + s\phi_p\left(p^k\right), \\
p^{k+1} &= p^k - s\psi_q\left(q^{k+1}\right).
\end{aligned}
$$

Both A_1^s, A_2^s are of 1st order. Their composite, e.g., $A_2^{s/2} \circ A_1^{s/2} =: A^s$ is 2nd order:

$$
\begin{aligned}
z^k &\rightarrow z^{k+1} =: A_2^{s/2} \circ A_1^{s/2} z^k, \\
p^{k+\frac{1}{2}} &= p^k - \frac{s}{2}\psi_q\left(q^k\right), \\
q^{k+1} &= q^k + s\phi_p\left(p^{k+\frac{1}{2}}\right), \\
p^{k+1} &= p^{k+\frac{1}{2}} - \frac{s}{2}\psi_q\left(q^{k+1}\right).
\end{aligned}
$$

The algorithms C^s, A^s are *revertible* in the sense that

$$
g^{-s} = \left(g^s\right)^{-1}.
$$

Revertibility is a desirable property of numerical methods, since it is always satisfied by the true solution of the system and it implies that the order of accuracy is necessarily even. It also provides a practical means for monitoring the inevitable loss of significant digits due to round-off in long-term numerical integration. Moreover, for any revertible Hamiltonian algorithm g^s of order 2, a revertible Hamiltonian algorithm $\tilde{g}^s$ of order 4 can be easily constructed by splitting the time-step of size s into 3 substeps of sizes $\alpha s, \beta s, \alpha s$

$$
\tilde{g}^s = g^{\alpha s} \circ g^{\beta s} \circ g^{\alpha s}
$$

with $2\alpha + \beta = 1, 2\alpha^3 + \beta^3 = 0$, i.e., $\alpha = (2 - \sqrt[3]{2})^{-1} > 0, \beta = 1 - 2\alpha < 0$, [3].

It is important to note that, although Hamiltonian algorithms emerge as the proper methodology for solving Hamiltonian systems, they are applicable as well to non-Hamiltonian systems with advantage. Hamiltonian algorithms are *neutral* in the sense that they give the proper amount of dissipation as determined by the systems themselves, without overcharge or discount. The advantage is indiscernible under strong dissipation but becomes ever more prominent with decreasing dissipation. The phenomenal stability property of Hamiltonian algorithms makes them good for solving stiff problems too. So Hamiltonian algorithms are recommented as a safe and high-performance methodology for general usage.

The study of numerical methods for Hamiltonian systems grossly neglected thus far is a new promising and vital field of research at the common frontier of computational physics, computational mechanics and computational mathematics. Owing to the universality of Hamiltonian formalism, Hamiltonian algorithms are promising for wide applications and developments and to have impact on many fields of scientific computing.

Acknowledgements This research program is supported by the National Natural Science Foundation of China. The author expresses his gratitude to his collaborators Prof. Li Wangyao for supplying the results on leap-frog scheme and Huygens oscillator, Mr. Tang Yi-fa for the results on geodesic flows on surfaces and Kepler motion, Mr. Jiang Li-xin for the Runge-Kutta and Adams calculation of harmonic oscillator, Dr. Wang Dao-liu, Dr. Shang Zai-jiu, Mr. Zhang Mei-qing, Mr. Zhu Wen-jie for the help in preparing the manuscript.

References

[1] Feng K. On difference schemes and symplectic geometry. In Feng K, ed. Proc 1984 Beijing Symp Diff Geometry and Diff Equations. Beijing: Science Press, 1985. 42-58

[2] Feng K, Qin M Z. Hamiltonian algorithms and a comparative numerical study. Comput Phys Comm. 1991 65: 173-187

[3] Qin M Z, Wang D L, Zhang M Q. Explicit sympiectic difference schemes for separable Hamiltonian systems. J Comp Math. 1991 9(3): 211-221

[4] Feng K. The calculus of generating functions and the formal energy for Hamiltonian algorithms: [Preprint]. Beijing: CAS. Computing Center, 1991

[5] Feng K. The calculus of formal power series for diffeomorphisms and vector fields: [Preprint]. Beijing: CAS. Computing Center, 1991

[6] Shang Z J. On KAM theorem for symplectic algorithms for Hamiltonian systems: [dissertation]. Beijing: CAS. Computing Center, 1991

[7] Tang Y F. Hamiltonian systems and algorithms for geodesic flows on compact Riemannian manifolds: [Master Thesis]. Beijing: CAS. Computing Center, 1990

[8] Feng K, Wang D L. A note on conservation laws of symplectic difference schemes for Hamiltonian systems. J Comp Math. 1991 9(3): 229-237

16 Formal Power Series and Numerical Algorithms for Dynamical Systems[①]

形式幂级数与动力系统的数值算法

Abstract

We give a brief survey of our results on the formal theory of vector fields and flows in dynamical systems based on the apparatus of formal power series. For any formal vector field $\boldsymbol{a}^s$ as a formal power series in s we obtain formal flow $\boldsymbol{e}^t_{\boldsymbol{a}^s}$ in two parameters s, t whose diagonal flow $\boldsymbol{e}^t_{\boldsymbol{a}^s}|_{t=s}$ is a near-1 formal map $\boldsymbol{f}^s$ as a formal power series in s. Conversely, any near-1 formal map $\boldsymbol{f}^s$ is the diagonal flow of a uniquely determined formal field $\boldsymbol{a}^s$. We give a series of dual properties between formal maps and fields. In particular, for the classical cases of dynamical systems with structures, i.e., for Hamiltonian, Liouville and contact systems we give the dual properties between the structure preserving formal maps and formal fields. Since every algorithm is characterized by its step transition map $\boldsymbol{f}^s$ which is a near-1 formal map depending on step-size s and approximating the flow $\boldsymbol{e}^t_{\boldsymbol{a}}$ of the original system, so this formal theory has important implications for the construction, analysis, assessment and understanding of numerical algorithms.

§1 Preliminaries

We study vector fields and their flows in $\boldsymbol{R}^N$ together with their approximations, specifically from the *formal power series* approach [1]. We use coordinate description and matrix notation. The point-vector $\boldsymbol{x} \in \boldsymbol{R}^N$ and vector function $\boldsymbol{a} : \boldsymbol{R}^N \longrightarrow \boldsymbol{R}^N$ are denoted by column matrix $\boldsymbol{x} = (x_1, \cdots, x_N)'$, $\boldsymbol{a}(\boldsymbol{x}) = (a_1(\boldsymbol{x}), \cdots, a_N(\boldsymbol{x}))'$. Prime $'$ denotes matrix transpose. Jacobian matrix of $\boldsymbol{b} = (b_1, \cdots, b_M)' : \boldsymbol{R}^N \longrightarrow \boldsymbol{R}^M$ is denoted by $\boldsymbol{b}_*$. Our considerations are always local in $\boldsymbol{x}$ and t and $\boldsymbol{C}^\infty$ smooth.

We identify all vector functions $\boldsymbol{R}^N \to \boldsymbol{R}^N$ as vector fields. When diffeomorphism conditions for the mapping are satisfied, we identify them as diffeomorphisms. The totality of vector fields on $\boldsymbol{R}^N$ form a Lie algebra $\boldsymbol{V}^N$ under Lie bracket

$$[\boldsymbol{a}, \boldsymbol{b}] = \boldsymbol{a}_*\boldsymbol{b} - \boldsymbol{b}_*\boldsymbol{a}.$$

① In Chan T, Shi Z C, eds. Proc Conf on Scientific Computation. Hangzhou, 1991. Singapore: World Scientific. 1992: 28 -35

For each $\boldsymbol{a} \in \boldsymbol{V}^N$ we define a linear differential operator of order 1 (and vice versa):

$$\boldsymbol{a}^* = \sum_{i=1}^{N} a_i \frac{\partial}{\partial x_i},$$

which operates on the space $\boldsymbol{S}^N$ of smooth functions on $\boldsymbol{R}^N$,

$$\phi \in \boldsymbol{S}^N \longrightarrow \boldsymbol{a}^*\phi = \sum_{i=1}^{N} a_i \frac{\partial}{\partial x_i}\phi \in \boldsymbol{S}^N.$$

This operation can be naturally extended to vector functions $\boldsymbol{b} = (b_1, \cdots, b_N)' \in \boldsymbol{V}^N$ component-wise as $\boldsymbol{a}^*\boldsymbol{b} := (\boldsymbol{a}^*b_1, \cdots, \boldsymbol{a}^*b_N)' \in \boldsymbol{V}^N$. We have

$$\boldsymbol{a}^*\phi = \phi_*\boldsymbol{a}, \quad \boldsymbol{a}^*\boldsymbol{b} = \boldsymbol{b}_*\boldsymbol{a}, \quad \boldsymbol{a}^*1_N = \boldsymbol{a},$$

where 1_N is the identity vector function $\in \boldsymbol{V}^N, 1_N(\boldsymbol{x}) = \boldsymbol{x} = (x_1, \cdots, x_N)'$.

The products and multiple products

$$\boldsymbol{a}^*\boldsymbol{b}^*, \quad \boldsymbol{a}^{*2} := \boldsymbol{a}^*\boldsymbol{a}^*, \quad \boldsymbol{a}^*\boldsymbol{b}^*\boldsymbol{c}^*, \quad \boldsymbol{a}^{*3} = \boldsymbol{a}^*\boldsymbol{a}^*\boldsymbol{a}^*, \cdots$$

of linear differential operators of order 1 are naturally defined. They are multi-linear, associative but non-commutative, they become linear differential operators of higher orders. However,

$$[\boldsymbol{a}, \boldsymbol{b}] = \boldsymbol{b}^*\boldsymbol{a} - \boldsymbol{a}^*\boldsymbol{b} = (\boldsymbol{b}^*\boldsymbol{a}^* - \boldsymbol{a}^*\boldsymbol{b}^*)\,1_N = [\boldsymbol{a}, \boldsymbol{b}]^*1_N,$$
$$\{\boldsymbol{a}^*, \boldsymbol{b}^*\} := \boldsymbol{b}^*\boldsymbol{a}^* - \boldsymbol{a}^*\boldsymbol{b}^* = [\boldsymbol{a}, \boldsymbol{b}]^*$$

so the totality $\boldsymbol{D}_1^N$ of linear differential operators of order 1 in N variables forms a Lie algebra under this bracket and is isomorphic to the Lie algebra $\boldsymbol{V}^N$.

It is well-known that the dynamical evolution of the system

$$\frac{d\boldsymbol{x}}{dt} = \boldsymbol{a}(x)$$

defined by the *vector field* $\boldsymbol{a} : \boldsymbol{R}^N \to \boldsymbol{R}^N$ is given by the *flow* $\boldsymbol{e}_{\boldsymbol{a}}^t$ of $\boldsymbol{a}$, as a diffeomorphism with parameter t, representable as a convergent power series

$$\boldsymbol{e}_{\boldsymbol{a}}^t = \boldsymbol{e}^t = \sum_{k=0}^{\infty} t^k \boldsymbol{e}_k, \quad \boldsymbol{e}_k = \boldsymbol{e}_{\boldsymbol{a},k} : \boldsymbol{R}^N \to \boldsymbol{R}^N$$

satisfying the one-parameter *group property in* t

$$\boldsymbol{e}^0 = 1_N,$$
$$\boldsymbol{e}^{t+s} = \boldsymbol{e}^t \circ \boldsymbol{e}^s, \quad \forall t, s \in \boldsymbol{R};$$

so $\boldsymbol{e}^t$ is a *near-identity diffeomorphism*, or *near-1 map* for short, it generates the solution by $\boldsymbol{x}(0) \to \boldsymbol{e}^t_{\boldsymbol{a}}\boldsymbol{x}(0) = \boldsymbol{x}(t)$. The coefficients can be determined recursively by the differential equation

$$\frac{d}{dt}\boldsymbol{e}^t = \boldsymbol{a} \circ \boldsymbol{e}^t,$$
$$\boldsymbol{e}_0 = 1_N, \boldsymbol{e}_k = \frac{1}{k}\boldsymbol{a}^*\boldsymbol{e}_{k-1} = \frac{1}{k!}\boldsymbol{a}^{*k}1_N = \frac{1}{k!}\boldsymbol{a}^k, \quad k \geqslant 1.$$

Then

$$\boldsymbol{e}^t_{\boldsymbol{a}} = 1_N + \sum_1^\infty \frac{t^k}{k!}\boldsymbol{a}^{*k}1_N = 1_N + \sum_1^\infty \frac{1}{k!}(t\boldsymbol{a}^*)^k 1_N,$$

i.e.,

$$\boldsymbol{e}^t_{\boldsymbol{a}}(\boldsymbol{x}) = \boldsymbol{x} + \sum_1^\infty \frac{t^k}{k!}\boldsymbol{a}^{*k}\boldsymbol{x}.$$

Let $I = \boldsymbol{a}^{*0}$ be the identity operator operating on $\boldsymbol{S}^N$ or $\boldsymbol{V}^N$, then we may write

$$\boldsymbol{e}^t_{\boldsymbol{a}} = \left(I + \sum_1^\infty \frac{t^k}{k!}\boldsymbol{a}^{*k}\right)1_N = \left(\sum_0^\infty \frac{t^k}{k!}\boldsymbol{a}^{*k}\right)1_N.$$

Similarly for $\phi \in \boldsymbol{S}^N, \boldsymbol{b} \in \boldsymbol{V}^N$

$$\boldsymbol{e}^{\lambda t}_{\boldsymbol{a}} = \boldsymbol{e}^t_{\lambda \boldsymbol{a}}, \quad \lambda \in \boldsymbol{R}$$
$$\phi \circ \boldsymbol{e}^t_{\boldsymbol{a}} = \left(\sum_0^\infty \frac{t^k}{k!}\boldsymbol{a}^{*k}\right)\phi, \quad \forall \phi \in \boldsymbol{S}^N,$$
$$\boldsymbol{b} \circ \boldsymbol{e}^t_{\boldsymbol{a}} = \left(\sum_0^\infty \frac{t^k}{k!}\boldsymbol{a}^{*k}\right)\boldsymbol{b}, \quad \forall \boldsymbol{b} \in \boldsymbol{V}^N.$$

The *exponential* transform is defined as

$$\operatorname{Exp}\boldsymbol{a} := \boldsymbol{e}^t_{\boldsymbol{a}}\Big|_{t=1}, \quad \boldsymbol{e}^t_{\boldsymbol{a}} = \operatorname{Exp} t\boldsymbol{a}.$$

Exp carries near-nullity, i.e. near-0 elements of the Lie algebra $\boldsymbol{V}^N$ of all vector fields to near-1 elements of the Lie group $\boldsymbol{D}^N$ of all diffeomorphisms.

§2 Formal Vector Fields and Formal Flows

Now consider a formal *vector field* on $\boldsymbol{R}^N$ expressed as a formal power series in some real parameter s

$$\boldsymbol{a}^s = \sum_{i=1}^\infty s^{i-1}\boldsymbol{a}_i, \quad \boldsymbol{a}_i : \boldsymbol{R}^N \to \boldsymbol{R}^N,$$

letting aside the problem of convergence. The totality of formal vector fields on $\boldsymbol{R}^N$ form a Lie algebra $F\boldsymbol{V}^N$ under the bracket

$$[\boldsymbol{a}^s, \boldsymbol{b}^s] = \left[\sum_{i=1}^{\infty} s^{i-1}\boldsymbol{a}_i, \sum_{j=1}^{\infty} s^{j-1}\boldsymbol{b}_j\right] = \sum_{k=1}^{\infty} s^{k-1} \sum_{i+j=k+1} [\boldsymbol{a}_i, \boldsymbol{b}_j],$$

and moreover

$$(\boldsymbol{a}^s)^* = \sum_{k=1}^{\infty} s^{k-1}\boldsymbol{a}_k^*,$$

$$(\boldsymbol{a}^s)^*(\boldsymbol{b}^s)^* = \left(\sum_{i=1}^{\infty} s^{i-1}\boldsymbol{a}_i\right)^* \left(\sum_{j=1}^{\infty} s^{j-1}\boldsymbol{b}_j\right)^* = \sum_{k=1}^{\infty} s^{k-1} \sum_{i+j=k+1} \boldsymbol{a}_i^*\boldsymbol{b}_j^*.$$

The *formal dynamical system*

$$\frac{d\boldsymbol{x}}{dt} = \boldsymbol{a}^s(\boldsymbol{x})$$

has *formal flow* $\boldsymbol{e}_{\boldsymbol{a}^s}^t$ in two parameters (t, s) satisfying group property in t

$$\begin{aligned}
\boldsymbol{e}_{\boldsymbol{a}^s}^t &= \mathrm{Exp}\, t\boldsymbol{a}^s = \boldsymbol{1}_N + \sum_{1}^{\infty} \frac{t^m}{m!}(\boldsymbol{a}^s)^{*m} 1_N \\
&= 1_N + \sum_{m=1}^{\infty} t^m \sum_{i=0}^{\infty} s^{i-m} \sum_{\substack{i_1+\cdots+i_m=i \\ i_p \geqslant 1}} \frac{1}{m!}\boldsymbol{a}_{i_1}^* \cdots \boldsymbol{a}_{i_m}^* 1_N.
\end{aligned}$$

Define the *formal diagonal flow* of the formal vector field $\boldsymbol{a}^s$ as

$$\begin{aligned}
\boldsymbol{e}_{\boldsymbol{a}^s}^t|_{t=s} &= \boldsymbol{e}_{\boldsymbol{a}^s}^s = \mathrm{Exp}\, s\boldsymbol{a}^s =: f^s = \sum_{0}^{\infty} s^k \boldsymbol{f}_k, \\
&= 1_N + \sum_{k=1}^{\infty} s^k \sum_{m=1}^{k} \sum_{\substack{i_1+\cdots+i_m=k \\ i_p \geqslant 1}} \frac{1}{m!}\boldsymbol{a}_{i_1}^* \cdots \boldsymbol{a}_{i_m}^* 1_N,
\end{aligned}$$

$$\begin{aligned}
\boldsymbol{f}_1 &= \boldsymbol{a}_1^* 1_N = \boldsymbol{a}_1, \\
\boldsymbol{f}_k &= \boldsymbol{a}_k + \sum_{m=2}^{\infty} \sum_{\substack{k_1+\cdots+k_m=k \\ k_p \geqslant 1}} \frac{1}{m!}\boldsymbol{a}_{k_1}^* \cdots \boldsymbol{a}_{k_m}^* 1_N, \quad k \geqslant 1.
\end{aligned}$$

$\boldsymbol{f}^s$ is a near-1 formal map. Conversely, *given any near-1 formal map* $f^s = 1_N + \sum_1^{\infty} s^k \boldsymbol{f}_k$, *there exists a unique formal field* $\boldsymbol{a}^s$ *whose formal diagonal flow is* $\boldsymbol{f}^s$. In fact, the coefficients $\boldsymbol{a}_k$ can be solved back recursively in terms of $\boldsymbol{f}_k$ and $\boldsymbol{a}_1, \cdots, \boldsymbol{a}_{k-1}$

$$\begin{aligned}
\boldsymbol{a}_1 &= \boldsymbol{f}_1 \\
\boldsymbol{a}_k &= \boldsymbol{f}_k - \sum_{m=2}^{k} \sum_{\substack{i_1+\cdots+i_m=k-m \\ i_p \geqslant 0}} \frac{1}{m!}\boldsymbol{a}_{i_1}^* \cdots \boldsymbol{a}_{i_m}^* 1_N, \quad k \geqslant 2.
\end{aligned}$$

The inversion from $\boldsymbol{f}^s$ to $\boldsymbol{a}^s$ can also be obtained by applying the *logarithm* transform for near-1 maps.

$$s\boldsymbol{a}^s = \sum_{k=1}^{\infty} s^k \boldsymbol{a}_k = \mathrm{Log}\boldsymbol{f}^s,$$

where Log is defined by

$$\mathrm{Log}\boldsymbol{f}^s = \sum_{m=1}^{\infty} \frac{(-1)^{m-1}}{m} \boldsymbol{h}_m^s,$$

$$\boldsymbol{h}_1^s = \boldsymbol{f}^s - 1_N = \sum_{k=1}^{\infty} s^k \boldsymbol{f}_k^* 1_N$$

$$\boldsymbol{h}_m^s = \boldsymbol{h}_{m-1}^s \circ \boldsymbol{f}^s - \boldsymbol{h}_{m-1}^s = \sum_{k_1,\cdots,k_m=1}^{\infty} s^{k_1+\cdots+k_m} \boldsymbol{f}_{k_1}^* \cdots \boldsymbol{f}_{k_m}^* 1_N, \quad m \geqslant 2$$

satisfying

$$\mathrm{Log}\,\mathrm{Exp}\,\boldsymbol{a}^s = \boldsymbol{a}^s, \quad \mathrm{Exp}\,\mathrm{Log}\boldsymbol{f}^s = \boldsymbol{f}^s$$

For details, see [2].

The existence of formal vector fields for arbitrary near-1 maps makes the latter to be respresentable in exponential form. The classical Campbell-Hausdorff-Dynkin formula is in fact a special case of formal vector fields. Apart from linear systems, the formal vector field may diverge even when the near-1 formal map converges. The convergence problem seems to be difficult and is related to the KAM-type theories.

§3 Dual Properties between Formal Fields and Flows

The near-1 formal map and the formal vector field completely characterize each other. We give a series of dual properties between them, $\boldsymbol{f}^s \sim \boldsymbol{a}^s, \boldsymbol{e}^s \sim \boldsymbol{b}^s$, [1].

1. $\boldsymbol{f}^s \circ \boldsymbol{e}^t = \boldsymbol{e}^t \circ \boldsymbol{f}^s, \forall s, t \Longleftrightarrow [\boldsymbol{a}^s, \boldsymbol{b}^t] = 0, \forall s, t; \Longleftrightarrow [\boldsymbol{a}_i, \boldsymbol{b}_j] = 0, \forall i, j.$

2. $\boldsymbol{f}^s$ is a one parameter group $\Longleftrightarrow \boldsymbol{a}^s$ is independent of s, i.e. $\boldsymbol{a}^s = \boldsymbol{a}_0$.

Note that the one parameter group property is a strong structural property, all coefficients $\boldsymbol{f}_k$ are uniquely determined by the leading one $\boldsymbol{f}_1$, and $\boldsymbol{f}^s = e^s_{\boldsymbol{f}_1}$, hence convergent.

3. $\boldsymbol{f}^s$ is *revertible*, i.e. $\boldsymbol{f}^{-s} \circ \boldsymbol{f}^s = 1_N, \Longleftrightarrow \boldsymbol{a}^s$ is even in s, i.e. all $\boldsymbol{a}_{2k+1} = 0$.

Note that the revertibility property is weaker than the one-parameter group property, all coefficients $\boldsymbol{f}_{2k}$ are determined by $\boldsymbol{f}_1, \boldsymbol{f}_3, \ldots, \boldsymbol{f}_{2k-1}$; so the order of approximation between two revertible maps is always even. They have important implications in numerical algorithms.

$$4.\quad \boldsymbol{f}_k = \boldsymbol{e}_{\boldsymbol{a}_0,k}, \quad k = 1,2,\cdots,p \qquad\qquad \boldsymbol{a}_0 = \boldsymbol{f}_1,$$

$$\boldsymbol{f}_{p+1} \neq \boldsymbol{e}_{\boldsymbol{a}_0,p+1} \qquad \Longleftrightarrow \qquad \boldsymbol{a}_1 - \cdots - \boldsymbol{a}_{p-1} = 0,$$

$$\boldsymbol{a}_p = \boldsymbol{f}_{p+1} - \boldsymbol{e}_{\boldsymbol{a}_0,p+1} \neq 0$$

This means that $\boldsymbol{f}^s$ approximates the flow $e^s_{\boldsymbol{a}_0}$ to the order p if and only if $\boldsymbol{a}^s$ approximates the field $\boldsymbol{a}_0$ to the order $p-1$.

5. Many physically important dynamical systems possess *structures*. The three classical cases with structures are *Hamiltonian*, *Liouville* and *contact* systems, [8].

For Hamiltonian systems, $N = 2n$, the vector fields are *infinitesimal symplectic* transformation, i.e., of the form

$$\boldsymbol{a} = J_{2n}\nabla H, \quad J_{2n} = \begin{pmatrix} O_n & -I_n \\ I_n & O_n \end{pmatrix}$$

for some $H \in \boldsymbol{S}^{2n}$. The flows $\boldsymbol{e}^t_{\boldsymbol{a}}$ are *symplectic* maps satisfying

$$\left(\boldsymbol{e}^t_{\boldsymbol{a}}(\boldsymbol{x})\right)'_* J_{2n} \left(\boldsymbol{e}^t_{\boldsymbol{a}}(\boldsymbol{x})\right)_* = J_{2n}, \quad \forall \boldsymbol{x}, t.$$

For Liouville systems in $\boldsymbol{R}^N$, the vector fields $\boldsymbol{a}$ are *source-free*, i.e., $\operatorname{div} \boldsymbol{a}(x) = 0, \forall \boldsymbol{x}$. The flows $\boldsymbol{e}^t_{\boldsymbol{a}}$ are *volume preserving*, i.e., $\det\ (\boldsymbol{e}^t_{\boldsymbol{a}}(\boldsymbol{x}))_* = 1, \quad \forall \boldsymbol{x}, t$. Source-free fields are also called infinitesimal volume preserving transformations.

For contact systems, $N = 2n+1$, the vector fields are *infinitesimal contact* transformations, i.e., of the form

$$\boldsymbol{a} = J_{2n+1}\nabla K + \boldsymbol{e}_{2n+1}K, \quad J_{2n+1} = \begin{pmatrix} O_n & -I_n & \tilde{\boldsymbol{x}} \\ I_n & O_n & 0_n \\ \tilde{\boldsymbol{x}}' & 0'_n & 0 \end{pmatrix}, \quad \boldsymbol{e}_{2n+1} = \begin{pmatrix} 0_n \\ 0_n \\ 1 \end{pmatrix}$$

for some $K \in \boldsymbol{S}^{2n+1}$; $\tilde{\boldsymbol{x}}$ is the column vector of length n, consisting of the first n coordinates of the entire column vector $\boldsymbol{x}$ of length $2n+1$. The flow $\boldsymbol{e}^t_{\boldsymbol{a}}$ are *contact* maps, satisfying

$$\left(\boldsymbol{e}^t_{\boldsymbol{a}}(\boldsymbol{x})\right)'_* \begin{pmatrix} 0_n \\ \tilde{e}^t_a(\boldsymbol{x}) \\ 1 \end{pmatrix} = \mu_{\boldsymbol{e}^t_a}(\boldsymbol{x}) \begin{pmatrix} 0_n \\ \tilde{\boldsymbol{x}} \\ 1 \end{pmatrix}, \quad \forall \boldsymbol{x}, t.$$

for some $\mu_{\boldsymbol{e}^t_a} \in \boldsymbol{S}^{2n+1}$ satisfying $\mu_{\boldsymbol{e}^t_a}|_{t=0} = 1$

Formal Hamiltonian systems are given by formal vector fields $\boldsymbol{a}^s$ which are formally infinitesimal symplectic transformations, i.e., of the form,

$$\boldsymbol{a}_k = J_{2n}\nabla H_k, \quad k = 0,1,2,\cdots.$$

The formal flow $\boldsymbol{e}_{\boldsymbol{a}^s}^t$, are formally symplectic, i.e.,

$$\left(\boldsymbol{e}_{\boldsymbol{a}^s}^t(\boldsymbol{x})\right)_*' J_{2n}\left(\boldsymbol{e}_{\boldsymbol{a}^s}^t(\boldsymbol{x})\right)=J_{2n}, \quad \forall \boldsymbol{x}, s, t$$

understood in the context of formal power series in s, t. Formal Lionville and formal contact systems are understood in analogous formal contexts.

We have proved the following theorems on structure preservation properties for near-1 formal maps $\boldsymbol{f}^s=\boldsymbol{e}_{\boldsymbol{a}^s}^s$ and their formal fields $\boldsymbol{a}^s$ for Hamiltonian, Liouville and contact systems respectively.

$$\text{(I) } \boldsymbol{f}^s=\boldsymbol{e}_{\boldsymbol{a}^s}^s \text{ formally}\left(\begin{array}{l}\text{symplectic} \\ \text{volume preserving} \\ \text{contact}\end{array}\right), \quad \forall s$$

$$\Longleftrightarrow \text{(II) } \boldsymbol{e}_{\boldsymbol{a}^s}^{s,t} \text{ formally}\left(\begin{array}{l}\text{symplectic} \\ \text{volume preserving} \\ \text{contact}\end{array}\right), \quad \forall s, t$$

$$\Longleftrightarrow \text{(III) } \boldsymbol{a}_i \text{ infinitesimal}\left(\begin{array}{l}\text{symplectic} \\ \text{volume preserving} \\ \text{contact}\end{array}\right) \text{transformation}, \forall i.$$

The crucial step is (I) $\Longrightarrow$ (II). The two-parameter family $\boldsymbol{e}_{\boldsymbol{a}^s}^t$ span a surface. The one parameter family $e_{a^s}^t|_{s=t}$ traces a diagonal curve on that surface. One needs to prove that once the diagonal curve belongs to the relevant structure preserving group, then the whole surface does too, [4, 5].

The above correspondences can be extended and unified to the general case of the *structure* defined by a *unique* differential k-form ω (as in the Hamiltonian and Liouville cases) and that of the *conformal* structure defined by differential k-form ω *unique up to a non-vanishing multiplier function* (as in the contact case). These correspondences are connected with the associations of various kinds of ∞-dimensional Lie algebras of vector fields on manifold with the corresponding ∞-dimensional local Lie groups of structure-preserving diffeomorphisms.

§4 Applications to Numerical Algorithms

The formal theory has important implications in numerical methods for ODE's. Any numerical algorithm for systems has its own *step-transition map* $\boldsymbol{f}^s$ depending on the vector field $\boldsymbol{a}$ defining the system, on the form of the scheme and on the step-size parameter s [2]. The near-1 map $\boldsymbol{f}^s=1_N+\sum_1^{\infty} s^k \boldsymbol{f}_k$ can be uniquely solved from the defining equation, in fact all coefficients $\boldsymbol{f}_1, \boldsymbol{f}_2, \cdots$ can be recursively determined, and the series is convergent. The marching of the numerical process is characterized by $\boldsymbol{f}^s$. We give some examples as follows.

Explicit Euler method E

$$\boldsymbol{x}_1 - \boldsymbol{x}_0 = s\boldsymbol{a}(\boldsymbol{x}_0), \quad \boldsymbol{f}^s - 1_N = s\boldsymbol{a},$$
$$\boldsymbol{f}_E^s = 1_N + s\boldsymbol{a}$$

Implicit Euler method I

$$\boldsymbol{x}_1 - \boldsymbol{x}_0 = s\boldsymbol{a}(\boldsymbol{x}_1), \quad \boldsymbol{f}^s - 1_N = s\boldsymbol{a} \circ \boldsymbol{f}^s,$$
$$\boldsymbol{f}_I^s = (1_N - s\boldsymbol{a})^{-1}.$$

Trapezoidal method T

$$\boldsymbol{x}_1 - \boldsymbol{x}_0 = \frac{s}{2}(\boldsymbol{a}(\boldsymbol{x}_1) + \boldsymbol{a}(\boldsymbol{x}_0)), \quad \boldsymbol{f}^s - 1_N = \frac{s}{2}\boldsymbol{a} \circ \boldsymbol{f}^s + \frac{s}{2}\boldsymbol{a},$$
$$\boldsymbol{f}_T^s = \left(1_N - \frac{s}{2}\boldsymbol{a}\right)^{-1} \circ \left(1_N + \frac{s}{2}\boldsymbol{a}\right).$$

Centered Euler method C

$$\boldsymbol{x}_1 - \boldsymbol{x}_0 = s\boldsymbol{a}\left(\frac{1}{2}\boldsymbol{x}_1 + \frac{1}{2}\boldsymbol{x}_0\right), \quad \boldsymbol{f}^s - 1_N = s\boldsymbol{a} \circ \left(\frac{1}{2}\boldsymbol{f}^s + \frac{1}{2}1_N\right),$$
$$\boldsymbol{f}_C^s = \left(1_N + \frac{s}{2}\boldsymbol{a}\right) \circ \left(1_N - \frac{s}{2}\boldsymbol{a}\right)^{-1}.$$

Linear multi-step method

$$\sum_{n=0}^{m} \alpha_n \boldsymbol{x}_n = s\sum_{n=1}^{m} \beta_n \boldsymbol{a}(\boldsymbol{x}_n), \quad \sum_{i=0}^{m} \alpha_i (\boldsymbol{f}^s)^i = s\sum_{i=0}^{m} \beta_i \boldsymbol{a} \circ (\boldsymbol{f}^s)^i,$$
$$(\boldsymbol{f}^s)^0 = 1_N, (\boldsymbol{f}^s)^k = \underbrace{\boldsymbol{f}^s \circ \boldsymbol{f}^s \circ \cdots \circ \boldsymbol{f}^s}_{k \text{ times}}.$$

The order of a method is measured by the difference of $\boldsymbol{f}^s$ from the flow $\boldsymbol{e}_{\boldsymbol{a}}^s$

$$\begin{aligned} \text{ord } \boldsymbol{f}^s = p &\Longleftrightarrow \boldsymbol{f}_k = \boldsymbol{e}_k \quad k = 1, \cdots, p, \boldsymbol{f}_{p+1} \neq \boldsymbol{e}_{p+1} \\ &\Longleftrightarrow \boldsymbol{a}_0 = \boldsymbol{a}, \boldsymbol{a}_i = 0, k = 1, \cdots, p-1, \quad \boldsymbol{a}_p = \boldsymbol{f}_{p+1} - \boldsymbol{e}_{p+1} \neq 0. \end{aligned}$$

Now the existence of formal vector field $\boldsymbol{a}^s$ for the near-1 step transition map $\boldsymbol{f}^s$ means that, although the algorithm gives only approximate solution of the original system, it gives the exact solution at discrete times of a *perturbed formal system* defined by the *formal vector field* $\boldsymbol{a}^s$ whose leading term $\boldsymbol{a}_0$ is the original vector field, the order of approximation of the perturbed system to the original one corresponds to the order of the algorithm. What is more important is the qualitative side of the problem. For dynamical systems with specific geometric structure, it is natural and mandatory to require the algorithm to be *structure-preserving*. Then the theorem on structure preservation in §3, 5 ensures that the perturbation of vector field is not arbitrary, but kept within the same governing structure. This was the one of the guidelines for the author's development of *Hamiltonian* (i.e., *symplectic*) *algorithms* for Hamiltonian systems with remarkable performance, extensive computer experimentation shows decisively their overwhelming superiority over the conventional non-symplectic algorithms, especially in global, structural and qualitative aspects and in long-term tracking capabilities, [3, 6, 7]. The related algebraic and geometric considerations seem to promise more.

Acknowledgements The author is grateful to Dr. Wang Dao-liu for discussions and help in preparing the manuscript.

References

[1] Feng K. The calculus of formal power series for diffeomorphisms and vector fields: [Preprint]. Beijing: CAS. Computing Center, 1991

[2] Feng K. The step transition operators for multi-step methods of ODE's: [Preprint]. Beijing: CAS. Computing Center, 1991

[3] Qin M Z, Wang D L, Zhang M Q. Explicit symplectic difference schemes for separable Hamiltonian systems. J Comp Math. 1991 9(3): 211-221

[4] Feng K. The calculus of generating functions and the formal energy for Hamiltonian algorithms: [Preprint]. Beijing: CAS. Computing Center, 1991

[5] Feng K. Volume preserving algorithms for systems with divergence-free vector fields: [Preprint]. Beijing: CAS. Computing Center, 1991

[6] Feng K. On difference schemes and symplectic geometry. In Feng K, ed. Proc 1984 Beijing Symp Diff Geometry and Diff Equations. Beijing: Science Press, 1985. 42-58

[7] Feng K, Qin M Z. Hamiltonian algorithms and a comparative numerical study. Comput Phys Comm. 1991 65: 173-187

[8] Arnold V I. Mathematical Methods of Classical Mechanics. New York: Springer. 1978

17 Symplectic, Contact and Volume-Preserving Algorithms①

辛算法、切触算法和保体积算法

§1 Introduction

We study structure-preserving algorithms for dynamical systems preserving certain geometric structure. Most important classical cases are Hamiltonian dynamics preserving the symplectic structure, volume-preserving dynamics and contact dynamics preserving the contact structure. We present a brief outlook on symplectic algorithms for Hamiltonian systems, some general methodologies for the construction of structure-preserving algorithms and results on contact algorithms for contact systems and volume-preserving algorithms for source-free systems. We use coordinate description and matrix notation, the coordinate vector in $\mathbf{R}^m$ and vector functions $a : \mathbf{R}^m \to \mathbf{R}^m$ are denoted by column matrices. The identity vector function 1_m is given by $1_m(x) = x$. For vector function $a = (a_1, \cdots, a_m)^T : \mathbf{R}^m \to \mathbf{R}^m$,

$$a_* := \left(\frac{\partial a_i}{\partial x_j}\right) = \text{ Jacobian matrix of } a =: \frac{\partial a}{\partial x}$$

$$a^* := \sum a_i \frac{\partial}{\partial x_i} = \text{ linear differential operator of first order associated to a.}$$

The association $a \to a^*$ is linear, a^* operates on scalar functions $\phi : \mathbf{R}^m \to \mathbf{R}$ and on vector functions $b : \mathbf{R}^m \to \mathbf{R}^m$ as

$$a^*\phi = \sum a_i \frac{\partial \phi}{\partial x_i}$$

$$a^*b = a^*\left(b_1, \cdots, b_m\right)^T = \left(a^*b_1, \cdots, a^*b_m\right)^T = b_*a, \quad a^*1_m = a.$$

Multiple applications of linear differential operators are naturally defined such as a^*b^*, $\left(a^*b^*\right)c^*, a^*\left(b^*c^*\right)$, etc. The operations are multilinear, associative but non-commutative; thus powers can be defined

$$a^{*k} = a^*a^* \cdots a^*(k \text{ times}), \quad a^k := a^{*k}1_m.$$

---

① In Shi Z C, Ushijima T, ed. Proc 1st Chinar-Japan Conf on Numer Math. Singapore: World Scientific, 1993. 1-28

The identity operator I operates on scalar and vector functions ϕ and b as $I\phi = \phi, Ib = b$.

We identify all vector functions $a : \mathbf{R}^m \to \mathbf{R}^m$ as vector fields. All vector fields in $\mathbf{R}^m$ form a (∞-dimensional) real Lie algebra $\mathbf{V}_m$ under Lie bracket

$$[a, b] := a_* b - b_* a = b^* a - a^* b = (b^* a^* - a^* b^*)\, 1_m.$$

The Lie algebra $\mathbf{V}_m$ is associated to the (∞-dimensional) local Lie group $\mathbf{D}_m$ of near-identity diffeomorphisms — or simply near-1 maps — of $\mathbf{R}^m$.

Define an operator power series in parameter t

$$\operatorname{Exp} ta^* := I + \sum_{k=1}^{\infty} \frac{1}{k!}(ta)^{*k} \tag{1.1}$$

which operates on scalar functions ϕ and vector functions b just term by term. We also define

$$\exp ta := (\operatorname{Exp} ta^*)\, 1_m := 1_m + \sum_{k=1}^{\infty} \frac{1}{k!} t^k a^{*k} 1_m = 1_m + \sum_{k=1}^{\infty} \frac{t^k}{k!} a^k \tag{1.2}$$

which is a family of near-1 maps of $\mathbf{R}^m$, having important properties

1. one parameter group property $\exp(t+s)a = (\exp ta) \circ (\exp sa)$
2. $\phi \circ (\exp ta) = (\operatorname{Exp} ta^*)\, \phi$, for all scalar function ϕ.
3. $b \circ (\exp ta) = (\operatorname{Exp} ta^*)\, b$, for all vector function b.

Each dynamical system $\dot{x} = a(x)$ in $\mathbf{R}^m$ defined by the vector function $a \in \mathbf{V}_m$ possesses a phase flow $e_a^t = e^t$ which is a one-parameter (in t) group of near-1 maps of $\mathbf{R}^m$, i.e., $e_a^t \in \mathbf{D}_m$, satisfying

$$e_a^{t+s} = e_a^t \circ e_a^s,$$

represented as an exponential series

$$e_a^t = 1_m + \sum_1^{\infty} t^k e_k = \exp ta \tag{1.3}$$

and generating the solution $x(0) \to e_a^t x(0) = x(t)$. Each numerical method for the given system $\dot{x} = a(x)$ possesses a step transition map $f_a^s = f^s$ which is a one-parameter (in step-size s) family of near-1 maps $\in \mathbf{D}_m$, in general not satisfying the group property in s, represented as a formal power series

$$f_a^t = 1_m + \sum_1^{\infty} s^k f_k, \quad f_k \in \mathbf{V}_m, \quad f_1 = a, \tag{1.4}$$

and generating the numerical solution $x(0) \to (f_a^s)^N x(0) \approx x(Ns)$[9, 10, 14]. We study the approximation problem $f_a^s \approx e_a^s$, specially for dynamical systems in $\mathbf{R}^m$ with vector fields a belonging to a Lie subalgebra $\mathbf{L} \subset \mathbf{V}_m$, then phase flows e_a^t belong to the associated group $\mathbf{G} \subset \mathbf{D}_m$. We want to construct approximations $f_a^s \approx e_a^s$ within the same group $\mathbf{G}$, called *structure-preserving* algorithms.

§2 Explicit Structure-preserving Algorithms

Let $\mathbf{G}$ be a local Lie group consisting of all near-1 maps $\in \mathbf{D}_m$ preserving a specific geometric structure (usually defined by a differential form on $\mathbf{R}^m$, for classical cases, see §3, 4 5, $\mathbf{G}$ is an ∞-dimensional Lie subgroup of $\mathbf{D}_m$. Then, all vector fields $a \in \mathbf{V}_m$ whose phase flow e_a^t belongs to $\mathbf{G}$ form a Lie algebra $\mathbf{L}$ under Lie bracket. $\mathbf{L}$ is an ∞-dimensional Lie subalgebra of $\mathbf{V}_m$, called *the* Lie algebra of Lie group $\mathbf{G}$. For emphasizing their relation we say $\mathbf{L}$ and $\mathbf{G}$ associate to each other.

Consider a dynamical system

$$\dot{x} = a(x), \quad a \in \text{ Lie algebra } \mathbf{L}. \tag{2.1}$$

Then the phase flow e_a^t belongs to the Lie group $\mathbf{G}$. For the algorithmic approximations $g^s \approx e_a^s$, we require, first of all, g^s belong to the same Lie group $\mathbf{G}$, then it is called structure-preserving or $\mathbf{L}$-preserving. Moreover, e_a^s is a one-parameter subgroup of $\mathbf{G}$, satisfying $e^s \circ e^t = e^{s+t}$. So one might be tempted to require g^s to satisfy the same group property $g^s \circ g^t = g^{s+t}$, then it turns out $g^s = e^s$, so this requirement is too stringent in general apart from certain important special cases to be mentioned below. However, phase flow satisfies the weaker condition $e^s \circ e^{-s} = 1_m$, then it is reasonable and practicable to require g^s to satisfy the same weaker condition $g^s \circ g^{-s} = 1_m$, then g^s is called *revertible*, that means g^s always generates coincident forward and backward orbits. So our aim is to construct $\mathbf{L}$-preserving revertible algorithms. For each specific $\mathbf{L}, \mathbf{G}$, for example, symplectic geometry for Hamiltonian systems, contact geometry for contact systems, volume-preserving geometry for source-free systems, it is quite different from case to case, see §3,4,5.

We indicate a general method of construction valid for all $\mathbf{L}, \mathbf{G}$[6, 22]. Suppose a vector field a in $\mathbf{L}$ admits a *decomposition* $a = a^1 + \cdots + a^k$, where each field a^i is in $\mathbf{L}$ and so simple that a simple algorithm in $\mathbf{G}$ approximating its phase flow e_a^s is available. Using these elementary maps as basic components, ..., etc. Since every subgroup $\mathbf{G} \subset \mathbf{D}_m$ is closed under composition (group operation in $\mathbf{D}_m$), then through suitable patterns of successive *composition* one gets various revertible algorithms in $\mathbf{G}$ approximating the phase flow e_a^s with various orders of accuracy. The process makes no use of specific properties of $\mathbf{L}, \mathbf{G}$. In order to be precise we list some basic properties of composition of near-1 maps including revertible maps.

We write $g^s \approx e_a^s$ if $g^s = e_a^s + O(s^2), g^s \approx e_a^s$, ord p if $g^s = e_a^s + O(s^{p+1}), g^s \approx e_a^s$, ord ∞ if $g^s = e_a^s$.

1. $\forall g^s$ define $\check{g}^s = (g^{-s})^{-1}$. Then $\check{\check{g}}^s = g^s, (f^s \circ g^s)^\vee = \check{g}^s \circ \check{f}^s, \{g^s \text{ revertible }\} \Longleftrightarrow \{g^s \circ g^{-s} = 1_m \Longleftrightarrow \{G^s = \check{g}^s\}, g^s \circ \check{g}^s$ and $\check{g}^s \circ g^s$ always revertible.
2. f^s, g^s revertible $\Longrightarrow$ symmetrical composites $f^s \circ g^s \circ f^s$ revertible, while simple composite $f^s \circ g^s$ may not be revertible, $\{f^s \circ g^s \text{ revertible }\} \Longleftrightarrow \{f^s \circ g^s = g^s \circ f^s\}$.
3. $g^s \approx e_a^s \Longrightarrow \check{g}^s \approx e_a^s, g^{s/2} \circ \check{g}^{s/2}$ and $\check{g}^{s/2} \circ g^{s/2} \approx e_a^s$, ord 2, always revertible. (2.2)

   $g^s \approx e_a^s$ revertible $\Longrightarrow g^s \approx e_a^s$, ord even $\geqslant 2$ or ∞.
4. $g^s \approx e_a^s$, ord 2, revertible $\Longrightarrow f^s := G^{\alpha s} \circ g^{\beta s} \circ g^{\alpha s} \approx e_a^s$, ord 4 for

   $$2\alpha + \beta = 1, 2\alpha^3 + \beta^3 = 0, \text{ i.e., } \alpha = \left(2 - 2^{1/3}\right)^{-1}, \beta = 1 - 2\alpha \leqslant 0. \tag{2.3}$$
5. $g^s \approx e_a^s$, ord $2l$, revertible $\Longrightarrow g^{\alpha s} \circ g^{\beta s} \circ g^{\alpha s} \approx e_a^s$, ord $2l+2$, for

   $$2\alpha + \beta = 1, 2\alpha^{2l+1} + \beta^{2l+1} = 0, \text{ i.e., } \alpha = \left(2 - 2^{1/(2l+1)}\right)^{-1}, \beta = 1 - 2\alpha \leqslant 0. \tag{2.4}$$

   See [23, 24], the practical utility seems to be restricted to order $\leqslant 6$.

   For $a = a_1 + \cdots + a_k$, write $e_{a_i}^s = e_i^s, \forall i$.
6. $g_i^s \approx e_i^s, \forall i \Longrightarrow g_1^s \circ \cdots \circ g_k^s \approx e_a^s$. (2.5)
7. $g_i^s \approx e_i^s, \forall i \Longrightarrow g_1^{s/2} \circ \cdots \circ g_k^{s/2} \circ g_k^{s/2} \circ \cdots \circ g_1^{s/2} \approx e_a^s$, ord 2. (2.6)
8. $g_i^s \approx e_i^s$, ord 2 revertible $\forall i \Longrightarrow g_1^{s/2} \circ \cdots \circ g_k^{s/2} \circ g_k^{s/2} \circ \cdots \circ g_1^{s/2} \approx e_a^s$, ord 2. (2.7)

We give some elementary algorithms, i.e., the explicit Euler method E, the implicit Euler method I, the centered Euler method C. E, I, C, especially the oldest and simplest E, will be the basic components for the composition of structure preserving algorithms. Then transition maps for step-size s for $x \to$ next step $\hat{x}$ will be denoted by E^s, I^s, C^s.

$$E: \hat{x} = x + sa(x), \quad E_a^s = 1_m + sa \approx e_a^s, \text{ ord } 1, \text{ non-revertible.} \tag{2.8}$$

$$I: \hat{x} = x + sa(\hat{x}), \quad I_a^s = (1_m - sa)^{-1} = \check{E}_a^s \approx e_a^s,$$

$$\text{ord } 1, \text{ non-revertible.} \tag{2.9}$$

$$C: \hat{x} = x + sa\left(\frac{1}{2}\hat{x} + \frac{1}{2}x\right), \quad \text{or in equivalent 2 -stage form}$$

$$\bar{x} = x + \frac{s}{2}a(\bar{x}), \quad \hat{x} = 2\bar{x} - x. \tag{2.10}$$

$$C_a^s = E_a^{s/2} \circ I_a^{s/2} = E_a^{s/2} \circ \check{E}_a^{s/2} = 2I_a^{s/2} - 1_m \approx e_a^s,$$

$$\text{ord } 2, \text{ revertible.}$$

A vector field a is called *nilpotent of degree 2* if

$$a^2 = a_* a = a^* a = 0 \Longleftrightarrow e_a^s = 1_m + sa = E_a^s, \tag{2.11}$$

so with such "simple" vector fields, the method E gives exact solutions.

Formula (2.6),(2.3),(2.4) lead immediately to the construction of revertible **L**-preserving algorithms by decomposition-composition: Let

$$a=\sum_{i=1}^{k} a^i, \quad a^i \in \mathbf{L} \quad f_i^s \approx e_i^s, \quad \mathbf{L}\text{-preserving} \tag{2.12}$$

ord $\geqslant 1, f_i^s \in \mathbf{G}, \forall i=1,\cdots,k$. Then

$$f^s := f_1^s \circ \cdots \circ f_k^s \approx e_a^s, \quad \text{ord } 1, \mathbf{L}\text{-preserving}, \tag{2.13}$$

$$g^s := f^{s/2} \circ \check{f}^{s/2} = f_1^{s/2} \circ \cdots \circ f_k^{s/2} \circ \check{f}_k^{s/2} \circ \cdots \circ \check{f}_1^{s/2} \approx e_a^s,$$
$$\text{ord 2, revertible } \mathbf{L}\text{-preserving}, \tag{2.14}$$

$$h^s := g^{\alpha s} \circ g^{\beta s} \circ g^{\alpha s} \approx e_a^s, \quad \text{with the } \alpha, \beta \text{ in (2.3)}$$
$$\text{ord 4, revertible } \mathbf{L}\text{-preserving; etc.} \tag{2.15}$$

A case with $a=a^1+a^2+a^3$ for construction of 3D volume preserving algorithms will be shown in §5. In particular, as an ideal case, take $f_i^s=e_{a^i}^s=e_i^s, i=1,\cdots k$, then

$$f^s := e_1^s \circ \cdots \circ e_k^s \approx e_a^s, \quad \text{ord } 1, \mathbf{L}\text{-preserving},$$
$$g^s := e_1^{s/2} \circ \cdots \circ e_{k-1}^{s/2} \circ e_k^s \circ e_{k-1}^{s/2} \circ \cdots \circ e_1^s \approx e_a^s,$$
$$\text{ord 2, revertible } \mathbf{L}\text{-preserving}. \tag{2.16}$$

This construction gives actually a *numerical algorithm* only when the component phase flows e_i^s are available and algorithmically implementable, this is quite often the case.

We call a vector field $a \in \mathbf{L}$ to be **L**-separable if it admits a decomposition

$$a=\sum_{i=1}^{k} a^i, \quad a^i \in \mathbf{L} \text{ and nilpotent, i.e., } a_*^i a^i = 0. \tag{2.17}$$

In this case, by (2.11), $e_{a^i}^s = E_{a^i}^s =: E_i^s$, so f^s, g^s become

$$f^s := E_1^s \circ \cdots \circ E_k^s \approx e_a^s, \quad \text{ord } 1, \mathbf{L}\text{-preserving},$$
$$g^s := E_1^{s/2} \circ \cdots \circ E_{k-1}^{s/2} \circ E_k^s \circ E_{k-1}^{s/2} \circ \cdots \circ E_1^{s/2} \approx e_a^s,$$
$$\text{ord 2, revertible } \mathbf{L}\text{-preserving}, \tag{2.18}$$

and all the related revertible **L**-preserving algorithms are explicit. A simple **L**-separable case with decomposition $a=a^1+a^2$ is important for the construction of explicit symplectic method for Hamiltonian systems. See immediately below §3.

§3 Symplectic Algorithms for Hamiltonian Systems

Hamiltonian systems are defined exclusively in even-dimensional phase space $\mathbf{R}^{2n}$ in variables $z_1 = p_1, \ldots, z_n = p_n, z_{n+1} = q_1, \cdots, z_{2n} = q_n$ in the form

$$\dot{a} = JH_z =: c(z), \quad J = J_{2n} = \begin{pmatrix} 0 & -I_n \\ I_n & 0 \end{pmatrix} \tag{3.1}$$

or, in 2-symbol notation

$$\begin{aligned} \dot{p} &= -H_q(p,q) =: a(p,q) \\ \dot{q} &= H_p(p,q) =: b(p,q) \end{aligned} \quad z = \begin{pmatrix} p \\ q \end{pmatrix}, \quad c = \begin{pmatrix} a \\ b \end{pmatrix}, \tag{3.2}$$

where $H(z) = H(p,q)$ is a function in $\mathbf{R}^{2n}$, unique up to a constant to define the system, called the Hamiltonian or energy function. For convenience we also write

$$z = (p,q), \quad c = JH_z = (-H_q, H_p) = (a,b). \tag{3.3}$$

All *Hamiltonian* vector fields c, i.e.,

$$c = JH_z \text{ for some } H(z) \Longleftrightarrow \left(\frac{\partial c}{\partial z}\right)^T J_{2n} + J_{2n}\left(\frac{\partial c}{\partial z}\right) = 0 \quad \text{identically in } z \tag{3.4}$$

form a Lie algebra $\mathbf{L} = SpV_{2n} \subset \mathbf{V}_{2n}$ under Lie bracket. $\mathbf{L}$ is associated to the local Lie group $\mathbf{G} = SpD_{2n} \subset \mathbf{D}_{2n}$, consisting of all near-identity-*symplectic* maps $g \in \mathbf{D}_{2n}$, i.e.,

$$g \text{ preserves the differential 2-form } \omega = \sum_{i=1}^{n} dp_i \wedge dq_i$$

$$\Longleftrightarrow \left(\frac{\partial g}{\partial z}\right)^T J\left(\frac{\partial g}{\partial z}\right) = 0, \quad \text{identically in } z. \tag{3.5}$$

The 2-form ω specifies the *symplectic* structure of the phase space, intrinsic to all Hamiltonian systems so that the phase flow $e_c^t, c = JH_z$, of any Hamiltonian function H is always symplectic. We require the algorithmic approximation $f^s \approx e_c^s$ to be symplectic too, i.e.,

$$\left(\frac{\partial f}{\partial z}\right)^T J\left(\frac{\partial f}{\partial z}\right) = 0. \tag{3.6}$$

This *proper* requirement had been overlooked by and large up to the very recent past, see e.g. [2 – 6].

We start with the simplest case $H_1(p,q) = \phi(p)$, independent of q. Then

$$c^1 := J\phi_z = \begin{pmatrix} 0 \\ \phi_p(p) \end{pmatrix}, \quad \left(\frac{\partial c^1}{\partial z}\right) c^1 = \begin{pmatrix} 0 & 0 \\ \phi_{pp} & 0 \end{pmatrix}\begin{pmatrix} 0 \\ \phi_p \end{pmatrix} = \begin{pmatrix} 0 \\ 0 \end{pmatrix}$$

$$\Longrightarrow c^1 \text{ nilpotent of degree } 2, e_{c^1}^s = E_{c^1}^s. \tag{3.7}$$

Using notational convention (3.3), we have

$$\begin{aligned} &c^1 = J\phi_z = (0, \phi_p(p)) = (0, b(p)), \\ &e^s_{c^1} = E^s(0, b) = E^s_b : \quad \hat{p} = p, \hat{q} = q + sb(q). \end{aligned} \tag{3.8}$$

Similarly for $H_2(p, q) = \psi(q)$, independent of p, we have

$$\begin{aligned} &c^2 = JH_{2,z} = (-\psi_q(q), 0) = (a(q), 0), \\ &e^s_{c^2} = E^s(a, 0) = E^s_a : \quad \hat{p} = p + sa(q), \hat{q} = q. \end{aligned} \tag{3.9}$$

In classical mechanics, most (but not all) Hamiltonians are of the form

$$H(p, q) = H_1 + H_2 = \phi(p) + \psi(q). \tag{3.10}$$

Then we have a natural decomposition

$$\begin{aligned} &c := JH_z = J\phi_z + J\psi_z = (-\psi(q), \phi(p)) \\ &(a(q), b(p)) = (a(q), 0) + (0, b(p)). \end{aligned} \tag{3.11}$$

The two simplest Hamiltonian fields on the right possess explicit Eulerian algorithms

$$E^s_a : \hat{p} = p + sa(q), \quad \hat{q} = q, \quad \hat{p} = p, \quad \hat{q} = q + sb(p), \tag{3.12}$$

which are Hamiltonian phase flows, hence symplectic and revertible and will be used as basic algorithmic components to form successive compositions giving the following explicit symplectic algorithms for Hamiltonians of the form $H = \phi(p) + \psi(q)$:

1. Non-revertible ord 1,

$$\begin{array}{ll} P^s_c := E^s_b \circ E^s_a, \quad \check{P}^s_c = Q^s_c & P^s_c := E^s_a \circ E^s_b, \quad \check{Q}^s_c = P^s_c \\ \hat{p} = p + sa(q) & \hat{q} = q + sb(p) \\ \hat{q} = q + sb(\hat{p}) & \hat{p} = p + sa(\hat{q}) \end{array} \tag{3.13}$$

2. Revertible ord 2,

$$\begin{array}{ll} F^s := E^{s/2}_b \circ E^s_a \circ E^{s/2}_b = P^{s/2}_c \circ \check{P}^{s/2}_c & G^s := E^{s/2}_a \circ E^s_b \circ E^{s/2}_a = Q^{s/2}_c \circ \check{Q}^{s/2}_c \\ q_1 = q + \frac{1}{2}sb(p) & p_1 = p + \frac{1}{2}sa(q) \\ \hat{p} = p + sa\,(q_1) & \hat{q} = q + sb\,(p_1) \\ \hat{q} = q_1 + \frac{1}{2}sb(\hat{p}) & \hat{p} = p_1 + \frac{1}{2}sa(\hat{q}) \end{array} \tag{3.14}$$

3. Revertible ord 4,

$$
\begin{aligned}
U_c^s &= F_c^{\alpha s} \circ F_c^{\beta s} \circ F_c^{\alpha s} & V_c^s &= G_c^{\alpha s} \circ G_c^{\beta s} \circ G_c^{\alpha s} \\
&= E_b^{\lambda s} E_a^{\alpha s} E_b^{\mu s} E_a^{\beta s} E_b^{\mu s} E_a^{\alpha s} E_b^{\lambda s}, & &= E_a^{\lambda s} E_b^{\alpha s} E_a^{\mu s} E_b^{\beta s} E_a^{\mu s} E_b^{\alpha s} E_a^{\lambda s}, \\
q_1 &= q + \lambda s b(p) & p_1 &= p + \lambda s a(q) \\
p_1 &= p + \alpha s a(q_1) & q_1 &= q + \alpha s b(p_1) \\
q_2 &= q_1 + \mu s a(p_1) & p_2 &= p_1 + \mu s a(q_1) \\
p_2 &= p_1 + \beta s a(q_2) & q_2 &= q_1 + \beta s b(p_2) \\
q_3 &= q_2 + \mu s b(p_2) & p_3 &= p_2 + \mu s a(q_2) \\
\hat{p} &= p_2 + \alpha s a(q_3) & \hat{q} &= q_2 + \alpha s b(p_3) \\
\hat{q} &= q_3 + \lambda s b(\hat{p}) & \hat{p} &= p_3 + \lambda s a(\hat{q})
\end{aligned}
\tag{3.15}
$$

Note that the 2nd order F^s, G^s need only 1 (instead of 2 in conventional methods) evaluations if $c = (a, b)$, the 4th order U^s, V^s need 3 (instead of conventional 4), so apart from being symplectic they are also fast.

We say a function H is nilpotent of degree 2, if its Hamiltonian vector field JH_z is nilpotent of degree 2, i.e., $H_{zz}JH_{zz} = 0$. A decomposition of a function H as a sum of functions is equivalent to a decomposition of the Hamiltonian fields JH_z as a sum of Hamiltonian fields. So in conformity with the definition (2.17) of **L**-separability of vector fields in **L** with $\mathbf{L} = SPV_{2n}$, we say a Hamiltonian function H is symplectically separable if it admits a decomposition $H = H_1 + \cdots + H_k$ where each function H_i is nilpotent of degree 2. The expression $H = \phi(p) + \psi(q)$ presents just one simple case of symplectically separable Hamiltonians. Symplectically separable Hamiltonians have wide coverage in applications.

It is easy to see that, the Hamiltonian of the form

$$H(p, q) = \phi(Ap + Bq), \quad AB^T = BA^T, \tag{3.16}$$

where $\phi(x)$ is a function of n variables, A and B are $n \times n$ matrices, is also nilpotent of degree 2. Then the explicit Euler scheme $E_H^s = E^s(\phi) = e_H^s$ is

$$E^s(\phi): \quad \begin{aligned} \hat{p} &= p - sB^T\phi_x(Ap + Bq) \\ \hat{q} &= q + sA^T\phi_x(Ap + Bq) \end{aligned} \tag{3.17}$$

so we get a class of symplectically separable Hamiltonians [7]

$$H(p, q) = \sum_{i=1}^{k} H_i(p, q), \quad H_i(p, q) = \phi_i(A_i p + B_i q), \quad A_i B_i^T = B_i A_i^T, \tag{3.18}$$

where $\phi_i(x)$ are functions of n variables, A_i, B_i are $n \times n$ matrices. Moreover, all polynomial Hamiltonians belong to this class; in fact, it has been proved in [7] that every polynomial $H(p,q)$ in $2n$ variables p,q can be decomposed in the form of (3.18), where $\phi_i(x)$ are polynomials of n variables x, $A_i=\operatorname{diag}(a_1^i,\cdots,a_n^i)$ and $B_i=\operatorname{diag}(b_1^i,\cdots,b_n^i)$.

For symplectically separable Hamiltonians of the above class, all explicit revertible symplectic schemes formed by compositions contain solely the Euler schemes $E^s(\phi_i)$ in the form (3.17) as basic components. A related conjecture is: this class already covers all the symplectically separable cases, i.e., H is nilpotent of degree 2 if and only if it is expressible in the form (3.16).

More generally, if we have a decomposition $H=H_1+\cdots+H_k$, for which each H_i is integrable and its phase flow $e_{H_i}^s$ is algorithmically implementable, then by (2.14),(2.15) the symmetrical composites

$$g_H^s=e_{H_1}^{s/2}\circ\cdots\circ e_{H_{k-1}}^{s/2}\circ e_{H_k}^s\circ e_{H_{k-1}}^{s/2}\circ\cdots\circ e_{H_1}^{s/2} \tag{3.19}$$

$$h_H^s=g_H^{\alpha s}\circ g_H^{\beta_s}\circ g_H^{\alpha s} \tag{3.20}$$

give explicit symplectic, revertible algorithms of order 2, 4, etc. This general approach is widely applicable to *many body problems* in different physical contexts for which the *2-body problem is solvable.* Since in such problems the Hamiltonian usually admits a natural decomposition $H=\sum_{i<j}H_{ij}$, each H_{ij} accounts for a 2-body problem.

An interesting result from this approach is a construction [24] of explicit symplectic method for computing the Hamiltonian system of N vortices $z_i=(x_i,y_i)$ with intensities k_i

$$k_i\frac{dx_i}{dt}=\frac{\partial H}{\partial y_i},\quad k_i\frac{dy_i}{dt}=-\frac{\partial H}{\partial x_i},\quad i=1,\cdots,N \tag{3.21}$$

with symplectic structure $\omega_k=\sum\frac{1}{k_i}dx_i\wedge dy_i$ and

$$H=\sum_{i<j}H_{ij},\quad H_{ij}=-\frac{1}{2\pi}k_ik_j\ln r_{ij},\quad r_{ij}=\left((x_i-x_j)^2+(y_i-y_j)^2\right)^{1/2}. \tag{3.22}$$

Each H_{ij} accounts for a solvable 2-vortex motion in which both vortices z_i,z_j rotate about their center of vorticity $z=(k_iz_i+k_jz_j)(k_i+k_j)^{-1}$ with angular velocity $a=(k_i+k_j)/2\pi r_{ij}^2$ if $k_i+k_j\neq0$ or translate with linear velocity $(b(y_i-y_j),-b(x_i-x_j))$, $b=k_i/2\pi r_{ij}^2$ if $k_i+k_j=0$. These simple phase flows $e_{H_{ij}}^s$ serve as the basic algorithmic components for successive compositions, resulting in efficient explicit revertible methods, symplectic for this specific system. They are promising in application to incompressible ideal flows for tracking the vortex particles, which ought to be done in the structure-preserving way.

We now give three basic *unconditional* symplectic methods C,P,Q and their composites. They are symplectic for all Hamiltonian systems, while almost all conventional methods are non-symplectic. Recall the notational convention

$$c(z) = JH_z(z) = (-H_q(z), H_p(z)) = (a(z), b(z))$$

C. 1-stage form: $\hat{z} = z + sJH_z\left(\frac{1}{2}\hat{z} + \frac{1}{2}z\right)$.

2-stage form: $\bar{z} = z + \frac{1}{2}sc(\bar{z}), \hat{z} = \pm\frac{1}{2}sc(\bar{z}) = 2\bar{z} - z$.

$C^s = E^{s/2} \circ I^{s/2} = \check{C}^s$, revertible, ord 2.

P. 1-stage form: $\hat{z} = z + sJH_z(\hat{p}, q)$, or $\hat{p} = p + sa(\hat{p}, q), \hat{q} = q + sb(\hat{p}, q)$

$P^s = E^s(0, b) \circ I^s(a, 0) = \check{Q}^s$, non-revertible, ord 1.

Q. 1-stage form: $\hat{z} = z + sJH_z(p, \hat{q})$, or $\hat{q} = q + sb(p, \hat{q}), \hat{p} = p + sa(p, \hat{q})$

$Q^s = E^s(a, 0) \circ I^s(0, b) = \check{P}^s$, non-revertible, ord 1.

For $H = \phi(p) + \psi(q)$, methods P, Q become explicit, already given in (3.13)

$$F^s := P^{s/2} \circ \check{P}^{s/2} : \begin{array}{l} \bar{q} = q + \frac{1}{2}sb(p, \bar{q}), \quad \bar{p} = p + \frac{1}{2}sa(p, \bar{q}); \\ \hat{p} = \bar{p} + \frac{1}{2}sa(\hat{p}, \bar{q}), \quad \hat{q} = \bar{q} + \frac{1}{2}sb(\hat{p}, \bar{q}). \end{array}$$

$F^s = E^{s/2}(0, b) \circ I^{s/2}(a, 0) \circ E^{s/2}(a, 0) \circ I^{s/2}(0, b) = \check{F}^s$, revertible, ord 2.

$$G^s := Q^{s/2} \circ \check{Q}^{s/2} : \begin{array}{l} \bar{p} = p + \frac{1}{2}sa(\bar{p}, q), \bar{q} = q + \frac{1}{2}sb(\bar{p}, q); \\ \hat{q} = \bar{q} + \frac{1}{2}sb(\bar{p}, \hat{q}), \hat{p} = \bar{p} + \frac{1}{2}sa(\bar{p}, \hat{q}). \end{array}$$

$G^s = E^{s/2}(a, 0) \circ I^{s/2}(0, b) \circ E^{s/2}(0, b) \circ I^{s/2}(a, 0) = \check{G}^s$, revertible, ord 2

Higher order revertible composites can be formed in the standard way.

For the symplecticity for all these methods one needs only to verify for C, P, Q. First differentiate the difference equations in 2-stage form, then solve for Jacobian matrix $\partial\hat{z}/\partial z$, then prove its symplecticity with the aid of some linear algebra on symplectic matrices. The method C preserves all quadratic invariants of the system, the methods P, Q and their composites preserve all quadratic invariants of the form $p^T Bq$(B arbitrary $n \times n$ matrix), including angular momenta. For general references, see [4 – 6].

Numerical evidences and theoretical analysis so far accumulated show unanimously and indisputably with the qualitative and the quantitative superiority of performance of symplectic methods over conventional non-symplectic ones for Hamiltonian systems. The constructs are overwhelming and striking especially in crucial aspects of global behavior, orbits stability and long-time tracking capabilities. For the first time one now knows *consciously* a *proper* or descent way to compute *Newton's equations of motion*, i.e., first put them in Hamiltonian form, then compute by symplectic algorithms [5,6,12].

As an illustration we give some results of symplectic computation with some comparisons against conventional non-symplectic ones in Fig. 1 – 4.

Fig. 1. Non-linear Huygens oscillator $H = p^2 - q^2 + q^4$, the phase portrait consists of Huygens ovals. The initial point for the outmost orbit of C and that for A and B are the same. The explicit Runge-Kutta methods produce two artificial attractors at the foci of the Huygens ovals. In A the orbit is attracted to the left, but with very minute change of step-size it is attracted to the right in B. C shows the surprising capability of very-long-time tracking and the fidelity in phase portrait structure for symplectic methods [12,15].

Fig. 2. Kepler motion, frequency ratio 11: 20, closed orbit [20]

Fig. 3, 4. Tracking of 3-vortex motion (3.21), $H = H_{12} + H_{13} + H_{23}$, by the explicit symplectic method based on 2-vortex motions. Fig.4 A,B,C are orbits of 4000 steps of 3 vortices with intensities $1, -2, 3$ and initial locations (0,0),(0,1),(1,0) respectively. Fig. 3 is the continuation of Fig. 4A to 8000 steps. The results are due to [24].

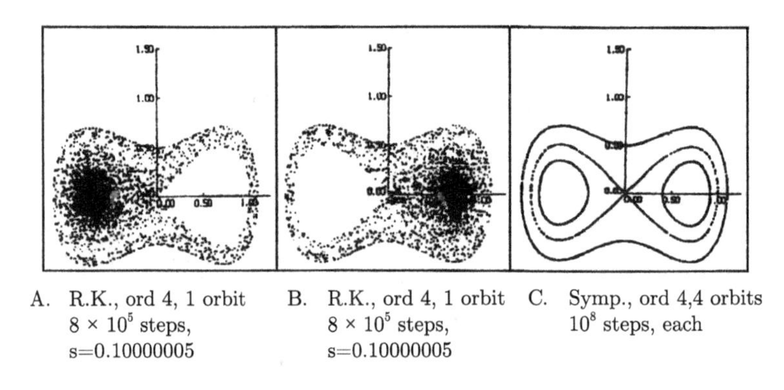

A. R.K., ord 4, 1 orbit 8×10^5 steps, s=0.10000005 B. R.K., ord 4, 1 orbit 8×10^5 steps, s=0.10000005 C. Symp., ord 4,4 orbits 10^8 steps, each

Fig. 1 Non-linear Huygens oscillator

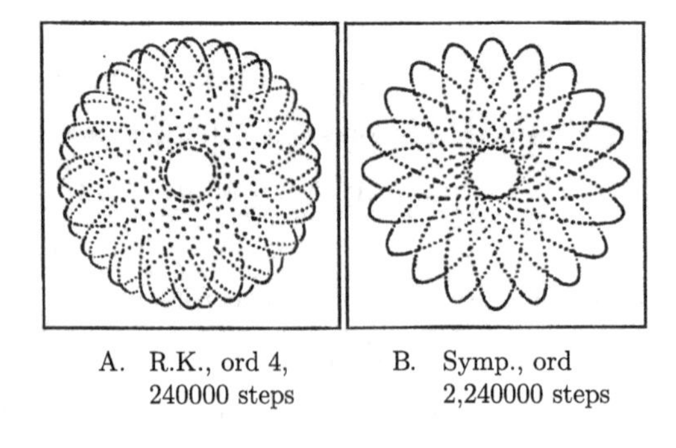

A. R.K., ord 4, 240000 steps B. Symp., ord 2,240000 steps

Fig. 2 Kepler motion, rational freg. ration, closed orbit.

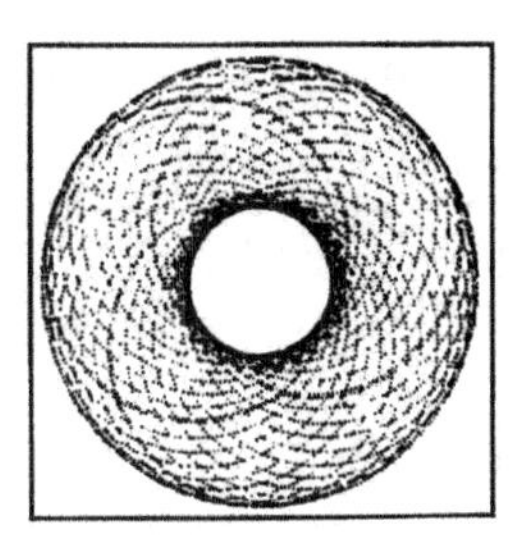

Fig. 3 One of 3 vortices

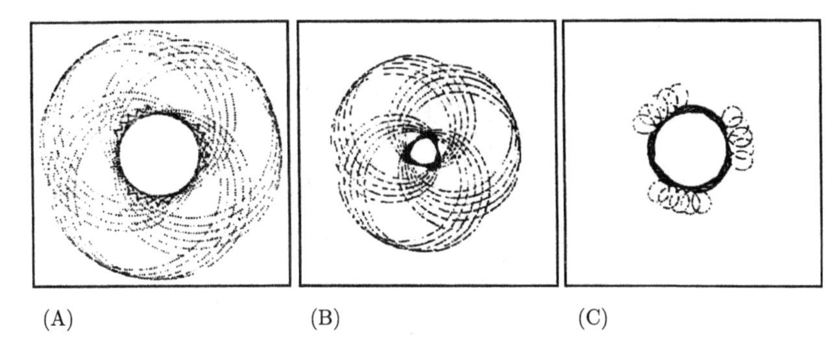

(A) (B) (C)

Fig. 4 Vortices, $s = 1$

§4 Contact Algorithms for Contact Systems

Contact systems occur only on space $\mathbf{R}^{2n+1}$ of odd dimensions. We use 3-symbol notation to denote the coordinates and vector fields on $\mathbf{R}^{2n+1}$

$$\begin{pmatrix} x \\ y \\ z \end{pmatrix}, \quad x = \begin{pmatrix} x_1 \\ \vdots \\ x_n \end{pmatrix}, \quad y = \begin{pmatrix} y_1 \\ \vdots \\ y_n \end{pmatrix}, \quad z = (z),$$

$$\begin{pmatrix} a(x,y,z) \\ b(x,y,z) \\ c(x,y,z) \end{pmatrix}, \quad a = \begin{pmatrix} a_1 \\ \vdots \\ a_n \end{pmatrix}, \quad b = \begin{pmatrix} b_1 \\ \vdots \\ b_n \end{pmatrix}, \quad c = (c).$$

A *contact* system is defined by a function $K(x,y,z)$, called contact Hamiltonian, on $\mathbf{R}^{2n+1}$ as follows

$$\begin{aligned} \frac{dx}{dt} &= -K_y + K_z x = a, \\ \frac{dy}{dt} &= K_x = b, \\ \frac{dz}{dt} &= K_e = c, \quad K_e(x,y,z) := K(x,y,z) - \langle x, K_x(x,y,z) \rangle . \end{aligned} \tag{4.1}$$

The right hand side is the general form of a *contact* vector field.

Intrinsic to all contact systems is the *contact* structure on $\mathbf{R}^{2n+1}$ defined by the differential 1-form

$$\alpha=\sum_{i=1}^{n} x_i dy_i+dz=\begin{pmatrix}0 & x^T & 1\end{pmatrix}\begin{pmatrix}dx\\ dy\\ dz\end{pmatrix} \tag{4.2}$$

up to an everywhere non-vanishing multiplier function. A *contact* transformation f is a diffeomorphism on $\mathbf{R}^{2n+1}$

$$f:\begin{pmatrix}x\\ y\\ z\end{pmatrix}\rightarrow\begin{pmatrix}\hat{x}(x,y,z)\\ \hat{y}(x,y,z)\\ \hat{z}(x,y,z)\end{pmatrix}$$

preserving the contact structure, that means

$$\sum_{i=1}^{n} \hat{x}_i d\hat{y}_i+d\hat{z}=\mu_f\left(\sum_{i=1}^{n} x_i dy_i+dz\right) \tag{4.3}$$

for some function $\mu_f \neq 0$ everywhere on $\mathbf{R}^{2n+1}$, called the multiplier of f. The explicit expression of (4.3) is

$$\begin{pmatrix}0 & \hat{x}^T & 1\end{pmatrix}\begin{pmatrix}\hat{x}_x & \hat{x}_y & \hat{x}_z\\ \hat{y}_x & \hat{y}_y & \hat{y}_z\\ \hat{z}_x & \hat{z}_y & \hat{z}_z\end{pmatrix}=\mu_f\begin{pmatrix}0 & x^T 1\end{pmatrix}.$$

All contact vector fields on $\mathbf{R}^{2n+1}$ form a Lie algebra $\mathbf{L}=\mathbf{KV}_{2n+1}$ under Lie bracket. The associated Lie group $\mathbf{G}=\mathbf{KD}_{2n+1}$ consists of all near-identity contact transformations on $\mathbf{R}^{2n+1}$. For $f\in\mathbf{G}$, $\mu_f>0$ everywhere. The phase flows e_K^t of contact systems defined by contact Hamiltonians K on $\mathbf{R}^{2n+1}$ are contact transformations. So the proper algorithms for contact systems should be *contact* too. For elements of contact geometry, see [1].

All the existing numerical methods including the symplectic ones are generically non-contact even for linear contact systems. A linear contact system is necessarily of the form,

$$\begin{aligned}\frac{dx}{dt}&=\left(\lambda I-L^T\right)x, \quad L\in gl(n), \quad \lambda\in\mathbf{R},\\ \frac{dy}{dt}&=Ly,\\ \frac{dz}{dt}&=\lambda z,\end{aligned} \tag{4.4}$$

the 3 equations are independent. The matrix of a linear contact transformation is necessarily of the block-diagonal form

$$\begin{pmatrix} \mu M^{-T} & 0 & 0 \\ 0 & M & 0 \\ 0 & 0 & \mu \end{pmatrix}, \quad M \in \mathbf{GL}(n), \quad \mu > 0. \tag{4.5}$$

Using the method C, e.g., one gets

$$\begin{aligned} \hat{x} &= \left(\left(1-\frac{s\lambda}{2}\right) I + \frac{s}{2} L^T\right)^{-1} \left(\left(1+\frac{s\lambda}{2}\right) I - \frac{s}{2} L^T\right) x =: Nx, \\ \hat{y} &= \left(I - \frac{s}{2}L\right)^{-1} \left(I + \frac{s}{2}L\right) =: My, \\ \hat{z} &= \left(1-\frac{s\lambda}{2}\right)^{-1} \left(1+\frac{s\lambda}{2}\right) z =: \mu z. \end{aligned}$$

The map is contact $\Longleftrightarrow N = \mu M^{-T}, \forall s \Longleftrightarrow \lambda = 0$ or $L^2 = \lambda L$; but this is highly exceptional.

Our main concern will be the construction of unconditional contact algorithms for general contact systems. The method is based on the well-known correspondence between contact geometry on $\mathbf{R}^{2n+1}$ and homogeneous (or conic) symplectic geometry on $\mathbf{R}^{2n+2}$.

We use 4-symbol notation for the coordinates on $\mathbf{R}^{2n+2}$

$$\begin{pmatrix} p_0 \\ p_1 \\ q_0 \\ q_1 \end{pmatrix} \in \mathbf{R}^{2n+2}, \quad p_0 = (p_0), q_0 = (q_0), \quad p_1 = \begin{pmatrix} p_{11} \\ \vdots \\ p_{1n} \end{pmatrix}, \quad q_1 = \begin{pmatrix} q_{11} \\ \vdots \\ q_{1n} \end{pmatrix}.$$

Consider *homogeneous* near-identity transformation $g : (p_0, p_1, q_0, q_1) \rightarrow (\hat{p}_0, \hat{p}_1, \hat{q}_0, \hat{q}_1)$ and *homogeneous* function $H(p_0, p_1, q_0, q_1)$ on $\mathbf{R}^{2n+2}$, they are defined for $p_0 \neq 0$ and satisfy the conditions

$$\begin{aligned} \forall \lambda \neq 0: \quad & \hat{p}_i (\lambda p_0, \lambda p_1, q_0, q_1) = \lambda \hat{p}_i (p_0, p_1, q_0, q_1), \\ & \hat{q}_i (\lambda p_0, \lambda p_1, q_0, q_1) = \hat{q}_i (p_0, p_1, q_0, q_1), \quad i = 1, 2, \\ & H (\lambda p_0, \lambda p_1, q_0, q_1) = \lambda H (p_0, p_1, q_0, q_1). \end{aligned}$$

They depend essentially only on $2n+1$ variables

$$\begin{aligned} p_0 \neq 0: \quad & \hat{p}_i (p_0, p_1, q_0, q_1) = p_0 \hat{p}_i \left(1, \frac{p_1}{p_0}, q_0, q_1\right), \\ & \hat{q}_i (p_0, p_1, q_0, q_1) = \hat{q}_i \left(1, \frac{p_1}{p_0}, q_0, q_1\right), \\ & H (p_0, p_1, q_0, q_1) = p_0 H \left(1, \frac{p_1}{p_0}, q_0, q_1\right). \end{aligned}$$

The phase flow e_H^t for Hamiltonian system defined by homogeneous function H on $\mathbf{R}^{2n+2}$ is homogeneous and symplectic.

Near-identity homogeneous symplectic transformation g and homogeneous function H on $\mathbf{R}^{2n+2}$ can be put in correspondence with near-identity contact transformation f and function K on $\mathbf{R}^{2n+1}$ as follows:

$$\begin{pmatrix} x \\ y \\ z \end{pmatrix} \rightarrow \begin{pmatrix} p_0 \\ p_1 \\ q_0 \\ q_1 \end{pmatrix} = \begin{pmatrix} p_0 \\ p_0 x \\ z \\ y \end{pmatrix} \xrightarrow{g} \begin{pmatrix} \hat{p}_0 \\ \hat{p}_1 \\ \hat{q}_0 \\ \hat{q}_1 \end{pmatrix} = \begin{pmatrix} \hat{p}_0 \\ \hat{p}_0 \hat{x} \\ \hat{z} \\ \hat{y} \end{pmatrix},$$

where $p_0 > 0$ is an arbitrary chosen parameter. Then $\hat{p}_0 > 0$ and the map

$$\begin{pmatrix} x \\ y \\ z \end{pmatrix} \xrightarrow{f} \begin{pmatrix} \hat{x} \\ \hat{y} \\ \hat{z} \end{pmatrix} = \begin{pmatrix} \hat{p}_1/\hat{p}_0 \\ \hat{q}_1 \\ \hat{q}_0 \end{pmatrix}, \quad \mu_f := \hat{p}_0/p_0 > 0$$

is a near-identity contact transformation independent of p_0 and μ_f is the multiplier of f.

$$\begin{aligned}
&(x, y, z) \longrightarrow K(x, y, z) := H(1, x, z, y), \\
&H(p_0, p_1, q_0, q_1) = p_0 K\left(\frac{p_1}{p_0}, q_1, q_0\right). \\
&H_{p_0}(p_0, p_1, q_0, q_1) = K(x, y, z) - \langle x, K_x(x, y, z)\rangle = K_e(x, y, z) \\
&H_{p_1}(p_0, p_1, q_0, q_1) = K_x(x, y, z), \\
&H_{q_0}(p_0, p_1, q_0, q_1) = p_0 K_z(x, y, z), \\
&H_{q_1}(p_0, p_1, q_0, q_1) = p_0 K_y(x, y, z), \\
&x = p_1/p_0, \quad y = q_1, \quad z = q_0.
\end{aligned}$$

The above process and its inverse are called *contactization* and *symplectization* respectively. Under this correspondence we have

1. $K(x, y, z) \longleftrightarrow H(p_0, p_1, q_0, q_1)$.
2. Contact system (4.1)$\longleftrightarrow$ homogeneous Hamiltonian system

$$\begin{aligned}
\frac{dp_0}{dt} &= -H_{q_0}, \quad \frac{dq_0}{dt} = H_{p_0}, \\
\frac{dp_1}{dt} &= -H_{q_1}, \quad \frac{dq_1}{dt} = H_{p_1}.
\end{aligned} \tag{4.6}$$

3. Contact phase flow e_K^t on $\mathbf{R}^{2n+1}$ $\longleftrightarrow$ homogeneous symplectic phase flow e_H^t on $\mathbf{R}^{2n+2}$.

4. Contact approximations $f_K^s \approx e_K^s \longleftrightarrow$ homogeneous symplectic approximations $g_H^s \approx e_H^s$. Note that the basic unconditional symplectic algorithms C, P, Q, when applied to homogeneous Hamiltonian systems (4.6) turn out to be unconditional homogeneous symplectic, so their contactizations $\widetilde{C}, \widetilde{P}, \widetilde{Q}$ provide the basic unconditional contact algorithms. We give the end results.

$\tilde{C}$, contact version of method C, two stage form (write $\bar{K}_x = K_x(\bar{x}, \bar{y}, \bar{z})$, etc.)

$$\begin{aligned} &\bar{x} = x + \frac{s}{2}\left(-\overline{K}_y + x\overline{K}_z\right), \quad \bar{y} = y + \frac{s}{2}\overline{K}_x, \quad z = z + \frac{s}{2}\overline{K}_e, \\ &\hat{x} = \bar{x} + \frac{1}{2}s\left(-\overline{K}_y + \hat{x}\overline{K}_z\right) = \left(\bar{x} - \frac{s}{2}\overline{K}_y\right)\left(1 - \frac{s}{2}\overline{K}_z\right)^{-1}, \\ &\hat{y} = \bar{y} + \frac{1}{2}s\overline{K}_x = 2\bar{y} - y, \quad \hat{z} = \bar{z} + \frac{1}{2}s\overline{K}_x = 2\bar{x} - z. \end{aligned} \tag{4.7}$$

$\widetilde{C}^s(a,b,c) = \widetilde{E}^{s/2}(a,b,c) \circ \widetilde{I}^{s/2}(a,b,c) = \widetilde{C}^{\vee s}(a,b,c)$, ord 2, revertible.

$\widetilde{P}$, contact analog of symplectic method P,

1-stage form:
$$\begin{aligned} &\hat{x} = x + s\left(-K_y(\hat{x}, y, z) + xK_z(\hat{x}, y, z)\right), \\ &\hat{y} = y + sK_x(\hat{x}, y, z), \\ &\hat{z} = z + sK_e(\hat{x}, y, z). \end{aligned}$$

2-stage form:
$$\begin{aligned} &\bar{x} = x + s\left(-\overline{K}_y + x\overline{K}_z\right), \bar{y} = u, \bar{z} = z, \\ &\hat{x} = \bar{x}, \quad \hat{y} = \bar{y} + s\overline{K}_x, \quad \hat{z} = \bar{z} + s\overline{K}_e. \end{aligned} \tag{4.8}$$

$\widetilde{P}^s(a,b,c) = \widetilde{E}^{s/2}(0,b,c) \circ \widetilde{I}^{s/2}(a,0,0) = \tilde{Q}^{\vee s}(a,b,c)$, ord 1, non-revertible. $\widetilde{P}^{s/2} \circ \widetilde{Q}^{s/2}$ revertible.

$\widetilde{Q}$, contact analog of symplectic method Q :

1-stage form:
$$\begin{aligned} &\hat{x} = x + s\left(-K_y(x, \hat{y}, \hat{z}) + \hat{x}K_z(x, \hat{y}, \hat{z})\right), \\ &\hat{y} = y + sK_x(x, \hat{y}, \hat{z}), \\ &\hat{z} = z + sK_e(x, \hat{y}, \hat{z}). \end{aligned}$$

2-stage form:
$$\begin{aligned} &\bar{x} = x, \quad \bar{y} = y + s\overline{K}_x, \quad \bar{z} = z + s\overline{K}_e, \\ &\hat{x} = \bar{x} + s\left(-\overline{K}_y + \hat{x}\overline{K}_z\right), \quad \hat{y} = \bar{y}, \quad \hat{z} = \bar{z}. \end{aligned} \tag{4.9}$$

$\widetilde{Q}^s(a,b,c) = \widetilde{E}^s(a,0,0) \circ \widetilde{I}^s(0,b,c) = \widetilde{P}^{\vee s}(a,b,c)$, ord 1, non-revertible. $\widetilde{Q}^{s/2} \circ \widetilde{P}^{s/2}$ revertible.

All three methods contain an almost explicit or almost implicit (or both) stages. The differences from the explicit and implicit Euler is minute and delicate. They consist only in the treatment of a single factor x in the term $x\overline{K}_z$ in component a, resulting in contact versions $\tilde{I}^s, \tilde{E}^s$ of I^s, E^s. Using ordinary versions, the methods $C^s = E^{s/2} \circ I^{s/2}, E^s(0,b,c) \circ I^s(a,0,0)$ and $E^s(a,0,0) \circ I^s(0,b,c)$, which are slightly different from $\tilde{C}^s, \tilde{P}^s, \tilde{Q}^s$, are noncontact! For further development of contact algorithms, see [11,13,19].

We give in Fig. 5, 6 comparisons of 3-D contact vs non-contact computations. The contact systems are those describing the geodesic lines on ellipsoidal and torus surface respectively. A geodesic line will be closed or dense on the surface according as a frequency ratio is rational or irrational. The contact Hamiltonains are of the form

$$K(x,y,z) = \left(\left(Gx^2 + 2Fx + E\right)/2D\right)^{1/2}, \quad D = EG - F^2,$$

with $E(y,z)(dy)^2 + 2F(y,z)dydx + F(y,z)(dz)^2$ as the first fundamental form on the surface (in parameters $y = \psi =$ latitude angle, $z = \phi =$ longitude angle) in question. Contact algorithm is the 2nd order $\widetilde{C}(4.7)$, non-contact algorithm is a 3rd order implicit R.K. The results are due to [21].

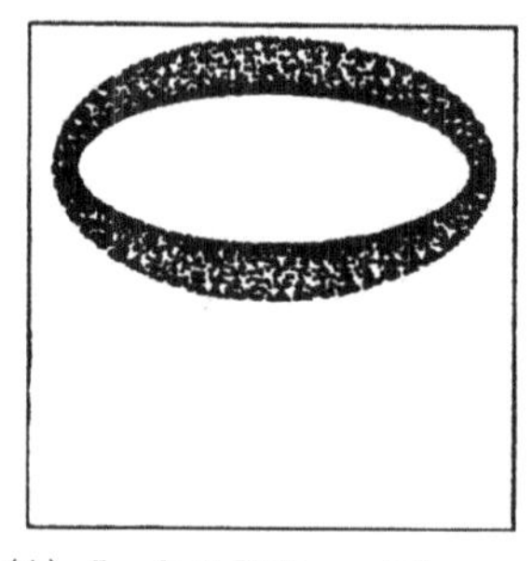

(A) Implicit R.K., ord 3, non-contact, 640000 steps

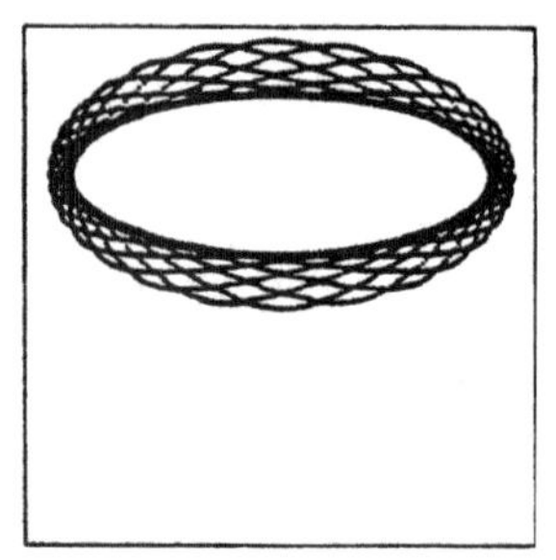

(B) Contact method $\widetilde{C}$, ord 2, 640000 steps

Fig. 5 Geodesics on ellipsoid, rational frequency ratio 9: 16

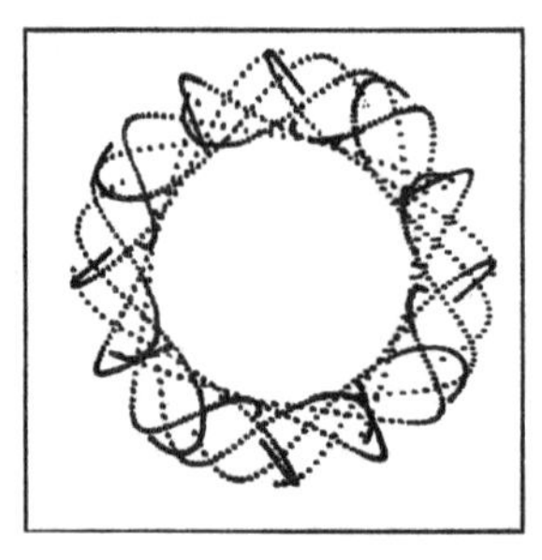

(A) Implicit R.K., ord 3, non-contact, 800000 steps

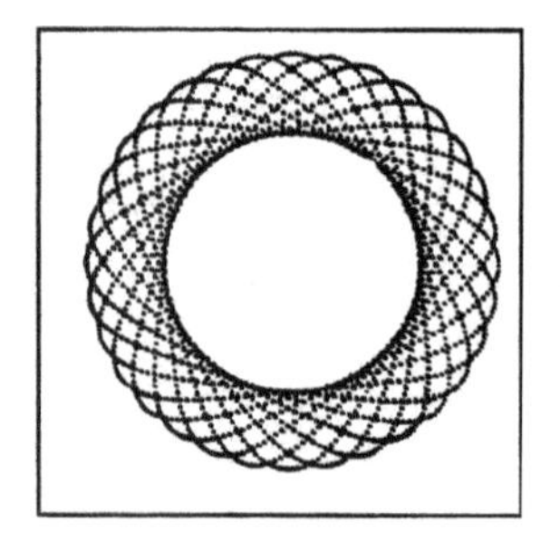

(B) Contact method $\widetilde{C}$, ord 2, 800000 steps

Fig. 6 Geodesics on torus, rational frequency ratio 11: 16

§5 Volume-preserving Algorithms for Source-free Systems

The source-free systems on $\mathbf{R}^m$ are governed by the m-D volume structure, the underlying Lie algebra $\mathbf{L} = \mathbf{SV}_m$ consists of all *source-free* vector fields $a, \operatorname{div} a = \operatorname{tr} a_* = 0$, the underlying Lie group $\mathbf{G} = \mathbf{SD}_m$ consists of all near-identity *volume-preserving* transformations $g, \det g_* = 1$.

The phase flow e_a^t of source-free field a is always volume preserving. Hence the proper algorithms for source-free systems should be also *volume-preserving*, since otherwise the dynamics would be polluted by artificial sources and sinks.

For $m = 2$, source-free fields = Hamiltonian fields, area-preserving maps = symplectic maps; so the problem for area-preserving algorithms has been solved in principle.

For $m \geqslant 3$, the problem is new, since all the conventional methods plus even the symplectic methods are generically not volume-preserving, even for linear source-free systems. As an illustration, solve on $\mathbf{R}^3$

$$\frac{dx}{dt} = Ax, \quad \operatorname{tr} A = 0$$

by the method C, we get

$$x \rightarrow \hat{x} = G^s x, \quad G^s = \left(I - \frac{s}{2}A\right)^{-1}\left(I + \frac{s}{2}A\right).$$

$\det G^s = 1 \Longleftrightarrow \det A = 0$, which is exceptional.

The construction of explicit Euler type structure-preserving algorithms by composition is applicable here when the vector field is source-free separable, i.e., decomposable as

$$a = \sum_{i=1}^{n} a^i, \quad \operatorname{div} a^i = 0, \quad a_*^i a^i = 0, \quad i = 1, \cdots, n. \tag{5.1}$$

A special case is

$$\begin{aligned} & a = (a_1, \cdots, a_m), \quad \frac{\partial a_k}{\partial x_k} = 0, \quad a^k = (0, \cdots, 0, a_k, 0, \cdots, 0), \quad k = 1, \cdots, m; \\ & a = \sum_{k=1}^{m} a^k. \end{aligned} \tag{5.2}$$

Example 1. Euler equation for free rigid body and Jacobian elliptic functions of modulus k,

$$\begin{aligned} \frac{dx_1}{dt} &= c_1 x_2 x_3 = a_1, \\ \frac{dx_2}{dt} &= c_2 x_3 x_1 = a_2, \\ \frac{dx_3}{dt} &= c_3 x_1 x_2 = a_3, \end{aligned} \qquad \begin{pmatrix} a_1 \\ a_2 \\ a_3 \end{pmatrix} = \begin{pmatrix} c_1 x_2 x_3 \\ 0 \\ 0 \end{pmatrix} + \begin{pmatrix} 0 \\ c_2 x_3 x_1 \\ 0 \end{pmatrix} + \begin{pmatrix} 0 \\ 0 \\ c_3 x_1 x_2 \end{pmatrix}$$

where (x_1, x_2, x_3) are angular momenta along principle axes of inertia with diagonal elements $I_1, I_2, I_3, c_1 = (I_2 - I_3)/I_2 I_3$, cyclic. Take $c_1 = 1, c_2 = -1, c_3 = -k^2$, it gives elliptic functions for $x_1(0) = 0, x_2(0) = x_3(0) = 1, x_1(t) = \operatorname{sn}(t,k), x_2(t) = \operatorname{cn}(t,k), x_3(t) = \operatorname{dn}(t,k)$.

Example 2. ABC flows [1]

$$\frac{dx_1}{dt} = A\sin x_3 + C\cos x_2 = a_1,$$
$$\frac{dx_2}{dt} = B\sin x_1 + A\cos x_3 = a_2,$$
$$\frac{dx_3}{dt} = C\sin x_2 + B\cos x_1 = a_3,$$

$$\begin{pmatrix} a_1 \\ a_2 \\ a_3 \end{pmatrix} = \begin{pmatrix} A\sin x_3 + C\cos x_2 \\ 0 \\ 0 \end{pmatrix} + \begin{pmatrix} 0 \\ B\sin x_1 + A\cos x_3 \\ 0 \end{pmatrix} + \begin{pmatrix} 0 \\ 0 \\ C\sin x_2 + B\cos x_1 \end{pmatrix}$$
$$= \begin{pmatrix} A\sin x_3 \\ A\cos x_3 \\ 0 \end{pmatrix} + \begin{pmatrix} 0 \\ B\sin x_1 \\ B\cos x_1 \end{pmatrix} + \begin{pmatrix} C\sin x_2 \\ 0 \\ C\cos x_2 \end{pmatrix}.$$

All decompositions on the right satisfy (5.1), then (2.18) gives corresponding volume-preserving revertible algorithms.

In $\mathbf{R}^m$ all source-free fields have "vector potential" representations: under such representation, a field always decomposes into a sum of essentially 2-D source-free fields, for which essentially area-preserving algorithms can always be constructed, then one can apply the decomposition-composition method in §2 to obtain m-D volume-preserving algorithms [8,22].

In $\mathbf{R}^2$, every source-free field (a_1, a_2) corresponds to a stream function or 2-D Hamiltonian ψ, unique up to a constant:

$$a_1 = -\frac{\partial\psi}{\partial x_2}, \quad a_2 = \frac{\partial\psi}{\partial x_1}, \tag{5.3}$$

then area-preserving algorithms = 2-D symplectic algorithms.

In $\mathbf{R}^3$, every source-free field (a_1, a_2, a_3) corresponds to a vector potential (b_1, b_2, b_3), unique up to a gradient:

$$a = \mathrm{curl} b, \quad a_1 = \frac{\partial b_3}{\partial x_2} - \frac{\partial b_2}{\partial x_3}, \quad a_2 = \frac{\partial b_1}{\partial x_3} - \frac{\partial b_3}{\partial x_1}, \quad a_3 = \frac{\partial b_2}{\partial x_1} - \frac{\partial b_1}{\partial x_2}, \tag{5.4}$$

then one gets source-free decomposition

$$
\begin{aligned}
a &= (a_1, a_2, a_3) \\
&= \left(0, \frac{\partial b_1}{\partial x_3}, -\frac{\partial b_1}{\partial x_2}\right) + \left(-\frac{\partial b_2}{\partial x_3}, 0, \frac{\partial b_2}{\partial x_1}\right) + \left(\frac{\partial b_3}{\partial x_2}, -\frac{\partial b_3}{\partial x_1}, 0\right) \\
&= a^1 + a^2 + a^3.
\end{aligned} \tag{5.5}
$$

Each field a^i is 2-D source-free and 0 in the 3rd dimension, apply area-preserving method P, say, with 3rd dimension fixed, then one gets three volume-preserving maps

$$
\begin{aligned}
&P_1^s \approx e^s\left(a^1\right) = e^s\left(0, \frac{\partial b_1}{\partial x_3}, -\frac{\partial b_1}{\partial x_2}\right), \quad \text{ord 1, volume-preserving },\\
&P_2^s \approx e^s\left(a^2\right) = e^s\left(-\frac{\partial b_2}{\partial x_3}, 0, \frac{\partial b_2}{\partial x_1}\right), \quad \text{ord 1, volume-preserving },\\
&P_3^s \approx e^s\left(a^3\right) = e^s\left(\frac{\partial b_3}{\partial x_2}, -\frac{\partial b_3}{\partial x_1}, 0\right), \quad \text{ord 1, volume-preserving },\\
&P_1^s = E^s\left(0, \frac{\partial b_1}{\partial x_3}, 0\right) I^s\left(0, 0, -\frac{\partial b_1}{\partial x_2}\right), \quad \check{P}_1^s = E^s\left(0, 0, -\frac{\partial b_1}{\partial x_2}\right) I^s\left(0, \frac{\partial b_1}{\partial x_3}, 0\right),\\
&P_2^s = E^s\left(0, 0, \frac{\partial b_2}{\partial x_1}\right) I^s\left(-\frac{\partial b_2}{\partial x_3}, 0, 0\right), \quad \check{P}_2^s = E^s\left(-\frac{\partial b_2}{\partial x_3}, 0, 0\right) I^s\left(0, 0, \frac{\partial b_2}{\partial x_1}\right), \quad (5.6)\\
&P_3^s = E^s\left(\frac{\partial b_3}{\partial x_2}, 0, 0\right) I^s\left(0, -\frac{\partial b_3}{\partial x_1}, 0\right), \quad \check{P}_3^s = E^s\left(0, -\frac{\partial b_3}{\partial x_1}, 0\right) I^s\left(\frac{\partial b_3}{\partial x_2}, 0, 0\right),
\end{aligned}
$$

then one gets volume-preserving algorithms for field a

$$
\begin{aligned}
&f^s = P_1^s P_2^s P_3^s \approx e^s(a), \quad \text{ord 1, non-revertible },\\
&g^s = P_1^{s/2} P_2^{s/2} P_3^{s/2} \circ \check{P}_3^{s/2} \check{P}_2^{s/2} \check{P}_1^{s/2} \approx e^s(a), \ \text{ord 2, revertible},\\
&h^s = g^{\alpha s} g^{\beta s} g^{\alpha s} \approx e^s(a), \quad \text{ord 4, revertible.}
\end{aligned} \tag{5.7}
$$

The basic elements of all these composites consist only in 6 explicit Euler and 6 implicit Euler in one variable appeared in (5.6).

On $\mathbf{R}^m$, to every source-free field (a_i), there exists, as a generalization of cases $m = 2, 3$, a skew-symmetric tensor field of order $2, b = (b_{ik}), b_{ik} = -b_{ki}$ so that

$$
a_i = \sum_{k=1}^{m} \frac{\partial b_{ik}}{\partial x_k}, \quad i = 1, \cdots, m. \tag{5.8}
$$

By (5.8) we can decompose

$$
a = \sum_{i<k} a_{(ik)}, a_{(ik)} = \left(0, \cdots, 0, \frac{\partial b_{ik}}{\partial x_k}, 0, \cdots, 0, -\frac{\partial b_{ik}}{\partial x_i}, 0, \cdots, 0\right)^T, \quad i < k. \tag{5.9}
$$

Every vector field $a_{(ik)}$ is 2-D source-free on the i-k plane and 0 in other dimensions so we can approximate the phase flow of $a_{(ik)}$ by area-preserving method which implies

m-D volume-preserving, similar to the cases discussed above for $m=3$. For more details see [8].

The tensor potential (b_{ik}) for a given (a_i) is by far not unique. For uniqueness one may impose normalizing conditions in many different ways. One way is to impose

$$N_0: \quad b_{ik}=0, |i-k| \geqslant 2, \tag{5.10}$$

(this condition is ineffective for $m=2$). The non-zero components are

$$\begin{aligned}
&b_{12}=-b_{21}, \quad b_{23}=-b_{32}, \quad \cdots \quad b_{m-1,m}=-b_{m,m-1}. \\
&N_1: \quad b_{12}\big|_{x_1=x_2=0}=0. \\
&N_k,\ 2 \leqslant k \leqslant m-1: \quad b_{k,k+1}\big|_{x_k=0}=0,
\end{aligned}$$

(this condition is ineffective for $m=2$). Then all $b_{k,k+1}$ are uniquely determined by quadrature

$$\begin{aligned}
b_{12} &= \int_0^{x_2} a_1 dx_2 - \int_0^{x_1} a_2\bigg|_{x_2=0} dx_1, \\
b_{k,k+1} &= -\int_0^{x_k} a_{k+1} dx_k, \quad 2 \leqslant k \leqslant m-1; \qquad (5.11) \\
a &= \sum_{k=1}^{m-1} a^k, \quad a^k=\left(0,\cdots,0,\frac{\partial b_{k,k+1}}{\partial x_{k+1}},-\frac{\partial b_{k,k+1}}{\partial x_k},0,\cdots,0\right)^T. \qquad (5.12)
\end{aligned}$$

When the vector field a is polynomial, then by (5.10) and (5.11) the tensor components b_{ik} are polynomials. Then the theorem on symplectic separability of all polynomial Hamiltonians (3.18) implies the source-free separability in the sense of (5.1) for all polynomial source-free systems.

We give in Fig 7 comparison of volume-preserving vs non-volume-preserving compu tations for source-free system Example 1 for Jacobian elliptic functions with modulus $k=\sqrt{2}$. For initial data cited there, the orbits will be closed curves as intersection of two cylinders $x_1^2+x_2^2=1, k^2x_1^2+x_3^2=1:$ The projection of this orbit on x_1x_3 plane is a closed ellipse, while both projections on x_2x_3 plane and x_1x_2 plane are *limited* elliptic arcs. The volume-preserving algorithm is a 2nd order composite from the decomposition of field. Both qualitative and quantitative properties are well preserved. The 4th order explicit R.K. producces artificial sources and sinks for source-free systems, the "beautiful" patterns in Fig 7(B) are fabricated. Fig 8 give similar comparisons for ABC flows for integrable case $A=B=1$, $C=0$. All results are due to [25].

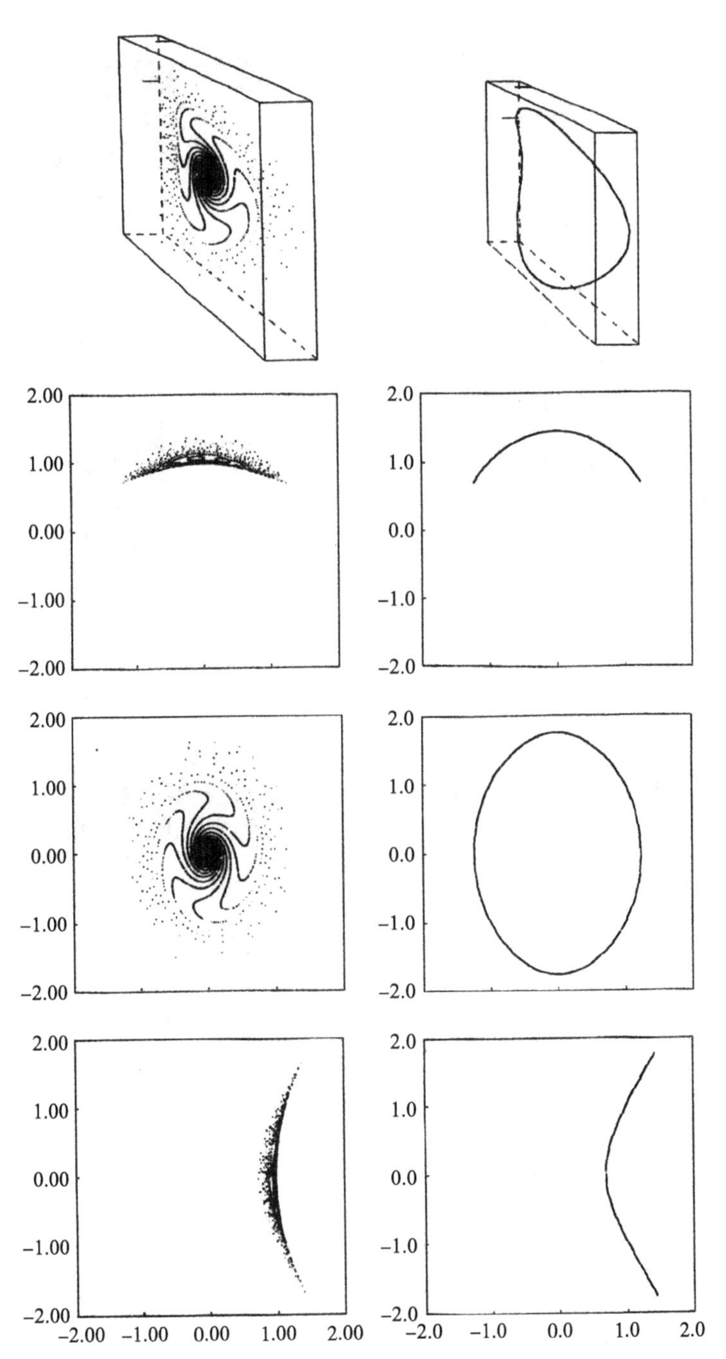

(A) Explicit R.K., ord 4, $s = 0.5$, 6000 steps (B) Exp vol-preserving, ord 2, $s = 0.25$,6000 steps

Fig. 7 Jacobian elliptic functions, $k = \sqrt{2}$.

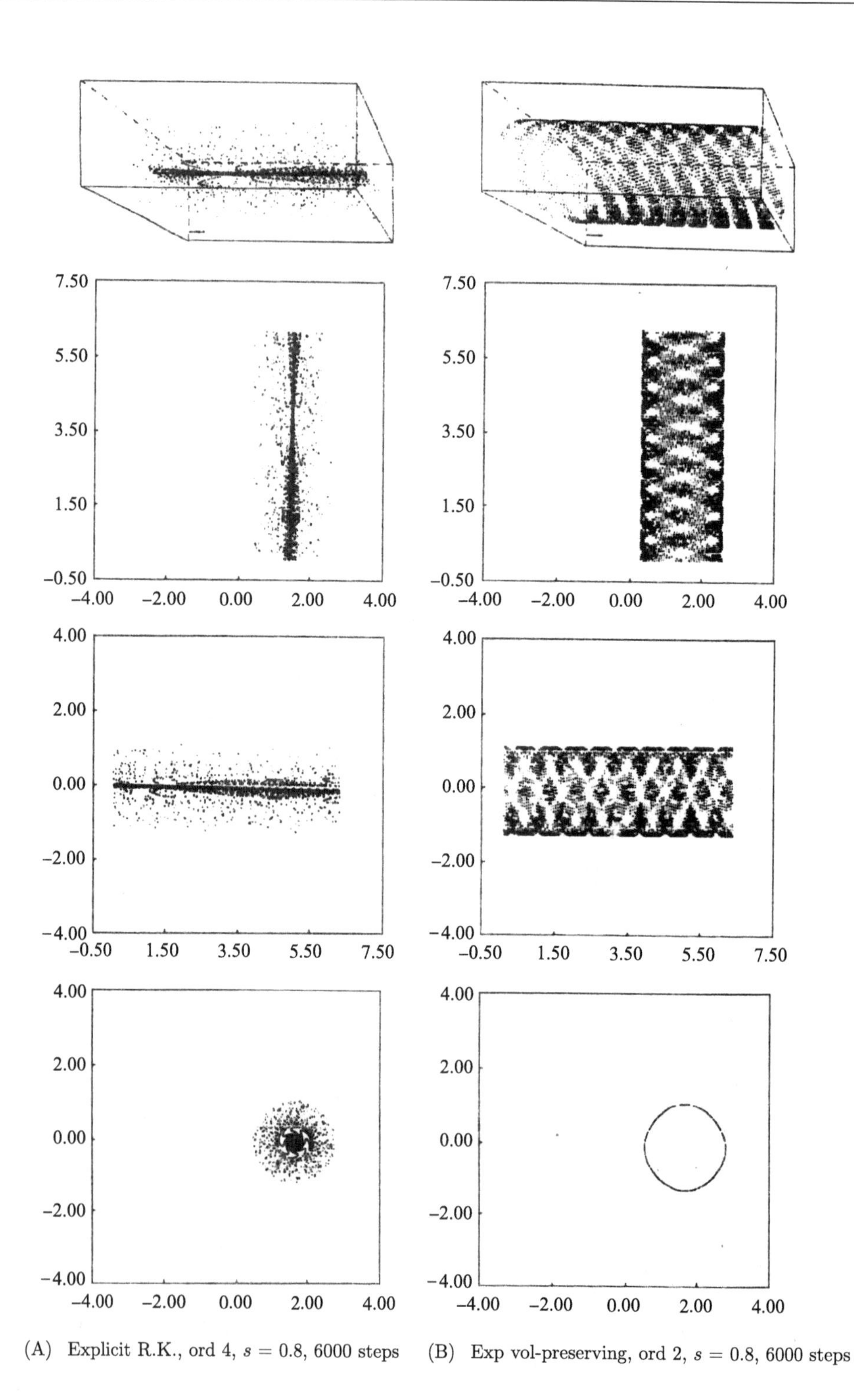

(A) Explicit R.K., ord 4, $s = 0.8$, 6000 steps (B) Exp vol-preserving, ord 2, $s = 0.8$, 6000 steps

Fig. 8 ABC flow , $A = B = 1, C = 0$

§6 Logarithms of Structure-preserving Maps

In order to facilitate the study of the problem $f^s \approx e_a^s$, the apparatus of formal power series is useful. Among the formal power series $\sum_0^\infty s^k a_k, a_k : \mathbf{R}^m \to \mathbf{R}^m$, we pick out two special classes. The first class consists of those with $a_0 = 0$, called near-0 formal vector fields; the second class consists of those with $a_0 = 1_m$, called near-1 formal maps (diffeomorphisms) [9,10,14].

All near-0 formal vector fields $a^s = \sum\limits_1^\infty s^k a_k$ form a (∞-dim.) real Lie algebra $\mathbf{FV}_m$ under the Lie bracket

$$[a^s, b^s] = \left[\sum_1^\infty s^k a_k, \sum_1^\infty s^k b_k\right] := \sum_{k=2}^\infty s^k \sum_{i+j=k} [a_i, b_j] .$$

The associated near-0 formal differential operators and their products are

$$(a^s)_* := \left(\sum_1^\infty s^k a_k\right)_* := \sum_1^\infty s^k a_{k*}$$

$$a^{s*} = \left(\sum_1^\infty s^k a_k\right)^* := \sum_1^\infty s^k a_k^*,$$

$$a^{s*} b^{s*} := \sum_2^\infty s^k \sum_{i+j=k} a_i^* b_j^*, \quad (a^{s*})^2 := a^{s*} a^{s*}, \text{ etc.}$$

The totality of near-1 formal maps $f^s = 1_m + \sum_1^\infty s^k f_k, f_k : \mathbf{R}^m \to \mathbf{R}^m$ is denoted by $\mathbf{FD}_m$, in which one can define a composition rule $f^s \circ g^s$ to make up a group. We omit the details here, it is simply adopted from substitution rule for formal power series and agrees with the usual composition of diffeomorphisms $f \circ g(x) = f(g(x))$ in case of convergence. Algebra $\mathbf{FV}_m$ associates to group $\mathbf{FD}_m$ just like $\mathbf{V}_m$ associates to $\mathbf{D}_m$.

There is an obvious one-one correspondence between the Lie algebra $\mathbf{FV}_m$ and the Lie group $\mathbf{FD}_m$, established simply by $+1_m$ and by -1_m. However, the more significant one-one correspondence between them is given by exp and its inverse log.

$$\exp : \mathbf{FV}_m \to \mathbf{FD}_m$$

$$a^s = \sum_1^\infty s^k a_k \to \exp a^s := 1_m + \sum_{k=1}^\infty \frac{1}{k!} (a^{s*})^k 1_m =: 1_m + \sum_1^\infty s^k f_k = f^s. \tag{6.1}$$

Note that

$$(a^{a*})^k = \left(\sum s^{i_1} a_{i_1}^*\right) \cdots \left(\sum s^{i_k} a_{i_k}^*\right) = \sum_{i_1,\cdots,i_k=1}^\infty s^{i_1+\cdots+i_k} a_{i_1}^* \cdots a_{i_k}^*,$$

so we get easily

$$f_k = \sum_{n=1}^{k} \frac{1}{n!} \sum_{k_1+\cdots+k_n=k} a^*_{k_1} \cdots a^*_{k_n} 1_m, \quad k \geqslant 1, \quad f_1 = a_1, \tag{6.2}$$

$$f_k = a_k + \sum_{n=2}^{k} \frac{1}{n!} \sum_{k_1+\cdots+k_n=k} a^*_{k_1} \cdots a^*_{k_n} 1_m, \quad k \geqslant 2, \quad f_2 = a_2 + \frac{1}{2} a_1^2.$$

Note that (6.2) provides a 2-way recursion formula from $a_1, \cdots, a_k$ to $f_1, \cdots, f_k$ and vice versa. Therefore exp maps $\mathbf{FV}_m$ one-one onto $\mathbf{FD}_m$ and its inverse, i.e., log is defined by the same (6.2):

$$\log = (\exp)^{-1} : \mathbf{FD}_m \to \mathbf{FV}_m, \quad \log\exp a^s = a^s, \quad \exp\log f^s = f^s.$$

In particular

$$\begin{aligned} &a_1 = f_1 \quad a_2 = f_2 - \frac{1}{2} a_1^2, \quad a_3 = f_3 - \frac{1}{2}(a_1^* a_2 + a_2^* a_1) - \frac{1}{3!} a_1^3, \\ &a_4 = f_4 - \frac{1}{2}\left(a_1^* a_3 + a_2^2 + a_3^* a_1\right) - \frac{1}{3!}\left(a_1^* a_1^* a_2 + a_1^* a_2^* a_1 + a_2^* a_1^* a_1\right) - \frac{1}{4!} a_1^4, \\ &a_k = f_k - \sum_{n=2}^{k-1} \frac{1}{n!} \sum_{k_1+\cdots+k_m=k} a^*_{k_1} \cdots a^*_{k_m} 1_m - \frac{1}{k!} a_1^k, \quad k \geqslant 3. \end{aligned} \tag{6.3}$$

An equivalent construction of $\log f^s = a^s$ is

$$\log f^s = \sum_{k=1}^{\infty} \frac{(-1)^{k-1}}{k} h_k^s, \text{ where } h_1^s = f^s - 1_m, \quad h_k^s = h_{k-1}^s \circ f^s - h_{k-1}^s. \tag{6.4}$$

It is easy to compute

$$\begin{aligned} h_1^s &= \sum_{k=1}^{\infty} s^k f_k = \sum_{k_1=1}^{\infty} s^{k_1} (1_m \circ f)_{k_1} \\ h_2^s &= \sum_{k_1=k_2=1}^{\infty} s^{k_1+k_2} \left((1_m \circ f)_{k_1} \circ f\right)_{k_2}, \\ h_3^s &= \sum_{k_1,k_2,k_3=1}^{\infty} s^{k_1+k_2+k_3} \left(\left((1_m \circ f)_{k_1} \circ f\right)_{k_2} \circ f\right)_{k_3}, \text{ etc.} \end{aligned}$$

Substituting in (6.4) and equating with $\sum\limits_{1}^{\infty} s^k a_k$, we get

$$a_k = \sum_{n=1}^{k} \frac{(-1)^{n-1}}{n} \sum_{k_1+\cdots+k_n=k} \left(\cdots\left((1_m \circ f)_{k_1} \circ f\right)_{k_2} \cdots \circ f\right)_{k_n}. \tag{6.5}$$

It is easy to verify log exp $a^s = a^s$ for this log, so this is precisely the inverse of exp, thus agrees with the previous one.

We use the above construction (6.5) to establish the *formal* Campbell-Hausdorff formula: For arbitrary near-1 formal maps f^s, g^s

$$\log\left(f^s \circ g^s\right) = \log f^s + \log g^s + \sum_{k=1}^{\infty} d_k\left(\log f^s, \log g^s\right), \tag{6.6}$$

where, for $\log f^s = a^s, \log g^s = b^s$,

$$d_k(a^s, b^s) = \frac{1}{k}\sum_{n=1}^{k}\frac{(-1)^{n-1}}{n}\sum_{\substack{p_1+q_1+\cdots+p_n+q_n=k\\ p_i+q_i\geqslant 1, p_i\geqslant 0, q_i\geqslant 0}}\frac{\left[(a^s)^{p_1}(b^s)^{q_1}\cdots(a^s)^{p_n}(b^s)^{q_n}\right]}{p_1!q_1!\cdots p_n!q_n!} \tag{6.7}$$

where

$$(x)^p = xx\cdots x\ (p\text{-times}), \quad [x_1x_2x_3\cdots x_n] = \left[\left[\cdots\left[\left[x_1, x_2\right], x_3\right], \cdots\right], x_n\right].$$

In particular

$$d_1 = \frac{1}{2}\left[a^s, b^s\right], \quad d_2 = \frac{1}{12}\left(\left[a^s b^s b^s\right] + \left[b^s a^s a^s\right]\right), \quad d_3 = -\frac{1}{24}\left[a^s b^s b^s a^s\right].$$

Let $\log\left(f^s \circ g^s\right) = c^s = \sum\limits_{1}^{\infty} s^k c_k$. Then

$$\begin{aligned}
c_1 =& a_1 + b_1, \quad c_2 = a_2 + b_2 + \frac{1}{2}\left[a_1 b_1\right],\\
c_3 =& a_3 + b_3 + \frac{1}{2}\left(\left[a_1 b_2\right] + \left[a_2 b_1\right]\right) + \frac{1}{12}\left(\left[a_1 b_1 b_1\right] + \left[b_1 a_1 a_1\right]\right),\\
c_4 =& a_4 + b_4 + \frac{1}{12}\left(\left[a_1 b_3\right] + \left[a_2 b_2\right] + \left[a_3 b_1\right]\right)\\
&+ \frac{1}{12}\left(\left[a_1 b_1 b_2\right] + \left[a_1 b_2 b_1\right] + \left[a_2 b_1 b_1\right] + \left[b_1 a_1 a_2\right] + \left[b_1 a_2 a_1\right] + \left[b_2 a_1 a_1\right]\right)\\
&- \frac{1}{24}\left[a_1 b_1 b_1 a_1\right], \quad \text{etc.}
\end{aligned}$$

Note that the classical CH formula is restricted to the composition of two one-parameter groups, where $\log f^s = sa_1, \log g^s = sb_1$.

The log transform reduces matters at the Lie group level to those at the easier level of Lie algebra. All properties of near-1 formal maps have their logarithmic interpretations. We list some of them, let $\log f^s = a^s = \sum\limits_{s}^{\infty} s^k a_k$:

1. f^s is a phase flow, i.e., $f^{s+t} = f^s \circ f^t \Longleftrightarrow \log f^s = sa_1$.
2. f^s is revertible, i.e., $f^s \circ f^{-s} = 1_m \Longleftrightarrow \log f^s$ is odd in s.
3. f^s raised to real μ th power $(f^s)^\mu \Longleftrightarrow \log(f^s)^\mu = \mu \log f^s$. In particular $\log(f^s)^{-1} = -\log f^s, \log\sqrt{f^s} = \frac{1}{2}\log f^s$.
4. f^s scaled to $f^{\alpha s} \Longleftrightarrow \log\left(f^{\alpha s}\right) = (\log f)^{\alpha s}$. In particular $\log\left(f^{-s}\right) = (\log f)^{-s}$.
5. $f^s - g^s = O\left(s^{p+1}\right) \Longleftrightarrow \log f^s - \log g^s = O\left(s^{p+1}\right)$.

6. $f^s \circ g^s = g^s \circ f^s \Longleftrightarrow [\log f^s, \log g^s] = 0 \Longleftrightarrow \log(f^s \circ g^s) = \log f^s + \log g^s$.

7. $(f^s \circ g^s) = h^s \Longleftrightarrow \log h^s = \log(f^s \circ g^s) = \log f^s + \log g^s + \sum_1^\infty d_k(\log f^s, \log g^s)$.

8. f^s symplectic $\Longleftrightarrow$ all a_k are Hamiltonian fields. .

9. f^s contact $\Longleftrightarrow$ all a_k are contact fields.

10. f^s volume-preserving $\Longleftrightarrow$ all a_k are source-free fields.

The log transform has important bearing on dynamical systems with Lie algebra structure, the structure-preserving property of maps f^s at the Lie group ($\mathbf{G} \subset \mathbf{D}_m$) level can be characterized through their logarithms at the associated Lie algebra ($\mathbf{L} \subset \mathbf{V}_m$) level.

We return to the main problem of approximation to the phase flow for dynamical system $dx/dt = a(x)$.

$$f_a^s = f^s = 1_m + \sum s^k f_k \approx e_a^s = 1_m + \sum s^k e_k, \quad e_k = a^k/k!. \tag{6.8}$$

If $f_k = e_k, 1 \leqslant k \leqslant p$, we say f_a^s is accurate to order $\geqslant p$, if moreover, $f_{p+1} \neq e_{p+1}$, we say it is accurate to order p.

Let $\log f^s = a^s = \sum s^k a_k$. Note that the first $p+1$ equations in (10.2) completely determine $a_1, a_2, \cdots, a_{p+1}$ and $f_1, f_2, \cdots, f_{p+1}$ each other. It is then easy to establish

$$\begin{aligned} &f_k = e_k, \quad 1 \leqslant k \leqslant p; \quad f_{p+1} \neq e_{p+1} \Longleftrightarrow \\ &a = a_1 = e_1; \quad a_k = 0, \quad 1 < k \leqslant p; \quad a_{p+1} = f_{p+1} - e_{p+1} \neq 0. \end{aligned} \tag{6.9}$$

So the orders of approximation for $f_a^s \approx e_a^s$ and for $\log f_a^s - sa$ are the same.

Moreover, note that we have a *formal field*

$$s^{-1} \log f^s = s^{-1} a^s = a + \sum_1^\infty s^k a_{k+1} = a + O(s^p) \tag{6.10}$$

which is equal to the original field a up to a near-0 perturbation and defines a *formal dynamical system*

$$\frac{dx}{dt} = \left(s^{-1} \log f^s\right)(x) = a(x) + \sum_1^\infty s^k a_{k+1}(x) \tag{6.11}$$

having a *formal phase flow* (in two parameters t and s with group property in t) $e_{s^{-1}a^s}^t = \exp ts^{-1}a^s$ whose *diagonal formal flow* $e_{s^{-1}a^s}^t|_{t=s}$ is exactly f^s. This means that any compatible algorithm f_a^s of order p gives perturbed solution of a right equation with field a; however, it gives the right solution of a perturbed equation with field $s^{-1} \log f_a^s = a + O(s^p)$. There could be many methods with the same formal order of accuracy but with quite different qualitative behavior. The problem is to choose among them those leading to allowable perturbations in the equation. For systems with geometric structure, the propositions 8, 9 and 10 provide guidelines for a proper choice. The structure-preservation requirement for the algorithms precludes all unallowable perturbations alien to the pertinent type of dynamics. Take, for example, Hamiltonian systems. A transition map f_a^s for Hamiltonian field a is symplectic if and only if all fields a_k are Hamiltonian, i.e., the induced perturbations in the equation are Hamiltonian [5,6]. So symplectic algorithms are clean, inherently free from all kinds of perturbations alien to Hamiltonian dynamics (such as artificial dissipation inherent in the vast majority of conventional methods), this accounts for their superior performance. The situations are analogous for contact and volume-preserving algorithms.

Acknowledgements The authors is grateful to Dr. Wang Dao-liu for discussions and valuable help in preparation the manuscript and also to his colleagues, Li Wang-yao, Qin Meng-zhao, Shang Zai-jiu, Shu Hai-bing, Tang Yi-fa, Zhang Mei-qing and Zhu Wen-jie for collaboration and provision of computing results.

References

[1] Arnold V I. Mathematical Methods of Classical Mechanics. New York: Springer. 1978

[2] Feng K. On difference schemes and symplectic geometry. In Feng K, ed. Proc 1984 Beijing Symp Diff Geometry and Diff Equations. Beijing: Science Press, 1985: 42-58

[3] Feng K. Difference schemes for Hamiltonian formalism and symplectic geometry. J Comp Math. 1986 4(3): 279-289

[4] Feng K, Wu H M, Qin M Z, Wang D L. Construction of canonical difference schemes for Hamiltonian formalism via generating functions. J Comp Math. 1989 7(1): 71-96

[5] Feng K. The Hamiltonian way for computing Hamiltonian dynamics. In Spigler R, ed. Applied and Industrial Mathematics. Netherlands: Kluwer, 1991: 17-35

[6] Feng K, Qin M Z. Hamiltonian algorithms and a comparative numerical study. Comput Phys Comm. 1991 65: 173-187

[7] Feng K, Shang Z J. Volume preserving algorithms for source-free dynamical systems: [Preprint]. Beijing: CAS. Computing Center, 1991

[8] Feng K. Formal power series and numerical methods for differential equations. In Chan T, Shi Z C, ed. Proc of Inter Conf on Scientific Computation. Singapore: World Scientific, 1992: 28-35

[9] Feng K. The step transition operators for multi-step methods of ODE's: [Preprint]. Beijing: CAS. Computing Center, 1991

[10] Feng K. The calculus of generating functions and the formal energy for Hamiltonian algorithms: [Preprint]. Beijing: CAS. Computing Center, 1991

[11] Feng K. The calculus of formal power series for diffeomorphisms and vector fields: [Preprint]. Beijing: CAS. Computing Center, 1991

[12] Feng K. Symplectic, contact and volume-preserving algorithms. In Shi Z C, Ushijima T, ed. Proc 1st China-Japan Conf on Numer Math. Singapore: World Scientific, 1993: 1-28

[13] Feng K. Formal dynamical systems and numerical algorithms. In Feng K, Shi Z C, ed. Proc Inter Conf on Computation of Differential Equations and Dynamical Systems. Singapore: World Scientific, 1993: 1-10

[14] Feng K. Contact algorithms for contact dynamical systems: [Preprint]. Beijing: CAS. Computing Center, 1992

[15] Feng K. How to compute properly Newton's equation of motion. In Ying L A, Guo B Y, ed. Proc of 2nd Conf on Numerical Methods for Partial Differential Equations. Singapore: World Scientific, 1992: 15-22

[16] Feng K. Theory of contact generating functions: [Preprint]. Beijing: CAS. Computing Center, 1992.

[17] Feng K, Qin M Z. The symplectic methods for computation of Hamiltonian equations. In Zhu Y L, Guo Ben-yu, ed. Proc Conf on Numerical Methods for PDE's. Berlin: Springer, 1987: 1-37. Lect Notes in Math 1297

[18] Feng K, Tang Y F. Non-symplecticity of linear multi-step methods: [Preprint]. Beijing: CAS. Computing Center, 1991

[19] Feng K, Wang D L. A note on conservation laws of symplectic difference schemes for Hamiltonian systems. J Comp Math. 1991 9(3): 229-237

[20] Feng K, Wang D L. Variations on a theme by Euler: [Preprint]. Beijing: CAS. Computing Center, 1991

[21] Qin M Z, Wang D L, Zhang M Q. Explicit symplectic difference schemes for separable Hamiltonian systems. J Comp Math. 1991 9(3): 211-221

[22] Qin M Z, Zhang M Q. Explicit Runge-Kutta like unitary schemes to solve certain quantum operator equations of motion. J Stat Phys. 1990 60: 837-843

[23] Shang Z J. On KAM theorem for symplectic algorithms for Hamiltonian systems: [dissertation]. Beijing: CAS. Computing Center, 1991

[24] Shu H B. A new approach to generating functions for contact systems: [Preprint]. Beijing: CAS. Computing Center, 1992

[25] Tang Y F. Hamiltonian systems and algorithms for geodesic flows on compact Riemannian manifolds: [Master Thesis]. Beijing: CAS. Computing Center, 1990

[26] Tang Y F. Private communications

[27] Wang D L. Decomposition of vector fields and composition of algorithms. In Feng K, Shi Z C, ed. Proc Inter Conf on Computation of Differential Equations and Dynamical Systems. Singapore: World Scientific, 1993: 179-184

[28] Yoshida H. Construction of higher order symplectic integrators. Phys Letters A. 1990 150: 262-268

[29] Zhang M Q, Qin M Z. Explicit symplectic schemes to solve vortex systems: [Preprint]. Beijing: CAS. Computing Center, 1992

[30] Zhu W J. Private communications

18 Formal Dynamical Systems and Numerical Algorithms[①]

形式动力系统与数值算法

§1 Preliminaries

We study vector fields, their associated dynamical systems, phase flows together with their algorithmic approximations in $\mathbf{R}^N$ from the formal power series approach. Our considerations will be local in both space and time, all related objects are $\mathbf{C}^\infty$ smooth. We use coordinate description and matrix notation, the coordinate vectors in $\mathbf{R}^N$ and vector functions $a : \mathbf{R}^N \to \mathbf{R}^N$ are denoted by column matrices. The identity vector function 1_N is given by $1_N(x) = x$. For vector function $a = (a_1, \cdots, a_N)^T : \mathbf{R}^N \to \mathbf{R}^N$,

$$a_* := \left(\frac{\partial a_i}{\partial x_j}\right) = \text{Jacobian matrix of } a$$

$$a^* := \sum a_i \frac{\partial}{\partial x_i} = \text{linear differential operator of first order associated to } a.$$

The association $a \to a^*$ is linear, a^* operates on scalar functions $\phi : \mathbf{R}^N \to \mathbf{R}$ and on vector functions $b : \mathbf{R}^N \to \mathbf{R}^N$ as

$$a^*\phi = \sum a_i \frac{\partial \phi}{\partial x_i}$$

$$a^*b = a^*(b_1, \cdots, b_N)^T = (a^*b_1, \cdots, a^*b_N)^T = b_*a, \quad a^*1_N = a.$$

Multiple application of linear differential operators are naturally defined such as a^*b^*, $(a^*b^*)c^*, a^*(b^*c^*)$, etc. The operations are multilinear, associative but non-commutative; thus powers can be defined

$$a^{*k} = a^*a^* \cdots a^*(k \text{ times }), \quad a^k := a^{*k}1_N$$

The identity operator I operates on scalar and vector functions ϕ and b as $I\phi = \phi, Ib = b$.

We identify all vector functions $a : \mathbf{R}^N \to \mathbf{R}^N$ as vector fields. All vector fields in

① In Feng K, Shi Z C, ed. Proc Conf on Computation of Differential Equations and Dynamical Systems. Singapore: World Scientific, 1993. 1-10

$\mathbf{R}^N$ form a (∞-dimensional) real Lie algebra $\mathbf{V}^N$ under Lie bracket

$$[a,b] := a_* b - b_* a = b^* a - a^* b = (b^* a^* - a^* b^*)\, 1_N.$$

The Lie algebra $\mathbf{V}^N$ is associated to the (∞-dimensional) local Lie group $\mathbf{D}^N$ of near-identity diffeomorphisms—or simply near-1 maps—of $\mathbf{R}^N$.

Consider the dynamical system in $\mathbf{R}^N$

$$\frac{dx}{dt} = a(x) \tag{1}$$

defined by a vector field a. It possesses a phase flow $e_a^t = e^t$, which is a one-parameter (in t) group of near-1 maps of $\mathbf{R}^N$,

$$e^0 = 1_N, \quad e^{t+s} = e^t \circ e^s,$$

and generates the solution by $x(0) \to e_a^t x(0) = x(t)$. The phase flow is expressible as a convergent power series in t,

$$e_a^t = 1_N + \sum_{k=1}^{\infty} t^k e_k,$$
$$e_0 = 1_N, \quad e_k = \frac{1}{k} a^* e_{k-1} = \frac{1}{k!}(a^*)^k 1_N = \frac{1}{k!} a^k.$$

We define

$$\operatorname{Exp} ta^* := I + \sum_{k=1}^{\infty} \frac{1}{k!}(ta)^{*k}, \quad I \text{ is the identity operator.} \tag{2}$$

This is an operator power series operating on scalar functions and vector functions, and define

$$\exp ta := (\operatorname{Exp} ta^*)\, 1_N = 1_N + \sum_{k=1}^{\infty} \frac{1}{k!}(ta^*)^k 1_N = 1_N + \sum_{k=1}^{\infty} \frac{t^k}{k!} a^k. \tag{2$'$}$$

then

$$\begin{cases} e_a^t = (\operatorname{Exp}\ ta^*)\, 1_N = \exp ta, \\ \phi \circ e_a^t = \phi \circ \exp ta = (\operatorname{Exp}\ ta^*)\phi, \quad \forall \text{ scaler function } \phi, \\ b \circ e_a^t = b \circ \exp ta = (\operatorname{Exp}\ ta^*)\, b = (\operatorname{Exp} ta^*)\, b^* 1_N, \quad \forall \text{ vector function } b. \end{cases} \tag{3}$$

Each numerical algorithm solving the system (1) possesses the *step transition map* f_a^s which is one-parameter (in step-size s) family (in general not a one-parameter group in s) of near-1 maps on R^N, expressible as a convergent power series in s

$$f_a^s = 1_N + \sum_1^{\infty} s^k f_k, \tag{4}$$

the coefficients can be determined recursively from the defining difference equation. The transition generates the numerical solution $x(0) \to (f_a^s)^N x(0) \approx x(Ns)$ by iterations with step-size s chosen fixed in general.

The main problem is to construct and analyse the algorithmic approximations $f_a^s \approx e_a^t|_{t=s} = e_a^s$ in a *proper* way. For this purpose we propose a *unified* framework based on the apparatus of formal power series, Lie algebra of vector fields and the corresponding Lie group of diffeomorphisms [1,2].

§2 Near-0 and Near-1 Formal Power Series

Among the formal power series $\sum_0^\infty s^k a_k, a_k : \mathbf{R}^N \to \mathbf{R}^N$, we pick out two special classes. The first class consists of those with $a_0 = 0$, called near-0 formal vector fields; the second class consists of those with $a_0 = 1_N$, called, near-1 formal maps (diffeomorphisms).

All near-0 formal vector fields $a^s = \sum_1^\infty s^k a_k$ form a (∞-dim.) real Lie algebra $\mathbf{FV}^N$ under the Lie bracket

$$[a^s, b^s] = \left[\sum_1^\infty s^k a_k, \sum_1^\infty s^k b_k\right] := \sum_{k=2}^\infty s^k \sum_{i+j=k} [a_i, b_j].$$

The associated near-0 formal differential operators and their products are

$$(a^s)_* := \left(\sum_1^\infty s^k a_k\right)_* := \sum_1^\infty s^k a_{k*},$$

$$a^{s*} = \left(\sum_1^\infty s^k a_k\right)^* := \sum_1^\infty s^k a_k^*,$$

$$a^{s*} b^{s*} := \sum_2^\infty s^k \sum_{i+j=k} a_i^* b_j^*, \quad (a^{s*})^2 := a^{s*} a^{s*}, \text{ etc.}$$

For any vector function $a = (a_1, \cdots, a_N)^T : \mathbf{R}^N \to \mathbf{R}^N$ and any near-1 formal map $g^s = 1_N + \sum s^k g_k$ we define the composition

$$(a \circ g^s)(x) = a(g^s(x)) = a(x) + \sum_1^\infty s^k (a \circ g)_k(x),$$

$$(a \circ g)_k = \sum_{m=1}^k \sum_{k_1+\cdots+k_m=k} \frac{1}{m!} (D^m a)(g_{k_1}, \cdots, g_{k_m}), \tag{5}$$

where

$$D^m a = (D^m a_1, \cdots, D^m a_N)^T,$$

$$D^m a_i(v_1, \cdots, v_m) = \sum_{j_1, \cdots, j_m = 1}^{N} \frac{\partial^m a_i}{\partial x_{j_1} \cdots \partial x_{j_m}} v_{1j_1} \cdots v_{mj_m}$$

is the usual m-th differential multi-linear form for m tangent vectors $v_i = (v_{i_1}, \cdots, v_{i_N})^T$, $i = 1, \cdots, m$ at point $x \in \mathbf{R}^N$ which is invariant under permutation of vectors. Using the identities

$$(D^1 a)(b) = b^* a,$$
$$(D^2 a)(b, c) = \left(c^* b^* - (c^* b)^*\right) a,$$
$$(D^3 a)(b, b, b) = \left(b^{*3} + 2b^{3*} - 3b^* b^{2*}\right) a,$$

we get in particular

$$(a \circ g)_1 = g_1^* a,$$
$$(a \circ g)_2 = g_2^* a + \frac{1}{2}\left(g_1^{*2} - g_1^{2*}\right) a,$$
$$(a \circ g)_3 = g_3^* a + \left(\left(g_2^* g_1^* - (g_2^* g_1)^*\right) a + \frac{1}{3!}\left(g_1^{*3} + 2g_1^{3*} - 3g_1^* g_1^{2*}\right) a.$$

For any two near-1 formal maps $f^s = 1_N + \sum s^k a_k$, $g^s = 1_N + \sum s^k g_k$, the composition $f^s \circ g^s$ is defined in a term by term way:

$$(f^s \circ g^s)(x) = f^s(g^s(x)) = 1_N(g^s(x)) + \sum_{k=1}^{\infty} s^k f_k(g^s(x))$$
$$=: 1_N(x) + \sum_{k=1}^{\infty} s^k (f \circ g)_k(x),$$

$$(f \circ g)_1 = f_1 + g_1, \tag{6}$$
$$(f \circ g)_k = f_k + g_k + \delta(f_1, \cdots, f_{k-1}; g_1, \cdots, g_k), \quad k \geqslant 2,$$
$$\delta(f_1, \cdots, f_{k-1}; g_1, \cdots, g_{k-1}) = \sum_{i=1}^{k-1} \sum_{m=1}^{i} \sum_{i_1 + \cdots + i_m = i} \frac{1}{m!}(D^m f_{k-i})(g_{i_1}, \cdots, g_{i_m}),$$

we get in particular,

$$(f \circ g)_2 = f_2 + g_2 + g_1^* f_1,$$
$$(f \circ g)_3 = f_3 + g_3 + g_1^* f_2 + g_2^* f_1 + \frac{1}{2}\left(g_1^{*2} - g_1^{2*}\right) f_1,$$
$$(f \circ g)_4 = f_4 + g_4 + g_1^* f_3 + g_2^* f_2 + g_3^* f_1 + \frac{1}{2}\left(g_1^{*2} - g_1^{2*}\right) f_2$$
$$+ \left(g_2^* g_1^* - (g_2^* g_1)^*\right) f_1 + \frac{1}{3!}\left(g_1^{*3} + 2g_1^{3*} - 3g_1^* g_1^{2*}\right) f_1.$$

Under this composition rule, all near-1 formal maps $f^s = 1_N + \sum_1^\infty s^k f_k$ form (∞-dim) formal Lie group $\mathbf{FD}^N$. In group $\mathbf{FD}^N$ inverse elements, square roots, rational powers, etc., always exist, their coefficients can always be determined recursively by the defining composition relations. For example, the inverse $(f^s)^{-1} := 1_N + \sum s^k h_k = h^s$ is defined by $(f^s \circ h^s) = 1_N$, hence

$$f_1 + h_1 = 0, \quad f_k + h_k + \delta(f_1, \cdots, f_{k-1}; h_1, \cdots, h_{k-1}) = 0, \quad k \geqslant 2.$$

In particular,

$$h_1 = -f_1; \quad h_2 = -f_2 + f_1^2, \quad h_3 = -f_3 + f_1^* f_2 + (f_2^* - f_1^{2*}) f_1 - \frac{1}{2} f_1^3 + \frac{1}{2} f_1^{2*} f_1.$$

There is an obvious one-one correspondence between the Lie algebra $\mathbf{FV}^N$ and the Lie group $\mathbf{FD}^N$, established simply by $+1_N$ and by -1_N. However, the more significant one-one correspondence between them is given by exp and its inverse log.

$$\exp : \mathbf{FV}^N \to \mathbf{FD}^N$$

$$a^s = \sum_1^\infty s^k a_k \to \exp a^s := 1_N + \sum_{m=1}^\infty \frac{1}{m!} (a^{s*})^m 1_N =: 1_N + \sum_1^\infty s^k f_k = f^s. \tag{7}$$

Note that

$$(a^{s*})^m = \left(\sum s^{k_1} a_{k_1}^*\right) \cdots \left(\sum s^{k_m} a_{k_m}^*\right) = \sum_{k_1, \cdots, k_m = 1}^\infty s^{k_1 + \cdots + k_m} a_{k_1}^* \cdots a_{k_m}^*,$$

so we get easily

$$f_k = \sum_{m=1}^k \frac{1}{m!} \sum_{k_1 + \cdots + k_m = k} a_{k_1}^* \cdots a_{k_m}^* 1_N, \quad k \geqslant 1, \quad f_1 = a_1, \tag{8}$$

$$f_k = a_k + \sum_{m=2}^k \frac{1}{m!} \sum_{k_1 + \cdots + k_m = k} a_{k_1}^* \cdots a_{k_m}^* 1_N, \quad k \geqslant 2, \quad f_2 = a_2 + \frac{1}{2} a_1^2.$$

Note that (8) provides a 2-way recursion formula from $a_1, \cdots, a_k$ to $f_1, \cdots, f_k$ and vice versa. Therefore exp maps $\mathbf{FV}^N$ one-one onto $\mathbf{FD}^N$ and its inverse, i.e., log is defined by the same (8):

$$\log = (\exp)^{-1} : \mathbf{FD}^N \to \mathbf{FV}^N, \quad \log \exp a^s = a^s, \quad \exp \log f^s = f^s.$$

In particular

$$a_1 = f_1, \quad a_2 = f_2 - \frac{1}{2} a_1^2, \quad a_3 = f_3 - \frac{1}{2} (a_1^* a_2 + a_2^* a_1) - \frac{1}{3!} a_1^3,$$

$$a_4 = f_4 - \frac{1}{2} (a_1^* a_3 + a_2^2 + a_3^* a_1) - \frac{1}{3!} (a_1^* a_1^* a_2 + a_1^* a_2^* a_1 + a_2^* a_1^* a_1) - \frac{1}{4!} a_1^4, \tag{9}$$

$$a_k = f_k - \sum_{m=2}^{k-1} \frac{1}{m!} \sum_{k_1 + \cdots + k_m = k} a_{k_1}^* \cdots a_{k_m}^* 1_N - \frac{1}{k!} a_1^k, \quad k \geqslant 3.$$

An equivalent construction of $\log f^s = a^s$ is

$$\log\ f^s = \sum_{m=1}^{\infty} \frac{(-1)^{m-1}}{m} h_m^s, \quad \text{where } h_1^s = f^s - 1_N, \quad h_m^s = h_{m-1}^s \circ f^s - h_{m-1}^s. \tag{10}$$

It is easy to compute

$$h_1^s = \sum_{k=1}^{\infty} s^k f_k = \sum_{k_1=1}^{\infty} s^{k_1} (1_N \circ f)_{k_1}$$

$$h_2^s = \sum_{k_1=k_2=1}^{\infty} s^{k_1+k_2} \left((1_N \circ f)_{k_1} \circ f\right)_{k_2},$$

$$h_3^s = \sum_{k_1,k_2,k_3=1}^{\infty} s^{k_1+k_2+k_3} \left(\left((1_N \circ f)_{k_1} \circ f\right)_{k_2} \circ f\right)_{k_3}, \text{ etc.}$$

Substituting in (10) and equating with $\sum\limits_{1}^{\infty} s^k a_k$, we get

$$a_k = \sum_{m=1}^{k} \frac{(-1)^{m-1}}{m} \sum_{k_1+\cdots+k_m=k} \left(\cdots\left((1_N \circ f)_{k_1} \circ f\right)_{k_2} \cdots \circ f\right)_{k_m}. \tag{11}$$

It is easy to verify $\log \exp\ a^s = a^s$ for this log, so this is precisely the inverse of exp, thus agrees with the previous one.

We use the above construction (11) to establish the *formal* Campbell-Haus-dorff formula:

For arbitrary near-1 formal maps f^s, g^s

$$\log(f^s \circ g^s) = \log\ f^s + \log\ g^s + \sum_{k=1}^{\infty} d_k(\log\ f^s, \log\ g^s), \tag{12}$$

where, for $\log f^s = a^s, \log g^s = b^s$,

$$d_k(a^s, b^s) = \frac{1}{k} \sum_{m=1}^{k} \frac{(-1)^{m-1}}{m} \sum_{\substack{p_1+q_1+\cdots+p_m+q_m=k \\ p_i+q_i \geqslant 1, p_i \geqslant 0, q_i \geqslant 0}} \frac{[(a^s)^{p_1}(b^s)^{q_1} \cdots (a^s)^{p_m}(b^s)^{q_m}]}{p_1! q_1! \cdots p_m! q_m!} \tag{13}$$

where

$$(x)^p = xx\cdots x(p\text{-times}), \quad [x_1 x_2 x_3 \cdots x_n] = [[\cdots[[x_1, x_2], x_3], \cdots], x_n].$$

In particular

$$d_1 = \frac{1}{2}[a^s, b^s], \quad d_2 = \frac{1}{12}\left([a^s b^s b^s] + [b^s a^s a^s]\right), \quad d_3 = -\frac{1}{24}[a^s b^s b^s a^s].$$

Let $\log(f^s \circ g^s) = c^s = \sum_1^\infty s^k c_k$. Then

$$
\begin{aligned}
c_1 &= a_1 + b_1, \quad c_2 = a_2 + b_2 + \frac{1}{2}[a_1 b_1], \\
c_3 &= a_3 + b_3 + \frac{1}{2}([a_1 b_2] + [a_2 b_1]) + \frac{1}{12}([a_1 b_1 b_1] + [b_1 a_1 a_1]), \\
c_4 &= a_4 + b_4 + \frac{1}{12}([a_1 b_3] + [a_2 b_2] + [a_3 b_1]) \\
&\quad + \frac{1}{12}([a_1 b_1 b_2] + [a_1 b_2 b_1] + [a_2 b_1 b_1] + [b_1 a_1 a_2] + [b_1 a_2 a_1] + [b_2 a_1 a_1]) \\
&\quad - \frac{1}{24}[a_1 b_1 b_1 a_1], \qquad \text{etc.}
\end{aligned}
$$

Note that the classical CH formula is restricted to the composition of two one-parameter groups, where $\log f^s = s a_1, \log g^s = s b_1$.

The log transform reduces matters at the Lie group level to those at the easier level of Lie algebra. All properties of near-1 formal maps have their logarithmic interpretations. We list some of them, let $\log f^s = a^s = \sum_1^\infty s^k a_k$:

1. f^s is a phase flow, i.e., $f^{s+t} = f^s \circ f^t \Longleftrightarrow \log f^s = s a_1$.
2. f^s is revertible, i.e., $f^s \circ f^{-s} = 1_N \Longleftrightarrow \log f^s$ is odd in s.
3. f^s raised to real μth power$(f^s)^\mu \Longleftrightarrow \log(f^s)^\mu = \mu \log f^s$. In particular log $(f^s)^{-1} = -\log f^s, \log\sqrt{f^s} = \frac{1}{2}\log f^s$.
4. f^s scaled to $f^{\alpha s} \Longleftrightarrow \log(f^{\alpha s}) = (\log f)^{\alpha s}$. In particular $\log(f^{-s}) = (\log f)^{-s}$.
5. $f^s - g^s = O(s^{p+1}) \Longleftrightarrow \log f^s - \log g^s = O(s^{p+1})$.
6. $f^s \circ g^s = g^s \circ f^s \Longleftrightarrow [\log f^s, \log g^s] = 0 \Longleftrightarrow \log(f^s \circ g^s) = \log f^s + \log g^s$.
7. $(f^s \circ g^s) = h^s \Longleftrightarrow \log h^s = \log(f^s \circ g^s) = \log f^s + \log g^s + \sum_1^\infty d_k(\log f^s, \log g^s)$.
8. f^s symplectic $\Longleftrightarrow$ all a_k are Hamiltonian fields.
9. f^s contact $\Longleftrightarrow$ all a_k are contact fields.
10. f^s volume-preserving $\Longleftrightarrow$ all a_k are source-free fields.

The log transform has important bearing on dynamical systems with Lie algebra structure, the structure-preserving property of maps f^s at the Lie group $(\mathbf{G} \subset \mathbf{D}_m)$ level can be characterized through their logarithms at the associated Lie algebra $(\mathbf{L} \subset \mathbf{V}_m)$ level, see [2,6].

§3 Algorithmic Approximations to Phase Flows

We return to the main problem of approximation to the phase flow for dynamical system $dx/dt = a(x)$.

$$
f_a^s = f^s = 1_N + \sum s^k f_k \approx e_a^s = 1_N + \sum s^k e_k, \quad e_k = a^k/k!. \tag{14}
$$

If $f_k = e_k, 1 \leqslant k \leqslant p$, we say f_a^s is accurate to order $\geqslant p$, if moreover, $f_{p+1} \neq e_{p+1}$, we say it is accurate to order p.

Let $\log f^s = a^s = \sum s^k a_k$. Note that the first $p+1$ equations in (8) completely determine $a_1, a_2, \cdots, a_{p+1}$ and $f_1, f_2, \cdots, f_{p+1}$ each other. It is then easy to establish

$$\begin{aligned} &f_k = e_k, \quad 1 \leqslant k \leqslant p; \quad f_{p+1} \neq e_{p+1} \Longleftrightarrow \\ &a = a_1 = e_1; \quad a_k = 0, \quad 1 < k \leqslant p; \quad a_{p+1} = f_{p+1} - e_{p+1} \neq 0. \end{aligned} \tag{15}$$

So the orders of approximation for $f_a^s \approx e_a^s$ and for $\log f_a^s - sa$ are the same.

Moreover, note that we have a *formal field*

$$s^{-1} \log f^s = s^{-1} a^s = a + \sum_1^\infty s^k a_{k+1} = a + O\left(s^p\right) \tag{16}$$

which is the original field a up to a near-0 perturbation and defines a *formal dynamical system*

$$\frac{dx}{dt} = \left(s^{-1} \log f^s\right)(x) = a(x) + \sum_1^\infty s^k a_{k+1}(x) \tag{17}$$

having a *formal phase flow* (in two parameters t and s with group property in t) $e_{s^{-1}a^s}^t = \exp t s^{-1} a^s$ whose *diagonal formal flow* $e_{s^{-1}a^s}^t|_{t=s}$ is exactly f^s. This means that any compatible algorithm f_a^s of order p gives perturbed solution of a right equation with field a; however, it gives the right solution of a perturbed equation with field $s^{-1} \log f_a^s = a + O\left(s^p\right)$. There could be many methods with the same formal order of accuracy but with quite different qualitative behavior. The problem is to choose among them those leading to allowable perturbations in the equation. For systems with geometric structure, the propositions 8, 9 and 10 provide guidelines for a proper choice. The structure-preservation requirement for the algorithms precludes all unallowable perturbations alien to the pertinent type of dynamics. Take, for example, Hamiltonian systems. A transition map f_a^s for Hamiltonian field a is symplectic if and only if all fields a_k are Hamiltonian, i.e., the induced perturbations in the equation are Hamiltonian [2,6]. So symplectic algorithms are clean, inherently free from all kinds of perturbations alien to Hamiltonian dynamics (such as artificial dissipation inherent in the vast majority of conventional methods), this accounts for their superior performance. The situations are the same for contact and volume-preserving algorithms [3-6].

Finally we give, as an illustration, four simplest methods together with step transition maps and their logarithms.

$$e_a^s = 1_N + sa + \frac{1}{2} s^2 a^2 + \frac{1}{3!} s^3 a^3 + O\left(s^4\right).$$

1. Explicit Euler method E

$$
\begin{aligned}
&x_1 - x_0 = sa\left(x_0\right), \\
&f^s - 1_N = sa, \\
&f_E^s = 1_N + sa, \\
&\log f_E^s = sa - \frac{s^2}{2}a^2 + O\left(s^3\right),
\end{aligned}
$$

non-revertible, ord=1.

2. Implicit Euler method I

$$
\begin{aligned}
&x_1 - x_0 = sa\left(x_1\right), \\
&f^s - 1_N = sa \circ f^s, \\
&f_I^s = \left(1_N - sa\right)^{-1} = \left(f_E^{-s}\right)^{-1} = sa + s^2a^2 + O\left(s^3\right), \\
&\log f_I^s = sa + \frac{s^2}{2}a^2 + O\left(s^3\right),
\end{aligned}
$$

non-revertible, ord=1.

3. Trapezoidal method T

$$
\begin{aligned}
x_1 - x_0 &= \frac{s}{2}\left(a\left(x_1\right) + a\left(x_0\right)\right), \\
f^s - 1_N &= \frac{s}{2}\left(a \circ f^s + a\right), \\
f_T^s &= \left(1_N - \frac{s}{2}a\right)^{-1} \circ \left(1_N + \frac{s}{2}a\right) = f_I^{s/2} \circ f_E^{s/2}, \\
&= \left(f_E^{s/2}\right)^{-1} \circ f_C^s \circ f_E^{s/2} = 1_N + sa + \frac{s^2}{2}a^2 + \frac{s^3}{4}a^3 + O\left(s^4\right), \\
\log f_T^s &= sa + \frac{s^3}{12}a^3 + O\left(s^5\right),
\end{aligned}
$$

revertible, ord = 2, symplectic for linear Hamiltonian but non-symplectic for non-linear Hamiltonian systems.

4. Centered Euler method C

$$
\begin{aligned}
&x_1 - x_0 = sa\left(\frac{1}{2}\left(x_1 + x_0\right)\right), \\
&f^s - 1_N = sa \circ \left(\frac{1}{2}\left(f^s + 1_N\right)\right),
\end{aligned}
$$

2-stage version recommended for implementation

$$\begin{aligned}
\bar{x} &= x + \frac{s}{2}a(\bar{x}), \quad x_1 = 2\bar{x} - x_0, \\
\bar{x} &= f_I^{s/2}(x_0), \quad x_1 = 2f_I^{s/2}(x_0) - 1_N(x_0), \\
f_C^s &= 2f_I^{s/2} - 1_N = \left(1_N + \frac{s}{2}a\right)^{-1} \circ \left(1_N - \frac{s}{2}a\right) = f_E^{s/2} \circ f_I^{s/2} \\
&= 1_N + sa + \frac{s^2}{2}a^2 + \frac{s^3}{8}\left(a_* a^2 + a^3\right) + O\left(s^4\right), \\
\log f_C^s &= sa + s^3\left(\frac{1}{8}a_* a^2 - \frac{1}{24}a^3\right) + O\left(s^5\right),
\end{aligned}$$

revertible, ord = 2, unconditionally symplectic with preservation of all quadratic invariants for Hamiltonian systems.

Note the similarities and delicate differences between C and T : Both can be composed by a $s/2$ implicit and a $s/2$ explicit stages but in opposite orderings. Moreover, they are conjugate to each other, noted earlier independently in [7]. C is far less known than T, it becomes prominent only after the recent development of symplectic algorithms [3]. In crucial aspects C is superior.

Acknowledgements The author is grateful to Dr. Wang Dao-liu for discussions and valuable help in preparing the manuscript.

References

[1] Feng K. The calculus of formal power series for diffeomorphisms and vector fields: [Preprint]. Beijing: CAS. Computing Center, 1991

[2] Feng K. Formal dynamical systems and numerical algorithms. In Feng K, Shi Z C, ed. Proc Inter Conf on Computation of Differential Equations and Dynamical Systems. Singapore: World Scientific, 1993: 1-10

[3] Feng K. On difference schemes and symplectic geometry. In Feng K, ed. Proc 1984 Beijing Symp Diff Geometry and Diff Equations. Beijing: Science Press, 1985: 42-58

[4] Feng K. The Hamiltonian way for computing Hamiltonian dynamics. In Spigler R, ed. Applied and Industrial Mathematics. Netherlands: Kluwer, 1991: 17-35

[5] Feng K, Qin M Z. Hamiltonian algorithms and a comparative numerical study. Comput Phys Comm. 1991 65: 173-187

[6] Feng K. Symplectic, contact and volume-preserving algorithms. In Shi Z C, Ushijima T, ed. Proc 1st China-Japan Conf on Numer Math. Singapore: World Scientific, 1993: 1-28

[7] Qin M Z. Oral communications

19 Dynamical Systems and Geometric Construction of Algorithms[①②]

动力系统与算法的几何构造

Abstract

We give a brief survey on the research, undertaken by the senior author and his group, on the construction and analysis of numerical algorithms for ordinary differential equations, including specifically Hamiltonian systems and other systems with algebraic-geometric structures from the view-point of dynamical systems and ∞-dimensional Lie algebras of vector fields on manifolds and the associated ∞-dimensional Lie transformation groups.

§1 Introduction

Consider the autonomous system on $\mathbf{R}^m$

$$\dot{z} = a(z), \quad z = (z_1, \cdots, z_m)^T, \quad a(z) = (a_1(z), \cdots, a_m(z))^T \tag{1.1}$$

defined by its "right hand function", i.e., a smooth vector field $z \to a(z)$ on $\mathbf{R}^m$. The phase flow operator of dynamical system (1.1) defined by the vector field a is determined by the one parameter family $e_a^t = e^t$ of near-identity diffeomorphisms (or synonymously, transformations, operators) on $\mathbf{R}^m$ satisfying

$$\begin{cases} \dfrac{d}{dt} e_a^t = a \circ e_a^t \\ e_a^0 = \text{identity} := 1_m. \end{cases} \tag{1.2}$$

It completely determines the dynamical evolution of the given continuous-time dynamical system. The phase flow operator e_a^t satisfies the one parameter group property

$$\begin{cases} e_a^{t+s} = e_a^t \circ e_a^s, \quad \forall t, s \in \mathbf{R}, \\ e_a^0 = 1_m \end{cases} \tag{1.3}$$

and depends analytically on t, expressible as a convergent, at least locally in space and time, power series

$$e_a^t = \sum_{k=0}^{\infty} t^k e_k, \tag{1.4}$$

---

① Joint with Wang D L

② In Shi Z C, Yang C J, ed. Computational Mathematics of China. Contemporary Mathematics. 1994 163: 1-32

where the coefficients are smooth vector functions from $\mathbf{R}^m$ to $\mathbf{R}^m$ and can be determined recursively

$$e_0 = e^0 = 1_m, e_1 = a, e_k = \frac{1}{k}(e_{k-1})_* e_1, \quad k = 1, 2, \cdots, \tag{1.5}$$

here Jacobian matrix of $u : \mathbf{R}^m \to \mathbf{R}^m$ is denoted by u_*.

Now to any difference scheme solving the system (1.1) there corresponds a step-transition operator $g^s = g_a^s$ for step size s which completely (for single-step methods) or essentially (for multi-step methods) determines the dynamical evolution of the discrete-time dynamical system. Operator g^s is a one parameter family in step size s of near-identity diffeomorphisms

$$g_a^s = \sum_{k=0}^{\infty} s^k g_k, \quad g_0 = 1_m. \tag{1.6}$$

The algorithm g_a^s is said to be consistent with system (1.1) if $g_1 = a$; it is said to be of (accuracy) order $p(\geqslant 1)$ if

$$g_k = e_k, \quad k = 0, 1, \cdots, p; \quad g_{p+1} \neq e_{p+1}. \tag{1.7}$$

If g_a^s is consistent with system (1.1) and is a one-parameter group in s, then $g_a^s = e_a^s$, i.e., the difference scheme is exact. So, apart from some exceptional cases, g_a^s does not satisfy the group property $g_a^{s+t} = g_a^s \circ g_a^t$. Therefore, in the most cases, g_a^s is an approximation of e_a^s. The problem is to construct the algorithmic approximation g^s to the phase flow e_a^s in a *proper* way.

As is well known, all (smooth) vector fields on $\mathbf{R}^m$ form an ∞-dim Lie algebra, denoted by $\mathbf{V}(\mathbf{R}^m) = \mathbf{V}_m$, under the Lie bracket $[a, b] = a_* b - b_* a$, where a and b are two vector fields on $\mathbf{R}^m$. The Lie algebra $\mathbf{V}_m$ is associated with the ∞-dimensional Lie group $\mathbf{D}(\mathbf{R}^m) = \mathbf{D}_m$ of near-identity diffeomorphisms on $\mathbf{R}^m$. Every Lie subalgebra $\mathbf{L}$ of $\mathbf{V}_m$ is associated with a Lie subgroup $\mathbf{G}$ of $\mathbf{D}_m$. For any vector field a in $\mathbf{L}$, the phase flow e_a^t is a one-parameter subgroup in $\mathbf{G}$.

The three classical cases of dynamical systems with Lie-algebraic structures are: 1. Hamiltonian systems on $\mathbf{R}^{2n}$—with the symplectic structure. $\mathbf{L}$ is the Lie algebra $\mathbf{SpV}_{2n}$ of Hamiltonian vector fields, $\mathbf{G}$ is the Lie group $\mathbf{SpD}_{2n}$ of symplectic transformations. 2. Contact systems on $\mathbf{R}^{2n+1}$—with the contact structure. $\mathbf{L}$ is the Lie algebra $\mathbf{KV}_{2n+1}$ of contact vector fields, $\mathbf{G}$ is the Lie group $\mathbf{KD}_{2n+1}$ of contact transformations. 3. Source-free systems on $\mathbf{R}^m$—with the volume structure. $\mathbf{L}$ is the Lie algebra $\mathbf{SV}_m$ of source-free vector fields, $\mathbf{G}$ is the Lie group $\mathbf{SD}_m$ of volume-preserving transformations. It is natural and even mandatory to require that the numerical algorithms for solving dynamical systems with structures to be structure-preserving, i.e., g^s belongs to the associated group $\mathbf{G}$ for small $|s|$. This is the main theme of this survey.

In section 2, we discuss the revertibility property of a one-parameter family of operators. Since the phase flows of all autonomous systems are revertible, their algorithms should be revertible too. In section 3, we give a kind of explicit structure-preserving algorithms for systems with Lie algebraic structures. In section 4, based on explicit Euler schemes, a class of explicit symplectic schemes for separable Hamiltonian systems are given. Symplectic schemes are implicit in general. But for specific Hamiltonians explicit Euler schemes may be symplectic and even exact. Consequently, if a Hamiltonian can be decomposed as a sum of specific Hamiltonians, then explicit symplectic schemes can be constructed by composing explicit Euler schemes for these specific Hamiltonians and higher order explicit symplectic schemes can be obtained as well. This kind of Hamiltonians is not too rare but can cover many important cases. In section 5 we give simple families of algorithms which are symplectic for all Hamiltonian systems. The general and systematic method to construct unconditional symplectic schemes is the generating function method which is discussed in section 6. Section 7 is to construct Hamiltonian algorithms for Hamiltonian systems with a perturbation parameter. In sections 8 and 9 we consider contact algorithms for contact systems and volume preserving algorithms for source-free systems respectively. In section 10 we consider the general theory of numerical approximations for ordinary differential equations based on the apparatus of formal power series and their exponential and logarithmic transforms.

§2 Revertibility of Operators

Let g^t be a one-parameter family of near-1 diffeomorphisms expansible as a convergent power series in t,

$$g^t = \sum_0^\infty t^k g_k, \quad g_0 = 1_m. \tag{2.1}$$

g^t is called *revertible* if $g^t \circ g^{-t} = 1_m, \forall t$. We define the reversion of g^t as

$$\check{g}^t = \left(g^{-t}\right)^{-1}. \tag{2.2}$$

g^t is revertible if and only if $g^t = \check{g}^t$. We have

(1) $\check{g}^t \circ g^t$ and $g^t \circ \check{g}^t$ are always revertible for all g^t;

(2) $\check{\check{g}}^t = g^t, (f^t \circ g^t)^\vee = \check{g}^t \circ \check{f}^t$;

(3) f^t and g^t are revertible $\Longrightarrow f^{\alpha t} \circ g^{\beta t} \circ f^{\alpha t}$ and $f^{\alpha t} \circ g^{\beta t} \circ g^{\beta t} \circ f^{\alpha t}$ are revertible $\forall \alpha, \beta \in \mathbf{R}$;

(4) f_i^t are revertible, $i=1,\cdots,k \Longrightarrow f_1^{\alpha_1 t}\circ\cdots\circ f_k^{\alpha_k t}\circ f_k^{\alpha_k t}\circ\cdots\circ f_1^{\alpha_1 t}$ are revertible for $\forall\alpha_i\in\mathbf{R}, i=1,\cdots,k_i$;

(5) When f^t and g^t are revertible then $f^t\circ g^t$ is revertible if and only if $f^t\circ g^t=g^t\circ f^t$.

Since revertibility is a weaker form of the group property (1.3), all one-parameter groups of diffeomorphisms are always revertible.

For example, consider the system (1.1). The explicit and implicit Euler schemes are

$$z\to\hat{z}=E_a^s z:\quad \hat{z}=z+sa(z),\quad E_a^s=1_m+sa; \tag{2.3}$$

$$z\to\hat{z}=I_a^s z:\quad \hat{z}=z+sa(\hat{z}),\quad I_a^s=(1_m-sa)^{-1}, \tag{2.4}$$

where E_a^s and I_a^s are step transition operators of the explicit and implicit Euler schemes respectively. So $I_a^s=(E_a^{-s})^{-1}=\check{E}_a^s$, i.e., the reversion of the explicit Euler operator is the implicit Euler operator.

Two-leg and one-leg weighted Euler schemes are

$$\begin{aligned} z\to\hat{z}=&T_{a,c}^s z:\hat{z}=z+s(ca(\hat{z})+(1-c)a(z)),\\ &T_{a,c}^s=(1_m-sca)^{-1}\circ(1_m+s(1-c)a); \end{aligned} \tag{2.5}$$

$$\begin{aligned} z\to\hat{z}=&E_{a,c}^s z:\hat{z}=z+sa(c\hat{z}+(1-c)z),\\ &E_{a,c}^s=(1_m+s(1-c)a)\circ(1_m-sca)^{-1}. \end{aligned} \tag{2.6}$$

So

$$\check{T}_{a,c}^s=T_{a,1-c}^s,\quad \check{E}_{a,c}^s=E_{a,1-c}^s.$$

Consequently, $T_{a,c}^s$ and $E_{a,c}^s$ are revertible for any vector field a if and only if $c=\dfrac{1}{2}$, i.e., they are trapezoidal and centered Euler schemes respectively.

We write $g^s\approx e_a^s$ if $g^s=e_a^s+O(s^2), g^s\approx e_a^s$, ord p if $g^s=e_a^s+O(s^{p+1}), g^s\approx e_a^s$, ord ∞ if $g^s=e_a^s$. A consistent revertible step transition operator is always of even order $2l(l\geqslant 1)$. For $g^s\approx e_a^s$, ord $1, g^{s/2}\circ\check{g}^{s/2}$ and $\check{g}^{s/2}\circ g^{s/2}\approx e_a^s$ are revertible and of order 2. If $g^s\approx e_a^s$ is revertible and of order 2, then the revertible composite of g^s :

$$g^{\alpha s}\circ g^{\beta s}\circ g^{\alpha s}\approx e_a^s \tag{2.7}$$

is of order 4 when

$$2\alpha+\beta=1,\quad 2\alpha^3+\beta^3=0,$$

i.e.,

$$\alpha=\frac{1}{2-2^{1/3}}>0,\quad \beta=1-2\alpha<0. \tag{2.7'}$$

Generally, if $g^s \approx e_a^s$ is revertible and of order $2l$, then the revertible composite (2.7) of g^s is of order $2(l+1)$, when

$$2\alpha + \beta = 1, \quad 2\alpha^{2l+1} + \beta^{2l+1} = 0, \tag{2.8}$$

i.e.,

$$\alpha = \frac{1}{2 - 2^{1/(2l+1)}} > 0, \quad \beta = 1 - 2\alpha < 0. \tag{2.8$'$}$$

For more details, see [19, 28, 31, 44].

We give some elementary algorithms, i.e., the explicit Euler method E, the implicit Euler method I, the centered Euler method C. $E, I,\ C$, especially the oldest and simplest E, will be the basic components for the composition of structure preserving algorithms. Then transition maps for step-size s for $x \to$ next step $\hat{x}$ will be denoted by E^s, I^s, C^s.

$$E: \quad \hat{x} = x + sa(x), \quad E_a^s = 1_m + sa \approx e_a^s,$$

$$\text{ord } 1, \text{non-revertible.} \tag{2.9}$$

$$I: \quad \hat{x} = x + sa(\hat{x}), \quad I_a^s = (1_m - sa)^{-1} = \check{E}_a^s \approx e_a^s,$$

$$\text{ord } 1, \text{non-revertible.} \tag{2.10}$$

$$C: \quad \hat{x} = x + sa\left(\frac{1}{2}\hat{x} + \frac{1}{2}x\right), \quad \text{or in equivalent 2 -stage form}$$

$$\bar{x} = x + \frac{s}{2}a(\bar{x}), \quad \hat{x} = 2\bar{x} - x. \tag{2.11}$$

$$C_a^s = E_a^{s/2} \circ I_a^{s/2} = E_a^{s/2} \circ \check{E}_a^{s/2} = 2I_a^{s/2} - 1_m \approx e_a^s,$$

$$\text{ord } 2, \text{ revertible.}$$

A vector field a is called *nilpotent of degree 2* if

$$a^2 = a_* a = a^* a = 0 \Longleftrightarrow e_a^s = 1_m + sa = E_a^s, \tag{2.12}$$

so with such "simple" vector fields, the method E gives exact solutions.

§3 Explicit Structure-preserving Algorithms

Consider a dynamical system

$$\frac{dz}{dt} = a(z), \quad a \in \mathbf{L} \tag{3.1}$$

where $\mathbf{L} \subset \mathbf{V}_m$, a Lie algebra of vector fields under the Lie bracket. Then the phase flow e_a^t is a one parameter subgroup of $\mathbf{G} \subset \mathbf{D}_m$, the Lie group associated with the Lie algebra $\mathbf{L}$.

We want to construct algorithms g_a^s approximating to e_a^s in the same group $\mathbf{G}$, i.e., $g_a^s \in \mathbf{G}$, for $|s|$ small. Such algorithm g_a^s is called *structure-preserving*, or $\mathbf{L}$-preserving. We have the following general propositions.

Let

$$a = \sum_{i=1}^{k} a^i, \quad a^i \in \mathbf{L} \tag{3.2}$$

and let e_i^s be the phase flow of $a^i, e_i^s \in \mathbf{G}$, and $f_i^s \approx e_i^s$, ord $\geqslant 1, f_i^s \in \mathbf{G}, \forall i = 1, \cdots, k$. Then

$$f^s := f_1^s \circ \cdots \circ f_k^s \approx e_a^s, \quad \text{ord } 1, \ \mathbf{L}\text{-preserving}, \tag{3.3}$$

$$g^s := f^{s/2} \circ \check{f}^{s/2} = f_1^{s/2} \circ \cdots \circ f_k^{s/2} \circ \check{f}_k^{s/2} \circ \cdots \circ \check{f}_1^{s/2} \approx e_a^s,$$
$$\text{ord } 2, \text{revertible } \mathbf{L}\text{-preserving}, \tag{3.4}$$

$$h^s := g^{\alpha s} \circ g^{\beta s} \circ g^{\alpha s} \approx e_a^s, \quad \text{with the } \alpha, \beta \text{ in } (2.7')$$
$$\text{ord } 4, \ \text{revertible } \mathbf{L}\text{-preserving; etc.} \tag{3.5}$$

A case with $a = a^1 + a^2 + a^3$ for construction of 3D volume preserving algorithms will be shown in §9. In particular, as an ideal case, take $f_i^s = e_{a^i}^s = e_i^s, i = 1, \cdots k$, then

$$f^s := e_1^s \circ \cdots \circ e_k^s \approx e_a^s, \quad \text{ord } 1, \ \mathbf{L}\text{-preserving} , \tag{3.6}$$

$$g^s := e_1^{s/2} \circ \cdots \circ e_{k-1}^{s/2} \circ e_k^s \circ e_{k-1}^{s/2} \circ \cdots \circ e_1^{s/2} \approx e_a^s,$$
$$\text{ord } 2, \ \text{revertible } \mathbf{L}\text{-preserving.} \tag{3.7}$$

This construction gives actually a *numerical algorithm* only when the component phase flows e_i^s are available and algorithmically implementable, this is quite often the case.

We call a vector field $a \in \mathbf{L}$ to be $\mathbf{L}$-separable if it admits a decomposition

$$a = \sum_{i=1}^{k} a^i, \quad a^i \in \mathbf{L} \text{ and nilpotent, i.e., } a_*^i a^i = 0. \tag{3.8}$$

In this case, by (2.12), $e_{a^i}^s = E_{a^i}^s =: E_i^s$, so f^s, g^s become

$$f^s := E_1^s \circ \cdots \circ E_k^s \approx e_a^s, \quad \text{ord } 1, \ \mathbf{L}\text{-preserving}, \tag{3.9}$$

$$g^s := E_1^{s/2} \circ \cdots \circ E_{k-1}^{s/2} \circ E_k^s \circ E_{k-1}^{s/2} \circ \cdots \circ E_1^{s/2} \approx e_a^s,$$
$$\text{ord } 2, \ \text{revertible } \mathbf{L}\text{-preserving}, \tag{3.10}$$

and all the related revertible $\mathbf{L}$-preserving algorithms are explicit. A simple $\mathbf{L}$-separable case with decomposition $a = a^1 + a^2$ is important for the construction of explicit symplectic method for Hamiltonian systems. See immediately the next section.

§4 Explicit Euler-type Hamiltonian Algorithms

We consider now Hamiltonian systems on $\mathbf{R}^{2n}$. The Hamiltonian vector field is of the form

$$a = J\nabla H, \quad J = J_{2n} = \begin{pmatrix} 0 & -I_n \\ I_n & 0 \end{pmatrix} \tag{4.1}$$

defined by an energy function $H(z)$ on $\mathbf{R}^{2n}$, which defines the Hamiltonian system

$$\frac{dz}{dt} = a(z) = J\nabla H(z), \quad z \in \mathbf{R}^{2n}, \tag{4.2}$$

where $\nabla H(z) = H_z = (H_{z_1}(z), \cdots, H_{z_{2n}}(z))^T$ is the gradient of H, or

$$\begin{aligned} \frac{dp}{dt} &= -H_q, \\ \frac{dq}{dt} &= H_p, \end{aligned} \quad \begin{pmatrix} p \\ q \end{pmatrix} = z \in \mathbf{R}^{2n}, \quad p = \begin{pmatrix} p_1 \\ \vdots \\ p_n \end{pmatrix}, \quad q = \begin{pmatrix} q_1 \\ \vdots \\ q_n \end{pmatrix}.$$

Intrinsic to all Hamiltonian systems is the symplectic structure defined by the differential 2-form

$$\omega = \sum_{i=1}^{n} dp_i \wedge dq_i.$$

A *symplectic* (or *canonical*) transformation is a diffeomorphism $g : z \to \hat{z}$ on $\mathbf{R}^{2n}$ preserving the symplectic structure

$$\sum d\hat{p}_i \wedge d\hat{q}_i = \sum dp_i \wedge dq_i, \quad \text{i.e.} \quad (g_*(z))^T J_{2n} (g_*(z)) \equiv J_{2n}.$$

All Hamiltonian vector fields form a Lie algebra $\mathbf{L} = \mathbf{SpV}_{2n}$ under Lie bracket. The associated Lie group $\mathbf{G} = \mathbf{SpD}_{2n}$ consists of all near-identity symplectic transformations on $\mathbf{R}^{2n}$. The phase flows $e^t_{J\nabla H}$, abbreviated as e^t_H, of Hamiltonian systems are symplectic transformations. So it is natural and mandatory to require the algorithms with transition operator $g^s_{J\nabla H}$, abbreviated as g^s_H, to be symplectic too, such algorithms are called *symplectic* (or *Hamiltonian or canonical*). Non-symplectic algorithms inevitably bring in distortions alien to Hamiltonian dynamics, such as artificial dissipation and other parasitic effects. See [2-5, 14-15, 20].

The phase flow e^t_H of (4.2) is a one parameter group of symplectic maps

$$e^t_H = e_0 + te_1 + t^2 e_2 + \cdots, \tag{4.3}$$

where

$$\begin{aligned} e_0 &= 1_{2n}, e_1 = J\nabla H, e_2 = \frac{1}{2}(J\nabla H)_* J\nabla H, \\ e_k &= \frac{1}{k}(e_{k-1})_* J\nabla H, k \geqslant 3. \end{aligned} \tag{4.4}$$

We need to construct *symplectic* algorithms.

By (2.12), a Hamiltonian vector field $J\nabla H$ is nilpotent of degree 2 if

$$(J\nabla H)_* J\nabla H = 0 \quad \text{or} \quad H_{zz}JH_z = 0, \tag{4.5}$$

for simplicity we say H is nilpotent of degree 2. For the Hamiltonian H nilpotent of degree 2, the explicit Euler scheme E_H^s is just the phase flow of the system (4.2) at $t = s$

$$e_H^s(z) = z + se_1(z) = z + sJ\nabla H(z) \equiv E_H^s(z).$$

So it must be symplectic.

If

$$H(z) = \sum_{i=1}^{k} H_i(z), \tag{4.6}$$

and every $H_i(z)$ is nilpotent of degree 2, then $H(z)$ is said to be *symplectically separable*. The following composite

$$z \to \hat{z} = f^s z, \quad f^s := E_{H_1}^s \circ E_{H_2}^s \circ \cdots \circ E_{H_k}^s, \tag{4.7}$$

is explicit, symplectic and of order 1. The composite

$$\begin{aligned} & z \to \hat{z} = g^s z, \\ & g^s := f^{s/2} \circ \check{f}^{s/2} = E_{H_1}^{s/2} \circ \cdots \circ E_{H_{k-1}}^{s/2} \circ E_{H_k}^s \circ E_{H_{k-1}}^{s/2} \circ \cdots \circ E_{H_1}^{s/2} \end{aligned} \tag{4.8}$$

is explicit, symplectic and revertible of order 2. High order revertible symplectic schemes can be constructed from these schemes by the procedure (2.7) and (2.8).

Examples If

$$H(z) = H(p,q) = \phi(p) \quad \text{or} \quad \psi(q),$$

where ϕ and ψ are functions of n variables, then $H(z)$ is nilpotent of degree 2

$$\begin{pmatrix} H_p(p,q) \\ H_q(p,q) \end{pmatrix} = \begin{pmatrix} \phi_p(p) \\ 0 \end{pmatrix} \quad \text{or} \quad \begin{pmatrix} 0 \\ \psi_q(q) \end{pmatrix}.$$

The corresponding explicit Euler schemes

$$\begin{aligned} \hat{p} &= p, \\ \hat{q} &= q + s\phi_p(p), \end{aligned}$$

or

$$\begin{aligned} \hat{p} &= p - s\psi_q(q), \\ \hat{q} &= q, \end{aligned}$$

are symplectic. If

$$H(p,q)=\phi(p)+\psi(q), \tag{4.9}$$

which is symplectically separable, then the composite schemes

$$E_\psi^s\circ E_\phi^s:\quad \hat{p}=p-s\psi_q(\hat{q}),\quad \hat{q}=q+s\phi_p(p), \tag{4.10}$$

$$E_\phi^s\circ E_\psi^s:\quad \hat{p}=p-s\psi_q(q),\quad \hat{q}=q+s\phi_p(\hat{p}), \tag{4.11}$$

are explicit, symplectic and of order 1. The revertible scheme

$$\begin{aligned}&g^s:=\left(E_\psi^{s/2}\circ E_\phi^{s/2}\right)^\vee\circ\left(E_\psi^{s/2}\circ E_\phi^{s/2}\right)=E_\phi^{s/2}\circ E_\psi^s\circ E_\phi^{s/2},\\&q^1=q+\frac{s}{2}\phi_p(p),\quad \hat{p}=p-s\psi_q\left(q_1\right),\quad \hat{q}=q^1+\frac{s}{2}\phi_p(\hat{p}).\end{aligned} \tag{4.12}$$

is symplectic and of order 2. Its revertible composite

$$h^s:=g^{\alpha s}\circ g^{\beta s}\circ g^{\alpha s}$$

gives a symplectic scheme of order 4 with the parameters (2.7′) , i.e.,

$$\begin{aligned}p^1&=p-c_1s\psi_q(q),\quad q^1=q+d_1s\phi_p\left(p^1\right),\\p^2&=p^1-c_2s\psi_q\left(q^1\right),\quad q^2=q^1+d_2s\phi_p\left(p^2\right),\\p^3&=p^2-c_3s\psi_q\left(q^2\right),\quad q^3=q^2+d_3s\phi_p\left(p^3\right),\\\hat{p}&=p^3-c_4s\psi_q\left(q^3\right),\quad \hat{q}=q^3+d_4s\phi_p(\hat{p}),\end{aligned} \tag{4.13}$$

with the parameters $\alpha=\left(2-2^{1/3}\right)^{-1}$, $\beta=1-2\alpha$ and either

$$c_1=0,\quad c_2=c_4=\alpha,\quad c_3=\beta,\quad d_1=d_4=\alpha/2,\quad d_2=d_3=(\alpha+\beta)/2,$$

or

$$c_1=c_4=\alpha/2,\quad c_2=c_3=(\alpha+\beta)/2,\quad d_1=d_3=\alpha,\quad d_2=\beta,\quad d_4=0.$$

Symplectically separable Hamiltonians have wide coverage in applications. It is easy to see that, the Hamiltonian of the form

$$H(p,q)=\phi(Ap+Bq),\quad AB^T=BA^T, \tag{4.14}$$

where $\phi(x)$ is a function of n variables, A and B are $n\times n$ matrices, is also nilpotent of degree 2. Then the explicit Euler scheme $E_H^s=E^s(\phi)=e_H^s$ is

$$E^s(\phi):\quad \begin{aligned}\hat{p}&=p-sB^T\phi_x(Ap+Bq)\\\hat{q}&=q+sA^T\phi_x(Ap+Bq)\end{aligned} \tag{4.15}$$

so we get a class of symplectically separable Hamiltonians [19]

$$H(p,q)=\sum_{i=1}^{k} H_i(p,q),\quad H_i(p,q)=\phi_i\left(A_i p+B_i q\right),\quad A_i B_i^T=B_i A_i^T, \tag{4.16}$$

where $\phi_i(x)$ are functions of n variables, A_i, B_i are $n\times n$ matrices. Moreover, all polynomial Hamiltonians belong to this class; in fact, it has been proved in [19] that every polynomial $H(p,q)$ in $2n$ variables p,q can be decomposed in the form of (4.16), where $\phi_i(x)$ are polynomials of n variables x, $A_i=\operatorname{diag}(a_1^i,\cdots,a_n^i)$ and $B_i=\operatorname{diag}(b_1^i,\cdots,b_n^i)$.

For symplectically separable Hamiltonians of the above class, all explicit revertible symplectic schemes formed by compositions contain solely the Euler schemes $E^s(\phi_i)$ in the form (4.15) as basic components. A related conjecture is: this class already covers all the symplectically separable cases, i.e., H is nilpotent of degree 2 if and only if it is expressible in the form (4.14).

More generally, if we have a decomposition $H=H_1+\cdots+H_k$, for which each H_i is integrable and its phase flow $e_{H_i}^s$ is algorithmically implementable, then by (3.4), (3.5) the symmetrical composites

$$g_H^s=e_{H_1}^{s/2}\circ\cdots\circ e_{H_{k-1}}^{s/2}\circ e_{H_k}^s\circ e_{H_{k-1}}^{s/2}\circ\cdots\circ e_{H_1}^{s/2} \tag{4.17}$$

$$h_H^s=g_H^{\alpha s}\circ g_H^{\beta s}\circ g_H^{\alpha s} \tag{4.18}$$

give explicit symplectic, revertible algorithms of order 2, 4, etc. This general approach is widely applicable to *many body problems* in different physical contexts for which the 2-*body problem* is *solvable*. Since in such problems the Hamiltonian usually admits a natural decomposition $H=\sum\limits_{i<j} H_{ij}$, each H_{ij} accounts for a 2-body problem.

An interesting result from this approach is a construction [47] of explicit symplectic method for computing the Hamiltonian system of N vortices $z_i=(x_i,y_i)$ with intensities k_i

$$k_i\frac{dx_i}{dt}=\frac{\partial H}{\partial y_i},\quad k_i\frac{dy_i}{dt}=-\frac{\partial H}{\partial x_i},\quad i=1,\cdots,N \tag{4.19}$$

with symplectic structure $\omega_k=\sum\frac{1}{k_i}dx_i\wedge dy_i$ and

$$\begin{aligned}H=\sum_{i<j}H_{ij},\quad H_{ij}&=-\frac{1}{2\pi}k_ik_j\ln r_{ij},\\ r_{ij}&=\left((x_i-x_j)^2+(y_i-y_j)^2\right)^{1/2}.\end{aligned} \tag{4.20}$$

Each H_{ij} accounts for a solvable 2 -vortex motion in which both vortices z_i, z_j rotate about their center of vorticity $z=(k_iz_i+k_jz_j)(k_i+k_j)^{-1}$ with angular velocity $a=$

$(k_i + k_j)/2\pi r_{ij}^2$ if $k_i + k_j \neq 0$ or translate with linear velocity $(b(y_i - y_j), -b(x_i - x_j))$, $b = k_i/2\pi r_{ij}^2$ if $k_i + k_j = 0$. These simple phase flows $e_{H_{ij}}^s$ serve as the basic algorithmic components for successive compositions, resulting in efficient explicit revertible methods, symplectic for this specific system. They are promising in application to incompressible ideal flows for tracking the vortex particles, which ought to be done in the structure-preserving way.

§5 Unconditional Hamiltonian Algorithms

Now we want to construct *unconditional* Hamiltonian algorithms, i.e., they are symplectic for *all* Hamiltonian systems.

It can be shown that

(a) The two-leg weighted Euler schemes (2.5) are generically non-symplectic for any real number c or real $2n \times 2n$ matrix c.

(b) The one-leg weighted Euler schemes (2.6), i.e.,

$$\hat{z} = g_H^s z: \quad \hat{z} = z + sJH_z(c\hat{z} + (1-c)z), \tag{5.1}$$

with real number c is unconditionally symplectic if and only if $c = \frac{1}{2}$, which corresponds to the *centered Euler scheme*

$$\hat{z} = z + sJH_z\left(\frac{\hat{z} + z}{2}\right). \tag{5.2}$$

These simple propositions illustrate a general situation: apart from some very rare exceptions, the vast majority of conventional schemes are non-symplectic. However, if we allow c in (5.1) to be a real matrix of order $2n$, we get a far-reaching generalization: (5.1) is symplectic iff

$$c = \frac{1}{2}(I_{2n} + J_{2n}B), \quad B^T = B, \quad c^T J + Jc = J. \tag{5.3}$$

The simplest and important cases are

$$\begin{aligned}
&C: \quad c = \frac{1}{2}I_{2n}, \qquad \hat{z} = z + sJH_z\left(\frac{\hat{z}+z}{2}\right),\\
&P: \quad c = \begin{pmatrix} I & 0 \\ 0 & 0 \end{pmatrix}, \quad \begin{aligned} \hat{p} &= p - sH_q(\hat{p}, q),\\ \hat{q} &= q + sH_p(\hat{p}, q), \end{aligned}\\
&Q: \quad c = \begin{pmatrix} 0 & 0 \\ 0 & I \end{pmatrix}, \quad \begin{aligned} \hat{p} &= p - sH_q(p, \hat{q}),\\ \hat{q} &= q + sH_p(p, \hat{q}), \end{aligned}
\end{aligned} \tag{5.4}$$

For $H(p,q) = \phi(p) + \psi(q)$, the above schemes P and Q reduce to explicit schemes (4.11) and (4.10).

Note that we may split the Hamiltonian vector field as

$$J\nabla H = cJ\nabla H + (I-c)J\nabla H \tag{5.5}$$

and it is more convenient to implement (5.1) as the composite

$$\begin{aligned} &z \to \bar{z} = z + scJ\nabla H(\bar{z}) = I^s_{cJ\nabla H}(z), \\ &\bar{z} \to \hat{z} = \bar{z} + s(I-c)J\nabla H(\bar{z}) = E^s_{(I-c)J\nabla H}(\bar{z}), \\ &g^s_{H,c} = E^s_{(I-c)J\nabla H} \circ I^s_{cJ\nabla H}. \end{aligned} \tag{5.6}$$

The first step is the implicit Euler method for the vector field $cJ\nabla H$, the second step is the explicit Euler method for the vector field $(I-c)J\nabla H$, both maps are not symplectic, but their composite is symplectic.

Scheme (5.1) is revertible of order 2 for $c = \frac{1}{2}I$, this is (5.2). (5.1) is of order 1 for $c \neq \frac{1}{2}I$, from §2 it follows that $\check{g}^s_{H,c} = g^s_{H,I-c}$. Then the composites

$$g^{s/2}_{H,c} \circ g^{s/2}_{H,I-c} \quad \text{and} \quad g^{s/2}_{H,I-c} \circ g^{s/2}_{H,c}$$

are symplectic, revertible and of order 2 for step size s.

So we get a great variety of simple symplectic schemes of order 1 or 2, classified according to type matrices $B \in sm(2n) :=$ space of symmetric matrices of order $2n$, which is a linear space of dimension $2n^2+n$.

The conservation properties of the above symplectic schemes are well-understood [4, 15, 18, 43]. In fact we have proved:

Let $\phi_A(z) = \frac{1}{2}z^TAz, A = A^T$ be a quadratic form and invariant under the system with Hamiltonian H. Then ϕ_A is invariant under symplectic difference scheme $g^s_{H,c}$ iff A commutes symplectically with the type matrix B in (5.3), i.e., $AJB = BJA$.

We list some of the most important weight matrices c, the type matrices B, together with the corresponding form of symmetric matrices A of the conserved quadratic invariants ϕ_A :

$$c = I - c = \frac{1}{2}I, \quad B = 0, \quad A \text{ arbitrary}.$$

$$c = \begin{pmatrix} I_n & 0 \\ 0 & 0 \end{pmatrix}, \quad B = \begin{pmatrix} 0 & -I_n \\ -I_n & 0 \end{pmatrix}, \qquad c = \begin{pmatrix} 0 & 0 \\ 0 & I_n \end{pmatrix}, \quad B = \begin{pmatrix} 0 & I_n \\ I_n & 0 \end{pmatrix},$$

$$A = \begin{pmatrix} 0 & b \\ b^T & 0 \end{pmatrix}, \quad b \text{ arbitrary angular momentum type}.$$

$$c=\frac{1}{2}\begin{pmatrix} I_n & \pm I_n \\ \mp I_n & I_n \end{pmatrix}, \quad B=\mp I_{2n}, \quad A=\begin{pmatrix} a & b \\ -b & a \end{pmatrix}, \quad \begin{matrix} a^T=a, b^T=-b; \\ \text{Hermitian type.} \end{matrix}$$

$$c=\frac{1}{2}\begin{pmatrix} I & \pm I \\ \pm I & I \end{pmatrix}, \quad B=\pm\begin{pmatrix} I_n & 0 \\ 0 & -I_n \end{pmatrix}, \quad A=\begin{pmatrix} a & b \\ -b & -a \end{pmatrix}, \quad \begin{matrix} a^T=a, \\ b^T=-b. \end{matrix}$$

Numerical evidences and theoretical analysis so far accumulated show unanimously and indisputably with the qualitative and the quantitative superiority of performance of symplectic methods over conventional non-symplectic ones for Hamiltonian systems. The contrasts are overwhelming and striking especially in crucial aspects of global behavior, orbital stability and long-time tracking capabilities. For the first time one now knows *consciously* a *proper* or descent way to compute *Newton's equations* of *motion*, i.e., first put them in Hamiltonian form, then compute by symplectic algorithms [4-5, 10, 15].

§6 Generating Function Methods

The construction of unconditional symplectic algorithms in §5 presents a special case of a general methodology based on generating functions explained in this section.

A matrix α of order $4n$ is called a *Darboux matrix* if

$$\alpha^T J_{4n}\alpha=\tilde{J}_{4n}, \quad J_{4n}=\begin{pmatrix} 0 & -I_{2n} \\ I_{2n} & 0 \end{pmatrix}, \quad \tilde{J}_{4n}=\begin{pmatrix} J_{2n} & 0 \\ 0 & -J_{2n} \end{pmatrix},$$

$$\alpha=\begin{pmatrix} a & b \\ c & d \end{pmatrix}, \quad \alpha^{-1}=\begin{pmatrix} a_1 & b_1 \\ c_1 & d_1 \end{pmatrix}.$$

Every Darboux matrix induces a (linear) *fractional transform* between symplectic and symmetric matrices

$$\begin{aligned} &\sigma_\alpha : Sp(2n)\to sm(2n), \\ &\qquad \sigma_\alpha(S)=(aS+b)(cS+d)^{-1}=A, \quad \text{for} \quad |cS+d|\neq 0 \end{aligned}$$

with the inverse transform $\sigma_\alpha^{-1}=\sigma_{\alpha^{-1}}$

$$\begin{aligned} &\sigma_\alpha^{-1}: sm(2n)\to Sp(2n), \\ &\qquad \sigma_\alpha^{-1}(A)=(a_1A+b_1)(c_1A+d_1)^{-1}=S, \quad \text{for } |c_1A+d_1|\neq 0, \end{aligned}$$

where $Sp(2n)=\left\{S\in GL(2n,\mathbf{R})|S^TJ_{2n}S=J_{2n}\right\}$ is the group of symplectic matrices.

The above machinery can be extended to generally non-linear operators on $\mathbf{R}^{2n}$. Denote $\mathbf{SpD}_{2n}$ the totality of symplectic operators, and *symm*(2n) the totality of symmetric operators (not necessary one-one). Every $f\in symm(2n)$ corresponds, at least locally, a real function ϕ (unique up to a constant) such that f is the gradient of $\phi: f(w) = \nabla\phi(w)$, where $\nabla\phi(w) = (\phi_{w_1}(w), \cdots, \phi_{w_{2n}}(w))^T = \phi_w(w)$. Then we have

$$\begin{aligned}&\sigma_\alpha : \mathbf{SpD}_{2n} \to symm(2n),\\&\sigma_\alpha(g) = (a \circ g + b) \circ (c \circ g + d)^{-1} = \nabla\phi, \quad \text{for} \quad |cg_z + d| \neq 0,\end{aligned}$$

or alternatively

$$ag(z) + bz = (\nabla\phi)(cg(z) + dz),$$

where ϕ is called the *generating function* of Darboux type α for the symplectic operator g. Then

$$\begin{aligned}&\sigma_\alpha^{-1}: \quad symm(2n) \to \mathbf{SpD}_{2n},\\&\sigma_\alpha^{-1}(\nabla\phi) = (a_1 \circ \nabla\phi + b_1) \circ (c_1 \circ \nabla\phi + d_1)^{-1} = g, \quad \text{for } |c_1\phi_{ww} + d_1| \neq 0\end{aligned} \tag{6.1}$$

or alternatively

$$a_1\nabla\phi(w) + b_1(w) = g\left(c_1\nabla\phi(w) + d_1 w\right), \tag{6.2}$$

where g is called the symplectic operator of Darboux type α for the generating function ϕ.

For the study of symplectic difference scheme we may narrow down the class of Darboux matrices to the subclass of *normal Darboux matrices*, i.e., those satisfying $a + b = 0, c + d = I_{2n}$. The normal Darboux matrices α can be characterized as

$$\alpha = \begin{pmatrix} a & b \\ c & d \end{pmatrix} = \begin{pmatrix} J & -J \\ c & I - c \end{pmatrix}, \quad c = \frac{1}{2}(I + JB), \quad B^T = B, \tag{6.3}$$

$$\alpha^{-1} = \begin{pmatrix} a_1 & b_1 \\ c_1 & d_1 \end{pmatrix} = \begin{pmatrix} (c - I)J & I \\ cJ & I \end{pmatrix}. \tag{6.4}$$

The fractional transform induced by a normal Darboux matrix establishes a 1-1 correspondence between *symplectic operators near identity* and *symmetric operators near nullity*. Then the determinantal conditions could be taken for granted. Those B's listed in section 5 correspond to the most important normal Darboux matrices.

For every Hamiltonian H with its phase flow e_H^t and for every normal Darboux matrix α, we get the *generating function* $\phi(w,t) = \phi_H^t(w) = \phi_{H,\alpha}^t(w)$ of *normal Darboux type* α for the *phase flow* of H by

$$\nabla\phi_{H,\alpha}^t = \left(Je_H^t - J\right) \circ \left(ce_H^t + I - c\right)^{-1}, \quad \text{for small} \quad |t|. \tag{6.5}$$

$\phi^t_{H,\alpha}$ satisfies the *Hamilton-Jacobi* equation

$$\frac{\partial}{\partial t}\phi(w,t) = -H\left(w + a_1\nabla\phi(w,t)\right) = -H\left(w + c_1\nabla\phi(w,t)\right) \tag{6.6}$$

and can be expressed by Taylor series in t

$$\phi(w,t) = \sum_{k=1}^{\infty}\phi^{(k)}(w)t^k, \quad |t| \text{ small}. \tag{6.7}$$

The coefficients can be determined recursively

$$\phi^{(1)}(w) = -\,H(w),$$

and for $k \geqslant 0,\ a_1 = (c-I)J:$

$$\phi^{(k+1)}(w) = \frac{-1}{k+1}\sum_{m=1}^{k}\frac{1}{m!}\sum_{\substack{j_1+\cdots+j_m=k\\ j_l\geqslant 1}} D^m H(w)\times$$
$$\times\left(a_1\nabla\phi^{(j_1)}(w),\cdots,a_1\nabla\phi^{(j_m)}(w)\right), \tag{6.8}$$

where we use the notation of the m-linear form

$$D^m H(w)\left(a_1\nabla\phi^{(j_1)}(w),\cdots,a_1\nabla\phi^{(j_m)}(w)\right)$$
$$:= \sum_{i_1,\cdots,i_m=1}^{2n} H_{z_{i_1}\cdots z_{i_m}}(w)\left(a_1\nabla\phi^{(j_1)}(w)\right)_{i_1}\cdots\left(a_1\nabla\phi^{(j_m)}(w)\right)_{i_m}.$$

Let ψ^s be a truncation of $\phi^s_{H,\alpha}$ up to a certain power s^m, say. Using inverse transform σ_α^{-1} we get the symplectic operator

$$g^s = \sigma_\alpha^{-1}\left(\nabla\psi^s\right), \quad |s| \text{ small}, \tag{6.9}$$

which depends on s, H, α (or equivalently B) and the mode of truncation. It is a symplectic approximation to the phase flow e^s_H and can serve as the transition operator of a symplectic difference scheme (for the Hamiltonian system (4.2))

$$z \to \hat{z} = g^s z: \quad \hat{z} = z - J\nabla\psi^s(c\hat{z} + (I-c)z), \quad c = \frac{1}{2}(I + JB). \tag{6.10}$$

Thus, using the machinery of phase flow generating functions we have constructed, for every H and every normal Darboux matrix, a hierarchy of symplectic schemes by truncation. The simple symplectic schemes (5.1, 5.3) correspond to the lowest truncation. The conservation properties of all these higher order schemes are the same as stated in section 5.

§7 Hamiltonian Algorithms for Hamiltonian Systems with a Perturbation Parameter

The machinery above can also be applied to construct symplectic algorithms for perturbed Hamiltonian systems defined by the perturbed Hamiltonian

$$H(z;\epsilon)=\sum_{k=0}^{\infty}\epsilon^k H_k(z)=H_0(z)+\sum_{k=1}^{\infty}\epsilon^k H_k(z), \tag{7.1}$$

where ϵ is the small perturbation parameter. $H_0(z)$ is usually an integrable Hamiltonian. The corresponding perturbed Hamiltonian system is

$$\frac{dz}{dt}=JH_z(z,\epsilon), \quad z\in\mathbf{R}^{2n}. \tag{7.2}$$

Its phase flow, denoted by $e_\epsilon^t=e_{H,\epsilon}^t$, depends on the parameter ϵ. The Hamilton-Jacobi equation and generating function are also parameterized by ϵ. That means, the parameterized generating function $\phi_\epsilon^t(w)=\phi^t(w,\epsilon)=\phi(w,t,\epsilon)$ satisfies the parameterized Hamilton-Jacobi equation

$$\frac{\partial}{\partial t}\phi(w,t,\epsilon)=-H\left(w+a_1\nabla\phi(w,t,\epsilon),\epsilon\right). \tag{7.3}$$

$\phi(w,t,\epsilon)$ can be expanded as a power series in ϵ in stead of t

$$\phi(w,t;\epsilon)=\sum_{k=0}^{\infty}\epsilon^k\phi^{(k)}(w,t). \tag{7.4}$$

The coefficients $\phi^{(k)}(w,t)$ satisfy the following equations:

$$\phi_t^{(0)}(w,t)=-\,H_0\left(w+a_1\nabla\phi^{(0)}(w,t)\right), \tag{7.5}$$

$$k\geqslant 1:\phi_t^{(k)}(w,t)=-\,H_k\left(w^*\right)-\sum_{i=1}^{k}\sum_{m=1}^{i}\frac{1}{m!}\sum_{\substack{i_1+\cdots+i_m=i\\ i_j\geqslant 1}}D^mH_{k-i}\left(w^*\right)$$
$$\times\left(a_1\nabla\phi^{(i_1)}(w,t),\cdots,a_1\nabla\phi^{(i_m)}(w,t)\right) \tag{7.6}$$

with the initial values $\phi^{(i)}(w,0)=0$, where $w^*=w+a_1\nabla\phi^{(0)}(w,t)$.

(7.5) is just the Hamilton-Jacobi equation for the unperturbed Hamiltonian system with Hamiltonian $H_0(z)$. The right hand side of (7.6) can be written as

$$-DH_0\left(w^*\right)\cdot a_1\nabla\phi^{(k)}(w,t)+R^{(k)}(w,t),$$

where the remainder $R^{(k)}(w,t)$ depends only on $H_i\left(w^*\right), i=0,1,\cdots,k$ and $\phi^{(i)}(w,t)$, $i=0,1,\cdots,k-1$. Once $\phi^{(0)}(w,t),\cdots,\phi^{(k-1)}(w,t)$ are known, $R^{(k)}(w,t)$ is also known. Therefore, if $\phi^{(0)}(w,t)$ can be solved from (7.5), then for $k\geqslant 1$,

$$\phi_t^{(k)}(w,t)=-DH_0\left(w^*\right)\cdot a_1\nabla\phi^{(k)}(w,t)+R^{(k)}(w,t)$$

are the linear partial differential equations for $\phi^{(k)}(w,t)$. They have the same coefficients, only those for $R^{(k)}(w,t)$ are different.

In some cases, we can solve (7.5) easily, refer to [38]. In general, it is difficult to solve (7.5) by analytical method. Nevertheless we can always give an approximative solution, for example, using the methods discussed above.

Let now $\psi_\epsilon^s(w) = \psi(w,s,\epsilon)$ be a truncation of $\phi(w,s,\epsilon)$ up to a certain power ϵ^m. Using the inverse transform σ_α^{-1} we get the symplectic operator

$$g_\epsilon^s = \sigma_\alpha^{-1}\left(\nabla\psi_\epsilon^s\right), \quad |s| \text{ small.} \tag{7.7}$$

It is a symplectic approximation to the phase flow e_ϵ^s of order m in ϵ and can serve as the transition operator of a symplectic difference scheme for the perturbed Hamiltonian system (7.2)

$$z \to \hat{z} = g_\epsilon^s z: \quad \hat{z} = z - J\nabla\psi(c\hat{z} + (I-c)z, s, \epsilon), \quad c = \frac{1}{2}(I + JB). \tag{7.8}$$

For general perturbed Hamiltonians $H(z,\epsilon)$, these schemes are only consistent in the time stepsize s. But for the perturbed Hamiltonians with the form

$$H(z,\epsilon) = H_0(z) + \epsilon H_1(z), \tag{7.9}$$

the order of the schemes (7.8) in the time stepsize s is the same as in ϵ. More precisely, for the m-th order scheme g_ϵ^s for the perturbed Hamiltonian (7.9),

$$g_\epsilon^s = e_\epsilon^s + O\left((s\epsilon)^{m+1}\right). \tag{7.10}$$

Therefore, for small ϵ the time stepsize s can be taken quite large. For more details, refer to [38].

§8 Contact Algorithms for Contact Systems

Contact systems occur only on space $\mathbf{R}^{2n+1}$ of odd dimensions. We use 3-symbol notation to denote the coordinates and vectors on $\mathbf{R}^{2n+1}$

$$\begin{pmatrix} x \\ y \\ z \end{pmatrix}, \quad x = \begin{pmatrix} x_1 \\ \vdots \\ x_n \end{pmatrix}, \quad y = \begin{pmatrix} y_1 \\ \vdots \\ y_n \end{pmatrix}, \quad z = (z),$$

$$\begin{pmatrix} a(x,y,z) \\ b(x,y,z) \\ c(x,y,z) \end{pmatrix}, \quad a = \begin{pmatrix} a_1 \\ \vdots \\ a_n \end{pmatrix}, \quad b = \begin{pmatrix} b_1 \\ \vdots \\ b_n \end{pmatrix}, \quad c = (c).$$

A *contact* system is defined by a function $K(x, y, z)$, called contact Hamiltonian, on $\mathbf{R}^{2n+1}$ as follows

$$\begin{aligned} \frac{dx}{dt} &= -K_y + K_z x = a, \\ \frac{dy}{dt} &= K_x = b, \\ \frac{dz}{dt} &= K_e = c, \quad K_e(x, y, z) := K(x, y, z) - \langle x, K_x(x, y, z)\rangle . \end{aligned} \tag{8.1}$$

The right hand side is the general form of a *contact* vector field.

Intrinsic to all contact systems is the *contact* structure on $\mathbf{R}^{2n+1}$ defined by the differential 1-form

$$\alpha = \sum_{i=1}^{n} x_i dy_i + dz = \begin{pmatrix} 0 & x^T & 1 \end{pmatrix} \begin{pmatrix} dx \\ dy \\ dz \end{pmatrix} \tag{8.2}$$

up to an everywhere non-vanishing multiplier function. A *contact* transformation f is a diffeomorphism on $\mathbf{R}^{2n+1}$

$$f: \begin{pmatrix} x \\ y \\ z \end{pmatrix} \to \begin{pmatrix} \hat{x}(x, y, z) \\ \hat{y}(x, y, z) \\ \hat{z}(x, y, z) \end{pmatrix}$$

preserving the contact structure, that means

$$\sum_{i=1}^{n} \hat{x}_i d\hat{y}_i + d\hat{z} = \mu_f \left(\sum_{i=1}^{n} x_i dy_i + dz \right) \tag{8.3}$$

for some function $\mu_f \neq 0$ everywhere on $\mathbf{R}^{2n+1}$, called the multiplier of f. The explicit expression of (8.3) is

$$\begin{pmatrix} 0 & \hat{x}^T & 1 \end{pmatrix} \begin{pmatrix} \hat{x}_x & \hat{x}_y & \hat{x}_z \\ \hat{y}_x & \hat{y}_y & \hat{y}_z \\ \hat{z}_x & \hat{z}_y & \hat{z}_z \end{pmatrix} = \mu_f \begin{pmatrix} 0 & x^T & 1 \end{pmatrix} .$$

All contact vector fields on $\mathbf{R}^{2n+1}$ form a Lie algebra $\mathbf{L} = \mathbf{KV}_{2n+1}$ under Lie bracket. The associated Lie group $\mathbf{G} = \mathbf{KD}_{2n+1}$ consists of all near-identity contact transformations on $\mathbf{R}^{2n+1}$. For $f \in \mathbf{G}$, $\mu_f > 0$ everywhere. The phase flows e_K^t of contact systems defined by contact Hamiltonians K on $\mathbf{R}^{2n+1}$ are contact transformations. So the proper algorithms for contact systems should be *contact* too. For elements of contact geometry, see [1].

All the existing numerical methods including the symplectic ones are generically non-contact even for linear contact systems. A linear contact system is necessarily of the form,

$$\begin{aligned}\frac{dx}{dt} &= \left(\lambda I - L^T\right) x, \quad L \in gl(n), \quad \lambda \in \mathbf{R},\\ \frac{dy}{dt} &= Ly,\\ \frac{dz}{dt} &= \lambda z,\end{aligned} \tag{8.4}$$

the 3 equations are independent. The matrix of a linear contact transformation is necessarily of the block-diagonal form

$$\begin{pmatrix} \mu M^{-T} & 0 & 0 \\ 0 & M & 0 \\ 0 & 0 & \mu \end{pmatrix}, \quad M \in \mathbf{GL}(n), \quad \mu > 0. \tag{8.5}$$

Using the trapezoidal method (identical with the centered Euler method in linear cases), one gets

$$\begin{aligned}\hat{x} &= \left(\left(1 - \frac{s\lambda}{2}\right) I + \frac{s}{2} L^T\right)^{-1} \left(\left(1 + \frac{s\lambda}{2}\right) I - \frac{s}{2} L^T\right) x =: Nx,\\ \hat{y} &= \left(I - \frac{s}{2} L\right)^{-1} \left(I + \frac{s}{2} L\right) =: My,\\ \hat{z} &= \left(1 - \frac{s\lambda}{2}\right)^{-1} \left(1 + \frac{s\lambda}{2}\right) z =: \mu z.\end{aligned}$$

The map is contact $\Longleftrightarrow N = \mu M^{-T}, \forall s \Longleftrightarrow \lambda = 0$ or $L^2 = \lambda L$; but this is highly exceptional.

The construction of explicit Euler-type structure preserving algorithms in §3 is applicable to contact systems when the vector field is contact-separable, i.e., decomposable into

$$a = \sum_{i=1}^{n} a^i, \quad a^i \in \mathbf{KL}_{2n+1}, \quad a_*^i a^i = 0, \quad i = 1, \cdots, n. \tag{8.6}$$

Our main concern will be the construction of unconditional contact algorithms for general contact systems. The method is based on the well-known correspondence between contact geometry on $\mathbf{R}^{2n+1}$ and homogeneous (or conic) symplectic geometry on $\mathbf{R}^{2n+2}$.

We use 4-symbol notation for the coordinates on $\mathbf{R}^{2n+2}$

$$\begin{pmatrix} p_0 \\ p_1 \\ q_0 \\ q_1 \end{pmatrix} \in \mathbf{R}^{2n+2}, \quad p_0 = (p_0)\,, q_0 = (q_0)\,, \quad p_1 = \begin{pmatrix} p_{11} \\ \vdots \\ p_{1n} \end{pmatrix}, \quad q_1 = \begin{pmatrix} q_{11} \\ \vdots \\ q_{1n} \end{pmatrix}.$$

Consider *homogeneous* near-identity transformation $g : (p_0, p_1, q_0, q_1) \to (\hat{p}_0, \hat{p}_1, \hat{q}_0, \hat{q}_1)$ and *homogeneous* function $H(p_0, p_1, q_0, q_1)$ on $\mathbf{R}^{2n+2}$, they are defined for $p_0 \neq 0$ and satisfy the conditions

$$\begin{aligned} \forall \lambda \neq 0: \qquad & \hat{p}_i(\lambda p_0, \lambda p_1, q_0, q_1) = \lambda \hat{p}_i(p_0, p_1, q_0, q_1)\,, \\ & \hat{q}_i(\lambda p_0, \lambda p_1, q_0, q_1) = \hat{q}_i(p_0, p_1, q_0, q_1)\,, \quad i = 1, 2, \\ & H(\lambda p_0, \lambda p_1, q_0, q_1) = \lambda H(p_0, p_1, q_0, q_1)\,. \end{aligned}$$

They depend essentially only on $2n+1$ variables

$$\begin{aligned} p_0 \neq 0: \qquad & \hat{p}_i(p_0, p_1, q_0, q_1) = p_0 \hat{p}_i\left(1, \frac{p_1}{p_0}, q_0, q_1\right), \\ & \hat{q}_i(p_0, p_1, q_0, q_1) = \hat{q}_i\left(1, \frac{p_1}{p_0}, q_0, q_1\right), \\ & H(p_0, p_1, q_0, q_1) = p_0 H\left(1, \frac{p_1}{p_0}, q_0, q_1\right). \end{aligned}$$

The phase flow e_H^t for Hamiltonian system defined by homogeneous function H on $\mathbf{R}^{2n+2}$ is homogeneous and symplectic.

Near-identity homogeneous symplectic transformation g and homogeneous function H on $\mathbf{R}^{2n+2}$ can be put in correspondence with near-identity contact transformation f and function K on $\mathbf{R}^{2n+1}$ as follows:

$$\begin{pmatrix} x \\ y \\ z \end{pmatrix} \to \begin{pmatrix} p_0 \\ p_1 \\ q_0 \\ q_1 \end{pmatrix} = \begin{pmatrix} p_0 \\ p_0 x \\ z \\ y \end{pmatrix} \xrightarrow{g} \begin{pmatrix} \hat{p}_0 \\ \hat{p}_1 \\ \hat{q}_0 \\ \hat{q}_1 \end{pmatrix} = \begin{pmatrix} \hat{p}_0 \\ \hat{p}_0 \hat{x} \\ \hat{z} \\ \hat{y} \end{pmatrix}, \tag{8.7}$$

where $p_0 > 0$ is an arbitrary chosen parameter. Then $\hat{p}_0 > 0$ and the map

$$\begin{pmatrix} x \\ y \\ z \end{pmatrix} \xrightarrow{f} \begin{pmatrix} \hat{x} \\ \hat{y} \\ \hat{z} \end{pmatrix} = \begin{pmatrix} \hat{p}_1 / \hat{p}_0 \\ \hat{q}_1 \\ \hat{q}_0 \end{pmatrix}, \quad \mu_f := \hat{p}_0 / p_0 > 0 \tag{8.8}$$

is a near-identity contact transformation independent of p_0 and μ_f is the multiplier of f.

$$
\begin{aligned}
&(x, y, z) \longrightarrow K(x, y, z) := H(1, x, z, y), \\
&H(p_0, p_1, q_0, q_1) = p_0 K\left(\frac{p_1}{p_0}, q_1, q_0\right). \\
&H_{p_0}(p_0, p_1, q_0, q_1) = K(x, y, z) - \langle x, K_x(x, y, z)\rangle = K_e(x, y, z) \\
&H_{p_1}(p_0, p_1, q_0, q_1) = K_x(x, y, z), \\
&H_{q_0}(p_0, p_1, q_0, q_1) = p_0 K_z(x, y, z), \\
&H_{q_1}(p_0, p_1, q_0, q_1) = p_0 K_y(x, y, z), \\
&x = p_1/p_0, \quad y = q_1, \quad z = q_0.
\end{aligned}
$$

The above process and its inverse are called *contactization* and *symplectization* respectively. Under this correspondence we have

1. $K(x, y, z) \longleftrightarrow H(p_0, p_1, q_0, q_1)$.
2. Contact system (8.1)$\longleftrightarrow$ homogeneous Hamiltonian system

$$
\begin{aligned}
\frac{dp_0}{dt} &= -H_{q_0}, \quad \frac{dq_0}{dt} = H_{p_0}, \\
\frac{dp_1}{dt} &= -H_{q_1}, \quad \frac{dq_1}{dt} = H_{p_1}.
\end{aligned} \tag{8.9}
$$

3. Contact phase flow e_K^t on $\mathbf{R}^{2n+1}$ $\longleftrightarrow$ homogeneous symplectic phase flow e_H^t on $\mathbf{R}^{2n+2}$

Now symplectic phase flow e_H^t can be approximated by various unconditional symplectic algorithms, e.g., $g_{H,c}^s$ generated by (5.4) and (6.10) on $\mathbf{R}^{2n+2}$. For weight matrices c on $\mathbf{R}^{2n+2}$ compatible with homogeneity conditions of functions and mappings, $g_{H,c}^s$ are homogeneous symplectic, the generating relations (5.4) and (6.10) are contactized to give the contactization $f_{K,c}^s$ on $\mathbf{R}^{2n+2}$ of $g_{H,c}^s$. They provide unconditional contact algorithms for contact systems. We give some basic contact algorithms in the following. For more details and the theory of contact generating functions see [12-13].

$\tilde{C}$. $c = \dfrac{1}{2} I_{2n+2}$, contact version of method C, two stage form (write $\bar{K}_x = K_x(\bar{x}, \bar{y}, \bar{z})$, etc.)

$$
\begin{aligned}
&\bar{x} = x + \frac{s}{2}\left(-\overline{K}_y + x\overline{K}_z\right), \quad \bar{y} = y + \frac{s}{2}\overline{K}_x, \quad \bar{z} = z + \frac{s}{2}\overline{K}_e, \\
&\hat{x} = \left(\bar{x} - \frac{s}{2}\overline{K}_y\right) / \left(1 - \frac{s}{2}\overline{K}_z\right), \quad \hat{y} = 2\bar{y} - y, \quad \hat{z} = 2\bar{x} - z.
\end{aligned} \tag{8.10}
$$

$\tilde{P} \cdot c = \begin{pmatrix} I_{n+1} & 0 \\ 0 & 0_{n+1} \end{pmatrix}$, contact analog of symplectic method P :

$$
\begin{aligned}
&& \hat{x} =& x + s\left(-K_y(\hat{x}, y, z) + xK_z(\hat{x}, y, z)\right), \\
&\text{1-stage form :} & \hat{y} =& y + sK_x(\hat{x}, y, z), \\
&& \hat{z} =& z + sK_e(\hat{x}, y, z). \\
&\text{2-stage form :} & \bar{x} =& x + s\left(-\overline{K}_y + x\overline{K}_z\right), \quad \bar{y} = y, \quad \bar{z} = z, \\
&& \hat{x} =& \bar{x}, \quad \hat{y} = \bar{y} + s\overline{K}_x, \quad \hat{z} = \bar{z} + s\overline{K}_e.
\end{aligned}
\tag{8.11}
$$

$\widetilde{Q} \cdot c = \begin{pmatrix} 0_{n+1} & 0 \\ 0 & I_{n+1} \end{pmatrix}$, contact analog of symplectic method Q :

$$
\begin{aligned}
&& \hat{x} =& x + s\left(-K_y(x, \hat{y}, \hat{z}) + \hat{x}K_z(x, \hat{y}, \hat{z})\right), \\
&\text{1-stage form:} & \hat{y} =& y + sK_x(x, \hat{y}, \hat{z}), \\
&& \hat{z} =& z + sK_e(x, \hat{y}, \hat{z}). \\
&\text{2-stage form:} & \bar{x} =& x, \quad \bar{y} = y + s\overline{K}_x, \quad z = z + s\overline{K}_e, \\
&& \hat{x} =& \bar{x} + s\left(-\overline{K}_y + \hat{x}\overline{K}_z\right), \quad \hat{y} = \bar{y}, \quad \hat{z} = \bar{z}.
\end{aligned}
\tag{8.12}
$$

One might suggest, by analogy with (5.4), for example, the following scheme for (8.1):

$$
\hat{x} = x + sa(\hat{x}, y, z), \quad \hat{y} = y + sb(\hat{x}, y, z), \quad \hat{z} = z + sc(\hat{x}, y, z).
$$

It differs from (8.11) only in one term for $\hat{x}$, i.e., $\hat{x}K(\hat{x}, y, z)$ instead of $xK(\hat{x}, y, z)$. This minute but delicate difference makes (8.11) contact and other non-contact! For further developments of contact algorithms, see [10, 12-13]. For numerical comparisons of 3D contact vs non-contact computations, see [10].

§9 Volume-preserving Algorithms for Source-free Systems

The source-free systems on $\mathbf{R}^m$ are governed by the m-D volume structure, the underlying Lie algebra $\mathbf{L} = \mathbf{SV}_m$ consists of all *source-free* vector fields a, $\operatorname{div} a = \operatorname{tr} a_* = 0$, the underlying Lie group $\mathbf{G} = \mathbf{SD}_m$ consists of all near-identity *volume-preserving* transformations g, $\det g_* = 1$.

The phase flow e_a^t of source-free field a is always volume preserving. Hence the proper algorithms for source-free systems should be also *volume-preserving*, since otherwise the dynamics would be polluted by artificial sources and sinks.

For $m = 2$, source-free fields = Hamiltonian fields, area-preserving maps = symplectic maps; so the problem for area-preserving algorithms has been solved in principle.

For $m \geqslant 3$, the problem is new, since all the conventional methods plus even the symplectic methods are generically not volume-preserving, even for linear source-free systems. As an illustration, solve on $\mathbf{R}^3$

$$\frac{dx}{dt} = Ax, \quad \operatorname{tr} A = 0$$

by the method C, we get

$$x \to \hat{x} = G^s x, \quad G^s = \left(I - \frac{s}{2}A\right)^{-1}\left(I + \frac{s}{2}A\right).$$

$\det G^s = 1 \Longleftrightarrow \det A = 0$, which is exceptional.

The construction of explicit Euler type structure-preserving algorithms by composition is applicable here when the vector field is source-free separable, i.e., decomposable as

$$a = \sum_{i=1}^{n} a^i, \quad \operatorname{div} a^i = 0, \quad a^i_* a^i = 0, \quad i = 1, \cdots, n. \tag{9.1}$$

A special case is

$$a = (a_1, \cdots, a_m), \quad \frac{\partial a_k}{\partial x_k} = 0, \quad a^k = (0, \cdots, 0, a_k, 0, \cdots, 0), \quad k = 1, \cdots, m;$$
$$a = \sum_{k=1}^{m} a^k. \tag{9.2}$$

Example 1. Euler equation for free rigid body and Jacobian elliptic functions of modulus k,

$$\frac{dx_1}{dt} = c_1 x_2 x_3 = a_1, \quad \frac{dx_2}{dt} = c_2 x_1 x_3 = a_2, \quad \frac{dx_3}{dt} = c_3 x_1 x_2 = a_3, \qquad \begin{pmatrix} a_1 \\ a_2 \\ a_3 \end{pmatrix} = \begin{pmatrix} c_1 x_2 x_3 \\ 0 \\ 0 \end{pmatrix} + \begin{pmatrix} 0 \\ c_2 x_1 x_3 \\ 0 \end{pmatrix} + \begin{pmatrix} 0 \\ 0 \\ c_3 x_1 x_2 \end{pmatrix}$$

where (x_1, x_2, x_3) are angular momenta along principle axes of inertia with diagonal elements $I_1, I_2, I_3, c_1 = (I_2 - I_3)/I_2 I_3$, cyclic. Take $c_1 = 1, c_2 = -1, c_3 = -k^2$, it gives elliptic functions for $x_1(0) = 0, x_2(0) = x_3(0) = 1, x_1(t) = \operatorname{sn}(t,k), x_2(t) = \operatorname{cn}(t,k), x_3(t) = \operatorname{dn}(t,k)$.

Example 2. ABC flows [1]

$$\frac{dx_1}{dt} = A\sin x_3 + C\cos x_2 = a_1,$$
$$\frac{dx_2}{dt} = B\sin x_1 + A\cos x_3 = a_2,$$

$$
\begin{aligned}
\frac{dx_3}{dt} =& C\sin x_2 + B\cos x_1 = a_3,\\
\begin{pmatrix} a_1\\ a_2\\ a_3 \end{pmatrix} =& \begin{pmatrix} A\sin x_3 + C\cos x_2\\ 0\\ 0 \end{pmatrix} + \begin{pmatrix} 0\\ B\sin x_1 + A\cos x_3\\ 0 \end{pmatrix}\\
&+ \begin{pmatrix} 0\\ 0\\ C\sin x_2 + B\cos x_1 \end{pmatrix}\\
=& \begin{pmatrix} A\sin x_3\\ A\cos x_3\\ 0 \end{pmatrix} + \begin{pmatrix} 0\\ B\sin x_1\\ B\cos x_1 \end{pmatrix} + \begin{pmatrix} C\sin x_2\\ 0\\ C\cos x_2 \end{pmatrix}.
\end{aligned}
$$

All decompositions on the right satisfy (9.1), then (3.10) gives corresponding volume-preserving revertible algorithms.

In $\mathbf{R}^m$ all source-free fields have "vector potential" representations: under such representation, a field always decomposes into a sum of essentially 2-D source-free fields, for which essentially area-preserving algorithms can always be constructed, then one can apply the decomposition-composition method in §3 to obtain m-D volume-preserving algorithms [16, 39].

In $\mathbf{R}^2$, every source-free field (a_1, a_2) corresponds to a stream function or 2-D Hamiltonian ψ, unique up to a constant:

$$
a_1 = -\frac{\partial \psi}{\partial x_2}, \quad a_2 = \frac{\partial \psi}{\partial x_1}, \tag{9.3}
$$

then area-preserving algorithms = 2-D symplectic algorithms.

In $\mathbf{R}^3$, every source-free field (a_1, a_2, a_3) corresponds to a vector potential (b_1, b_2, b_3), unique up to a gradient:

$$
a = \mathrm{curl} b, \quad a_1 = \frac{\partial b_3}{\partial x_2} - \frac{\partial b_2}{\partial x_3}, \quad a_2 = \frac{\partial b_1}{\partial x_3} - \frac{\partial b_3}{\partial x_1}, \quad a_3 = \frac{\partial b_2}{\partial x_1} - \frac{\partial b_1}{\partial x_2}, \tag{9.4}
$$

then one gets source-free decomposition

$$
\begin{aligned}
a =& (a_1, a_2, a_3)\\
=& \left(0, \frac{\partial b_1}{\partial x_3}, -\frac{\partial b_1}{\partial x_2}\right) + \left(-\frac{\partial b_2}{\partial x_3}, 0, \frac{\partial b_2}{\partial x_1}\right) + \left(\frac{\partial b_3}{\partial x_2}, -\frac{\partial b_3}{\partial x_1}, 0\right)\\
=& a^1 + a^2 + a^3.
\end{aligned} \tag{9.5}
$$

Each field a^i is 2-D source-free and 0 in the 3rd dimension, apply area-preserving method P, say, with 3rd dimension fixed, then one get three volume-preserving maps

$$
P_1^s \approx e^s\left(a^1\right) = e^s\left(0, \frac{\partial b_1}{\partial x_3}, -\frac{\partial b_1}{\partial x_2}\right), \quad \text{ord } 1, \ \text{volume-preserving},
$$

$$
\begin{aligned}
P_2^s &\approx e^s\left(a^2\right)=e^s\left(-\frac{\partial b_2}{\partial x_3},0,\frac{\partial b_2}{\partial x_1}\right), \quad \text{ord 1, volume-preserving,}\\
P_3^s &\approx e^s\left(a^3\right)=e^s\left(\frac{\partial b_3}{\partial x_2},-\frac{\partial b_3}{\partial x_1},0\right), \quad \text{ord 1, volume-preserving,}\\
P_1^s &= E^s\left(0,\frac{\partial b_1}{\partial x_3},0\right)I^s\left(0,0,-\frac{\partial b_1}{\partial x_2}\right),\\
\check{P}_1^s &= E^s\left(0,0,-\frac{\partial b_1}{\partial x_2}\right)I^s\left(0,\frac{\partial b_1}{\partial x_3},0\right),\\
P_2^s &= E^s\left(0,0,\frac{\partial b_2}{\partial x_1}\right)I^s\left(-\frac{\partial b_2}{\partial x_3},0,0\right),\\
\check{P}_2^s &= E^s\left(-\frac{\partial b_2}{\partial x_3},0,0\right)I^s\left(0,0,\frac{\partial b_2}{\partial x_1}\right),\\
P_3^s &= E^s\left(\frac{\partial b_3}{\partial x_2},0,0\right)I^s\left(0,-\frac{\partial b_3}{\partial x_1},0\right),\\
\check{P}_3^s &= E^s\left(0,-\frac{\partial b_3}{\partial x_1},0\right)I^s\left(\frac{\partial b_3}{\partial x_2},0,0\right),
\end{aligned}
\tag{9.6}
$$

then one get volume-preserving algorithms for field a

$$
\begin{aligned}
f^s &= P_1^s P_2^s P_3^s \approx e^s(a), \quad \text{ord 1, non-revertible,}\\
g^s &= P_1^{s/2} P_2^{s/2} P_3^{s/2} \circ \check{P}_3^{s/2} \check{P}_2^{s/2} \check{P}_1^{s/2} \approx e^s(a), \ \text{ord 2, revertible,}\\
h^s &= g^{\alpha s} g^{\beta s} g^{\alpha s} \approx e^s(a), \quad \text{ord 4, revertible.}
\end{aligned}
\tag{9.7}
$$

The basic elements of all these composites consist only in 6 explicit Euler and 6 implicit Euler in one variable appeared in (9.6).

On $\mathbf{R}^m$, to every source-free field (a_i), there exists, as a generalization of cases $m=2,3$, a skew-symmetric tensor field of order 2, $b=(b_{ik})$, $b_{ik}=-b_{ki}$ so that

$$
a_i=\sum_{k=1}^{m}\frac{\partial b_{ik}}{\partial x_k}, \quad i=1,\cdots,m. \tag{9.8}
$$

By (9.8) we can decompose

$$
a=\sum_{i<k} a_{(ik)},\ a_{(ik)}=\left(0,\cdots,0,\frac{\partial b_{ik}}{\partial x_k},0,\cdots,0,-\frac{\partial b_{ik}}{\partial x_i},0,\cdots,0\right)^T, \quad i<k. \tag{9.9}
$$

Every vector field $a_{(ik)}$ is 2-D source-free on the i-k plane and 0 in other dimensions so we can approximate the phase flow of $a_{(ik)}$ by area-preserving method which implies m-D volume-preserving, similar to the cases discussed above for $m=3$. For more details see [16].

The tensor potential (b_{ik}) for a given (a_i) is by far not unique. For uniqueness one may impose normalizing conditions in many different ways. One way is to impose

$$
N_0: \quad b_{ik}=0, |i-k|\geqslant 2, \tag{9.10}
$$

(this condition is ineffective for $m=2$). The non-zero components are

$$b_{12}=-b_{21},\quad b_{23}=-b_{32},\quad \cdots \quad b_{m-1,m}=-b_{m,m-1}.$$

$$N_1: b_{12}|_{x_1=x_2=0}=0.$$

$$N_k, 2\leqslant k\leqslant m-1:\ b_{k,k+1}|_{x_k=0}=0,$$

(this condition is ineffective for $m=2$). Then all $b_{k,k+1}$ are uniquely determined by quadrature

$$b_{12}=\int_0^{x_2} a_1 dx_2-\int_0^{x_1} a_2\Big|_{x_2=0} dx_1,$$

$$b_{k,k+1}=-\int_0^{x_k} a_{k+1} dx_k,\quad 2\leqslant k\leqslant m-1; \tag{9.11}$$

$$a=\sum_{k=1}^{m-1} a^k,\quad a^k=\left(0,\cdots,0,\frac{\partial b_{k,k+1}}{\partial x_{k+1}},-\frac{\partial b_{k,k+1}}{\partial x_k},0,\cdots,0\right)^T. \tag{9.12}$$

When the vector field a is polynomial, then by (9.10) and (9.11) the tensor components b_{ik} are polynomials. Then the theorem on symplectic separability of all polynomial Hamiltonians (4.16) implies the source-free separability in the sense of (9.1) for all polynomial source-free systems. For numerical comparisons of 3D volume-preserving vs non-volume-preserving computations, see [10].

§10 Formal Dynamical Systems and Numerical Algorithms

In order to facilitate the study of the problem $f^s\approx e_a^s$, the apparatus of formal power series is useful. We first introduce some notation. For $a=(a_1,\cdots,a_m)^T:\mathbf{R}^m\to\mathbf{R}^m$,

$$a^*:=\sum a_i\frac{\partial}{\partial x_i}=\text{linear differential operator of first order associated to } a.$$

The association $a\to a^*$ is linear, a^* operates on scalar functions $\phi:\mathbf{R}^m\to\mathbf{R}$ and on vector functions $b:\mathbf{R}^m\to\mathbf{R}^m$ as

$$a^*\phi=\sum a_i\frac{\partial\phi}{\partial x_i}$$

$$a^*b=a^*(b_1,\cdots,b_m)^T=(a^*b_1,\cdots,a^*b_m)^T=b_*a,\quad a^*1_m=a.$$

Multiple applications of linear differential operators are naturally defined such as a^*b^*, $(a^*b^*)c^*, a^*(b^*c^*)$, etc. The operations are multilinear, associative but non-commutative; thus powers can be defined

$$a^{*k}=a^*a^*\cdots a^*(k\text{ times }),\quad a^k:=a^{*k}1_m$$

The identity operator I operates on scalar and vector functions ϕ and b as $I\phi=\phi, Ib=b$. Lie bracket of vector functions a, b on $\mathbf{R}^m$ can be written as

$$[a,b]:=a_*b-b_*a=b^*a-a^*b=(b^*a^*-a^*b^*)\,1_m.$$

Among the formal power series $\sum\limits_0^\infty s^k a_k$, $a_k:\mathbf{R}^m\rightarrow\mathbf{R}^m$, we pick out two special classes. The first class consists of those with $a_0=0$, called near-0 formal vector fields; the second class consists of those with $a_0=1_m$, called near-1 formal maps (diffeomorphisms).

All near-0 formal vector fields $a^s=\sum\limits_1^\infty s^k a_k$ form a (∞-dim.) real Lie algebra $\mathbf{FV}_m$ under the Lie bracket

$$[a^s,b^s]=\left[\sum_1^\infty s^k a_k,\sum_1^\infty s^k b_k\right]:=\sum_{k=2}^\infty s^k\sum_{i+j=k}[a_i,b_j].$$

The associated near-0 formal differential operators and their products are

$$(a^s)_*:=\left(\sum_1^\infty s^k a_k\right)_*:=\sum_1^\infty s^k a_{k*}$$

$$a^{s*}=\left(\sum_1^\infty s^k a_k\right)^*:=\sum_1^\infty s^k a_k^*,$$

$$a^{s*}b^{s*}:=\sum_2^\infty s^k\sum_{i+j=k}a_i^*b_j^*,\quad (a^{s*})^2:=a^{s*}a^{s*},\ \text{etc.}$$

For any vector function $a=(a_1,\cdots,a_m)^T:\mathbf{R}^m\rightarrow\mathbf{R}^m$ and any near-1 formal map $g^s=1_m+\sum s^k g_k$ we define the composition

$$(a\circ g^s)(x)=a\left(g^s(x)\right)=a(x)+\sum_1^\infty s^k(a\circ g)_k(x),$$

$$(a\circ g)_k=\sum_{n=1}^k\sum_{j_1+\cdots+j_n=k}\frac{1}{n!}\left(D^n a\right)\left(g_{j_1},\cdots,g_{j_n}\right),$$

where

$$D^n a=\left(D^n a_1,\cdots,\ D^n a_m\right)^T,$$

$$D^n a_i\left(v_1,\cdots,v_n\right)=\sum_{j_1,\cdots,j_n=1}^m\frac{\partial^n a_i}{\partial x_{j_1}\cdots\partial x_{j_n}}v_{1j_1}\cdots v_{nj_n}$$

is the usual n-th differential multi-linear form for n tangent vectors $v_j=\left(v_{j1},\cdots,v_{jm}\right)^T$, $j=1,\cdots,n$ at point $x\in\mathbf{R}^m$ which is invariant under permutation of vectors. Using the identities

$$
\begin{aligned}
&(D^1a)(b) = b^*a,\\
&(D^2a)(b,c) = \left(c^*b^* - (c^*b)^*\right)a,\\
&(D^3a)(b,b,b) = \left(b^{*3} + 2b^{3*} - 3b^*b^{2*}\right)a,
\end{aligned}
$$

we get in particular

$$
\begin{aligned}
&(a\circ g)_1 = g_1^*a,\\
&(a\circ g)_2 = g_2^*a + \frac{1}{2}\left(g_1^{*2} - g_1^{2*}\right)a,\\
&(a\circ g)_3 = g_3^*a + \left((g_2^*g_1^* - (g_2^*g_1)^*\right)a + \frac{1}{3!}(g_1^{*3} + 2g_1^{3*} - 3g_1^*g_1^{2*})a.
\end{aligned}
$$

For any two near-1 formal maps $f^s = 1_m + \sum s^k a_k, g^s = 1_m + \sum s^k g_k$, the composition $f^s\circ g^s$ is defined in a term by term way:

$$
\begin{aligned}
(f^s\circ g^s)(x) = f^s\left(g^s(x)\right) &= 1_m\left(g^s(x)\right) + \sum_{k=1}^{\infty} s^k f_k\left(g^s(x)\right)\\
&=: 1_m(x) + \sum_{k=1}^{\infty} s^k (f\circ g)_k(x),
\end{aligned}
$$

$$
\begin{aligned}
&(f\circ g)_1 = f_1 + g_1,\\
&(f\circ g)_k = f_k + g_k + \delta\left(f_1,\cdots,f_{k-1};g_1,\cdots,g_{k-1}\right),\quad k\geqslant 2,\\
&\delta\left(f_1,\cdots,f_{k-1};g_1,\cdots,g_{k-1}\right) = \sum_{n=1}^{k-1}\sum_{j=1}^{n}\sum_{i_1+\cdots+i_j=n}\frac{1}{j!}\left(D^j f_{k-n}\right)\left(g_{i_1},\cdots,g_{i_j}\right),
\end{aligned}
$$

we get in particular,

$$
\begin{aligned}
(f\circ g)_2 =& f_2 + g_2 + g_1^*f_1,\\
(f\circ g)_3 =& f_3 + g_3 + g_1^*f_2 + g_2^*f_1 + \frac{1}{2}\left(g_1^{*2} - g_1^{2*}\right)f_1,\\
(f\circ g)_4 =& f_4 + g_4 + g_1^*f_3 + g_2^*f_2 + g_3^*f_1 + \frac{1}{2}\left(g_1^{*2} - g_1^{2*}\right)f_2\\
&+ \left(g_2^*g_1^* - (g_2^*g_1)^*\right)f_1 + \frac{1}{3!}\left(g_1^{*3} + 2g_1^{3*} - 3g_1^*g_1^{2*}\right)f_1.
\end{aligned}
$$

Under this composition rule, all near-1 formal maps $f^s = 1_m + \sum_1^\infty s^k f_k$ form a (∞-dim) formal Lie group $\mathbf{FD}_m$. In group $\mathbf{FD}_m$ inverse elements, square roots, rational powers, etc., always exist, their coefficients can always be determined recursively by the defining composition relations. For example, the inverse $(f^s)^{-1} := 1_m + \sum s^k h_k = h^s$ is defined by $(f^s\circ h^s) = 1_m$, hence

$$
f_1 + h_1 = 0,\quad f_k + h_k + \delta\left(f_1,\cdots,f_{k-1};h_1,\cdots,h_{k-1}\right) = 0,\quad k\geqslant 2.
$$

In particular,

$$
\begin{aligned}
h_1 &= -f_1, \\
h_2 &= -f_2 + f_1^2, \\
h_3 &= -f_3 + f_1^* f_2 + \left(f_2^* - f_1^{2*}\right) f_1 - \frac{1}{2} f_1^3 + \frac{1}{2} f_1^{2*} f_1.
\end{aligned}
$$

There is an obvious one-one correspondence between the Lie algebra $\mathbf{FV}_m$ and the Lie group $\mathbf{FD}_m$, established simply by $+1_m$ and by -1_m. However, the more significant one-one correspondence between them is given by exp and its inverse log.

$$
\begin{aligned}
&\exp : \mathbf{FV}_m \to \mathbf{FD}_m \\
a^s = \sum_1^{\infty} s^k a_k \to \exp a^s &:= 1_m + \sum_{k=1}^{\infty} \frac{1}{k!} (a^{s*})^k 1_m \\
&=: 1_m + \sum_1^{\infty} s^k f_k = f^s. \qquad (10.1)
\end{aligned}
$$

Note that

$$
(a^{s*})^k = \left(\sum s^{i_1} a_{i_1}^*\right) \cdots \left(\sum s^{i_k} a_{i_k}^*\right) = \sum_{i_1,\cdots,i_k=1}^{\infty} s^{i_1+\cdots+i_k} a_{i_1}^* \cdots a_{i_k}^*,
$$

so we get easily

$$
\begin{aligned}
f_k &= \sum_{n=1}^{k} \frac{1}{n!} \sum_{k_1+\cdots+k_n=k} a_{k_1}^* \cdots a_{k_m}^* 1_m, \quad k \geqslant 1, \quad f_1 = a_1, \qquad (10.2) \\
f_k &= a_k + \sum_{n=2}^{k} \frac{1}{n!} \sum_{k_1+\cdots+k_n=k} a_{k_1}^* \cdots a_{k_n}^* 1_m, \quad k \geqslant 2, \quad f_2 = a_2 + \frac{1}{2} a_1^2.
\end{aligned}
$$

Note that (10.2) provides a 2-way recursion formula from $a_1, \cdots, a_k$ to $f_1, \cdots, f_k$ and vice versa. Therefore exp maps $\mathbf{FV}_m$ one-one onto $\mathbf{FD}_m$ and its inverse, i.e., log is defined by the same (10.2) :

$$
\log = (\exp)^{-1} : \mathbf{FD}_m \to \mathbf{FV}_m, \quad \log \exp a^s = a^s, \quad \exp \log f^s = f^s.
$$

In particular

$$
\begin{aligned}
a_1 &= f_1, \quad a_2 = f_2 - \frac{1}{2} a_1^2, \quad a_3 = f_3 - \frac{1}{2}\left(a_1^* a_2 + a_2^* a_1\right) - \frac{1}{3!} a_1^3, \\
a_4 &= f_4 - \frac{1}{2}\left(a_1^* a_3 + a_2^2 + a_3^* a_1\right) \\
&\quad - \frac{1}{3!}\left(a_1^* a_1^* a_2 + a_1^* a_2^* a_1 + a_2^* a_1^* a_1\right) - \frac{1}{4!} a_1^4, \qquad (10.3) \\
a_k &= f_k - \sum_{n=2}^{k-1} \frac{1}{n!} \sum_{k_1+\cdots+k_n=k} a_{k_1}^* \cdots a_{k_n}^* 1_m - \frac{1}{k!} a_1^k, \quad k \geqslant 3.
\end{aligned}
$$

An equivalent construction of $\log f^s = a^s$ is

$$\log f^s = \sum_{k=1}^{\infty} \frac{(-1)^{k-1}}{k} h_k^s, \tag{10.4}$$

where

$$h_1^s = f^s - 1_m, \quad h_k^s = h_{k-1}^s \circ f^s - h_{k-1}^s.$$

It is easy to compute

$$\begin{aligned} h_1^s &= \sum_{k=1}^{\infty} s^k f_k = \sum_{k_1=1}^{\infty} s^{k_1} (1_m \circ f)_{k_1} \\ h_2^s &= \sum_{k_1=k_2=1}^{\infty} s^{k_1+k_2} \left((1_m \circ f)_{k_1} \circ f \right)_{k_2}, \\ h_3^s &= \sum_{k_1,k_2,k_3=1}^{\infty} s^{k_1+k_2+k_3} \left(\left((1_m \circ f)_{k_1} \circ f \right)_{k_2} \circ f \right)_{k_3}, \text{ etc.} \end{aligned}$$

Substituting in (10.4) and equating with $\sum\limits_{1}^{\infty} s^k a_k$, we get

$$a_k = \sum_{n=1}^{k} \frac{(-1)^{n-1}}{n} \sum_{k_1+\cdots+k_n=k} \left(\cdots \left((1_m \circ f)_{k_1} \circ f \right)_{k_2} \cdots \circ f \right)_{k_n}. \tag{10.5}$$

It is easy to verify log exp $a^s = a^s$ for this log, so this is precisely the inverse of exp, thus agrees with the previous one.

We use the above construction (10.5) to establish the *formal* Campbell-Hausdorff formula:

For arbitrary near-1 formal maps f^s, g^s

$$\log (f^s \circ g^s) = \log f^s + \log g^s + \sum_{k=1}^{\infty} d_k (\log f^s, \log g^s), \tag{10.6}$$

where, for $\log f^s = a^s, \log g^s = b^s$,

$$d_k(a^s, b^s) = \frac{1}{k} \sum_{n=1}^{k} \frac{(-1)^{n-1}}{n} \sum_{\substack{p_1+q_1+\cdots+p_n+q_n=k \\ p_i+q_i \geqslant 1, p_i \geqslant 0, q_i \geqslant 0}} \frac{[(a^s)^{p_1} (b^s)^{q_1} \cdots (a^s)^{p_n} (b^s)^{q_n}]}{p_1! q_1! \cdots p_n! q_n!} \tag{10.7}$$

where

$$(x)^p = xx \cdots x (p\text{ -times }), \quad [x_1 x_2 x_3 \cdots x_n] = [[\cdots\cdots [[x_1, x_2], x_3], \cdots], x_n].$$

In particular

$$d_1 = \frac{1}{2} [a^s, b^s], \quad d_2 = \frac{1}{12} ([a^s b^s b^s] + [b^s a^s a^s]), \quad d_3 = -\frac{1}{24} [a^s b^s b^s a^s].$$

Let $\log(f^s \circ g^s) = c^s = \sum_1^{\infty} s^k c_k$. Then

$$\begin{aligned}
c_1 &= a_1 + b_1, \quad c_2 = a_2 + b_2 + \frac{1}{2}[a_1 b_1], \\
c_3 &= a_3 + b_3 + \frac{1}{2}([a_1 b_2] + [a_2 b_1]) + \frac{1}{12}([a_1 b_1 b_1] + [b_1 a_1 a_1]), \\
c_4 &= a_4 + b_4 + \frac{1}{12}([a_1 b_3] + [a_2 b_2] + [a_3 b_1]) \\
&\quad + \frac{1}{12}([a_1 b_1 b_2] + [a_1 b_2 b_1] + [a_2 b_1 b_1] + [b_1 a_1 a_2] \\
&\quad + [b_1 a_2 a_1] + [b_2 a_1 a_1]) - \frac{1}{24}[a_1 b_1 b_1 a_1], \\
&\text{etc.}
\end{aligned}$$

Note that the classical CH formula is restricted to the composition of two one-parameter groups, where $\log f^s = s a_1, \log g^s = s b_1$.

The log transform reduces matters at the Lie group level to those at the easier level of Lie algebra. All properties of near-1 formal maps have their logarithmic interpretations. We list some of them, let $\log f^s = a^s = \sum_1^{\infty} s^k a_k$:

1. f^s is a phase flow, i.e., $f^{s+t} = f^s \circ f^t \Leftrightarrow \log f^s = s a_1$.
2. f^s is revertible, i.e., $f^s \circ f^{-s} = 1_m \Leftrightarrow \log f^s$ is odd in s.
3. f^s raised to real μth power $(f^s)^\mu \Leftrightarrow \log(f^s)^\mu = \mu \log f^s$. In particular $\log(f^s)^{-1} = -\log f^s, \log\sqrt{f^s} = \frac{1}{2}\log f^s$.
4. f^s scaled to $f^{\alpha s} \Leftrightarrow \log(f^{\alpha s}) = (\log f)^{\alpha s}$. In particular $\log(f^{-s}) = (\log f)^{-s}$.
5. $f^s - g^s = O(s^{p+1}) \Leftrightarrow \log f^s - \log g^s = O(s^{p+1})$.
6. $f^s \circ g^s = g^s \circ f^s \Leftrightarrow [\log f^s, \log g^s] = 0 \Leftrightarrow \log(f^s \circ g^s) = \log f^s + \log g^s$.
7. $(f^s \circ g^s) = h^s \Leftrightarrow \log h^s = \log(f^s \circ g^s) = \log f^s + \log g^s + \sum_1^{\infty} d_k(\log f^s, \log g^s)$.
8. f^s symplectic $\Leftrightarrow$ all a_k are Hamiltonian fields.
9. f^s contact $\Leftrightarrow$ all a_k are contact fields.
10. f^s volume-preserving $\Leftrightarrow$ all a_k are source-free fields.

The log transform has important bearing on dynamical systems with Lie algebra structure, the structure-preserving property of maps f^s at the Lie group ($\mathbf{G} \subset \mathbf{D}_m$) level can be characterized through their logarithms at the associated Lie algebra ($\mathrm{L} \subset \mathbf{V}_m$) level.

We return to the main problem of approximation to the phase flow for dynamical system $dx/dt = a(x)$.

$$f_a^s = f^s = 1_m + \sum s^k f_k \approx e_a^s = 1_m + \sum s^k e_k, \quad e_k = a^k/k!. \tag{10.8}$$

If $f_k = e_k, 1 \leqslant k \leqslant p$, we say f_a^s is accurate to order $\geqslant p$, if moreover, $f_{p+1} \neq e_{p+1}$, we say it is accurate to order p.

Let $\log f^s = a^s = \sum s^k a_k$. Note that the first $p+1$ equations in (10.2) completely determine $a_1, a_2, \cdots, a_{p+1}$ and $f_1, f_2, \cdots, f_{p+1}$ each other. It is then easy to establish

$$\begin{aligned} & f_k = e_k, \quad 1 \leqslant k \leqslant p; \quad f_{p+1} \neq e_{p+1} \Longleftrightarrow \\ & a = a_1 = e_1; \quad a_k = 0, \quad 1 < k \leqslant p; \quad a_{p+1} = f_{p+1} - e_{p+1} \neq 0. \end{aligned} \tag{10.9}$$

So the orders of approximation for $f_a^s \approx e_a^s$ and for $\log f_a^s - sa$ are the same.

Moreover, note that we have a *formal field*

$$s^{-1} \log f^s = s^{-1} a^s = a + \sum_1^{\infty} s^k a_{k+1} = a + O\left(s^p\right) \tag{10.10}$$

which is equal to the original field a up to a near-0 perturbation and defines a *formal dynamical system*

$$\frac{dx}{dt} = \left(s^{-1} \log f^s\right)(x) = a(x) + \sum_1^{\infty} s^k a_{k+1}(x) \tag{10.11}$$

having a *formal phase flow* (in two parameters t and s with group property in t)$e_{s^{-1}a^s}^t = \exp ts^{-1}a^s$ whose *diagonal formal flow* $e_{s^{-1}a^s}^t\big|_{t=s}$ is exactly f^s. This means that any compatible algorithm f_a^s of order p gives perturbed solution of a right equation with field a; however, it gives the right solution of a perturbed equation with field $s^{-1} \log f_a^s = a + O\left(s^p\right)$. There could be many methods with the same formal order of accuracy but with quite different qualitative behavior. The problem is to choose among them those leading to allowable perturbations in the equation. For systems with geometric structure, the propositions 8, 9 and 10 provide guidelines for a proper choice. The structure-preservation requirement for the algorithms precludes all unallowable perturbations alien to the pertinent type of dynamics. Take, for example, Hamiltonian systems. A transition map f_a^s for Hamiltonian field a is symplectic if and only if all fields a_k are Hamiltonian, i.e., the induced perturbations in the equation are Hamiltonian [4,15]. So symplectic algorithms are clean, inherently free from all kinds of perturbations alien to Hamiltonian dynamics (such as artificial dissipation inherent in the vast majority of conventional methods), this accounts for their superior performance. The situations are analogous for contact and volume-preserving algorithms. Propositions 8, 9 lead to the existence of formal energy for symplectic algorithms and that of formal contact energy for contact algorithms, see [8-9, 12-13, 37, and also 45]. For the incompatibility of symplectic structure preservation with energy conservation for algorithms, see [24,26].

Appended Remarks Our results on structure-preserving algorithms for dynamical systems and the constructive theory of generating functions for symplectic maps as surveyed above has been extended and further developed in various aspects as follows: For Poisson manifolds and Poisson maps, see [21–23, 25, 36]. For infinite-dimensional Hamiltonian systems, see [24, 27, 29, 35]. For quantum mechanical systems and unitary algorithms, see [30, 46]. For Hamiltonization of ordinary differential equations, see [39, 41, 42]. For comparative numerical studies for Hamiltonian vs non-Hamiltonian algorithms, see [4, 5, 14, 15, 34]. For the problem of symplecticity of multi-step methods, see [7, 17]. For KAM theorem for symplectic algorithms for Hamiltonian systems, see [32].

Acknowledgements The authors would like to thank their collaborators Qin Meng-zhao, Li Wang-yao, Ge Zhong, Shang Zai-jiu, Wu Yu-hua, Li Chun-wang, Zhang Mei-qing, Tang Yi-fa, Zhu Wen-jie, Jiang Li-xin and Shu Hai-bing for fruitful cooperation.

References

[1] Arnold V I. Mathematical Methods of Classical Mechanics. New York: Springer, 1978

[2] Feng K. On difference schemes and symplectic geometry. In Feng K, ed. Proc 1984 Beijing Symp Diff Geometry and Diff Equations. Beijing: Science Press, 1985: 42-58

[3] Feng K. Difference schemes for Hamiltonian formalism and symplectic geometry. J Comp Math. 1986 4(3): 279-289

[4] Feng K. The Hamiltonian way for computing Hamiltonian dynamics. In Spigler R, ed. Applied and Industrial Mathematics. Netherlands: Kluwer, 1991: 17-35

[5] Feng K. How to compute properly Newton's equation of motion. In Ying L A, Guo B Y, ed. Proc of 2nd Conf on Numerical Methods for Partial Differential Equations. Singapore: World Scientific, 1992: 15-22

[6] Feng K. Formal power series and numerical methods for differential equations. In Chan T, Shi Z C, ed. Proc of Inter Conf on Scientific Computation. Singapore: World Scientific, 1992: 28-35

[7] Feng K. The step transition operators for multi-step methods of ODE's: [Preprint]. Beijing: CAS. Computing Center, 1991

[8] Feng K. The calculus of generating functions and the formal energy for Hamiltonian algorithms: [Preprint]. Beijing: CAS. Computing Center, 1991

[9] Feng K. The calculus of formal power series for diffeomorphisms and vector fields: [Preprint]. Beijing: CAS. Computing Center, 1991

[10] Feng K. Symplectic, contact and volume-preserving algorithms. In Shi Z C, Ushijima T, ed. Proc 1st China-Japan Conf on Numer Math. Singapore: World Scientific, 1993: 1-28

[11] Feng K. Formal dynamical systems and numerical algorithms. In Feng K, Shi Z C, ed. Proc Inter Conf on Computation of Differential Equations and Dynamical Systems. Singapore: World Scientific, 1993: 1-10

[12] Feng K. Contact algorithms for contact dynamical systems: [Preprint]. Beijing: CAS. Computing Center, 1992

[13] Feng K. Theory of contact generating functions: [Preprint]. Beijing: CAS. Computing Center, 1992

[14] Feng K, Qin M Z. The symplectic methods for computation of Hamiltonian equations. In Zhu Y L, Guo Ben-yu, ed. Proc Conf on Numerical Methods for PDE's. Berlin: Springer, 1987: 1-37. Lect Notes in Math 1297

[15] Feng K, Qin M Z. Hamiltonian algorithms and a comparative numericcil study. Comput Phys Comm. 1991 65: 173-187

[16] Feng K, Shang Z J. Volume preserving algorithms for source-free dynamical systems. Numerische Mathematik. 1995

[17] Feng K, Tang Y F. Non-symplecticity of linear multi-step methods: [Preprint]. Beijing: CAS. Computing Center， 1991

[18] Feng K, Wang D L. A note on conservation laws of symplectic difference schemes for Hamiltonian systems. J Comp Math. 1991 9(3): 229-237

[19] Feng K, Wang D L. Variations on a theme by Euler. Preprint, Acad. Sin. Comp. Ctr., 1991

[20] Feng K, Wu H M, Qin M Z, Wang D L. Construction of canonical difference schemes for Hamiltonian formalism via generating functions. J Comp Math. 1989 7(1): 71-96

[21] Ge Z. Generating functions for the Poisson map: [Preprint]. Beijing：CAS. Computing Center, 1987

[22] Ge Z. Geometry of symplectic difference schemes and generating functions: [Preprint]. Beijing: CAS. Computing Center, 1988

[23] Ge Z. Generating functions, Hamilton-Jacobi equations and symplectic groupsoids on Poisson manifolds: [Preprint]. Berkeley: MSRI, 1988

[24] Ge Z, Feng K. On the approximation of Hamiltonian systems. J Comp Math. 1988 6(1): 88-97

[25] Ge Z, Marsden J. Lie-Poisson Hamilton-Jacobi theory and Lie-Poisson integrators. Phys Lett A. 1988 133(3)：137-139

[26] Li C W. The incompatibility of symplectic structure preservation with energy conservation for numerical computation of Hamiltonian systems: [Preprint]. Beijing: CAS. Computing Center, 1987

[27] Li C W, Qin M Z. Symplectic difference schemes for infinite dimensional Hamiltonian systems. J Comp Math. 1988 6(2): 164-174

[28] Qin M Z, Wang D L, Zhang M Q. Explicit symplectic difference schemes for separable Hamiltonian systems. J Comp Math. 1991 9(3): 211-221

[29] Qin M Z, Zhang M Q. Multi-stage symplectic schemes of two kind of Hamiltonian systems of wave equations. Computer Math Applic. 1990 19(10): 51-62

[30] Qin M Z，Zhang M Q. Explicit Runge-Kutta like unitary schemes to solve certain quantum operator equations of motion. J Stat Phys. 1990 60: 837-843

[31] Qin M Z, Zhu W J. Construction of higher order symplectic schemes by composition. Computing. 1992 47: 309-321

[32] Shang Z J. On KAM theorem for symplectic algorithms for Hamiltonian systems: [dissertation]. Beijing: CAS. Computing Center, 1991

[33] Shu H B. A new approach to generating functions for contact systems: [Preprint]. Beijing: CAS. Computing Center, 1992

[34] Tang Y F. Hamiltonian systems and algorithms for geodesic flows on compact Riemannian manifolds: [Master Thesis]. Beijing: CAS. Computing Center, 1990

[35] Wang D L. Semi-discrete Fourier spectral approximations of infinite dimensional Hamiltonian systems and conservation laws. Computer Math Applic. 1991 21(4): 63-75

[36] Wang D L. Poisson difference schemes for Hamiltonian systems on Poisson manifolds. J Comp Math. 1991 9(2): 115-124

[37] Wang D L. Some aspects of Hamiltonian systems and symplectic algorithms. Physica D. 1994 73: 1-16

[38] Wang D L. Symplectic difference schemes for perturbed Hamiltonian systems: [Preprint]. Berlin: ZIB, 1990

[39] Wang D L. Decomposition of vector fields and composition of algorithms. In Feng K, Shi Z C, ed. Proc Inter Conf on Computation of Differential Equations and Dynamical Systems. Singapore: World Scientific, 1993, 179-184

[40] Wang D L, Wu Y H. Generating function methods for the construction of one-step schemes for ODE's: [Preprint]. Beijing: CAS. Computing Center, 1988

[41] Wu Y H. The generating function of the solution of ODE and its discrete methods. Computer Math Applic. 1988 15(12): 1041-1050

[42] Wu Y H. Symplectic transformations and symplectic difference schemes. Chinese J Numer Math Applic. 1990 12(1): 23-31

[43] Wu Y H. Discrete variational principle to the Euler-Lagrange equation. Computer Math Applic. 1990 20(8): 61-75

[44] Yoshida H. Construction of higher order symplectic integrators. Phys Letters A. 1990 150: 262-268

[45] Yoshida H. Conserved quantities of symplectic integrators for Hamiltonian systems: [Preprint]. Japan, 1990

[46] Zhang M Q. Explicit unitary schemes to solve quantum operator equations of motion. J Stat Phys. 1991 65(3/4): 793-799

[47] Zhang M Q, Qin M Z. Explicit symplectic schemes to solve vortex systems: [Preprint]. Beijing: CAS. Computing Center, 1992

20 Variations on a Theme by Euler[①]

欧拉型差分格式

Abstract

The oldest and simplest difference scheme is the explicit Euler method. Usually, it is not symplectic for general Hamiltonian systems. It is interesting to ask: under what conditions of Hamiltonians, the explicit Euler method becomes symplectic? In this paper, we give the class of Hamiltonians for which systems the explicit Euler method is symplectic. In fact, in these cases, the explicit Euler method is really the phase flow of the systems, therefore symplectic. Most of important Hamiltonian systems can be decomposed as the sum of these simple systems. Then composition of the Euler method acting on these systems yields a symplectic method, also explicit. These systems are called symplectically separable. Classical separable Hamiltonian systems are symplectically separable. Especially, we proved that any polynomial Hamiltonian is symplectically separable.

§1 Introduction

A Hamiltonian system of differential equations on $\mathbf{R}^{2n}$ is given by

$$\dot{p} = -H_q(p,q), \quad \dot{q} = H_p(p,q), \tag{1}$$

where $p = (p_1, \cdots, p_n)^T, q = (q_1, \cdots, q_n)^T \in \mathbf{R}^n$ are the generalized coordinates and momenta respectively and $H(p,q)$ is the energe of the system. The system (1) can be rewritten as the compact form

$$\dot{z} = JH_z(z), \quad J = \begin{bmatrix} 0 & -I_n \\ I_n & 0 \end{bmatrix}, \tag{2}$$

where $z = (z_1, \cdots, z_n, z_{n+1}, \cdots, z_{2n})^T = (p,q)^T \in \mathbf{R}^{2n}, H(z) = H(p,q)$. The phase flow, denoted by e_H^t, of the Hamiltonian system is symplectic, i.e., it preserves the differential 2-form on $\mathbf{R}^{2n}$

$$e_H^{t^*}\omega = \omega,$$

where

$$\omega = dp_1 \wedge dq_1 + \cdots + dp_n \wedge dq_n,$$

① Joint with Wang D L

or

$$\left(e_H^t\right)_z^T(z)J(e_H^t)_z(z) = J, \quad \forall z \in \mathbf{R}^{2n}.$$

For the Hamiltonian system (2), a single step numerical method can be characterized by a map g_H^τ, τ is the time step size,

$$z^{n+1} = g_H^\tau z^n, \qquad \text{or} \qquad \hat{z} = g_H^\tau z.$$

If g_H^τ is symplectic, i.e.,

$$\left(g_H^\tau\right)_z^T(z)J\left(g_H^\tau\right)_z(z) = J, \quad \forall z \in \mathbf{R}^{2n},$$

then, the method g_H^τ is called symplectic.

The oldest and simplest difference scheme for Hamiltonian system (2) is the explicit Euler method

$$\hat{z} = E_H^\tau z := z + \tau J H_z(z), \quad E_H^\tau = 1 + \tau J H_z. \tag{3}$$

Usually, it is not symplectic for general Hamiltonian systems. But, it is symplectic for a kind of specific Hamiltonian systems, i.e., systems with nilpotent of degree 2 (see Section 2). In fact, it is the exact phase flow for these systems, therefore is symplectic. Many important Hamiltonian systems can be decomposed as the sum of Hamiltonian systems with nilpotent of degree 2, which are called symplectically separable. Then explicit symplectic schemes can be derived by composition of explicit Euler methods acting on these systems (exact phase flows). This kind of Hamiltonians is not too rare but can cover most important cases. Usual Hamiltonian systems in classical mechanics are symplectically separable. Especially, classical separable Hamiltonian systems are symplectically separable. At last, we proved that any polynomial Hamiltonian is symplectic separable.

§2 Systems with Nilpotent of Degree 2

Definition 1 A Hamiltonian H is nilpotent of degree 2 if H satisfies

$$JH_{zz}(z)JH_z(z) = 0, \quad \forall z \in \mathbf{R}^{2n}. \tag{4}$$

Evidently, $H(p,q) = \phi(p)$ or $H(p,q) = \psi(q)$, which presents inertial flow and standing flow, are nilpotent of degree 2 since for $H(p,q) = \phi(p)$,

$$H_{zz}(z)JH_z(z) = \begin{bmatrix} \phi_{pp} & 0 \\ 0 & 0 \end{bmatrix}\begin{bmatrix} 0 & -I \\ I & 0 \end{bmatrix}\begin{bmatrix} \phi_p \\ 0 \end{bmatrix} = \begin{bmatrix} \phi_{pp} & 0 \\ 0 & 0 \end{bmatrix}\begin{bmatrix} 0 \\ \phi_p \end{bmatrix} = 0$$

and for $H(p,q)=\psi(q)$,

$$H_{zz}(z)JH_z(z)=\begin{bmatrix}0 & 0\\ 0 & \phi_{qq}\end{bmatrix}\begin{bmatrix}0 & -I\\ I & 0\end{bmatrix}\begin{bmatrix}0\\ \psi_q\end{bmatrix}=\begin{bmatrix}0 & 0\\ 0 & \phi_{qq}\end{bmatrix}\begin{bmatrix}-\psi_q\\ 0\end{bmatrix}=0.$$

Theorem 1 *If H is nilpotent of degree 2 then the explicit Euler method E_H^τ is the exact phase flow of the Hamiltonian, therefore symplectic.*

Proof. Let $z=z(0)$. From the condition (4) it follows that

$$\ddot{z}(t)=\frac{d}{dt}JH_z(z(t))=(JH_z(z(t)))_z\,\dot{z}(t)=JH_{zz}(z(t))JH_z(z(t))=0.$$

Thorefore,

$$\dot{z}(t)=\dot{z}(0)=JH_z(z(0)).$$

Hence

$$z(t)=z(0)+tJH_z(z(0))=z+tJH_z(z)=E_H^t(z).$$

It is just the explicit Euler method E_H^t. This shows that for such a system, explicit Euler method E_H^τ is the exact phase flow, therefore symplectic.

Theorem 2 *Let $\phi:\mathbf{R}^n\to\mathbf{R}$ be a scalar function on n variables u, $\phi(u)=\phi(u_1,\cdots,u_n)$. Let $C_{n\times 2n}=(A,B)$ be a linear transformation from $\mathbf{R}^{2n}$ to $\mathbf{R}^n$. Then the Hamiltonian $H(z)=\phi(Cz)$ satisfies*

$$JH_{zz}(z)JH_z(z)=0,\quad \forall\phi,z \tag{5}$$

if and only if

$$CJC^T=0. \tag{6}$$

Proof. Since

$$JH_{zz}(z)JH_z(z)=JC^T\phi_{uu}(Cz)CJC^T\phi_u(Cz), \tag{7}$$

the sufficient condition is trivial.

We now prove the necessity. If

$$JH_{zz}(z)JH_z(z)=0,\quad \forall\phi,z$$

then from (7) it follows that

$$JC^T\phi_{uu}(Cz)CJC^T\phi_u(Cz)=0,\quad \forall\phi,z.$$

Especially, take $\phi(u) = \frac{1}{2}u^T u$. Then

$$JC^TCJC^TCz = 0, \quad \forall z.$$

i.e.,

$$JC^TCJC^TC = 0.$$

Left multiplying C and right multiplying JC^T this equation, we get

$$\left(CJC^T\right)^3 = 0.$$

The anti-symmetry of CJC^T implies $CJC^T = 0$.

Lemma 1 *Let $C = (A, B)$. Then $CJC^T = 0$ if and only if $AB^T = BA^T$.*

Theorem 3 *For any Hamiltonian*

$$H(z) = H(p, q) = \phi(Cz) = \phi(Ap + Bq), \quad AB^T = BA^T,$$

where $\phi(u)$ is any n variable function, the explicit Euler method

$$\hat{z} = E_H^\tau z = E_\phi^\tau z = z + \tau JH_z(z) = z + \tau JC^T\phi_u(Cz)$$

is the exact phasc flow, i.e.,

$$e_\phi^\tau = E_\phi^\tau = 1 + \tau JH_z = 1 + \tau JC^T\phi_u \circ C.$$

Hence, E_ϕ^τ is symplectic.

§3 Symplectically Separable Hamiltonian Systems

Definition 2 Hamiltonian $H(z)$ is symplectically separable if

$$H(z) = \sum_{i=1}^{m} H_i(z), \quad H_i(z) = \phi_i\left(C_i z\right) = \phi\left(A_i p + B_i q\right), \tag{8}$$

where ϕ_i are the functions on n variables and $C_i = (A_i, B_i)$ with the condition $AB^T = BA^T$, $i = 1, \cdots, m$.

Proposition 1 *Linear combination of symplectic separable Hamiltonians is symplectically separable.*

For symplectically separable Hamiltonian (8), the explicit composition scheme

$$\begin{aligned} g_H^\tau &= E_m^\tau \circ E_{m-1}^\tau \circ \cdots \circ E_2^\tau \circ E_1^\tau \\ &:= E_{H_m}^\tau \circ E_{H_{m-1}}^\tau \circ \cdots \circ E_{H_2}^\tau \circ E_{H_1}^\tau \end{aligned} \tag{9}$$

is symplectic and of order 1. As a matter of fact

$$
\begin{aligned}
E_{H_2}^{\tau}\circ E_{H_1}^{\tau} &= (1+\tau JH_{2,z})\circ(1+\tau JH_{1,z})\\
&= 1+\tau JH_{2,z}+\tau JH_{1,z}+O\left(\tau^2\right)\\
&= 1+\tau J\left(H_{2,z}+H_{1,z}\right)+O\left(\tau^2\right)\\
&\cdots\\
g_H^{\tau} &= E_m^{\tau}\circ E_{m-1}^{\tau}\circ\cdots\circ E_2^{\tau}\circ E_1^{\tau}\\
&= (1+\tau JH_{m,z})\circ\left(1+\tau J\sum_{i=1}^{m-1}H_{i,z}+O\left(\tau^2\right)\right)\\
&= 1+\tau J\sum_{i=1}^{m}H_{i,z}+O\left(\tau^2\right)\\
&= 1+\tau JH_z+O\left(\tau^2\right).
\end{aligned}
$$

The symplecticity of g_H^{τ} follows from the fact that symplectic maps on $\mathbf{R}^{2n}$ forms a group under composition.

Similarly,

$$
\check{g}_H^{\tau}=E_1^{\tau}\circ E_2^{\tau}\circ\cdots\circ E_{m-1}^{\tau}\circ E_m^{\tau}
$$

is symplectic and of order 1. Composite schemes of arbitrary ordering

$$
E_{i_m}^{\tau}\circ E_{i_{m-1}}^{\tau}\circ\cdots\circ E_{i_2}^{\tau}\circ E_{i_1}^{\tau}
$$

and

$$
E_{i_1}^{\tau}\circ E_{i_2}^{\tau}\circ\cdots\circ E_{i_{m-1}}^{\tau}\circ E_{i_m}^{\tau},
$$

where $i_1,i_2,\cdots,i_m$ is a permutation of $1,2,\cdots,m$, are also symplectic and of order 1.

§4 Construction of Explicit Symplectic Methods with High Order

Let g^t be a one-parameter family of diffeomorphisms, $g^0=1_{2n}$. Define its reversion as

$$
\check{g}^t=\left(g^{-t}\right)^{-1}.
$$

g^t is revertible (or self-adjoint, or symmetric) if $g^t=\check{g}^t$, i.e., $g^t\circ g^{-t}=$ identity. Evidently, the phase flow of Hamiltonian systems is revertible.

The step transition operator g^τ of a difference scheme is said to be of order r if it is an r -th approximation to the phase flow e^t, i.e.,

$$g^\tau = e^\tau + O\left(\tau^{r+1}\right).$$

If g^τ is revertible, then g^τ is of even order of accuracy.

Let g^τ be revertible and of order $2r$, then its revertible composition

$$\boldsymbol{g}^{\alpha\tau} \circ \boldsymbol{g}^{\beta\tau} \circ \boldsymbol{g}^{\alpha\tau}$$

is of order $2(r+1)$ if

$$2\alpha + \beta = 1, \quad \beta^{2r+1} + 2\alpha^{2r+1} = 0, \tag{10}$$

i.e.,

$$\alpha = \frac{1}{2 - 2^{1/(2r+1)}} > 0, \quad \beta = 1 - 2\alpha < 0.$$

The proof can refer to Yoshida [23] and Qin / Zhu [19]. Here we give a more direct proof.

Since g^τ is of order $2r$, then it has the expansion

$$g^\tau = e^\tau + O\left(\tau^{2r+1}\right) = e^\tau + \tau^{2r+1} g_{2r+1} + O\left(\tau^{2r+2}\right).$$

Therefore,

$$\begin{aligned}
\boldsymbol{g}^{\alpha\tau} \circ \boldsymbol{g}^{\beta\tau} = &\left(e^\tau + \alpha^{2r+1}\tau^{2r+1} g_{2r+1} + O\left(\tau^{2r+2}\right)\right) \circ \left(e^\tau + \beta^{2r+1}\tau^{2r+1} g_{2r+1} + O\left(\tau^{2r+2}\right)\right) \\
= &\, e^\tau \circ \left(e^\tau + \beta^{2r+1}\tau^{2r+1} g_{2r+1} + O\left(\tau^{2r+2}\right)\right) \\
&+ \alpha^{2r+1}\tau^{2r+1} g_{2r+1} \circ \left(e^\tau + \beta^{2r+1}\tau^{2r+1} g_{2r+1} + O\left(\tau^{2r+2}\right)\right) + O\left(\tau^{2r+2}\right) \\
= &\, e^{\alpha\tau} e^{\beta\tau} + \beta^{2r+1}\tau^{2r+1} g_{2r+1} + \alpha^{2r+1}\tau^{2r+1} g_{2r+1} + O\left(\tau^{2r+2}\right) \\
= &\, e^{(\alpha+\beta)\tau} + \left(\beta^{2r+1} + \alpha^{2r+1}\right)\tau^{2r+1} g_{2r+1} + O\left(\tau^{2r+2}\right)
\end{aligned}$$

Similarly,

$$g^{\alpha\tau} \circ g^{\beta\tau} \circ g^{\alpha\tau} = e^{(2\alpha+\beta)\tau} + \left(\beta^{2r+1} + 2\alpha^{2r+1}\right)\tau^{2r+1} g_{2r+1} + O\left(\tau^{2r+2}\right).$$

Therefore, $g^{\alpha\tau} \circ g^{\beta\tau} \circ g^{\alpha\tau}$ is of order $2r+1$ if and only if

$$2\alpha + \beta = 1, \quad \beta^{2r+1} + 2\alpha^{2r+1} = 0.$$

The revertibility of $g^{\alpha\tau} \circ g^{\beta\tau} \circ g^{\alpha\tau}$ implies that it is of order $2(r+1)$.

Since the phase flow of a Hamiltonian system is revertible, the Euler method E_H^τ for Hamiltonians with nilpotent of degree 2 is also revertible. Hence,

$$\breve{g}_H^\tau = E_1^\tau \circ E_2^\tau \circ \cdots \circ E_{m-1}^\tau \circ E_m^\tau$$

is the reversion of

$$g_H^\tau = E_m^\tau \circ E_{m-1}^\tau \circ \cdots \circ E_2^\tau \circ E_1^\tau.$$

The composition

$$g_2^\tau = \breve{g}_H^{\tau/2} \circ g_H^{\tau/2} = E_{H_1}^{\tau/2} \circ E_{H_2}^{\tau/2} \circ \cdots \circ E_{H_{m-1}}^{\tau/2} \circ E_{H_m}^\tau \circ E_{H_{m-1}}^{\tau/2} \circ \cdots \circ E_{H_2}^{\tau/2} \circ E_{H_1}^{\tau/2}$$

is revertible and of order 2. The composition

$$g_4^\tau = g_2^{\alpha\tau} \circ g_2^{\beta\tau} \circ g_2^{\alpha\tau}$$

gives a revertible explicit symplectic scheme of order 4 when

$$2\alpha + \beta = 1, \quad 2\alpha^3 + \beta^3 = 0,$$

i.e.,

$$\alpha = \frac{1}{2 - 2^{1/3}} > 0, \quad \beta = 1 - 2\alpha < 0, \tag{11}$$

which was derived by Qin/Wang/Zhang [18] and Yoshida [23] etc. by different ways. Similarly, we can get high order symplectic schemes by this procedure.

Example 1. Since $\phi(p)$ and $\psi(q)$ are nilpotent of degree 2, the classical separable Hamiltonian $H(p, q) = \phi(p) + \psi(q)$ is also symplectically separable. The composition of E_ϕ^τ and E_ψ^τ gives explicit symplectic schemes of order 1

$$E_\psi^\tau \circ E_\phi^\tau : \hat{p} = p - \tau\psi_q(\hat{q}), \quad \hat{q} = q + \tau\phi_p(p),$$
$$E_\phi^\tau \circ E^\tau\psi : \hat{p} = p - \tau\psi_q(q), \quad \hat{q} = q + \tau\phi_p(\hat{p}).$$

The revertible schemes

$$g^\tau := \left(E_\psi^{\tau/2} \circ E_\phi^{\tau/2}\right)^\vee \circ \left(E_\psi^{\tau/2} \circ E_\phi^{\tau/2}\right) = E_\phi^{\tau/2} \circ E_\psi^\tau \circ E_\phi^{\tau/2} :$$
$$q^1 = q + \frac{\tau}{2}\phi_p(p), \quad \hat{p} = p - \tau\psi_q\left(q_1\right), \quad \hat{q} = q^1 + \frac{\tau}{2}\phi_p(\hat{p}).$$

and

$$f^s := \left(E_\phi^{\tau/2} \circ E_\psi^{\tau/2}\right)^\vee \circ \left(E_\phi^{\tau/2} \circ E_\psi^{\tau/2}\right) = E_\psi^{\tau/2} \circ E_\phi^\tau \circ E_\psi^{\tau/2} :$$
$$p^1 = p - \frac{\tau}{2}\psi_q(q), \quad \hat{q} = q + \tau\phi_p\left(p_1\right), \quad \hat{p} = p^1 - \frac{\tau}{2}\psi_q(\hat{q}).$$

are symplectic and of order 2. The revertible composition

$$g^{\alpha\tau} \circ g^{\beta\tau} \circ g^{\alpha\tau} \quad f^{\alpha\tau} \circ f^{\beta\tau} \circ f^{\alpha\tau}$$

give symplectic schemes of order 4 with the parameters (11), i.e. ([18], [23]),

$$\begin{aligned} p^1 &= p - c_1\tau\psi_q(q), & q^1 &= q + d_1\tau\phi_p\left(p^1\right), \\ p^2 &= p^1 - c_2\tau\psi_q\left(q^1\right), & q^2 &= q^1 + d_2\tau\phi_p\left(p^2\right), \\ p^3 &= p^2 - c_3\tau\psi_q\left(q^2\right), & q^3 &= q^2 + d_3\tau\phi_p\left(p^3\right), \\ \hat{p} &= p^3 - c_4\tau\psi_q\left(q^3\right), & \hat{q} &= q^3 + d_4\tau\phi_p(\hat{p}), \end{aligned}$$

with the parameters $\alpha = \left(2 - 2^{1/3}\right)^{-1}, \beta = 1 - 2\alpha$ and either

$$c_1 = 0, \quad c_2 = c_4 = \alpha, \quad c_3 = \beta, \quad d_1 = d_4 = \alpha/2, \quad d_2 = d_3 = (\alpha+\beta)/2,$$

or

$$c_1 = c_4 = \alpha/2, \quad c_2 = c_3 = (\alpha+\beta)/2, \quad d_1 = d_3 = \alpha, \quad d_2 = \beta, \quad d_4 = 0.$$

Example 2. The Hamiltonian

$$H_k(p,q) = \sum_{i=0}^{k-1} \cos\left(p\cos\frac{2\pi i}{k} + q\sin\frac{2\pi i}{k}\right)$$

with k -fold rotational symmetry in phase plane [2, 4] are not separable in the conventional sence for $k \neq 1, 2, 4$, but symplectically separable, since every term

$$\cos\left(p\cos\frac{2\pi i}{k} + q\sin\frac{2\pi i}{k}\right)$$

is nilpotent of degree 2 according to Theorem 2. Such as, for $k = 3$,

$$\begin{aligned} H_3(p,q) &= \cos p + \cos\left(p\cos\frac{2\pi}{3} + q\sin\frac{2\pi}{3}\right) + \cos\left(p\cos\frac{4\pi}{3} + q\sin\frac{4\pi}{3}\right) \\ &= \cos p + \cos\left(\frac{1}{2}p - \frac{\sqrt{3}}{2}q\right) + \cos\left(-\frac{1}{2}p - \frac{\sqrt{3}}{2}q\right) \end{aligned}$$

The explicit symplectic scheme of order 1 is

$$\begin{aligned} q^1 &= q - \frac{1}{2}\tau\sin\left(\frac{1}{2}p + \frac{\sqrt{3}}{2}q\right) \\ p^1 &= p + \frac{\sqrt{3}}{2}\tau\sin\left(\frac{1}{2}p + \frac{\sqrt{3}}{2}q\right) \end{aligned}$$

$$\begin{aligned} q^2 &= q^1 - \frac{1}{2}\tau\sin\left(\frac{1}{2}p^1 - \frac{\sqrt{3}}{2}q^1\right) \\ \hat{p} &= p^1 - \frac{\sqrt{3}}{2}\tau\sin\left(\frac{1}{2}p - \frac{\sqrt{3}}{2}q\right) \\ \hat{q} &= q^2 - \tau\sin\hat{p}. \end{aligned}$$

The explicit revertible symplectic scheme of order 2 is

$$
\begin{aligned}
q^1 &= q - \frac{1}{4}\tau\sin\left(\frac{1}{2}p + \frac{\sqrt{3}}{2}q\right) \\
p^1 &= p + \frac{\sqrt{3}}{4}\tau\sin\left(\frac{1}{2}p + \frac{\sqrt{3}}{2}q\right) \\
q^2 &= q^1 - \frac{1}{4}\tau\sin\left(\frac{1}{2}p^1 - \frac{\sqrt{3}}{2}q^1\right) \\
p^2 &= p^1 - \frac{\sqrt{3}}{4}\tau\sin\left(\frac{1}{2}p^1 - \frac{\sqrt{3}}{2}q^1\right) \\
q^3 &= q^2 - \tau\sin p^2 \\
p^3 &= p^2 - \frac{\sqrt{3}}{4}\tau\sin\left(\frac{1}{2}p^2 - \frac{\sqrt{3}}{2}q^2\right) \\
q^4 &= q^3 - \frac{1}{4}\tau\sin\left(\frac{1}{2}p^3 - \frac{\sqrt{3}}{2}q^3\right) \\
\hat{p} &= p^3 + \frac{\sqrt{3}}{4}\tau\sin\left(\frac{1}{2}p^3 + \frac{\sqrt{3}}{2}q^4\right) \\
\hat{q} &= q^4 - \frac{1}{4}\tau\sin\left(\frac{1}{2}p^3 + \frac{\sqrt{3}}{2}q^4\right)
\end{aligned}
$$

Similarly, we can get explicit symplectic schemes of order 4.

§5 Separability of All Polynomials in $\mathbf{R}^{2n}$

Theorem 4 *Every monomial $x^{n-k}y^k$ of degree n in 2 variables x and $y, n \geqslant 2, 0 \leqslant k \leqslant n$ can be expanded as a linear combination of $n+1$ terms*

$$\left\{(x+y)^n, (x+2y)^n, \cdots, \left(x+2^{n-2}y\right)^n, x^n, y^n\right\}$$

Proof. Using binomial expansion

$$(x+y)^n = x^n + C_n^1 x^{n-1}y^1 + C_n^2 x^{n-2}y^2 + \cdots + C_n^2 x^2 y^{n-2} + C_n^1 x^1 y^{n-1} + y^n,$$

define

$$
\begin{aligned}
P_1(x,y) &= (x+y)^n - x^n - y^n \\
&= C_n^1 x^{n-1}y^1 + C_n^2 x^{n-2}y^2 + \cdots + C_n^2 x^2 y^{n-2} + C_n^1 x^1 y^{n-1},
\end{aligned}
$$

which is separable, and the right hand side consists of "mixed terms". P_1 is a linear combination of 3 terms $(x+y)^n, x^n$ and y^n. Then

$$P_1(x,2y) = 2C_n^1x^{n-1}y^1 + 2^2C_n^2x^{n-2}y^2 + \cdots + 2^{n-2}C_n^2x^2y^{n-2} + 2^{n-1}C_n^1x^1y^{n-1},$$
$$2P_1(x,y) = 2C_n^1x^{n-1}y^1 + 2C_n^2x^{n-2}y^2 + \cdots + 2C_n^2x^2y^{n-2} + 2C_n^1x^1y^{n-1}.$$

Define

$$\begin{aligned} P_2(x,y) &= P_1(x,2y) - 2P_1(x,y) \\ &= \left(2^2-2\right)C_n^2x^{n-2}y^2 + \cdots + \left(2^{n-2}-2\right)C_n^2x^2y^{n-2} \\ &\quad + \left(2^{n-1}-2\right)C_n^1x^1y^{n-1}. \end{aligned}$$

It is separable in 4 terms $(x+y)^n, (x+2y)^n, x^n$ and y^n. Define

$$\begin{aligned} P_3(x,y) &= P_2(x,2y) - 2^2P_2(x,y) \\ &= \left(2^3-2^2\right)\left(2^3-2\right)C_n^3x^{n-3}y^3 + \cdots + \left(2^{n-2}-2^2\right)\left(2^{n-2}-2\right)C_n^2x^2y^{n-2} \\ &\quad + \left(2^{n-1}-2^2\right)\left(2^{n-1}-2\right)C_n^1x^1y^{n-1}. \end{aligned}$$

It is separable in 5 terms $(x+y)^n, (x+2y)^n, \left(x+2^2y\right)^n, x^n$ and y^n.

Define

$$\begin{aligned} P_{n-2}(x,y) &= P_{n-3}(x,2y) - 2^{n-3}P_{n-3}(x,y) \\ &= \left(2^{n-2}-2^{n-3}\right)\cdots\left(2^{n-2}-2\right)C_n^2x^2y^{n-2} \\ &\quad + \left(2^{n-1}-2^{n-3}\right)\cdots\left(2^{n-1}-2\right)C_n^1x^1y^{n-1}, \end{aligned}$$

separable in n terms $(x+y)^n, (x+2y)^n, \cdots, \left(x+2^{n-3}y\right)^n, x^n$ and y^n. Finally, we get

$$\begin{aligned} P_{n-1}(x,y) &= P_{n-2}(x,2y) - 2^{n-2}P_{n-2}(x,y) \\ &= \left(2^{n-1}-2^{n-2}\right)\left(2^{n-1}-2^{n-3}\right)\cdots\left(2^{n-1}-2\right)C_n^1x^1y^{n-1} \\ &= \gamma_{n-1}x^1y^{n-1}, \quad \gamma \neq 0 \end{aligned}$$

separable in $n+1$ terms $(x+y)^n, (x+2y)^n, \cdots, \left(x+2^{n-2}y\right)^n, x^n$ and y^n.

Hence, the mixed term xy^{n-1} is separable in $n+1$ terms. Then from the separability of $P_{n-2}(x,y)$ and xy^{n-1} we know that x^2y^{n-2} is separable in $n+1$ terms. Similarly, x^3y^{n-3}, $x^4y^{n-4}, \cdots, x^{n-2}y^2$ and $x^{n-1}y$ are separable in $n+1$ terms.

Remark We can also work with formula

$$
\begin{aligned}
&\frac{1}{2}(x+y)^{2m+1}+\frac{1}{2}(x-y)^{2m+1}-x^{2m+1}\\
=&C_{2m+1}^{2}x^{2m-1}y^{2}+C_{2m+1}^{4}x^{2m-3}y^{4}+\cdots+C_{2m+1}^{2m}xy^{2m}\\
&\frac{1}{2}(x+y)^{2m+1}-\frac{1}{2}(x-y)^{2m+1}-y^{2m+1}\\
=&C_{2m+1}^{1}x^{2m}y+C_{2m+1}^{3}x^{2m-2}y^{3}+\cdots+C_{2m+1}^{2m-1}x^{2}y^{2m-1}\\
&\frac{1}{2}(x+y)^{2m}+\frac{1}{2}(x-y)^{2m}-x^{2m}-y^{2m}\\
=&C_{2m}^{2}x^{2m-2}y^{2}+C_{2m}^{4}x^{2m-4}y^{4}+\cdots+C_{2m}^{2m-2}x^{2}y^{2m-2}\\
&\frac{1}{2}(x+y)^{2m}-\frac{1}{2}(x-y)^{2m}\\
=&C_{2m}^{1}x^{2m-1}y+C_{2m}^{3}x^{2m-3}y^{3}+\cdots+C_{2m}^{2m-1}xy^{2m-1}
\end{aligned}
$$

by means of elimination to get more symmetric and economic expansions, e.g.,

$$
xy=\frac{1}{2}(x+y)^{2}-\frac{1}{2}x^{2}-\frac{1}{2}y^{2}=\frac{1}{4}(x+y)^{2}-\frac{1}{4}(x-y)^{2}.
$$

Theorem 5 *Every polynomial $P(p,q)$ of degree n in variables p and q can be expanded as $n+1$ terms*

$$
P_{1}(x,y),P_{2}(x,y),\cdots,P_{n-1}(x,y),P_{n}(x),P_{n+1}(y),
$$

where each $P_i(u)$ is a polynomial of degree n in one variable or more generally every polynomial $P(p,q)$ can be expanded as

$$
P(p,q)=\sum_{i=1}^{m}P_{i}\left(a_{i}p+b_{i}q\right),\quad m\leqslant n+1,
$$

where $P_i(u)$ are polynomial of degree n in one variable.

Theorem 6 *Every monomial in 2n variables is of the form*

$$
f(p,q)=\left(p_{1}^{m_{1}-k_{1}}q_{1}^{k_{1}}\right)\left(p_{2}^{m_{2}-k_{2}}q_{2}^{k_{2}}\right)\cdots\left(p_{n}^{m_{n}-k_{n}}q_{n}^{k_{n}}\right)
$$

and can be expanded as a linear combination of the terms in the form

$$
\phi(Ap+Bq)=\left(a_{1}p_{1}+b_{1}q_{1}\right)^{m_{1}}\left(a_{2}p_{2}+b_{2}q_{2}\right)^{m_{2}}\cdots\left(a_{n}p_{n}+b_{n}q_{n}\right)^{m_{n}},
$$

where $\phi(u)=\phi\left(u_{1},\cdots,u_{n}\right)=u_{1}^{m_{1}}u_{2}^{m_{2}}\cdots u_{n}^{m_{n}}$ is the monomial in n variables with total degree $m=\sum\limits_{i=1}^{m}m_{i}$ and with degree m_i in varable u_i, A and B are diagonal matrices of *order n*

$$A=\begin{pmatrix} a_1 & 0 & \cdots & 0 \\ 0 & a_2 & \cdots & 0 \\ \cdots & \cdots & \cdots & \cdots \\ 0 & 0 & \cdots & a_n \end{pmatrix}, \quad B=\begin{pmatrix} b_1 & 0 & \cdots & 0 \\ 0 & b_2 & \cdots & 0 \\ \cdots & \cdots & \cdots & \cdots \\ 0 & 0 & \cdots & b_n \end{pmatrix},$$

which automatically satisfies $AB^T = BA^T$. *The elements* a_i, b_i *can be chosen integers.*

Theorem 7 *Every polynomial* $P(p_1, q_1, \cdots, p_n, q_n)$ *of degree* m *in* $2n$ *variables can be expanded as*

$$P(p,q) = \sum_{i=1}^{m} P_i\left(A_i p + B_i q\right),$$

each P_i *is a polynomial of degree* m *in* n *variables,* A_i *and* B_i *are diagonal matrices,* $\left(\text{satisfying } A_i B_i^T = B_i A_i^T\right)$. *So for polynomial Hamiltonians, the symplectic explicit Euler composite schemes of order 1, 2 or 4 can be easily constructed.*

References

[1] Arnold V I. Mathematical Methods of Classical Mechanics. New York: Springcr. 1978

[2] Beloshapkin V V, Chernikov A A, Natenzon M, Petrovichev B A, Sagdeev R Z, Zaslavsky G M. Chaotic streamlines in pre-turbulent states. Nature. 1989 337(12): 133-137

[3] Channell P J, Scovel J C. Symplectic Integration of Hamiltonian Systems. Nonlinearity. 1990 3: 231-259

[4] Chernikov A A, Sagdeev R Z, Zaslavsky G M. Stochastic Webs. Physica D. 1988 33: 65-76

[5] Feng K. On difference schemes and symplectic geometry. In Feng K, ed. Proc 1984 Beijing Symp Diff Geometry and Diff Equations. Beijing: Science Press, 1985. 42-58

[6] Feng K. Difference schemes for Hamiltonian formalism and symplectic geometry. J Comp Math. 1986 4(3): 279-289

[7] Feng K. The Hamiltonian way for computing Hamiltonian dynamics. In Spigler R, ed. Applied and Industrial Mathematics. Netherlands: Kluwer, 1991. 17-35

[8] Feng K. How to compute properly Newton's equation of motion. In Ying L A, Guo B Y, ed. Proc of 2nd Conf on Numerical Methods for Partial Differential Equations. Singapore: World Scientific, 1992. 15-22

[9] Feng K. Formal dynamical systems and numerical algorithms. In Feng K, Shi Z C, ed. Proc Inter Conf on Computation of Differential Equations and Dynamical Systems. Singapore: World Scientific, 1993. 1-10

[10] Feng K. The step transition operators for multi-step methods of ODE's: [Preprint]. Beijing: CAS. Computing Center, 1991

[11] Feng K. The calculus of generating functions and the formal energy for Hamiltonian algorithms: [Preprint]. Beijing: CAS. Computing Center, 1991

[12] Feng K, Qin M Z. The symplectic methods for computation of Hamiltonian equations. In Zhu Y L, Guo Ben-yu, ed. Proc Conf on Numerical Methods for PDE' s. Berlin: Springer, 1987. 1-37. Lect Notes in Math 1297

[13] Feng K, Qin M Z. Hamiltonian algorithms and a comparative numerical study. Comput Phys Comm. 1991 65: 173-187

[14] Feng K, Wang D L. A note on conservation laws of symplectic difference schemes for Hamiltonian systems. J Comp Math. 1991 9(3): 229-237

[15] Feng K, Wu H M, Qin M Z, Wang D L. Construction of canonical difference schemes for Hamiltonian formalism via generating functions. J Comp Math. 1989 7(1): 71-96

[16] Ge Z, Feng K. On the approximation of Hamiltonian systems. J Comp Math. 1988 6(1): 88-97

[17] Ge Z, Marsden J. Lie-Poisson Hamilton-Jacobi theory and Lie-Poisson integrators. Phys Lett A. 1988 133(3): 137-139

[18] Qin M Z, Wang D L, Zhang M Q. Explicit symplectic difference schemes for separable Hamiltonian systems. J Comp Math. 1991 9(3): 211-221

[19] Qin M Z, Zhu W J. Construction of higher order symplectic schemes by composition. Computing. 1992 47: 309-321

[20] Wang D L. Some aspects of Hamiltonian systems and symplectic algorithms. Physica D. 1994 73: 1-16

[21] Wu Y H. The generating function of the solution of ODE and its discrete methods. Computer Math Applic. 1988 15(12): 1041-1050

[22] Wu Y H. Symplectic transformations and symplectic difference schemes. Chinese J Numer Math Applic. 1990 12(1): 23-31

[23] Yoshida H. Construction of higher order symplectic integrators. Phys Letters A. 1990 150: 262-268

21 The Step-Transition Operators for Multi-Step Methods of ODE's①

常微分方程多步法的步进算子

Abstract

In this paper, we propose a new definition of symplectic multistep methods. This definition differs from the old ones in that it is given via the one step method defined directly on M which is corresponding to the m step scheme defined on M while the old definitions are given out by defining a corresponding one step method on $M \times M \times \cdots \times M = M^m$ with a set of new variables. The new definition gives out a steptransition operator $g: M \longrightarrow M$. Under our new definition, the Leap-frog method is symplectic only for linear Hamiltonian systems. The transition operator g will be constructed via continued fractions and rational approximations.

§1 Introduction

The disadvantage of symplectic methods in using the information from past time steps leads to their needing more function evaluation than nonsymplectic methods. This disadvantage can be overcome if one could construct symplectic multi-step methods. But the first problem should be solved is to give out the definition of symplectic multi-step method. Until now, a popular idea is that an m-step method on M may be written as a one-step method on M^m. In paper [2, 7], the authors have investigated the circumstance under which a difference scheme can preserve the product symplectic structure on M^m. In this paper, a completely different criterion is given because the induced one-step method corresponding to the original multi-step method is defined, it gives out a transition operator $g: M \longrightarrow M$.

Consider the autonomous ODE's on R^n

$$\frac{dz}{dt} = a(z). \tag{1.1}$$

For equation (1.1), we define a linear m step method(LMM) in standard form by

$$\sum_{j=0}^{m} \alpha_j z_j = \tau \sum_{j=0}^{m} \beta_j a(z_j), \tag{1.2}$$

① 由秦孟兆整理

where α_j and β_j are constants subject to the conditions

$$\alpha_m = 1, \quad |\alpha_0| + |\beta_0| \neq 0.$$

If $m = 1$, we call (1.2) a one step method. In other cases, we call it a multi-step method. Here linearity means the right hand of (1.2) linearly dependent on the value of $a(z)$ on integral points. For the compatibility of (1.2) with equation (1.1), it must at least of order one and thus satisfies

1°. $\alpha_1 + \alpha_2 + \cdots + \alpha_m = 0.$

2°. $\beta_0 + \beta_2 + \cdots + \beta_m = \sum_{j=0}^{m} j\alpha_j \neq 0.$

LMM method (1.2) has two characteristic polynomials

$$\zeta(\lambda) = \sum_{i=0}^{m} \alpha_i \lambda^i, \quad \sigma(\lambda) = \sum_{i=0}^{m} \beta_i \lambda^i. \tag{1.3}$$

Equation (1.2) can be written as

$$\zeta(E) y_n = \tau a \left(\sigma(E) y_n\right). \tag{1.4}$$

In section 2, we will study symplectic multi-step methods for linear Hamiltonian systems. We will give a new definition via transition operators which are corresponding to the multi-step methods. We will point out that if these operators are of exponential forms and their reverse maps are of Log forms then the original multi-step method are symplectic. In section 3, we will use continued fractions and rational approximations to approximate the transition operators. In section 4, we show that for non-linear Hamiltonian systems, there exists no symplectic multi-step methods in the sense of our new definition. Numerical examples are also presented.

§2 Symplectic LMM for Linear Hamiltonian Systems

First we consider a linear Hamiltonian system

$$\frac{dz}{dt} = az, \tag{2.1}$$

where a is an infinitesimal $n \times n$ symplectic matrix. Its phase flow is $z(t) = \exp(ta) z_0$. The LMM for (2.1) is

$$\alpha_m z_m + \cdots + \alpha_1 z_1 + \alpha_0 z_0 = \tau a \left(\beta_m z_m + \cdots + \beta_1 z_1 + \beta_0 z_0\right). \tag{2.2}$$

Our goal is to find a matrix g, i.e., a linear transformation $g : R^{2n} \longrightarrow R^{2n}$ which can satisfy (2.2)

$$\alpha_m g^m (z_0) + \cdots + \alpha_1 g (z_0) + \alpha_0 z_0 = \tau a \left(\beta_m g^m (z_0) + \cdots + \beta_1 g (z_0) + \beta_0 z_0\right). \tag{2.3}$$

Such a map g exists for sufficiently small τ and can be represented by continued fractions and rational approximations. We call this transformation step transition operator.

Definition 2.1 If g is a symplectic transformation, then we call its corresponding LMM (2.2) is symplectic. (We simply call this method a SLMM.)

From (2.3), we have

$$\tau a = \frac{\alpha_0 I + \alpha_1 g^1 + \cdots + \alpha_m g^m}{\beta_0 I + \beta_1 g^1 + \cdots + \beta_m g^m}. \tag{2.4}$$

The characteristic equation for LMM is

$$\zeta(\lambda) = \tau\mu\sigma(\lambda), \tag{2.5}$$

where μ is the eigenvalue of the infinitesimal symplectic matrix a and λ is the eigenvalue of g.

Let

$$\psi(\lambda) = \frac{\zeta(\lambda)}{\sigma(\lambda)}, \tag{2.6}$$

then (2.5) can be written as

$$\tau\mu = \psi(\lambda). \tag{2.7}$$

It's reverse function is

$$\lambda = \phi(\tau\mu). \tag{2.8}$$

To study the symplecticity of the LMM, one only needs to study the properties of functions ϕ and ψ. We will see if ϕ is of the exponential form or ψ is of logarithmic form, the corresponding LMM is symplectic. We first study the properties of the exponential functions and logarithmic functions.

Explike and Loglike functions:

First we give out the properties of exponential functions

1°. $\exp(x)|_{x=0} = 1$.

2°. $\frac{d}{dx}\exp(x)|_{x=0} = 1$.

3°. $\exp(x+y) = \exp(x)\cdot\exp(y)$.

If we substitute y by $-x$, we have

$$\exp(x)\exp(-x) = 1. \tag{2.9}$$

Definition 2.2 If function $\phi(x)$ satisfies $\phi(0) = 1, \phi'(0) = 1$ and $\phi(x)\phi(-x) = 1$, we call this function is an *explike* function.

It's well known, the inverse function of an exponential function is a logarithmic function $x \longrightarrow \log(x)$. It has the following properties

1°. $\log(x)|_{x=1} = 0$.

2°. $\frac{d}{dx}\log(x)|_{x=1} = 1$.

3°. $\log(xy) = \log(x) + \log(y)$.

If we take $y = 1/x$, we get

$$\log(x) + \log\left(\frac{1}{x}\right) = 0. \tag{2.10}$$

Definition 2.3 If a function ψ satisfies $\psi(1) = 0, \psi'(1) = 1$ and

$$\psi(x) + \psi\left(\frac{1}{x}\right) = 0, \tag{2.11}$$

we call it a *loglike* function.

Obviously, polynomials can not be explike functions or loglike functions, so we try to find explike and loglike functions in the form of rational functions.

Theorem 2.1 ([3]) *LMM is symplectic for linear Hamiltonian systems iff its step transition operator $g = \phi(\tau a)$ is explike, i.e., $\phi(\mu)\cdot\phi(-\mu) = 1, \phi(0) = 1, \phi'(0) = 1$.*

Theorem 2.2 ([4]) *LMM is symplectic for linear Hamiltonian systems iff $\psi(\lambda) = \frac{\zeta(\lambda)}{\sigma(\lambda)}$ is a loglike function, i.e., $\psi(\lambda) + \psi\left(\frac{1}{\lambda}\right) = 0, \psi(1) = 0, \psi'(1) = 1$.*

Proof. From Theorem 2.1, we have $\phi(\mu)\phi(-\mu) = 1$, so $\lambda = \phi(\mu), \frac{1}{\lambda} = \phi(-\mu)$. The inverse function of ϕ satisfies $\psi(\lambda) = \mu, \psi\left(\frac{1}{\lambda}\right) = -\mu$, i.e., $\psi(\lambda) + \psi\left(\frac{1}{\lambda}\right) = 0, \psi(0) = 1, \psi'(1) = 1$ follows from consistency condition 1°, 2°.

On the other side, if $\psi(\lambda) = -\psi\left(\frac{1}{\lambda}\right)$, let $\psi(\lambda) = \mu$, then its inverse function is $\phi(\mu) = \lambda$ and $\phi(-\mu) = \frac{1}{\lambda}$, we then have $\phi(\mu)\phi(-\mu) = 1$.

Theorem 2.3 *If $\xi(\lambda)$ is an antisymmetric polynomial, $\sigma(\lambda)$ is a symmetric one, then $\psi(\lambda) = \frac{\xi(\lambda)}{\sigma(\lambda)}$ satisfies*

$$\psi(1) = 0, \quad \psi\left(\frac{1}{\lambda}\right) + \psi(\lambda) = 0$$

Proof.

$$\tilde{\xi}(\lambda) = \lambda^m \xi\left(\frac{1}{\lambda}\right) = \sum_{i=0}^{m} \alpha_{m-i}\lambda^i = -\Sigma\alpha_i\lambda^i = -\xi(\lambda)$$

$$\tilde{\sigma}(\lambda) = \lambda^m \sigma\left(\frac{1}{\lambda}\right) = \sum_{i=0}^{m} \beta_{m-i}\lambda^i = \Sigma\beta_i\lambda^i = \sigma(\lambda)$$

$$\psi(\lambda) = \frac{\xi(\lambda)}{\sigma(\lambda)}, \quad \psi\left(\frac{1}{\lambda}\right) = \frac{\xi\left(\frac{1}{\lambda}\right)}{\sigma\left(\frac{1}{\lambda}\right)} = \frac{\lambda^m\xi\left(\frac{1}{\lambda}\right)}{\lambda^m\sigma\left(\frac{1}{\lambda}\right)} = -\frac{\xi(\lambda)}{\sigma(\lambda)}$$

we obtain $\psi(\lambda) + \psi\left(\frac{1}{\lambda}\right) = 0$. Now $\xi(1) = \sum\limits_{k=0}^{m} \alpha_k = 0, \sigma(1) = \sum\limits_{k=0}^{m} \beta_u \neq 0$, then $\psi(1) = \frac{\xi(1)}{\sigma(1)} = 0$.

Corollary 2.1 *If above generating polynomials is consistency with ODE* (1.1), *then* $\psi(\lambda)$ *is loglike function. i.e.* $\psi\left(\frac{1}{\lambda}\right) + \psi(\lambda) = 0, \psi(1) = 0, \psi'(1) = 1$.

Proof. $\psi'(1) = \frac{\xi'\sigma - \sigma'\xi}{\sigma^2} = \frac{\xi'(1)}{\sigma(1)} = 1$. This condition is not others just consistence condition.

Theorem 2.4 *Let* $\psi(\lambda) = \frac{\xi(\lambda)}{\sigma(\lambda)}$ *be irreducible loglike function, then* $\xi(\lambda)$ *is an autisymmetric polynomial while* $\sigma(\lambda)$ *is a symmetric one.*

Proof. We write formally

$$\xi(\lambda) = \alpha_m\lambda^m + \alpha_{m-1}\lambda^{m-1} + \cdots \alpha_1\lambda + \alpha_0$$

$$\sigma(\lambda) = \beta_m\lambda^m + \beta_{m-1}\lambda^{m-1} + \cdots \beta_1\lambda + \beta_0$$

(if $\deg\xi(\lambda) = p < m$, set $\alpha_i = 0$ for $i > p$, if $\deg Q(\lambda) = q < m$, set $\beta_i = 0$ for $i > q$). $\psi(1) = 0 \Longrightarrow \xi(1) = 0$, since otherwise, if $\xi(1) \neq 0$, then $\psi(1) = \frac{\xi(1)}{\sigma(1)} \neq 0$. Now $\xi(1) = 0 \Longleftrightarrow \sigma(1) \neq 0$, since otherwise $\xi(1) = \sigma(1) = 0 \Longrightarrow \xi(\lambda), \sigma(\lambda)$ would have common factor. So we have

$$\xi(1) = \sum_{k=0}^{m} \alpha_k = \sum_{k=0}^{p} \alpha_k = 0,$$

$$\sigma(1) = \sum_{k=0}^{m} \beta_k = \sum_{k=0}^{q} \beta_k \neq 0$$

If $m = \deg\xi = p$, then $\alpha_m = \alpha_p \neq 0$. If $m = \deg\sigma = q$, then $\beta_m = \beta_p \neq 0$

$$\psi\left(\frac{1}{\lambda}\right) = \frac{\xi\left(\frac{1}{\lambda}\right)}{\sigma\left(\frac{1}{\lambda}\right)} = \frac{\lambda^m\xi\left(\frac{1}{\lambda}\right)}{\lambda^m\sigma\left(\frac{1}{\lambda}\right)} = \frac{\tilde{\xi}(\lambda)}{\tilde{\sigma}(\lambda)}$$

Since $\psi(\lambda)+\psi\left(\dfrac{1}{\lambda}\right)=0$, we have

$$\begin{aligned}&\frac{\xi(\lambda)}{\sigma(\lambda)}=-\frac{\tilde{\xi}(\lambda)}{\tilde{\sigma}(\lambda)}\Longleftrightarrow \xi(\lambda)\tilde{\sigma}(\lambda)=-\tilde{\xi}(\lambda)\sigma(\lambda)\\&\Longrightarrow \xi(\lambda)|\tilde{\xi}(\lambda)\sigma(\lambda),\quad \sigma(\lambda)|\tilde{\sigma}(\lambda)\xi(\lambda).\end{aligned}$$

Since $\xi(\lambda),\sigma(\lambda)$ have no common factor, then $\xi(\lambda)|\tilde{\xi}(\lambda),\sigma(\lambda)|\tilde{\sigma}(\lambda)$. If $m=\deg\xi(\lambda)\Longrightarrow$ $\deg\tilde{\xi}\leqslant\deg\xi\Longrightarrow\exists c$ such that

$$\xi(\lambda)=c\tilde{\xi}(\lambda)\Longrightarrow\sigma(\lambda)=-c\tilde{\sigma}(\lambda)$$

since $\alpha_m\neq 0\Longrightarrow\alpha_m\lambda^m+\alpha_{m-1}\lambda^{m-1}+\cdots+\alpha_0=c\left(\alpha_m+\cdots\alpha_0\lambda^m\right)\Longrightarrow\alpha_m=c\alpha_0$, $\alpha_0=c\alpha_m\Longleftrightarrow\alpha_m=c^2\alpha_m$, therefore $c^2=1,c=\pm1$. Suppose $c=+1$, then $\sigma(\lambda)=-\tilde{\sigma}(\lambda)$, $\sum\limits_{r=0}^{m}\beta_k=\sigma(1)=-\tilde{\sigma}(1)=\sigma(1)\Longleftrightarrow\sigma(1)=0$, this leads to a contradiction with the assumption $\sigma(1)\neq0$. Therefore $c=-1$, i.e.

$$\begin{aligned}&\xi(\lambda)=-\tilde{\xi}(\lambda),\quad \alpha_j=-\alpha_{m-j},\quad j=0,1,\cdots,m\\&\sigma(\lambda)=\tilde{\sigma}(\lambda),\quad \beta_j=\beta_{m-j},\quad j=0,1,\cdots,m\end{aligned}$$

The proof for the case $m=\deg\sigma(\lambda)$ proceeds in exactly the same manner as above.

§3 Rational Approximations to Exponential and Logarithmic Functions

1. We first study a simple example, the Leap-frog scheme

$$z_2=z_0+2\tau a z_1.\tag{3.1}$$

Let $z_1=cz_0$, then $z_0=c^{-1}z_1$, insert this equation into (3.1), we get

$$\begin{aligned}&z_2=2\tau a z_1+\frac{1}{c}z_1=\left(2\tau a+\frac{1}{c}\right)z_1=d_1z_1,\quad z_1=\frac{1}{2\tau a+\frac{1}{c}}z_2=\frac{z_2}{d_1},\\&z_3=z_1+2\tau a z_2=\left(2\tau a+\frac{1}{2\tau a+\frac{1}{c}}\right)z_2=d_2z_2,\quad z_2=\frac{1}{2\tau a+\frac{1}{2\tau a+\frac{1}{c}}}z_3,\\&z_4=\left(2\tau a+\frac{1}{2\tau a+\frac{1}{2\tau a+\frac{1}{c}}}\right)=d_4z_3,\quad\cdots\end{aligned}$$

where d_k can be written in the form of continued fractions

$$d_k=2\tau a+\frac{1}{2\tau a+}\frac{1}{2\tau a+}\cdots\frac{1}{+2\tau a}+\cdots,\tag{3.2}$$

and

$$\lim_{k\to\infty} d_k = g = \tau a + \sqrt{1+(\tau a)^2}. \tag{3.3}$$

We assume the transition operator of Leap-frog to be g, from (3.1) we have

$$g^2 - 1 = 2\tau a g,$$

now we have $g = \tau a \pm \sqrt{1+(\tau a)^2}$. Here only sign + is meaningful, thus $g = \tau a + \sqrt{1+(\tau a)^2}$ which is just the limit of continued fraction (3.2). It is easy to verify that g is explike, i.e., $g(\mu)g(-\mu) = 1$. So the Leap-frog scheme is symplectic for linear Hamiltonian systems in the sense our new definition.

2. For the exponential function

$$\exp(z) = 1 + \sum_{k=1}^{\infty} \frac{z^k}{k!}, \tag{3.4}$$

we have Lagrange's continued function

$$\begin{aligned} \exp(z) &= 1 + \frac{z}{1+}\,\frac{-z}{2}+\cdots+\frac{z}{2n-1+}\,\frac{-z}{2}+\cdots \\ &= b_0 + \frac{a_1}{b_1+}\,\frac{a_2}{b_2}+\cdots+\frac{a_{2n-1}}{b_{2n-1}+}\,\frac{a_{2n}}{b_{2n}}+\cdots, \end{aligned} \tag{3.5}$$

where

$$a_1 = z,\quad a_2 = -z,\quad \cdots,\quad a_{2n-1} = z,\quad a_{2n} = -z,\quad n \geqslant 1,$$

$$b_0 = 1,\quad b_1 = 1,\quad b_2 = 2,\quad \cdots,\quad b_{2n-1} = 2n-1,\quad b_{2n} = 2,\quad n \geqslant 1,$$

and Euler's contract expansion

$$\begin{aligned} \exp(z) &= 1 + \frac{2z}{2-z+}\,\frac{z^2}{6}+\cdots+\frac{z^2}{2(2n-1)}+\cdots \\ &= B_0 + \frac{A_1}{B_1+}\,\frac{A_2}{B_2}+\cdots+\frac{A_n}{B_n}+\cdots, \end{aligned} \tag{3.6}$$

where

$$A_1 = 2z,\quad A_2 = z^2,\quad \cdots,\quad A_n = z^2,\quad n \geqslant 2,$$

$$B_0 = 1,\quad B_1 = 2-z,\quad B_2 = 6,\quad \cdots,\quad B_n = 2(2n-1),\quad n \geqslant 2.$$

We have

$$\begin{aligned} &\frac{P_0}{Q_0} = \frac{p_0}{q_0} = 1,\quad \frac{p_1}{q_1} = \frac{1+z}{1},\quad \frac{p_2}{q_2} = \frac{P_1}{Q_1} = \frac{2+z}{2-z},\quad \frac{p_3}{q_3} = \frac{6+4z+z^2}{6-2z} \\ &\frac{p_4}{q_4} = \frac{P_2}{Q_2} = \frac{12+6z+z^2}{12-6z+z^2} + \cdots \end{aligned} \tag{3.7}$$

In general $p_{2n-1}(z)$ is a polynomial of degree n, q_{2n-1} is a polynomial of degree $n-1$, so p_{2n-1}/q_{2n-1} is not explike. While $p_{2n}=P_n(x), q_{2n}=Q_n(z)$ are both polynomials of degree n and from the recursions

$$\begin{aligned}&P_0=1, \quad P=2+z, \quad P_n=z^2P_{n-2}+2(2n-1)P_{n-1},\\&Q_0=1, \quad Q=2-z, \quad Q_n=z^2Q_{n-2}+2(2n-1)Q_{n-1}.\end{aligned} \tag{3.8}$$

It's easy to check that for $n=0,1,\cdots$

$$Q_n(z)=P_n(-z), \quad P_n(0)>0.$$

So the rational function

$$\phi_n(z)=\frac{P_n(z)}{Q_n(z)}=\frac{P_n(z)}{P_n(-z)}$$

is explike and

$$\phi_n(z)-\exp(z)=O\left(|z|^{2n+1}\right),$$

where

$$P_0=1, \quad P_1=2+z, \quad P_n(z)=z^2P_{n-2}(z)+2(2n-1)P_{n-1}(z), \quad n\geqslant 2. \tag{3.9}$$

This is just the diagonal Padé approximation.

3. For the logarithmic function

$$\log(w)=\sum_{k=1}^{\infty}\frac{(w-1)^k}{kw^k}, \tag{3.10}$$

we have the Lagrange's continued fraction

$$\begin{aligned}\log(w)&=\frac{w-1}{1+}\frac{w-1}{2+}\frac{w-1}{3+}\frac{2(w-1)}{2}+\cdots\\&\quad+\frac{(n-1)(w-1)}{2n-1}+\frac{n(w-1)}{2}+\cdots\\&=\frac{a_1}{b_1+}\frac{a_2}{b_2+}\frac{a_3}{b_3+}\frac{a_4}{b_4+}\cdots+\frac{a_{2n-1}}{b_{2n-1}+}\frac{a_{2n}}{b_{2n}+}\cdots,\end{aligned} \tag{3.11}$$

where

$$a_1=w-1, \quad a_2=w-1, \quad a_3=w-1, \quad a_4=2(w-1),\cdots,$$

$$b_0=0, \quad b_1=1, \quad b_2=2, \quad b_3=3, \quad b_4=2,\cdots,$$

and

$$a_{2n-1}=(n-1)(w-1), \quad a_{2n}=n(w-1), \quad n\geqslant 2,$$

$$b_{2n-1}=2n-1, \quad b_{2n}=2, \quad n\geqslant 2,$$

and the Euler's contracted expansion

$$\begin{aligned}\log(w) =& \frac{2(w-1)}{w+1} - \frac{2(w-1)}{6(w+1)} - \frac{(2.2(w-1))^2}{2.5(w+1)} - \cdots \\ & - \frac{(2(n-1)(w-1))^2}{2(2n-1)(w+1)} - \cdots \\ =& \frac{A_1}{B_1 +} \frac{A_2}{B_2 +} \frac{A_3}{B_3 +} \cdots + \frac{A_n}{B_n +} \cdots,\end{aligned} \tag{3.12}$$

where

$$A_1 = 2(w-1), A_2 = -2(w-1), \cdots, A_n = -(2(n-1)(w-1))^2, n \geqslant 3,$$

$$B_0 = 0, B_1 = w+1, B_2 = 6(w+1), \cdots, B_n = 2(2n-1)(w+1), n \geqslant 2.$$

The followings can be get by recursion

$$\begin{aligned}&\frac{P_0}{Q_0} = \frac{p_0}{q_0} = 0, \quad \frac{p_1}{q_1} = w-1, \quad \frac{p_2}{q_2} = \frac{P_1}{Q_1} = \frac{2(w-1)}{w+1}, \\ &\frac{p_3}{q_3} = \frac{w^2+4w-5}{4w+2}, \quad \frac{p_4}{q_4} = \frac{P_2}{Q_2} = \frac{3\left(w^2-1\right)}{w^2+4w+1}, \cdots.\end{aligned} \tag{3.13}$$

In general

$$\frac{p_{2n-1}(w)}{q_{2n-1}(w)} - \log(w) = O\left(|w-1|^{2n}\right), \quad \frac{p_{2n}(w)}{q_{2n}(w)} - \log(w) = O\left(|w-1|^{2n+1}\right).$$

The rational function $\frac{p_{2n-1}(w)}{q_{2n-1}(w)}$ approximates $\log(w)$ only by odd order $2n-1$, it does not reach the even order $2n$, and is not loglike. However

$$R_n = \psi_n(w) = \frac{p_{2n}(w)}{q_{2n}(w)} = \frac{P_n(w)}{Q_n(w)}$$

is a loglike function. In fact, by recursion, it's easy to see that

$$\begin{aligned}P_n(w) &= -w^n P_n\left(\frac{1}{w}\right), \\ Q_n(w) &= w^n Q_n\left(\frac{1}{w}\right),\end{aligned} \tag{3.14}$$

and $\forall n, Q_n(1) \neq 0$. We also have

$$P_0 = 0, P_1(w) = 2(w-1), P_2(w) = 3\left(w^2-1\right),$$

$$Q_0 = 1, Q_1(w) = w+1, Q_2(w) = w^2+4w+1,$$

and for $n \geqslant 3$,

$$\begin{aligned}P_n(w) &= -(2(n-1)(w-1))^2 P_{n-2}(w) + 2(2n-1)(w-1)P_{n-2}(w), \\ Q_n(w) &= -((2n-1)(w-1))^2 Q_{n-2}(w) + 2(2n-1)(w-1)Q_{n-2}(w).\end{aligned} \tag{3.15}$$

So we see $R_1(\lambda)$ is just the Euler midpoint rule and $R_2(\lambda)=\dfrac{3\left(\lambda^2-1\right)}{\lambda^2+4\lambda+1}$ is just the Simpson scheme.

Conclusion: The odd truncation of the continued fraction of the Lagrange's approximation to $\exp(x)$ and $\log(x)$ is neither explike nor loglike, while the even truncation is explike and loglike. The truncation of the continued fraction got from Euler's contracted expansion is explike and loglike.

4. Another famous rational approximation to a given function is the Obreschkoff formula ([8]),

$$
\begin{aligned}
R_{m,n}(x):&\sum_{k=0}^{n}\frac{c_n^k}{c_{m+n}^k k!}\left(x_0-x\right)^k f^{(k)}(x)-\sum_{k=0}^{m}\frac{c_m^k}{c_{m+n}^k k!}\left(x-x_0\right)^k f^{(k)}\left(x_0\right)\\
&=\frac{1}{(m+n)!}\int_{x_0}^{x}(x-t)^m\left(x_0-t\right)^n f^{(m+n+1)}(t)dt.
\end{aligned}\tag{3.16}
$$

1°. Take $f(x)=e^x, x_0=0$, we obtain Padé approximation $\exp(x)\doteq R_{m,n}(x)$. If $m=n$, we obtain Padé diagonal approximation $R_{m,m}(x)$.

2°. Take $f(x)=\log(x), x_0=1$, we obtain $\log(x)\doteq R_{m,n}(x)$. If $m=n$, we obtain loglike function $R_m(x)$,

$$
R_m(\lambda)=\frac{1}{\lambda^m}\sum_{k=1}^{m}\frac{c_m^k}{c_{2m}^k k}(\lambda-1)^k\left(\lambda^{m-k}+(-1)^{k-1}\lambda^m\right),
$$

i.e.,

$$
R_m(\lambda)+R_m\left(\frac{1}{\lambda}\right)=0.
$$

We have

$$
\begin{aligned}
&R_m(\lambda)-\log(\lambda)=O\left(|\lambda|^{2n+1}\right),\\
&R_1=\frac{\lambda^2-1}{2\lambda},\\
&R_2(\lambda)=\frac{1}{12\lambda^2}\left(-\lambda^4+8\lambda^3-8\lambda+1\right),\\
&R_3(\lambda)=\frac{1}{60\lambda^3}\left(\lambda^6-9\lambda^5+45\lambda^4-45\lambda^2+9\lambda-1\right),\\
&\qquad\cdots\cdots
\end{aligned}
$$

where $R_1(\lambda)$ is just the leap-frog scheme.

§4 Nonexistence of SLMM for Nonlinear Hamiltonian Systems

For nonlinear Hamiltonian systems, there exists no symplectic LMM. When equation (1.1) is nonlinear, how to define a symplectic LMM? The answer is to find the step-transition operator $g : R^n \longrightarrow R^n$, let

$$
\begin{aligned}
& z = g^0(z), \\
& z_1 = g(z), \\
& z_2 = g(g(z)) = g \circ g(z) = g^2(z), \\
& \quad \cdots\cdots \\
& z_n = g(g(\cdots(g(z))\cdots) = g \circ g \circ \cdots \circ g \circ z) = g^n(z),
\end{aligned}
\tag{4.1}
$$

we get from (1.2)

$$
\sum_{i=0}^{k} \alpha_i g^i(z) = \tau \sum_{i=0}^{n} \beta_i a \circ g^i(z). \tag{4.2}
$$

It's easy to prove that if LMM(4.2) is consistent with equation (1.1), then for smooth a and sufficiently small step-size τ, the operator g defined by (4.1) exists and it can be represented as a power series in τ and is near identity. Consider the case that equation (1.1) is a Hamiltonian system, i.e., $a(z) = J\nabla H(z)$, we have the following definition.

Definition 4.1 LMM is symplectic if the trsnsition operator g defined by (4.1) is symplectic for all $H(z)$ and all step-size τ, i.e.,

$$
g_*(z)' J g_*(z) = J. \tag{4.3}
$$

This definition is a completely different criterion that can include the symplectic condition for one-step methods in the usual sense. But Tang in [5] has proven that non linear multistep method can satisfy such a strict criterion. Numerical experiments due to Li in [6] shows the explicit 3-level centered method (Leap-frog method) is symplectic for linear Hamiltonian systems $H = \frac{1}{2}(p^2 + 4q^2)$ (See Fig.1 of [6]) but is non-symplectic for nonlinear Hamiltonian system $H = \frac{1}{2}(p^2 + q^2) + \frac{2}{3}q^4$(See Fig 2$(a, b)$ of [6]).

References

[1] Feng K. On difference schemes and symplectic geometry. In Feng K, ed. Proc 1984 Beijing Symp Diff Geometry and Diff Equations. Beijing: Science Press, 1985. 42-58

[2] Feng K, Qin M Z. The symplectic methods for computation of Hamiltonian equations. In Zhu Y L, Guo Ben-yu, ed. Proc Conf on Numerical Methods for PDE's. Berlin: Springer, 1987. 1-37. Lect Notes in Math 1297

[3] Feng K, Wu H M, Qin M Z. Canonical difference schemes for linear Hamiltonian formalism. J Comp Math. 1990 8(4): 371-380

[4] Feng K. On approximation by algebraic functions. Proc of the Annual Meeting of Computational Mathematics in Tianjing, China. 1990

[5] Tang Y F. The symplecticity of multi-step methods. Computers Math Applic. 1993 25: 83-90

[6] Li W Y. Private communications

[7] Ge Z, Feng K. On the approximation of Hamiltonian systems. J Comp Math. 1988 6(1): 88-97

[8] Obreschkof F N. Neue Quadraturformeln. Abhandlungen pröß. Klasse: Akad Wiss, MathNaturwiss. 1940: 1-20

22 The Calculus of Generating Functions and the Formal Energy for Hamiltonian Algorithms①

生成函数的微积运算与哈密尔顿算法的形式能量

Abstract

In [2–4], symplectic schemes of arbitrary order are constructed by generating functions. However the construction of generating functions is dependent on the chosen coordinates. One would like to know that under what circumstance the construction of generating functions will be independent of the coordinates. The generating functions are deeply associated with the conservation laws, so it is important to study their properties and computations. This paper will begin with the study of Darboux transformation, then in section 2, a normalization Darboux transformation will be defined naturally. Every symplectic scheme which is constructed from Darboux transformation and compatible with the Hamiltonian equation will satisfy this normalization condition. In section 3, we will study transformation properties of generator maps and generating functions. Section 4 will be devoted to the study of the relationship between the invariance of generating functions and the generator maps. In section 5, formal symplectic energy of symplectic schemes are presented.

§1 Darboux Transformation

Consider cotangent hundle $T^*R^n \simeq R^{2n}$ with natural symplectic structure

$$J_{2n} = \begin{bmatrix} 0 & I_n \\ -I_n & 0 \end{bmatrix} \tag{1.1}$$

and the product of cotangent bundles $(T^*R^n) \times (T^*R^n) \simeq R^{4n}$ with natural product symplectic structure

$$\tilde{J}_{4n} = \begin{bmatrix} -J_{2n} & 0 \\ 0 & J_{2n} \end{bmatrix}. \tag{1.2}$$

---

① 由秦孟兆整理

Correspondingly, we consider the product space $R^n \times R^n \simeq R^{2n}$. Its cotangent bundle, $T^*(R^n \times R^n) = T^*R^{2n} \simeq R^{4n}$ has natural symplectic structure

$$J_{4n} = \begin{bmatrix} 0 & I_{2n} \\ -I_{2n} & 0 \end{bmatrix}. \tag{1.3}$$

Choose symplectic coordinates $z = (p, q)$ on the symplectic manifold, then for symplectic transformation $g : T^*R^n \longrightarrow T^*R^n$, we have

$$\operatorname{gra}(g) = \left\{ \begin{bmatrix} g_z \\ z \end{bmatrix}, \quad z \in T^*R^n \right\}, \tag{1.4}$$

it is a Lagrangian submanifold of $T^*R^n \times T^*R^n$ in $\tilde{R}^{4n} = \left(R^{4n}, \tilde{J}_{4n}\right)$. Note that on R^{4n} there is a standard symplectic structure (R^{4n}, J_{4n}). A generating map

$$\alpha : T^*R^n \times T^*R^n \longrightarrow T^*(R^n \times R^n)$$

maps the symplectic structure (1.2) to the standard one (1.3). In particular, α maps Lagrangian submanifolds in $\left(R^{4n}, \tilde{J}_{4n}\right)$ to Lagrangian submanifolds L_g in (R^{4n}, J_{4n}). Suppose that α satisfies the transversality condition of g, then

$$L_g = \left\{ \begin{bmatrix} d\phi_g(w) \\ w \end{bmatrix}, w \in R^{2n} \right\}. \tag{1.5}$$

ϕ_g is called generating function of g. We call this generating map α (linear case) or α_*(nonlinear case) Darboux transformation, in other words, we have the following definition.

Definition 1.1 A linear map

$$\alpha = \begin{bmatrix} A_\alpha & B_\alpha \\ C_\alpha & D_\alpha \end{bmatrix}, \tag{1.6}$$

which acts as the followings

$$\begin{bmatrix} z_0 \\ z_1 \end{bmatrix} \in R^{4n} \longmapsto \alpha \begin{bmatrix} z_0 \\ z_1 \end{bmatrix} = \begin{bmatrix} A_\alpha z_0 + B_\alpha z_1 \\ C_\alpha z_0 + D_\alpha z_1 \end{bmatrix} = \begin{bmatrix} w_0 \\ w_1 \end{bmatrix} \in R^{4n}$$

is called a Darboux transformation, if

$$\alpha' J_{4n} \alpha = \tilde{J}_{4n}. \tag{1.7}$$

Denote

$$E_\alpha = C_\alpha + D_\alpha, \quad F_\alpha = A_\alpha + B_\alpha. \tag{1.8}$$

We have

Definition 1.2

$$
\begin{aligned}
&Sp\left(\tilde{J}_{4n}, J_{4n}\right)=\left\{\alpha \in GL(4n) | \alpha' J_{4n}\, \alpha=\tilde{J}_{4n}\right\}=Sp(\tilde{J}, J);\\
&Sp\left(J_{4n}\right)=\left\{\beta \in GL(4n) | \beta' J_{4n}\, \beta=J_{4n}\right\}=Sp(4n);\\
&Sp\left(\tilde{J}_{4n}\right)=\left\{\gamma \in GL(4n) | \gamma' \tilde{J}_{4n} \gamma=\tilde{J}_{4n}\right\}=\widetilde{S}_p(4n).
\end{aligned}
$$

Definition 1.3 A special case of Darboux transformation $\alpha_0=\begin{bmatrix} J_{2n} & -J_{2n} \\ \frac{1}{2} I_{2n} & \frac{1}{2} I_{2n} \end{bmatrix}$ is called Poincare transformation.

Proposition 1.4 *If*

$$
\alpha \in Sp\left(\tilde{J}_{4n}, J_{4n}\right), \quad \beta \in Sp(4n), \quad \gamma \in \widetilde{Sp}(4n),
$$

then $\beta\alpha\gamma \in Sp\left(\tilde{J}_{4n}, J_{4n}\right)$.

Proposition 1.5 $Sp\left(\tilde{J}_{4n}, J_{4n}\right)=\widetilde{Sp}(4n) \alpha_0=\alpha_0 \widetilde{Sp}(4n)$ *and* $Sp\left(\tilde{J}_{4n}, J_{4n}\right)=Sp(4n) \alpha=\alpha \widetilde{Sp}(4n), \forall \alpha \in Sp\left(\tilde{J}_{4n}, J_{4n}\right)$.

Proposition 1.6 *If* $\alpha=\begin{bmatrix} A_\alpha & B_\alpha \\ C_\alpha & D_\alpha \end{bmatrix} \in Sp\left(\tilde{J}_{4n}, J_{4n}\right)$, *then*

$$
\alpha^{-1}=\begin{bmatrix} -J_{2n} C_\alpha' & J_{2n} A_\alpha' \\ J_{2n} D_\alpha' & -J_{2n} B_\alpha' \end{bmatrix}=\begin{bmatrix} A_{\alpha-1} & B_{\alpha^{-1}} \\ C_{\alpha^{-1}} & D_{\alpha^{-1}} \end{bmatrix}.
$$

We have the following well known theorem.

Theorem 1.7 ([2-5]) *If* $\alpha \in Sp\left(\tilde{J}_{4n}, J_{4n}\right)$ *satisfies transversality condition* $|C_\alpha+D_\alpha| \neq 0$, *then for all symplectic diffeomorphisms*, $z \longrightarrow g(z)$ *in* $R^{2n}, g \sim I_{2n}$ (*near identity*), $g_z \in Sp(2n)$ *there exists a generating function*

$$
\phi_{\alpha, g}: R^{2n} \longrightarrow R
$$

such that

$$
A_\alpha g(z)+B_\alpha z=\nabla \phi_{\alpha, g}\left(C_\alpha g(z)+D_\alpha z\right),
$$

i.e.,

$$
\left(A_\alpha \circ g+B_\alpha\right)\left(C_\alpha \circ g+D_\alpha\right)^{-1} z=\nabla \phi_{\alpha, g}(z)
$$

identically in z.

§2 Normalization of Darboux Transformation

Denote $M \equiv Sp\left(\tilde{J}_{4n}, J_{4n}\right)$ a submanifold in $GL(4n), \dim M = \frac{1}{2}4n(4n+1) = 8n^2+2n$. Denote $M^* \equiv \{\alpha \in M||E_\alpha| \neq 0\}$ an open submanifold of $M, \dim M^* = \dim M$. Denote $M' \equiv \{\alpha \in M|E_\alpha = I_n, F_\alpha = 0\} \subset M^* \subset M$.

Definition 2.1 Darboux transformation is called normalization Darboux transformation if $(1) E_\alpha = I_{2n}$ and $(2) F_\alpha = O_{2n}$.

The following theorem answers the question about how to construct a normalization Darboux transformation from a given one.

Theorem 2.2 *$\forall \alpha \in M^*$, there exists*

$$\beta_1 = \begin{bmatrix} I_{2n} & P \\ 0 & I_{2n} \end{bmatrix} \in Sp(4n), \quad |T| \neq 0, \quad \beta_2 = \begin{bmatrix} T'^{-1} & 0 \\ 0 & T \end{bmatrix} \in Sp(4n),$$

such that $\beta_2\beta_1\alpha \in M'$.

Proof. We only need take $P = -F_\alpha E_\alpha^{-1} = -\left(A_\alpha + B_\alpha\right)\left(C_\alpha + D_\alpha\right)^{-1}, T = E_\alpha^{-1}$, then

$$\beta_2 \cdot \beta_1 \cdot \alpha = \begin{bmatrix} E'_\alpha & 0 \\ 0 & E_\alpha^{-1} \end{bmatrix} \begin{bmatrix} I & -F_\alpha E_\alpha^{-1} \\ 0 & I \end{bmatrix} \begin{bmatrix} A_\alpha & B_\alpha \\ C_\alpha & D_\alpha \end{bmatrix} = \begin{bmatrix} A_{\beta_2\beta_1\alpha} & B_{\beta_2\beta_1\alpha} \\ C_{\beta_2\beta_1\alpha} & D_{\beta_2\beta_1\alpha} \end{bmatrix}$$

$$= \alpha_1 = \begin{pmatrix} E'_\alpha\left(A_\alpha - F_\alpha E_\alpha^{-1} C_\alpha\right) & E'_\alpha\left(B_\alpha - F_\alpha E_\alpha^{-1} D_\alpha\right) \\ E_\alpha^{-1} C_\alpha & E_\alpha^{-1} D_\alpha \end{pmatrix}. \tag{2.1}$$

It's easy to calculate that $A_{\beta_2\beta_{1_\alpha}} + B_{\beta_2\beta_1\alpha} = O_{2n}, C_{\beta_2\beta_1\alpha} + D_{\beta_2\beta_1\alpha} = I_{2n}$.

Later, we will assume α to be a normalization Darboux transformation if we do not point out specifically.

Theorem 2.3 *A Darboux transformation can be written in the standard form as*

$$\alpha = \begin{bmatrix} J_{2n} & -J_{2n} \\ \frac{1}{2}(I+V) & \frac{1}{2}(I-V) \end{bmatrix}, \quad V \in sp(2n).$$

It's not difficult to show

$$\forall \alpha_1 \in M \Longrightarrow \exists \beta \in Sp(4n), \quad \beta = \begin{bmatrix} A_\beta & B_\beta \\ C_\beta & D_\beta \end{bmatrix},$$

such that $\alpha_1 = \beta\alpha_0$, where α_0 is Poincare transformation. By computation, we get

$$\alpha_1 = \begin{bmatrix} A_\beta J + \frac{1}{2}B_\beta & -A_\beta J + \frac{1}{2}B_\beta \\ C_\beta J + \frac{1}{2}D_\beta & -C_\beta J + \frac{1}{2}D_\beta \end{bmatrix}.$$

Because $\alpha_1 \in M'$, we have $D_\beta = I_{2n}, B_\beta = 0$, i.e., $\beta = \begin{bmatrix} A_\beta & 0 \\ C_\beta & I_{2n} \end{bmatrix}$. Since $\beta \in Sp(4n)$, we have $\beta = \begin{bmatrix} I_{2n} & 0 \\ Q & I_{2n} \end{bmatrix}, \quad Q \in Sm(2n)$. Thus

$$\alpha_1 = \begin{bmatrix} I_{2n} & 0 \\ Q & I_{2n} \end{bmatrix} \begin{bmatrix} J & -J \\ \frac{1}{2}I & \frac{1}{2}I \end{bmatrix} = \begin{bmatrix} J & -J \\ \frac{1}{2}I + QJ & \frac{1}{2}I - QJ \end{bmatrix}$$
$$= \begin{bmatrix} J & -J \\ \frac{1}{2}(I+V) & \frac{1}{2}(I-V) \end{bmatrix},$$

where $Q' = Q, V = 2QJ$. We shall write

$$\alpha_V = \begin{bmatrix} J_{2n} & -J_{2n} \\ \frac{1}{2}(I+V) & \frac{1}{2}(I-V) \end{bmatrix}, \quad \alpha_V^{-1} = \begin{bmatrix} -\frac{1}{2}(I-V)J & I_{2n} \\ \frac{1}{2}(I+V)J & I_{2n} \end{bmatrix}.$$

Corollary 2.4 *Every* $\alpha = \begin{bmatrix} A_\alpha & B_\alpha \\ C_\alpha & D_\alpha \end{bmatrix} \in M^*$ *has a normalized form* $\alpha_V \in M'$ *with* $V = (C_\alpha + D_\alpha)^{-1}(C_\alpha - D_\alpha) \in sp(2n)$.

This result can be derived from (2.1).

From the following theorem, we can show that the normalization condition is natural.

Theorem 2.5 G^τ *is a consistent difference scheme for* $\dot{z} = J^{-1}H_z, \forall H$, *i.e.,*

(1) $G^\tau(z)|_{\tau=0} = z, \quad \forall z, H$;

(2) $\left.\dfrac{\partial G^\tau(z)}{\partial z}\right|_{\tau=0} = J^{-1}H_z(z) \quad \forall z, H$;

iff generating Darboux transformation is normalized with $A = -J$.

Proof. We take symplectic difference scheme of first order via generating function of type $\alpha = \begin{bmatrix} A & B \\ C & D \end{bmatrix}$. We have

$$AG^\tau(z) + Bz = -\tau H_w\left(CG^\tau(z) + Dz\right). \tag{2.2}$$

At first, we prove the only if part of the theorem. When take $\tau = 0$, we have

$$AG^0(z) + Bz = (A+B)z = 0, \quad \forall z \Longrightarrow A + B = 0,$$
$$A\left.\frac{\partial G^\tau(z)}{\partial \tau}\right|_{\tau=0} = -H_z((C+D)z).$$

Since we know $\left.\left(\dfrac{\partial G^\tau(z)}{\partial \tau}\right)\right|_{\tau=0} = J^{-1}H_z(z)$, then

$$AJ^{-1}H_z(z) = -H_z((C+D)z), \quad \forall H, z.$$

We have

$$H(z) = z'b \Rightarrow AJ^{-1}b = -b, \quad \forall b \Longrightarrow A = -J.$$

On the other hand since $H_z(z) = H_w((C+D)z), \forall H, z$, we have

$$H = \frac{1}{2}z'z \Longrightarrow z = (C+D)z, \quad \forall z \Rightarrow C+D = I.$$

Now we show the if part, take

$$A + B = 0, \quad A = -J, \quad C + D = I \tag{2.3}$$

then

$$A\left(G^{\tau}(z) - z\right) = -\tau H_z\left(CG^{\tau}(z) + Dz\right), \quad A = -J, \tau = 0 \Longrightarrow G^{\tau}(z)|_{\tau=0} = z.$$

On the other hand

$$A\left(\frac{\partial G^{\tau}(z)}{\partial \tau}\right)_{\tau=0} = -H_z((C+D)z) \Longrightarrow \left(\frac{\partial G^{\tau}(z)}{\partial \tau}\right)_{\tau=0} = J^{-1}H_z(z), \quad \forall z, H.$$

Theorem 2.6 *A normalized Darboux transformation with $A = -J$ iff can be written in the standard form*

$$\alpha = \begin{bmatrix} -J & J \\ \frac{1}{2}(I - V) & \frac{1}{2}(I + V) \end{bmatrix}, \quad \forall V \in sp(2n).$$

§3 Transform Properties of Generator Maps and Generating Functions

Let $\alpha = \begin{bmatrix} A_\alpha & B_\alpha \\ C_\alpha & D_\alpha \end{bmatrix} \in Sp\left(\tilde{J}_{4n}, J_{4n}\right)$. Define $E_\alpha = C_\alpha + D_\alpha, F_\alpha = A_\alpha + B_\alpha$. Suppose $g \in Sp\text{-}Diff, g \sim I_{2n}$. In the future, we assume that transversality condition $|E_\alpha| \neq 0$ is satisfied.

Theorem 3.1 *$\forall T \in GL(2n)$, define $\beta_T = \begin{bmatrix} T'^{-1} & 0 \\ 0 & T \end{bmatrix} \in Sp(4n), \beta_T\alpha \in Sp\left(\tilde{J}_{4n}, J_{4n}\right)$, we have*

$$\phi_{\beta_T \circ \alpha, g} \cong \phi_{\alpha, g} \circ T^{-1}. \tag{3.1}$$

Proof. Since

$$\beta_T\alpha = \begin{bmatrix} T'^{-1}A_\alpha & T'^{-1}B_\alpha \\ TC_\alpha & TD_\alpha \end{bmatrix} = \begin{bmatrix} A_{\beta_T\alpha} & B_{\beta_T\alpha} \\ C_{\beta_T\alpha} & D_{\beta_T\alpha} \end{bmatrix},$$

we have

$$A_\alpha g(z)+B_\alpha z=\nabla\phi_{\alpha,g}\circ(C_\alpha g(z)+D_\alpha z) \tag{3.2}$$

identically in z, and

$$T'^{-1}A_\alpha g(z)+T'^{-1}B_\alpha z=\nabla\phi_{\beta_T\alpha,g}\circ(TC_\alpha g(z)+TD_\alpha z)$$

$\Longleftrightarrow$

$$\begin{aligned}A_\alpha g(z)+B_\alpha z&=T'\left(\nabla\phi_{\beta_T\alpha,g}\right)\circ T\left(C_\alpha g(z)+D_\alpha z\right)\\&=\nabla\left(\phi_{\beta_T\alpha,g}\circ T\right)\left(C_\alpha g(z)+D_\alpha z\right)\end{aligned} \tag{3.3}$$

identically in z.

Compare (3.2) with (3.3), we find

$$\nabla\phi_{\alpha,g}\left(C_\alpha g(z)+D_\alpha z\right)=\nabla\left(\phi_{\beta_T\alpha,g}\circ T\right)\left(C_\alpha g(z)+D_\alpha z\right)$$

identically in z. Thus we obtain

$$\phi_{\alpha,g}\cong\phi_{\beta_T\alpha,g}\circ T$$

or

$$\phi_{\alpha,g}\circ T^{-1}\cong\phi_{\beta_T\alpha,g}$$

Theorem 3.2 *$\forall S\in Sp(2n)$, define $\gamma_S=\begin{bmatrix}S&0\\0&S\end{bmatrix}\in\tilde{S}p(4n)$, then we have*

$$\phi_{\alpha\gamma_{S,g}}\cong\phi_{\alpha,S\circ g\circ S^{-1}}. \tag{3.4}$$

Proof. Since

$$\alpha\gamma_S=\begin{bmatrix}A_\alpha S&B_\alpha S\\C_\alpha S&D_\alpha S\end{bmatrix}=\begin{bmatrix}A_{\alpha\gamma_S}&B_{\alpha\gamma_S}\\C_{\alpha\gamma_S}&D_{\alpha\gamma_S}\end{bmatrix}$$

and

$$A_\alpha S\circ g\circ S^{-1}(z)+B_\alpha z=\nabla\phi_{\alpha,S\circ g\circ S^{-1}}\left(C_\alpha S\circ g\circ S^{-1}(z)+D_\alpha z\right)$$

identically in z, because S nonsingular, we may replace z by $S(z)$, so we get

$$A_\alpha S\circ g(z)+B_\alpha Sz=\nabla\phi_{\alpha,S\circ g\circ S^{-1}}\left(C_\alpha Sg(z)+D_\alpha Sz\right),\quad\forall z. \tag{3.5}$$

On the other hand

$$(A_\alpha S)\,g(z)+(B_\alpha S)\,z=\nabla\phi_{\alpha\gamma_S},g\left[(C_\alpha S)\,g(z)+D_\alpha S)\,z\right],\quad\forall z. \tag{3.6}$$

Compare (3.5) with (3.6) and note that

$$|C_\alpha + D_\alpha| \neq 0 \Longleftrightarrow |C_\alpha S + D_\alpha S| \neq 0 \Longleftrightarrow |C_\alpha S g_z(z) + D_\alpha S| \neq 0,$$

we obtain

$$\nabla \phi_{\alpha\gamma s, g} = \nabla \phi_{\alpha, S\circ g \circ S^{-1}},$$

i.e.,

$$\phi_{\alpha\gamma s, g} \cong \phi_{\alpha, S\circ g\circ S^{-1}}.$$

Theorem 3.3 *Take* $\beta = \begin{bmatrix} I_{2n} & P \\ 0 & I_{2n} \end{bmatrix} \in Sp(4n), P \in Sm(2n), \alpha \in Sp\left(\tilde{J}_{4n}, J_{4n}\right)$, *then*

$$\phi_{\beta\alpha, g} \cong \phi_{\alpha, g} + \psi_P \tag{3.7}$$

where $\psi_P = \dfrac{1}{2} w' \ Pw$ *(function independent of* g*).*
Proof.

$$\beta\alpha = \begin{bmatrix} I_{2n} & P \\ 0 & I_{2n} \end{bmatrix} \begin{bmatrix} A_\alpha & B_\alpha \\ C_\alpha & D_\alpha \end{bmatrix} = \begin{bmatrix} A_\alpha + PC_\alpha & B_\alpha + PD_\alpha \\ C_\alpha & D_\alpha \end{bmatrix}.$$

Obviously, $E_{\beta\alpha} = E_\alpha, F_{\beta\alpha} = F_\alpha + PE_\alpha$, so

$$A_{\beta\alpha} g(z) + B_{\beta\alpha} z = \nabla \phi_{\beta\alpha, g} \left(C_{\beta\alpha} g(z) + D_{\beta\alpha} z\right), \tag{3.8}$$

$$A_\alpha g(z) + B_\alpha z + (PC_\alpha g(z) + PD_\alpha z) = \nabla \phi_{\beta\alpha, g} \left(C_\alpha g(z) + D_\alpha z\right). \tag{3.9}$$

On the other hand

$$\nabla \psi_P \left(C_\alpha g(z) + D_\alpha z\right) = P \left(C_\alpha g(z) + D_\alpha z\right). \tag{3.10}$$

Inserting (3.10) into (3.9), we obtain

$$\begin{aligned} A_\alpha g(z) + B_\alpha z &= \nabla \phi_{\beta\alpha, g} \left(C_\alpha g(z) + D_\alpha z\right) - \nabla \psi_P \left(C_\alpha g(z) + D_\alpha z\right) \\ &= \left(\nabla \left(\phi_{\beta\alpha, g} - \psi_P\right)\right) \left(C_\alpha g(z) + D_\alpha z\right). \end{aligned} \tag{3.11}$$

Compare (3.11) and

$$A_\alpha g(z) + B_\alpha z = \nabla \phi_{\alpha, g} \left(C_\alpha g(z) + D_\alpha z\right),$$

we obtain

$$\phi_{\beta\alpha, g} - \psi_P \cong \phi_{\alpha, g}.$$

Analogically, if we take $\beta = \begin{bmatrix} I_{2n} & 0 \\ Q & I_{2n} \end{bmatrix} \in Sp(4n), Q \in Sm(2n)$, we have the following result.

Theorem 3.4

$$\phi_{\alpha,g} + \frac{1}{2} (\nabla_w \phi_{\alpha,g}(w))' Q (\nabla_w \phi_{\alpha,g}(w)) \cong \phi_{\beta\alpha,g} (w + Q\nabla\phi_{\alpha,g}(w)). \tag{3.12}$$

Theorem 3.5

$$\phi_{\begin{bmatrix} A & B \\ C & D \end{bmatrix}, g^{-1}} \cong -\phi_{\begin{bmatrix} -B & -A \\ D & C \end{bmatrix}, g}. \tag{3.13}$$

Proof. Since

$$A_\alpha g^{-1}(z) + B_\alpha z = \nabla\phi_{\alpha,g^{-1}} \left(C_\alpha g^{-1}(z) + D_\alpha z\right),$$

replace z *by* $g(z)$, *we have*

$$A_\alpha z + B_\alpha g(z) = \nabla\phi_{\alpha,g^{-1}} (C_\alpha z + D_\alpha g(z)). \tag{3.14}$$

Comparing

$$-B_\alpha g(z) - A_\alpha z = \nabla\phi_{\begin{bmatrix} -B_\alpha & -A_\alpha \\ D_\alpha & C_\alpha \end{bmatrix}, g} (D_\alpha g(z) + C_\alpha z)$$

with (3.14) *concludes our proof.*

Theorem 3.5′ *If*

$$\phi_{\begin{bmatrix} A & B \\ C & D \end{bmatrix}, g^{-1}} = -\phi_{\begin{bmatrix} A & B \\ C & D \end{bmatrix}, g}, \quad \forall g,$$

then $A + B = 0, C = D$.

Proof. Since by Theorem 3.5 and uniqueness theorem 4.2, we have

$$\begin{bmatrix} A & B \\ C & D \end{bmatrix} = \pm \begin{bmatrix} -B & -A \\ D & C \end{bmatrix}.$$

We only consider the case "+", in this case we have $A + B = 0, C = D$.

Remark For Poincare map $\alpha_0 = \begin{bmatrix} J & -J \\ \frac{1}{2}I & \frac{1}{2}I \end{bmatrix}$, we have

$$\phi_{\alpha_0,g^{-1}} \cong -\phi_{\alpha_0,g}, \quad \forall g \in Sp\text{-}diff.$$

Theorem 3.6 *Let* g_H^t *be the phase flow of Hamiltonian system* $H(z)$, *then the generating function of* g_H^t *under Poincaré map* α_0 *is an odd in* t, *i.e.*,

$$\phi_{\alpha_0,g_H^t}(w,t) = -\phi_{\alpha_0,g_H^t}(w,-t), \quad \forall w \in R^{2n}, t \in R.$$

Proof.

$$g_H^{-t}=\left(g_H^t\right)^{-1},\quad \phi_{\alpha_0,g_H^{-t}}(w,t)=\phi_{\alpha_0,g_H^t}(w,-t)=\phi_{\alpha_0,\left(g_H^t\right)^{-1}}(w,t)=-\phi_{\alpha_0,g_H^t}(w,t).$$

Theorem 3.7 *If* $S\in Sp(2n),\ \alpha\in Sp\left(\tilde{J}_{4n},J_{4n}\right),\gamma_1=\begin{bmatrix}S&0\\0&I\end{bmatrix}$, *then*

$$\alpha\gamma_1=\begin{bmatrix}A_\alpha S&B_\alpha\\C_\alpha S&D_\alpha\end{bmatrix}.$$

Assume $|E_{\alpha\gamma}|=|C_\alpha S+D_\alpha|\neq 0$, we have

$$\phi_{\alpha,S\cdot g}\cong\phi_{\alpha\gamma_1,g},\tag{3.15}$$

i.e.,

$$\phi_{\begin{bmatrix}A&B\\C&D\end{bmatrix},S\cdot g}\cong\phi_{\begin{bmatrix}AS&B\\CS&D\end{bmatrix},g}.$$

Theorem 3.8 *If*

$$\gamma_2=\begin{bmatrix}I&0\\0&S\end{bmatrix},\quad \alpha\in Sp\left(\tilde{J}_{4n},J_{4n}\right),\quad \alpha\gamma_2=\begin{bmatrix}A_\alpha&B_\alpha S\\C_\alpha&D_\alpha S\end{bmatrix},$$

assume $|B_\alpha+D_\alpha S|\neq 0$, *we have*

$$\phi_{\alpha,g\circ S^{-1}}\cong\phi_{\alpha\gamma_2,g},\tag{3.16}$$

i.e.,

$$\phi_{\begin{bmatrix}A&B\\C&D\end{bmatrix},g\circ S^{-1}}\cong\phi_{\begin{bmatrix}A&BS\\C&DS\end{bmatrix},g}.$$

Proof. Since

$$Ag\left(S^{-1}z\right)+Bz=\nabla\phi_{\alpha,g\circ S^{-1}}\left(Cg\left(S^{-1}z\right)+Dz\right),\quad \forall z,$$

replace z by Sz, we get

$$Ag(z)+BSz=\nabla\phi_{\alpha,go S^{-1}}(Cg(z)+DSz)=\nabla\phi_{\begin{bmatrix}A&BS\\C&DS\end{bmatrix},g}(Cg(z)+DSz),$$

$$\phi_{\begin{bmatrix}A&BS\\C&DS\end{bmatrix},g}\cong\phi_{\begin{bmatrix}A&B\\C&D\end{bmatrix},g\circ S^{-1}}.$$

Proof of (3.7) is similar.

Theorem 3.9 *If*

$$\beta=\begin{bmatrix}\lambda I_{2n} & 0\\ 0 & I_{2n}\end{bmatrix}\in CSp(4n),\quad \alpha\in Sp\left(\tilde{J}_{4n},J_{4n}\right),\quad \lambda\neq 0,$$

$$\beta\alpha=\begin{bmatrix}\lambda A & \lambda B\\ C & D\end{bmatrix}\in CSp\left(\tilde{J}_{4n},J_{4n}\right),\quad \mu(\beta\alpha)=\lambda,$$

then we have

$$\phi_{\begin{bmatrix}\lambda A & \lambda B\\ C & D\end{bmatrix}}\cong\lambda\phi_{\begin{bmatrix}A & B\\ C & D\end{bmatrix},g}\tag{3.17}$$

Theorem 3.10 *Suppose*

$$\beta=\begin{bmatrix}I_{2n} & 0\\ 0 & \lambda I_{2n}\end{bmatrix}\in CSp\left(J_{4n}\right),\quad \lambda\neq 0,\quad \alpha\in Sp\left(\tilde{J}_{4n},J_{4n}\right),$$

$$\beta\alpha=\begin{bmatrix}A & B\\ \lambda C & \lambda D\end{bmatrix}\in CSp\left(\tilde{J}_{4n},J_{4n}\right),\quad \mu(\beta\alpha)=\lambda,$$

then we have

$$\phi_{\begin{bmatrix}A & B\\ \lambda C & \lambda D\end{bmatrix},g}\cong\lambda\phi_{\begin{bmatrix}A & B\\ C & D\end{bmatrix},g}\circ\lambda^{-1}I_{2n}\tag{3.18}$$

Proof of (3.9): Since

$$\begin{aligned}
&\alpha\in Sp\left(\tilde{J}_{4n},J_{4n}\right)\Longrightarrow Ag(z)+Bz=\nabla\phi_{\begin{bmatrix}A & B\\ C & D\end{bmatrix},g}(Cg(z)+Dz),\\
&\beta\in CSp\left(\tilde{J}_{4n},J_{4n}\right)\Longrightarrow\lambda Ag(z)+\lambda Bz=\nabla\phi_{\begin{bmatrix}\lambda A & \lambda B\\ C & D\end{bmatrix},g}(Cg(z)+Dz)\\
&\text{L.H.S}=\lambda Ag(z)+\lambda Bz=\lambda\nabla\phi_{\begin{bmatrix}A & B\\ C & D\end{bmatrix},g}(Cg(z)+Dz),\\
&\text{R.H.S}=\nabla\phi_{\begin{bmatrix}\lambda A & \lambda B\\ C & D\end{bmatrix},g}(Cg(z)+Dz),
\end{aligned}$$

then we have

$$\begin{aligned}
&\nabla\phi_{\begin{bmatrix}\lambda A & \lambda B\\ C & D\end{bmatrix},g}(Cg(z)+Dz)=\lambda\nabla\phi_{\begin{bmatrix}A & B\\ C & D\end{bmatrix},g}(Cg(z)+Dz),\\
&\phi_{\begin{bmatrix}\lambda A & \lambda B\\ C & D\end{bmatrix},g}\cong\lambda\phi_{\begin{bmatrix}A & B\\ C & D\end{bmatrix},g}.
\end{aligned}$$

Proof of (3.10): From

$$\begin{bmatrix}A & B\\ \lambda C & \lambda D\end{bmatrix}\in CSp\left(\tilde{J}_{4n},J_{4n}\right)$$

it follows that

$$\begin{aligned}Ag(z)+Bz&=\nabla\phi_{\left[\begin{smallmatrix}A & B\\ \lambda C & \lambda D\end{smallmatrix}\right],g}(\lambda Cg(z)+\lambda Dz).\\ \text{L.H.S}&=\nabla\phi_{\left[\begin{smallmatrix}A & B\\ C & D\end{smallmatrix}\right],g}(Cg(z)+Dz),\\ \text{R.H.S}&=(\phi_{\left[\begin{smallmatrix}A & B\\ \lambda C & \lambda D\end{smallmatrix}\right],g})\circ\lambda I_{2n}(Cg(z)+Dz)\\ &=\lambda^{-1}\nabla(\phi_{\left[\begin{smallmatrix}A & B\\ \lambda C & \lambda D\end{smallmatrix}\right],g}\circ\lambda I_{2n})(Cg(z)+Dz).\end{aligned}$$

Hence

$$\phi_{\left[\begin{smallmatrix}A & B\\ C & D\end{smallmatrix}\right],g}\cong\lambda^{-1}\phi_{\left[\begin{smallmatrix}A & B\\ \lambda C & \lambda D\end{smallmatrix}\right],g}\circ\lambda I_{2n}.$$

At the end of this section, we give out two conclusive theorems which can include the contents of the seven theorems given before. They are easy to prove and the proofs are omitted here.

Let

$$\alpha\in CSp\left(\tilde{J}_{4n},J_{4n}\right),\quad \beta\in CSp\left(J_{4n}\right),\quad \beta=\begin{bmatrix}a & b\\ c & d\end{bmatrix},$$

obviously,

$$\beta\alpha\in CSp\left(\tilde{J}_{4n},J_{4n}\right),\quad \mu(\beta\alpha)=\lambda(\beta)\mu(\alpha).$$

We have the following theorem.

Theorem 3.11

$$\begin{aligned}&\phi_{\beta\alpha,g}\left(c\nabla_w\phi_{\alpha,g}(w)+dw\right)\cong\lambda(\beta)\phi_{\alpha,g}(w)\\ &\quad+\left\{\frac{1}{2}w'\left(d'b\right)w+\left(\nabla_w\phi_{\alpha,g}(w)\right)'\left(c'b\right)w\frac{1}{2}\left(\nabla_w\phi_{\alpha,g}(w)\right)'\left(c'a\right)\left(\nabla_w\phi_{\alpha,g}(w)\right)\right\}.\end{aligned}\tag{3.19}$$

We now formulate the other one. Let $\alpha\in CSp\left(\tilde{J}_{4n},J_{4n}\right),\gamma\in CSp\left(\tilde{J}_{4n}\right)\Longleftrightarrow$ $\gamma'\tilde{J}_{4n}\gamma=v(\gamma)\tilde{J}_{4n}\Longrightarrow\alpha\gamma\in CSp\left(\tilde{J}_{4n},J_{4n}\right),\mu(\alpha\gamma)=\mu(\alpha)v(\gamma),\gamma=\begin{bmatrix}a & b\\ c & d\end{bmatrix}.$

Theorem 3.12

$$\phi_{\alpha\gamma,g}\cong\phi_{\alpha,(a\cdot g+b)\circ(c\cdot g+d)^{-1}}.\tag{3.20}$$

§4 Invariance of Generating Functions and Commutativity of Generator Maps

First we give out the uniqueness theorem of linear fractional transformation.

Theorem 4.1 *Suppose*

$$\alpha = \begin{bmatrix} A_\alpha & B_\alpha \\ C_\alpha & D_\alpha \end{bmatrix}, \bar{\alpha} = \begin{bmatrix} A_{\bar{\alpha}} & B_{\bar{\alpha}} \\ C_{\bar{\alpha}} & D_{\bar{\alpha}} \end{bmatrix} \in Sp\left(\tilde{J}_{4n}, J_{4n}\right), |E_\alpha| \neq 0, |E_{\bar{\alpha}}| \neq 0.$$

If
$(A_\alpha M+B_\alpha)(C_\alpha M+D_\alpha)^{-1} = (A_{\bar{\alpha}} M+B_{\bar{\alpha}})(C_{\bar{\alpha}} M+D_{\bar{\alpha}})^{-1}, \quad \forall M \sim I_{2n}, M \in Sp(2n)$, then $\bar{\alpha} = \pm\alpha$.

Proof. Let

$$N_0 = (A_\alpha I + B_\alpha)(C_\alpha I + D_\alpha)^{-1} = (A_{\bar{\alpha}} I + B_{\bar{\alpha}})(C_{\bar{\alpha}} I + D_{\bar{\alpha}})^{-1}.$$

Suppose $\beta \in Sp(4n)$, first we prove that

$$(A_\beta N + B_\beta)(C_\beta N + D_\beta)^{-1} = N, \quad \forall N \sim N_0, N \in Sm(2n),$$

then $\beta = \pm I_{4n}$.

1° : $(A_\beta N_0 + B_\beta)(C_\beta N_0 + D_\beta)^{-1} = N_0 \Longrightarrow A_\beta N_0 + B_\beta = N_0 C_\beta N_0 + N_0 D_\beta$.

2° : Take $N = N_0 + \varepsilon I \Longrightarrow A_\beta(N_0+\varepsilon I) + B_\beta = (N_0+\varepsilon I) C_\beta (N_0+\varepsilon I) + (N_0+\varepsilon I) D_\beta$.

From

$$1^\circ, 2^\circ \Longrightarrow \varepsilon A_\beta = \varepsilon N_0 C_\beta + \varepsilon C_\beta N_0 + \varepsilon D_\beta + \varepsilon^2 C_\beta, \forall \varepsilon$$
$$\Longrightarrow A_\beta - D_\beta - N_0 C_\beta - C_\beta N_0 = \varepsilon C_\beta \Longrightarrow C_\beta = 0,$$

then $A_\beta = D_\beta$.

From 1°, we have $B = \begin{bmatrix} A_\beta & B_\beta \\ 0 & A_\beta \end{bmatrix}, B_\beta = B'_\beta$. Therefore, from 1° we have

$$A_\beta N A_\beta^{-1} = N - B_\beta A_\beta^{-1}.$$

Subtract this formula by $A_\beta N_0 A_\beta^{-1} = N_0 - B_\beta A_\beta^{-1}$, we get

$$A_\beta (N - N_0) = (N - N_0) A_\beta.$$

Take $N - N_0 = \varepsilon S, S \in Sm(2n) \Longrightarrow A_\beta S = S A_\beta, \forall S \in Sm(2n) \Longrightarrow A_\beta = \lambda I_{2n}$. (This can be proved by mathematical induction).

Then from 1°, $A_\beta N_0 + B_\beta = N_0 A_\beta \Longrightarrow B_\beta = 0$, and

$$\beta = \begin{bmatrix} A_\beta & 0 \\ 0 & A_\beta \end{bmatrix} = \lambda I_{4n} \in Sp(4n) \Longrightarrow \lambda = \pm 1.$$

Let $\beta = \bar{\alpha}\alpha^{-1}$, then the fractional transformation of β leaves all symmetric $N \sim N_0$. Because $\alpha \in Sp(\tilde{J}, J), \alpha^{-1} \in Sp(J, \tilde{J}), \bar{\alpha}\alpha^{-1} \in Sp(J, J) = Sp(4n)$.

We now give out the uniqueness theorem for Darboux transformations.

Theorem 4.2 *Suppose* $\alpha, \bar{\alpha} \in Sp(\tilde{J}, J)$, *then*

$$\phi_{\alpha,g} \cong \phi_{\bar{\alpha},g}, \quad \forall g \in Sp - Diff, \quad g \sim I_{2n} \Longrightarrow \bar{\alpha} = \pm\alpha.$$

Proof. From hypothesis, we have

$$\phi_{\alpha,g} \cong \phi_{\bar{\alpha},g} \Longrightarrow \text{ Hessian of } \phi_{\alpha,g} = (\phi_{\alpha,g})_{ww} = (A_\alpha g(z) + B_\alpha)(C_\alpha g(z) + D_\alpha)^{-1},$$
$$(\phi_{\overline{\alpha},g})_{ww} = (A_{\bar{\alpha}} g(z) + B_{\overline{\alpha}})(C_{\bar{\alpha}} g(z) + D_{\bar{\alpha}})^{-1}, \quad \forall g_z \in Sp(2n) \sim I.$$

Then by uniqueness theorem of linear fractional transformation $\alpha = \pm\bar{\alpha}$. From the proof we know, Hessian of $\phi_{\alpha,g}$ = Hessian of $\phi_{-\alpha,g}, \forall g \in I, \alpha$.

The generating function $\phi_{\alpha,g}$ depends on Darboux transformation α, symplectic diffeomorphism g and coordinates. If we make a symplectic change of coordinates $w \longrightarrow S(z)$, then $\phi(S) \Longrightarrow \phi(S(z))$ while the symplectic diffeomorphism g is represented in z coordinates as $S^{-1} \circ g \circ S$.

For the invariance of generating function $\phi_g(S)$ under S, one would like to expect

$$\phi_{\alpha,S^{-1}\circ g\circ S} = \phi_{\alpha,g} \circ S, \quad \forall g \sim I.$$

This is in general not true. We shall study under what condition this is true for the normalized Darboux transformation α_V. The following theorem answers this question.

Theorem 4.3 *Let*

$$\alpha = \alpha_V = \begin{bmatrix} J_{2n} & -J_{2n} \\ \frac{1}{2}(I+V) & \frac{1}{2}(I-V) \end{bmatrix}, \quad V \in sp(2n), \alpha_V \in M'$$
$$S \in Sp(2n), \quad \beta_S = \begin{bmatrix} S'^{-1} & 0 \\ 0 & S \end{bmatrix} \in Sp(J_{4n}), \quad \gamma_S = \begin{bmatrix} S & 0 \\ 0 & S \end{bmatrix} \in Sp\left(\tilde{J}_{4n}\right).$$

Then the following conditions are equivalent:

(1) $\phi_{\alpha_V, S\circ g\circ S^{-1}} = \phi_{\alpha,g} \circ S^{-1}, \forall g \sim I$.

(2) $\phi_{\alpha_V\gamma_S,g} = \phi_{\beta_S\alpha_V,g}, \forall g \sim I$.

(3) $\alpha_V\gamma_S = \beta_S\alpha_V$.

(4) $SV = VS$.

Proof. (1)$\Longleftrightarrow$ (2) from theorems 3.1 and 3.2. (2) $\Rightarrow$ (3) using the uniqueness theorem on Darboux transformation 4.2. For

$$\alpha_V\gamma_S = \pm\beta_S\alpha_V,$$

since $JS = S'^{-1}J, (-)$ case is excluded. The rest of the proof is trivial.

There is a deep connection between the symmetry of a symplectic difference scheme and the preservation of first integrals.

Let $\mathcal{F}$ be the set of smooth functions defined on R^n.

Theorem 4.4 *Let* $H, F \in \mathcal{F}$,

$$F \circ g_H^t = F \Longleftrightarrow \{F, H\} = 0 \Longleftrightarrow H \circ g_F^t = H \Longleftrightarrow g_H^t = g_F^{-s} \circ g_H^t \circ g_F^s.$$

Theorem 4.5 *Let* $F \in \mathcal{F}, g \in Sp - Diff$,

$$g = g_F^{-t} \circ g \circ g_F^t \;(\text{ or } g_F^t = g^{-1} \circ g_F^t \circ g) \Longleftrightarrow F \circ g = F + C.$$

The "if" part of the proof is obvious. Since

$$\begin{aligned} F \circ g = F + C &\Longrightarrow \nabla F = \nabla F \circ g \Longrightarrow g_F^t = g_{Fog}^t = g^{-1} \circ g_F^t \circ g \\ &= g^{-1} g_F^t(g(z)) \Longleftrightarrow g = g_F^{-t} \circ g \circ g_F^t. \end{aligned}$$

On the other hand, take the derivative of both sides of the following equation by t at $t = 0$,

$$\left.\frac{d}{dt}\right|_{t=0} : \quad g_F^t(z) = g^{-1} g_F^t(g(z))$$

and notice that $g_*(z) \in Sp, g_*^{-1} J^{-1} = J^{-1} g_*'$, we get

$$J^{-1}\nabla F(z) = g_*^{-1}(z) J^{-1} \nabla F(g(z)) = J^{-1} g_*'(z) \nabla F(g(z)).$$

Then we have

$$\nabla F = \nabla(F \circ g) \Longrightarrow F \circ g = F + C.$$

§5 Formal Energy for Hamiltonian Algorithm

Let F^s be an analytic canonical transformation in s, i.e.,

(1) $F^s \in Sp\text{-}Diff$.

(2) $F^0 = id$.

(3) F^s analytic in s for $|s|$ small.

Then there exists a "formal" energy, i.e., a formal power series in s,

$$h^s(z) = h(s, z) = \sum h^i(z)$$

with the following property:

When $h^s(z)$ converges, the phase flow $g_{h^s}^t$, where $h^s(z)$ is considered as a time-independent Hamiltonian with s as a parameter which satisfies "equivalence condition"

$$g_{h^s}^t\Big|_{t=s}=F^s. \tag{5.1}$$

So that $h^s(z)=h^s\left(F^sz\right),\forall z\in R^{2n}$, thus $h^s(z)$ is invariant under F^s (or for those s,z in the domain of convergence of $h^s(z)$).

Let F^s be generated by $\psi(s,w)$ according to normal Darboux transformation α

$$\phi_{F^s,\alpha}(w)=:\psi(s,w)=\sum_{k=1}^{\infty}s^k\psi^{(k)}(w). \tag{5.2}$$

Introduce formal power series

$$h^s(z)=h(s,w)=\sum s^ih^i(w).$$

Assuming convergence, then we associate the phase flow with generating function

$$\begin{aligned}&h^s(z)\longrightarrow\psi_{h^s,\alpha}^t(w):=\chi(t,s,w)=\sum_{k=1}^{\infty}t^k\chi^{(k)}(s,w),\\&\chi^{(1)}(s,w)=-h(s,w).\end{aligned} \tag{5.3}$$

For $k>1$,

$$\begin{aligned}\chi^{(k+1)}(s,w)=&-\sum_{m=1}^{k}\frac{1}{(k+1)m!}\sum_{l_1,\cdots,l_m=1}^{2n}\sum_{k_1+\cdots+k_m=k}h_{w_{l_1},\cdots,w_{l_m}}(s,w)\\&\times\left(A_1\chi_w^{(k_1)}(s,w)\right)_{l_1}\cdots\left(A_1\chi_w^{(k_m)}(s,w)\right)_{l_m}\\=&\sum_{m=1}^{k}\frac{1}{(k+1)m!}\sum_{k_1+\cdots+k_m=k}\chi_{w_{l1},\cdots,w_{lm}}^{(1)}(s,w)\\&\times\left(A_1\chi_w^{(k_1)}(s,w)\right)_{l_1}\cdots\left(A_1\chi_w^{(k_m)}(s,w)\right)_{l_m}.\end{aligned} \tag{5.4}$$

Let $\chi^{(k)}(s,w)=\sum\limits_{i=0}^{\infty}s^i\chi^{(k,i)}(w)$, then $\chi(t,s,w)=\sum\limits_{k=1}^{\infty}\sum\limits_{i=0}^{\infty}t^ks^i\chi^{(k,i)}(w)$. Then

$$\begin{aligned}\sum_{i=0}^{\infty}s^i\chi^{(k+1,i)}(w)=&\sum_{i=0}^{\infty}s^i\sum_{m=1}^{k}\frac{1}{(k+1)m!}\sum_{\substack{i_0+i_1+\cdots+i_m=i\\k_1+\cdots+k_m=k}}\sum_{l_1,\cdots,l_m=1}^{2n}\chi_{w_{l_1},\ldots,w_{l_m}}^{(1,i_0)}(w)\\&\times\left(A_1\chi_w^{(k_1,i_1)}(w)\right)_{l_1}\cdots\left(A_1\chi_w^{(k_m,i_m)}(w)\right)_{l_m}.\end{aligned} \tag{5.5}$$

Thus

$$\begin{aligned}\chi^{(k+1,i)}(w)=&\sum_{m=1}^{k}\frac{1}{(k+1)m!}\sum_{\substack{i_0+i_1+\cdots+i_m=i\\k_1+\cdots+k_m=k}}\sum_{l_1,\cdots,l_m=1}^{2n}\chi_{w_{l_1},\cdots,w_{l_m}}^{(1,i_0)}(w)\\&\times\left(A_1\chi_w^{(k_1,i_1)}(w)\right)_{l_1}\cdots\left(A_1\chi_w^{(k_m,i_m)}(w)\right)_{l_m}.\end{aligned} \tag{5.6}$$

So the coefficient $\chi^{(k+1,i)}$ can be obtained by recursion, if

$$\chi^{(1)}(s,w)=\sum_{0}^{\infty}s^i\chi^{(1,i)}(w)=-h(s,w)=-\sum_{i=0}^{\infty}s^ih^i(w),$$

i.e.,

$$\chi^{(1,i)}=-h^{(i)},\quad i=0,1,2,\cdots. \tag{5.7}$$

Note that $\chi^{(k+1,i)}$ is determined only by the derivatives of $\chi^{(k',i')}, k'\leqslant k, i'\leqslant i$

$$\begin{array}{cccccc}
\chi^{(1,0)} & \chi^{(1,1)} & \chi^{(1,2)} & \cdots & \chi^{(1,i)} & \chi^{(1,i+1)}\\
\cdots & \cdots & \cdots & & \cdots & \cdots\\
\chi^{(k,0)} & \chi^{(k,1)} & \chi^{(k,2)} & \cdots & \chi^{(k,i)} & \chi^{(k,i+1)}\\
\chi^{(k+1,0)} & \chi^{(k+1,1)} & \chi^{(k+1,2)} & \cdots & \chi^{(k+1,i)} & \chi^{(k+1,i+1)}
\end{array} \tag{5.8}$$

The condition (5.1) can be now re-expressed as

$$\chi(t,s,w)|_{t=s}=\chi(s,s,w)=\psi(s,w),$$

i.e.,

$$\begin{aligned}
&\sum_{k=1}^{\infty}s^k\sum_{i=0}^{\infty}s^ix^{(k,i)}(w)=\sum_{k=1}^{\infty}s^k\psi^{(k)}(w),\\
&\sum_{i=0}^{\infty}s^i\sum_{j=0}\chi^{(k-j,j)}(w)=\sum_{i=1}^{\infty}s^i\psi^{(k)}(w),\\
&\sum_{j=0}^{k-1}\chi^{(k-j,j)}(w)=\psi^{(k)},\quad k=1,2,\cdots,\\
&\chi^{(1,0)}=\psi^{(1)},\\
&\sum_{i=0}^{k}\chi^{(k+1-i,i)}=\psi^{(k+1)},\quad k=0,1,2,\cdots,
\end{aligned} \tag{5.9}$$

so

$$\begin{array}{ccccccc}
 & -h^{(0)} & -h^{(1)} & -h^{(2)} & \cdots & -h^{(k-1)} & -h^{(k)}\\
\psi^{(1)} & \chi^{(1,0)} & \chi^{(1,1)} & \chi^{(1,2)} & \cdots & \chi^{(1,k-1)} & \chi^{(1,k)}\\
\psi^{(2)} & \chi^{(2,0)} & \chi^{(2,1)} & \chi^{(2,2)} & \cdots & \chi^{(2,k-1)} & \\
\cdots\cdots & & & & & & \\
\psi^{(k)} & \chi^{(k,0)} & \chi^{(k,1)} & & & & \\
\psi^{(k+1)} & \chi^{(k+1,0)} & & & & &
\end{array} \tag{5.10}$$

and

$$\begin{aligned}
&\chi^{(1,0)} = \psi^{(1)}, \\
&\chi^{(2,0)} + \chi^{(1,1)} = \psi^{(2)}, \\
&\chi^{(3,0)} + \chi^{(2,1)} + \chi^{(1,2)} = \psi^{(3)}, \\
&\cdots\cdots \\
&\chi^{(k+1,0)} + \chi^{(k,1)} + \cdots + \chi^{(2,k-1)} + \chi^{(1,k)} = \psi^{(k+1)}.
\end{aligned} \tag{5.11}$$

Now consider $\psi^{(1)}, \psi^{(2)}, \cdots, \psi^{(k)}, \psi^{(k+1)}, \cdots$ as known, then

$$h^{(0)} = -\chi^{(1,0)}, h^{(1)} = -\chi^{(1,1)}, \cdots, h^{(k-1)} = \chi^{(1,k-1)}, h^{(k)} = \chi^{(1,k)}, \cdots$$

have to be determined. We get

$$\begin{aligned}
&h^{(0)} = -\psi^{(1)}, \\
&h^{(1)} = -\psi^{(2)} + \chi^{(2,0)}, \\
&h^{(2)} = -\psi^{(3)} + \left(\chi^{(3,0)} + \chi^{(2,1)}\right), \\
&\cdots\cdots \\
&h^{(k)} = -\psi^{(k+1)} + \left(\chi^{(k+1,0)} + \chi^{(k,1)} + \cdots + \chi^{(2,k-1)}\right), \\
&\cdots\cdots
\end{aligned} \tag{5.12}$$

So $h^{(0)}, h^{(1)}, h^{(2)}, \cdots$ can be recursively determined by $\psi^{(1)}, \psi^{(2)}, \cdots$ So we get the formal power series $h^s = \sum\limits_{i=0}^{\infty} s^i h^{(i)}(z)$, and in case of convergence, satisfies

$$g_{h^s}^t\big|_{t=s} = F^s.$$

Now we give out a special example to show how to calculate the formal energy. Take normal Darboux transformation with

$$V = -E = \begin{bmatrix} 0 & -1 \\ -1 & 0 \end{bmatrix}, \quad \alpha_V^{-1} = \begin{bmatrix} A_1 & B_1 \\ C_1 & D_1 \end{bmatrix},$$

where

$$A_1 = \frac{1}{2}(JVJ - J) = \begin{bmatrix} 0 & 0 \\ -I & 0 \end{bmatrix}.$$

Suppose we just take the first term of the generating function of the generating map α_V, i.e., we just consider the first order scheme

$$F^s \sim \psi(s, w) = -sH(w) = \sum_1^{\infty} s^k \psi^{(k)}.$$

Here $\psi^{(1)}=-H(w), \psi^{(2)}=\psi^{(3)}=\cdots=0$. Obviously $\chi^{(1,0)}=\psi^{(1)}=-H$.

$$\chi_z^{(1,0)}=-\begin{pmatrix}H_p\\H_q\end{pmatrix},\quad A_1\chi_z^{(1,0)}=\begin{pmatrix}0&0\\-1&0\end{pmatrix}\begin{pmatrix}-H_p\\-H_q\end{pmatrix}=\begin{pmatrix}0\\H_p\end{pmatrix},$$

$$\chi_{zz}^{(1,0)}=-\begin{pmatrix}H_{pp}&H_{pq}\\H_{qp}&H_{qq}\end{pmatrix}.$$

Calculate by formula (5.6), we get

$$\begin{aligned}\chi^{(2,0)}&=\frac{1}{2(1!)}\sum_{i_0+i_1=0}\sum_{k_1=1}\sum_{l_1=1}^{2n}\chi_{zl_1}^{(1,i_0)}\left(A_1\chi_z^{(1,i_1)}\right)_{l_1}=\frac{1}{2}\left(\chi_z^{(1,0)}\right)'A_1\chi_z^{(1,0)}\\&=-\frac{1}{2}\begin{pmatrix}H_p\\H_q\end{pmatrix}'\begin{pmatrix}0\\H_p\end{pmatrix}=-\frac{1}{2}H_q'H_p.\end{aligned}$$

From formula (5.10), we get

$$\chi^{(2,0)}+\chi^{(1,1)}=\psi^{(2)}=0\Longrightarrow\chi^{(1,1)}=-\chi^{(2,0)}=\frac{1}{2}H_q'H_p,$$

$$\chi_z^{(1,1)}=\frac{1}{2}\begin{pmatrix}\dfrac{\partial}{\partial p}\sum\limits_{j=1}^{n}H_{q_j}H_{p_j}\\\dfrac{\partial}{\partial q}\sum\limits_{j=1}^{n}H_{q_j}H_{p_j}\end{pmatrix}=\frac{1}{2}\begin{pmatrix}H_{pq}H_p+H_{pp}H_q\\H_{qq}H_p+H_{qp}H_q\end{pmatrix}.$$

$$A_1\chi_z^{(1,1)}=\begin{pmatrix}0&0\\-1&0\end{pmatrix}\chi_z^{(1,1)}=-\frac{1}{2}\begin{pmatrix}0\\H_{pq}H_p+H_{pp}H_q\end{pmatrix},$$

$$A_1\chi_z^{(2,0)}=-A_1\chi_z^{(1,1)}=\frac{1}{2}\begin{pmatrix}0\\H_{pq}H_p+H_{pp}H_q\end{pmatrix}.$$

For $k=2, i=0$, we have

$$\begin{aligned}x^{(3,0)}=&\frac{1}{3}\left(\frac{1}{1!}\sum_{i_0+i_1=0}\sum_{k_1=2}\sum_{l_1=1}^{2n}\chi_{z_{l_1}}^{(1,0)}\left(A_1\chi_z^{(2,0)}\right)_{l_1}\right.\\&\left.+\sum_{l_1l_2=1}^{2n}\sum_{i_0+i_1+i_2=0}\sum_{k_1+k_2=2}\chi_{z_{l_1},z_{l_2}}^{(1,0)}\left(A_1\chi_z^{(k_1,0)}\right)_{l_1}\left(A_1\chi_z^{(k_2,0)}\right)_{l_2}\right)\\=&\frac{1}{3}\left(\chi_z^{(1,0)}\right)'A_1\chi_z^{(2,0)}+\frac{1}{6}\left(A_1\chi_z^{(1,0)}\right)'\chi_{zz}^{(1,0)}A_1\chi_z^{(1,0)}\\=&-\frac{1}{6}\left(H_q'H_{pq}H_p+H_q'H_{pp}H_q+H_p'H_{qq}H_p\right).\end{aligned}$$

For $k=1, i=1$, we have

$$\begin{aligned}
\chi^{(2,1)} &= \frac{1}{2}\sum_{i_0+i_1=1}\sum_{k_1=1}\sum_{l_1=1}^{2n}\chi_{z_{l_1}}^{(1,i_0)}\left(A_1\chi_z^{(1,i_1)}\right)_{l_1} \\
&= \frac{1}{2}\left\{\left(\chi_z^{(1,0)}\right)' A_1\chi_z^{(1,1)} + \left(\chi_z^{(1,1)}\right)' A_1\chi_z^{(1,0)}\right\} \\
&= \frac{1}{4}\left(H_q' H_{pp} H_q + H_p' H_{qq} H_p\right) + \frac{1}{2} H_q' H_{pq} H_p.
\end{aligned}$$

From (5.11), we have

$$\chi^{(3,0)} + \chi^{(2,1)} + \chi^{(1,2)} = \psi^{(3)} = 0 \Longrightarrow \chi^{(1,2)} = -\left(\chi^{(3,0)} + \chi^{(2,1)}\right)$$

$$\begin{aligned}
\chi^{(1,2)} &= -\left\{-\frac{1}{6}\left(H_q' H_{pp} H_q + H_p' H_{qq} H_p\right) - \frac{1}{6} H_q' H_{pq} H_p\right. \\
&\quad \left. + \frac{1}{4}\left(H_q' H_{pp} H_q + H_p' H_{qq} H_p\right) + \frac{1}{2} H_q' H_{pq} H_p\right\} \\
&= -\frac{1}{12}\left(H_q' H_{pp} H_q + H_p' H_{qq} H_p + 4 H_q' H_{pq} H_p\right).
\end{aligned}$$

Finally we get formal power series of energy

$$\begin{aligned}
h(s,z) &= -\left(\chi^{(1,0)} + s\chi^{(1,1)} + s^2\chi^{(1,2)}\right) + O\left(s^3\right) \\
&= H(z) - \frac{s}{2} H_q' H_p + \frac{s^2}{2}\left(H_q' H_{pp} H_q + H_p' H_{qq} H_p + 4 H_q' H_{pq} H_q\right) + O\left(s^3\right).
\end{aligned}$$

Now let $H(z)$ be a time-independent Hamiltonian, its phase flow is g_H^t, its generating function is

$$\phi_{g_H^t}(w) = \phi(t,w) = \sum_{k=1}^{\infty} t^k \phi^{(k)}(w).$$

We have

$$\begin{aligned}
\phi^{(1)}(w) &= -H(w), \\
\text{For } k \geqslant 1, \quad \phi^{(k+1)}(w) &= \sum_{m=1}^{k}\frac{1}{(k+1)m!}\sum_{l_1,\cdots,l_m=1}^{2n}\sum_{k_1+\cdots+k_m=k}\phi_{w_{l_1}\cdots w_{l_m}}^{(1)}(w) \\
&\quad \times (A_1\phi_w^{(k_1)}(w))_{l_1}\cdots(A_1\phi_w^{(k_m)}(w))_{l_m}).
\end{aligned} \tag{5.13}$$

Theorem 5.1 *Suppose F^s is the Sp-Diff operator of order m for Hamiltonian H, i.e., $\phi(s,w) - \psi(s,w) = O\left(|s|^{m+1}\right)$, i.e.,*

$$\begin{cases}
\psi^{(1)}(w) = \phi^{(1)}(w) = -H(w), \\
\psi^{(2)}(w) = \phi^{(2)}(w), \\
\cdots\cdots \\
\psi^{(m)}(w) = \phi^{(m)}(w).
\end{cases}$$

Then

$$h^{(0)}(w)=H(w),h^{(1)}(w)=h^{(2)}(w)=\cdots=h^{(m-1)}(w)=0,$$

i.e.,

$$h(s,w)-H(w)=o\left(|s|^{m}\right)$$

and

$$h^{(m)}(w)=\psi^{(m+1)}(w)-\phi^{(m+1)}(w).$$

First we show that $\chi^{(k+1,i)}$ dependent only on derivatives of $\chi^{\left(k',i'\right)},k'\leqslant k,i'\leqslant i$. The recursion for $i=0$ is the same for the recursion of phase flow generating function with Hamiltonian $\chi^{(1,0)}(w)$. For $i\geqslant 1,\chi^{(k+1,i)}=0$ if $\chi^{\left(k',i'\right)}=0$ for all i',k' such that $1\leqslant i'\leqslant i$, $1\leqslant k'\leqslant k$. We have

$$\begin{aligned}
\psi^{(1)}&=\chi^{(1,0)}\Longrightarrow\chi^{(1,0)}\\
&\overset{\text{recursion}}{\longrightarrow}\chi^{(2,0)},\chi^{(3,0)},\chi^{(4,0)},\cdots,\\
\psi^{(2)}&=\chi^{(1,1)}+\chi^{(2,0)}\Longrightarrow\chi^{(1,1)}\\
&\overset{\text{recursion}}{\longrightarrow}\chi^{(2,1)},\chi^{(3,1)},\chi^{(4,1)},\cdots,\\
\psi^{(3)}&=\chi^{(1,2)}+\chi^{(2,1)}+\chi^{(3,0)}\Longrightarrow\chi^{(1,2)}\\
&\overset{\text{recursion}}{\longrightarrow}\chi^{(2,2)},\chi^{(3,2)},\chi^{(4,2)},\cdots,\\
\psi^{(4)}&=\chi^{(1,3)}+\chi^{(2,2)}+\chi^{(3,1)}+\chi^{(4,0)}\Longrightarrow\chi^{(1,3)}\\
&\overset{\text{recursion}}{\longrightarrow}\chi^{(2,3)},\chi^{(3,3)},\chi^{(4,3)},\cdots,\\
&\cdots\cdots\\
\psi^{(k)}&=\chi^{(1,k-1)}+\chi^{(2,k-2)}+\cdots+\chi^{(k,0)}\Longrightarrow\chi^{(1,k-1)}\\
&\overset{\text{recursion}}{\longrightarrow}\chi^{(2,k-1)},\chi^{(3,k-1)},\chi^{(4,k-1)},\cdots.
\end{aligned}\tag{5.14}$$

So $\chi^{(k,i)}$ can be generated successively through (5.9), (5.6). Then

$$h(s,w)=\sum_{i=0}^{\infty}s^{i}\chi^{(1,i)}(w).$$

Using equation $H=\psi^{(1)}=\psi^{(1)}=\chi^{(1,0)}$ and (5.9), (5.14), we get

$$\chi^{(2,0)}=\phi^{(2)},\chi^{(3,0)}=\phi^{(3)},\cdots,\chi^{(k,0)}=\psi^{(k)},\cdots.$$

Using equation (5.14), we get

$$\psi^{(2)}=\phi^{(2)}=\chi^{(1,1)}+\phi^{(2)}\Longrightarrow\chi^{(1,1)}=0.$$

Applying equation (5.9), (5.14), we get

$$\chi^{(2,1)} = 0 \Longrightarrow \chi^{(3,1)} = \chi^{(4,1)} \doteq \cdots = \chi^{(k,1)} = \cdots = 0.$$

Applying equation

$$\psi^{(3)} = \phi^{(3)} = \chi^{(1,2)} + \chi^{(2,1)} + \chi^{(3,1)} = \chi^{(1,2)} + 0 + \phi^{(3)} \Longrightarrow \chi^{(1,2)} = 0,$$

then

$$\chi^{(2,2)} = \chi^{(3,2)} = \chi^{(4,2)} = \cdots = \chi^{(k,2)} = \cdots = 0.$$

Finally

$$\psi^{(m)} = \phi^{(m)} = \chi^{(1,m-1)} + \chi^{(2,m-2)} + \cdots + \chi^{(m-1,1)} + \phi^{(m)} \Longrightarrow \chi^{(1,m-1)} = 0,$$

then

$$\chi^{(2,m-2)} = \chi^{(3,m-2)} = \chi^{(4,m-2)} = \cdots = \chi^{(k,m-2)} = \cdots = 0.$$

So $\chi^{(k,i)} = 0$ for $i = 1, 2, \cdots, m-1$ and $k = 1, 2, 3 \cdots$. Then equation

$$\psi^{(m+1)} = \chi^{(1,m)} + \chi^{(2,m-1)} + \cdots + \chi^{(m,1)} + \chi^{(m+1,0)} \Longrightarrow \chi^{(1,m)} = \psi^{(m+1)} - \phi^{(m+1)},$$

so we get finally

$$h(s,z) = \sum_{i=0}^{\infty} s^i \chi^{(1,i)} = H(z) + s^m \left(\psi^{(m+1)} - \phi^{(m+1)}\right) + O\left(|s|^{m+1}\right),$$

i.e.,

$$h(s,z) - H(z) = s^m \left(\psi^{(m+1)}(z) - \phi^{(m+1)}(z)\right) + O\left(|s|^{m+1}\right).$$

So in particular, if $F^s \sim \psi(s,w)$ is given by the truncation of phase flow generating function, i.e.,

$$\psi^{(1)} = \phi^{(1)} = H, \quad \psi^{(2)} = \phi^{(2)}, \cdots, \psi^{(m)} = \phi^{(m)}, \quad \psi^{(m+1)} = \phi^{(m+2)} = 0,$$

then

$$h(s,z) = H(z) - s^m \phi^{(m+1)}(z) + O\left(|s|^{m+1}\right).$$

References

[1] Feng K. On difference schemes and symplectic geometry. In Feng K, ed. Proc 1984 Beijing Symp Diff Geometry and Diff Equations. Beijing: Science Press, 1985: 42-58

[2] Feng K. Difference schemes for Hamiltonian formalism and symplectic geometry. J Comp Math. 1986 4(3): 279-289

[3] Feng K, Qin M Z. The symplectic methods for computation of Hamiltonian equations. In Zhu Y L, Guo Ben-yu, ed. Proc Conf on Numerical Methods for PDE's. Berlin: Springer, 1987: 1-37. Lect Notes in Math 1297

[4] Feng K, Wu H M, Qin M Z, Wang D L. Construction of canonical difference schemes for Hamiltonian formalism via generating functions. J Comp Math. 1989 7(1): 71-96

[5] Feng K. The Hamiltonian way for computing Hamiltonian dynamics. In Spigler R, ed. Applied and Industrial Mathematics. Netherlands: Kluwer, 1991: 17-35

[6] Ge Z, Feng K. On the approximation of Hamiltonian systems. J Comp Math. 1988 6(1): 88-97

[7] Feng K, Wang D L. A note on conservation laws of symplectic difference schemes for Hamiltonian systems. J Comp Math. 1991 9(3): 229-237

23 Contact Algorithms for Contact Dynamical Systems[①]

切触动力系统的切触算法

§1 Introduction

Contact structure is an analog of a symplectic one for odd-dimensional manifolds and stems from manifolds of contact elements of configuration spaces in mechanics and, therefore, it is also of basic importance in physical and engineering sciences. We apply, in this paper, the ideas of preserving Lie group and Lie algebra structure of dynamical systems in constructing symplectic algorithms for Hamiltonian systems to the study of numerical algorithms for contact dynamical systems and present so-called contact algorithms, i.e., algorithms preserving contact structures, for solving numerically contact systems.

A contact structure on a manifold is defined as a nondegenerate field of tangent hyperplanes and, therefore, it is determined by a differential 1-form, uniquely up to an everywhere non-vanishing multiplier function, such that the zero set of the 1-form at a point on the manifold is the tangent hyperplane of the field at the point. So, contact structures occur only on manifolds of odd-dimensions. In this paper, we simply consider the Euclidean space $\mathbf{R}^{2n+1}$ of $2n+1$ dimensions as our basic manifold with the contact structure given by the normal form

$$\alpha = \sum_{i=1}^{n} x_i dy_i + dz =: xdy + dz = \left(0, x^T, 1\right) \begin{pmatrix} dx \\ dy \\ dz \end{pmatrix}, \tag{1.1}$$

here we have used 3-symbol notation to denote the coordinates and vectors on $\mathbf{R}^{2n+1}$

$$x = (x_1, \cdots, x_n)^T, \quad y = (y_1, \cdots, y_n)^T, \quad z = (z). \tag{1.2}$$

A contact dynamical system on $\mathbf{R}^{2n+1}$ is governed by a contact vector field $f =$

① 由尚在久整理

$\left(a^T, b^T, c^T\right) : \mathbf{R}^{2n+1} \to \mathbf{R}^{2n+1}$ through equations

$$\dot{x} = a(x, y, z), \quad \dot{y} = b(x, y, z), \quad \dot{z} = c(x, y, z), \qquad \cdot =: \frac{d}{dt}, \tag{1.3}$$

where the contactivity condition of the vector field f is

$$L_f \alpha = \lambda_f \alpha \tag{1.4}$$

with some function $\lambda_f : \mathbf{R}^{2n+1} \to \mathbf{R}$, called the multiplier of f. In (1.4), $L_f \alpha$ denotes the Lie derivative of α with respect to f and is usually calculated by the formula (see [9])

$$L_f \alpha = i_f d\alpha + d i_f \alpha. \tag{1.5}$$

It is easy to show from (1.4) and (1.5) that to any contact vector field f on $\mathbf{R}^{2n+1}$, there corresponds a function $K(x, y, z)$, called contact Hamiltonian, such that

$$a = -K_y + K_z x, \quad b = K_x, \quad c = K - x^T K_x =: K_e. \tag{1.6}$$

In fact, (1.6) represents the general form of a contact vector field. Its multiplier, denoted as λ_K from now, is equal to K_z.

A contact transformation g is a diffeomorphism on $\mathbf{R}^{2n+1}$

$$g : \begin{pmatrix} x \\ y \\ z \end{pmatrix} \to \begin{pmatrix} \hat{x}(x, y, z) \\ \hat{y}(x, y, z) \\ \hat{z}(x, y, z) \end{pmatrix}$$

conformally preserving the contact structure, i.e., $g^* \alpha = \mu_g \alpha$, that means

$$\sum_{i=1}^{n} \hat{x}_i d\hat{y}_i + d\hat{z} = \mu_g \left(\sum_{i=1}^{n} x_i dy_i + dz \right) \tag{1.7}$$

for some everywhere non-vanishing function $\mu_g : \mathbf{R}^{2n+1} \to \mathbf{R}$, called the multiplier of g. The explicit expression of (1.7) is

$$\left(0, \hat{x}^T, 1\right) \begin{pmatrix} \hat{x}_x & \hat{x}_y & \hat{x}_z \\ \hat{y}_x & \hat{y}_y & \hat{y}_z \\ \hat{z}_x & \hat{z}_y & \hat{z}_z \end{pmatrix} = \mu_g \left(0, x^T, 1\right).$$

A fundamental fact is that the phase flow g_K^t of a contact dynamical system associated with contact Hamiltonian $K : \mathbf{R}^{2n+1} \to \mathbf{R}$ is a one parameter (local) group of contact transformations on $\mathbf{R}^{2n+1}$, i.e., g_K^t satisfies

$$g_K^0 = \text{identity map on } \mathbf{R}^{2n+1}; \tag{1.8}$$

$$g_K^{t+s} = g_K^t \circ g_K^s, \quad \forall t, s \in \mathbf{R}; \tag{1.9}$$

$$\left(g_K^t\right)^* \alpha = \mu_{g_K^t} \alpha \tag{1.10}$$

for some everywhere non-vanishing function $\mu_{g_K^t}: \mathbf{R}^{2n+1} \to \mathbf{R}$. Moreover, we have the following relation between $\mu_{g_K^t}$ and the Hamiltonian K :

$$\mu_{g_K^t} = \exp \int_0^t (K_z \circ g_K^s)\, ds. \tag{1.11}$$

For general contact systems, condition (1.9) is stringent for algorithmic approximations to phase flows because only the phase flows themselves satisfy it. We will construct algorithms for contact systems such that the corresponding algorithmic approximations to the phase flows satisfy the condition (1.10), of course probably with different, but everywhere non-vanishing, multipliers from $\mu_{g_K^t}$. We call such algorithms contact ones.

§2　Contactization and Symplectization

There is a well known correspondence between contact geometry on $\mathbf{R}^{2n+1}$ and conic (or homogeneous) symplectic geometry on $\mathbf{R}^{2n+2}$. To establish this correspondence, we introduce two spaces $\mathbf{R}_+^{2n+2}$ and $\mathbf{R}_+ \times \mathbf{R}^{2n+1}$.

a. We use the 4-symbol notation for the coordinates on $\mathbf{R}^{2n+2}$

$$\begin{pmatrix} p_0 \\ p_1 \\ q_0 \\ q_1 \end{pmatrix}, \quad p_0 = (p_0), \quad q_0 = (q_0), \quad p_1 = \begin{pmatrix} p_{11} \\ \vdots \\ p_{1n} \end{pmatrix}, \quad q_1 = \begin{pmatrix} q_{11} \\ \vdots \\ q_{1n} \end{pmatrix}. \tag{2.1}$$

Consider

$$\mathbf{R}_+^{2n+2} = \left\{ (p_0, p_1, q_0, q_1) \in \mathbf{R}^{2n+2} | p_0 > 0 \right\} \tag{2.2}$$

as a conic symplectic space with the standard symplectic form

$$\omega = dp_0 \wedge dq_0 + dp_1 \wedge dq_1. \tag{2.3}$$

Definition 1　Function $\phi: \mathbf{R}_+^{2n+2} \to \mathbf{R}$ is called a conic function if it satisfies

$$\phi(p_0, p_1, q_0, q_1) = p_0 \phi\left(1, \frac{p_1}{p_0}, q_0, q_1\right), \quad \forall p_0 > 0. \tag{2.4}$$

So, a conic function on $\mathbf{R}^{2n+2}$ depends essentially only on $2n+1$ variables.

Definition 2　A map $F: \mathbf{R}_+^{2n+2} \to \mathbf{R}_+^{2n+2}$ is called a conic map if

$$F \circ T_\lambda = T_\lambda \circ F, \quad \forall \lambda > 0, \tag{2.5}$$

where T_λ is the linear transformation on $\mathbf{R}^{2n+2}$

$$T_\lambda \begin{pmatrix} p \\ q \end{pmatrix} = \begin{pmatrix} \lambda p \\ q \end{pmatrix}, \quad p = \begin{pmatrix} p_0 \\ p_1 \end{pmatrix}, \quad q = \begin{pmatrix} q_0 \\ q_1 \end{pmatrix}. \tag{2.6}$$

The conic condition (2.5) for the map $F:(p_0,p_1,q_0,q_1)\rightarrow(P_0,P_1,Q_0,Q_1)$ can be expressed as follows

$$\begin{aligned}
P_0\left(p_0,p_1,q_0,q_1\right)&=p_0P_0\left(1,\frac{p_1}{p_0},q_0,q_1\right)>0,\quad \forall p_0>0,\\
P_1\left(p_0,p_1,q_0,q_1\right)&=p_0P_1\left(1,\frac{p_1}{p_0},q_0,q_1\right),\\
Q_0\left(p_0,p_1,q_0,q_1\right)&=Q_0\left(1,\frac{p_1}{p_0},q_0,q_1\right),\\
Q_1\left(p_0,p_1,q_0,q_1\right)&=Q_1\left(1,\frac{p_1}{p_0},q_0,q_1\right).
\end{aligned}\tag{2.7}$$

So, a conic map is essentially depending only on $2n+2$ functions in $2n+1$ variables.

It should be noted that, in some cases, we also consider conic functions and conic maps defined on the whole Euclidean space. The following lemma gives a criterion of a conic symplectic map.

Lemma 1 *$F:(p_0,p_1,q_0,q_1)\rightarrow(P_0,P_1,Q_0,Q_1)$ is a conic symplectic map if and only if $\left(0,0,P_0^T,P_1^T\right)F_*-\left(0,0,p_0^T,p_1^T\right)=0$, where F_* is the Jacobi matrix of F at the point (p_0,p_1,q_0,q_1).*

Proof. For $F:(p_0,p_1,q_0,q_1)\rightarrow(P_0,P_1,Q_0,Q_1)$, the condition

$$\left(0,0,P_0^T,P_1^T\right)F_*-\left(0,0,p_0^T,p_1^T\right)=0\tag{2.8}$$

is equivalent to the condition

$$P_0dQ_0+P_1dQ_1=p_0dq_0+p_1dq_1\quad\text{or}\quad PdQ=pdq,\tag{2.9}$$

where $P=(P_0,P_1),Q=(Q_0,Q_1),p=(p_0,p_1),q=(q_0,q_1)$. Hence in matrix form, it can be written as

$$Q_p^T\cdot P=0,\quad Q_q^T\cdot P=p.\tag{2.10}$$

Notice that a function $f(x_1,x_2,\cdots,x_n)$ is homogeneous of degree k, i.e.,

$$f(\lambda x_1,\lambda x_2,\cdots,\lambda x_n)=\lambda^k f(x_1,x_2,\cdots,x_n),$$

if and only if

$$\sum_{i=1}^{n}x_if_{x_i}(x_1,x_2,\cdots,x_n)=kf(x_1,x_2,\cdots,x_n).$$

Therefore, the condition (2.7) is equivalent to

$$P_p(p,q)\cdot p=P(p,q),\quad Q_p(p,q)\cdot p=0.\tag{2.11}$$

If F is conic symplectic then

$$Q_p^T P_p - P_p^T Q_p = 0, \quad Q_q^T P_q - P_q^T Q_q = 0, \quad Q_q^T P_p - P_q^T Q_p = I.$$

Combining with (2.11) we get

$$\begin{aligned} p &= Q_q^T P_p p - P_q^T Q_p p = Q_q^T P \\ 0 &= Q_p^T P_p p - P_p^T Q_p p = Q_p^T P \end{aligned}$$

This proves the "only if" part.

Conversely, if F satisfies the condition (2.8), then it satisfies (2.9), which means that it is symplectic. We know that if a matrix is symplectic then its transpose is also symplectic. Therefore,

$$P_q P_p^T - P_p P_q^T = 0, \quad Q_q Q_p^T - Q_p Q_q^T = 0, \quad P_p Q_q^T - P_q Q_p^T = I.$$

Combining with (2.10), we get

$$\begin{aligned} P &= P_p Q_q^T P - P_q Q_p^T P = P_p p, \\ 0 &= Q_q Q_p^T P - Q_p Q_q^T P = Q_q p. \end{aligned}$$

This means that F is conic. This finishes the proof.

b. Consider $\mathbf{R}_+ \times \mathbf{R}^{2n+1}$ as the product of the positive real space $\mathbf{R}_+$ and the contact space $\mathbf{R}^{2n+1}$. We use (w, x, y, z) to denote the coordinates of $\mathbf{R}_+ \times \mathbf{R}^{2n+1}$ with $w > 0$ and with x, y, z as before.

Definition 3 A map $G : \mathbf{R}_+ \times \mathbf{R}^{2n+1} \to \mathbf{R}_+ \times \mathbf{R}^{2n+1}$ is called a positive product map if it is composed by a map $g : \mathbf{R}^{2n+1} \to \mathbf{R}^{2n+1}$ and a positive function $\gamma : \mathbf{R}^{2n+1} \to \mathbf{R}_+$ in the form

$$\begin{pmatrix} w \\ x \\ y \\ z \end{pmatrix} \to \begin{pmatrix} W \\ X \\ Y \\ Z \end{pmatrix}, \quad W = w\gamma(x, y, z), \quad (X, Y, Z) = g(x, y, z). \tag{2.12}$$

We denote by $\gamma \otimes g$ the positive product map composed by map g and function γ.

c. Define map $S : \mathbf{R}_+ \times \mathbf{R}^{2n+1} \to \mathbf{R}_+^{2n+2}$

$$\begin{pmatrix} w \\ x \\ y \\ z \end{pmatrix} \to \begin{pmatrix} p_0 = w \\ p_1 = wx \\ q_0 = z \\ q_1 = y \end{pmatrix}. \tag{2.13}$$

Then the inverse $S^{-1}: \mathbf{R}_+^{2n+2} \to \mathbf{R}_+ \times \mathbf{R}^{2n+1}$ is given by

$$\begin{pmatrix} p_0 \\ p_1 \\ q_0 \\ q_1 \end{pmatrix} \to \begin{pmatrix} w = p_0 \\ x = \frac{p_1}{p_0} \\ y = q_1 \\ z = q_0 \end{pmatrix}. \tag{2.14}$$

Lemma 2 *Given a transformation $F: (p_0, p_1, q_0, q_1) \to (P_0, P_1, Q_0, Q_1)$ on $\mathbf{R}_+^{2n+2}$ and let $G = S^{-1} \circ F \circ S$. Then we have*

(1) F is a conic map on $\mathbf{R}_+^{2n+2}$ if and only if G is a positive product map on $\mathbf{R}_+ \times \mathbf{R}^{2n+1}$; in this case, if we write $G = \gamma \otimes g$, then

$$\gamma(x, y, z) = P_0(1, x, z, y), \tag{2.15}$$

and $g: (x, y, z) \to (X, Y, Z)$ is given by

$$X = \frac{P_1(1, x, z, y)}{P_0(1, x, z, y)}, \quad Y = Q_1(1, x, z, y), \quad Z = Q_0(1, x, z, y). \tag{2.16}$$

(2) F is a conic symplectic map if and only if G is a positive product map, say $\gamma \otimes g$, on $\mathbf{R}_+ \times \mathbf{R}^{2n+1}$ with g also a contact map on $\mathbf{R}^{2n+1}$. Moreover, in this case, the multiplier of the contact map g is just equal to $\gamma^{-1} = P_0^{-1}(1, x, z, y)$.

Proof. The conclusion (1) is easily proved by some simple calculations. Below we devote to the proof of (2). Let F sends (p_0, p_1, q_0, q_1) to (P_0, P_1, Q_0, Q_1) and G sends (w, x, y, z) to (W, X, Y, Z). Then by using the conclusion (1), we have

$$P_0 \circ S = wP_0(1, x, z, y) = w\gamma, \quad P_1 \circ S = wP_1(1, x, z, y) = w\gamma X(x, y, z),$$

$$S_* = \begin{pmatrix} 1 & 0 & 0 & 0 \\ x & wI_n & 0 & 0 \\ 0 & 0 & 0 & 1 \\ 0 & 0 & I_n & 0 \end{pmatrix}, \quad G_* = \frac{\partial(W, X, Y, Z)}{\partial(w, x, y, z)} = \begin{pmatrix} \gamma & w\gamma_x & w\gamma_y & w\gamma_z \\ 0 & & & \\ 0 & & g_* & \\ 0 & & & \end{pmatrix},$$

$$S_* \circ G = \begin{pmatrix} 1 & 0 & 0 & 0 \\ X & WI_n & 0 & 0 \\ 0 & 0 & 0 & 1 \\ 0 & 0 & I_n & 0 \end{pmatrix}$$

and compute

$$\begin{aligned}&\left(\left[\left(0,0,P_0^T,P_1^T\right)F_*-\left(0,0,p_0^T,p_1^T\right)\right]\circ S\right)S_*\\&=\left(\left(0,0,P_0^T,P_1^T\right)\circ S\right)\left(F_*\circ S\right)S_*-\left(\left(0,0,p_0^T,p_1^T\right)\circ S\right)S_*\\&=\left(0,0,w\gamma,w\gamma X^T\right)\left(F_*\circ S\right)S_*-\left(0,0,w,wx^T\right)S_*\\&=\left(0,0,w\gamma,w\gamma X^T\right)\left(S_*\circ G\right)G_*-\left(0,0,w,wx^T\right)S_*\\&=w\gamma\left[0,\left(0,X^T,1\right)g_*\right]-w\gamma\left[0,\gamma^{-1}\left(0,x^T,1\right)\right].\end{aligned}$$

Noting that S is a diffeomorphism, S_* is non-singular, $w>0,\gamma>0$, we obtain

$$\left(0,0,P_0^T,P_1^T\right)F_*-\left(0,0,p_0^T,p_1^T\right)\equiv 0\Leftrightarrow\left(0,X^T,1\right)g_*-\gamma^{-1}\left(0,x^T,1\right)\equiv 0$$

which proves the conclusion (2).

Lemma 2 establishes correspondences between conic symplectic space and contact space and between conic symplectic maps and contact maps. We call the transform from F to $G=S^{-1}\circ F\circ S=\gamma\otimes g$ contactization of conic symplectic maps, the transform from $G=\gamma\otimes g$ to $F=S\circ G\circ S^{-1}$ symplectization of contact maps and call the transform $S:\mathbf{R}_+\times\mathbf{R}^{2n+1}\to\mathbf{R}_+^{2n+2}$ symplectization of contact space, and the transform $C=S^{-1}:\mathbf{R}_+^{2n+2}\to\mathbf{R}_+\times\mathbf{R}^{2n+1}$ contactization of conic symplectic space.

§3 Contact Generating Functions for Contact Maps

With the preliminaries of the last section, it is natural to derive contact generating function theory for contact maps from the well known symplectic analog.

The following two lemmas are easily proved.

Lemma 3 *Hamiltonian $\phi:\mathbf{R}^{2n+2}\to\mathbf{R}$ is a conic function if and only if the associated Hamiltonian vector field $a_\phi=J\nabla\phi$ is conic, i.e., $a(T_\lambda z)=T_\lambda a(z)$ for $\lambda\neq 0$ with $z\in\mathbf{R}^{2n+2}$, where $J=\begin{pmatrix}0&-I_n\\I_n&0\end{pmatrix}$.*

Lemma 4 *Linear map $\begin{pmatrix}p\\q\end{pmatrix}\to C\begin{pmatrix}p\\q\end{pmatrix}$ is a conic transformation on $\mathbf{R}^{2n+2}$, i.e., $C\circ T_\lambda=T_\lambda\circ C$, if and only if the matrix C has the diagonal form $C=\begin{pmatrix}C_0&0\\0&C_1\end{pmatrix}$ with $(n+1)\times(n+1)$ matrices C_0 and C_1.*

Noting that the matrix in $gl(2n+2)$

$$C=\frac{1}{2}(I+JB),\quad B=B^T\in sm(2n+2),^{①}\tag{3.1}$$

establishes a 1-1 correspondence between near-zero Hamiltonian vector fields $z\to$

① $sm(2n+2)$ denotes the set of all $(2n+2)\times(2n+2)$ real symmetric matrices.

$a(z) = J\nabla\phi(z)$ and near-identity symplectic maps $z \to g(z)$ via generating relation

$$g(z) - z = J\nabla\phi(Cg(z) + (I - C)z), \tag{3.2}$$

and combining Lemmas 3 and 4, we find that matrix

$$C = \begin{pmatrix} C_0 & 0 \\ 0 & I - C_0^T \end{pmatrix}, \quad C_0 \in gl(n+1) \tag{3.3}$$

establishes a 1-1 correspondence between near-zero conic Hamiltonian vector fields $z \to a(z) = J\nabla\phi(z)$ and near-identity conic symplectic maps $z \to g(z)$ via generating relation (3.2).

Write $C_0 = \begin{pmatrix} \alpha & \beta^T \\ \gamma & \delta \end{pmatrix}$ with $\alpha \in \mathbf{R}, \beta, \gamma \in \mathbf{R}^n$ and $\delta \in gl(n)$. Then the generating relation (3.2) with generating matrix C given by (3.3) can be expressed as

$$\begin{cases} \hat{p}_0 - p_0 = -\phi_{q_0}(\bar{p}, \bar{q}), \\ \hat{p}_1 - p_1 = -\phi_{q_1}(\bar{p}, \bar{q}), \\ \hat{q}_0 - q_0 = \phi_{p_0}(\bar{p}, \bar{q}), \\ \hat{q}_1 - q_1 = \phi_{p_1}(\bar{p}, \bar{q}), \end{cases} \tag{3.4}$$

where $\bar{p} = \begin{pmatrix} \bar{p}_0 \\ \bar{p}_1 \end{pmatrix}$ and $\bar{q} = \begin{pmatrix} \bar{q}_0 \\ \bar{q}_1 \end{pmatrix}$ are given by

$$\begin{cases} \bar{p}_0 = \alpha\hat{p}_0 + (1-\alpha)p_0 + \beta^T(\hat{p}_1 - p_1), \\ \bar{p}_1 = \delta\hat{p}_1 + (1-\delta)p_1 + \gamma(\hat{p}_0 - p_0), \\ \bar{q}_0 = (1-\alpha)\hat{q}_0 + \alpha q_0 - \gamma^T(\hat{q}_1 - q_1), \\ \bar{q}_1 = \left(1 - \delta^T\right)\hat{q}_1 + \delta^T q_1 - \beta(\hat{q}_0 - q_0). \end{cases} \tag{3.5}$$

Every conic function ϕ can be contactized as an arbitrary function $\psi(x, y, z)$ as follows

$$\psi(x, y, z) = \phi(1, x, z, y). \tag{3.6}$$

i.e., $\phi(p_0, p_1, q_0, q_1) = p_0\phi(1, p_1/p_0, q_0, q_1) = p_0\psi(p_1/p_0, q_1, q_0)$ for $p_0 \neq 0$ and we have the partial derivative relation

$$\begin{cases} \phi_{q_0}(p_0, p_1, q_0, q_1) = p_0\psi_z(x, y, z), \\ \phi_{q_1}(p_0, p_1, q_0, q_1) = p_0\psi_y(x, y, z), \\ \phi_{p_0}(p_0, p_1, q_0, q_1) = \psi(x, y, z) - x^T\psi_x(x, y, z) = \psi_e(x, y, z), \\ \phi_{p_1}(p_0, p_1, q_0, q_1) = \psi_x(x, y, z). \end{cases} \tag{3.7}$$

with $x=\dfrac{p_1}{p_0}, y=q_1, z=q_0$ on the right hand sides. So, under contactizing transforms

$$
\begin{aligned}
S: &\begin{pmatrix} w \\ x \\ y \\ z \end{pmatrix} \to \begin{pmatrix} p_0 \\ p_1 \\ q_0 \\ q_1 \end{pmatrix} = \begin{pmatrix} w \\ wx \\ z \\ y \end{pmatrix}, \begin{pmatrix} \hat{w} \\ \hat{x} \\ \hat{y} \\ \hat{z} \end{pmatrix} \to \begin{pmatrix} \hat{p}_0 \\ \hat{p}_1 \\ \hat{q}_0 \\ \hat{q}_1 \end{pmatrix} = \begin{pmatrix} \hat{w} \\ \hat{w}\hat{x} \\ \hat{z} \\ \hat{y} \end{pmatrix}, \\
&\begin{pmatrix} \bar{w} \\ \bar{x} \\ \bar{y} \\ \bar{z} \end{pmatrix} \to \begin{pmatrix} \bar{p}_0 \\ \bar{p}_1 \\ \bar{q}_0 \\ \bar{q}_1 \end{pmatrix} = \begin{pmatrix} \bar{w} \\ \bar{w}\bar{x} \\ \bar{z} \\ \bar{y} \end{pmatrix},
\end{aligned}
\tag{3.8}
$$

the generating relation (3.4) turns into

$$
\begin{cases}
\hat{w}-w=-\bar{w}\psi_z(\bar{x},\bar{y},\bar{z}), \\
\hat{w}\hat{x}-wx=-\bar{w}\psi_y(\bar{x},\bar{y},\bar{z}), \\
\hat{z}-z=\psi_e(\bar{x},\bar{y},\bar{z}), \\
\hat{y}-y=\psi_x(\bar{x},\bar{y},\bar{z}),
\end{cases}
\tag{3.9}
$$

and Eq. (3.5) turns into

$$
\begin{cases}
\bar{w}=\alpha\hat{w}+(1-\alpha)w+\beta^T(\hat{w}\hat{x}-wx), \\
\bar{w}\bar{x}=\delta\hat{w}\hat{x}+(1-\delta)wx+\gamma(\hat{w}-w), \\
\bar{z}=(1-\alpha)\hat{z}+\alpha z-\gamma^T(\hat{y}-y), \\
\bar{y}=\left(1-\delta^T\right)\hat{y}+\delta^T y-\beta(\hat{z}-z).
\end{cases}
\tag{3.10}
$$

Since the p_0-axis is distinguished for the contactization in which we should always take $p_0 \neq 0$, it is natural to require $\beta=0$ in Eq. (3.5). Let $\hat{\mu}=w/\hat{w}=p_0/\hat{p}_0$ and $\bar{\mu}=w/\bar{w}=p_0/\bar{p}_0$, we obtain from Eq. (3.9) and (3.10)

$$
\hat{\mu}=\frac{1+\alpha\psi_z(\bar{x},\bar{y},\bar{z})}{1-(1-\alpha)\psi_z(\bar{x},\bar{y},\bar{z})}, \quad \bar{\mu}=1+\alpha\psi_z(\bar{x},\bar{y},\bar{z}),
\tag{3.11}
$$

and the induced contact transformation on the contact (x,y,z) space $\mathbf{R}^{2n+1}$ is

$$
\begin{cases}
\hat{x}-x=-\psi_y(\bar{x},\bar{y},\bar{z})+\psi_x(\bar{x},\bar{y},\bar{z})((1-\alpha)\hat{x}+\alpha x), \\
\hat{y}-y=\psi_x(\bar{x},\bar{y},\bar{z}), \\
\hat{z}-z=\psi_e(\bar{x},\bar{y},\bar{z}),
\end{cases}
\tag{3.12}
$$

with the bar variables on the right hand sides given by

$$
\begin{cases}
\bar{x}=d_1\hat{x}+d_2x+d_0, \\
\bar{y}=\left(1-\delta^T\right)\hat{y}+\delta^T y, \\
\bar{z}=(1-\alpha)\hat{z}+\alpha z-\gamma^T(\hat{y}-y),
\end{cases}
\tag{3.13}
$$

where

$$d_1 = (1-(1-\alpha)\psi_z(\bar{x},\bar{y},\bar{z}))\,\delta, d_2 = (1+\alpha\psi_z(\bar{x},\bar{y},\bar{z}))\,(I-\delta), d_0 = -\psi_z(\bar{x},\bar{y},\bar{z})\gamma. \quad (3.14)$$

Summarizing the above discussions, we have

Theorem 1 *Relations* (3.12) – (3.14) *give a contact map* $(x,y,z) \to (\hat{x},\hat{y},\hat{z})$ *via contact generating function* $\psi(x,y,z)$ *under the type* $C_0 = \begin{pmatrix} \alpha & 0 \\ \gamma & \delta \end{pmatrix}$. *Conversely, any contact map* $(x,y,z) \to (\hat{x},\hat{y},\hat{z})$ *with the contact multiplier* $\mu = \hat{x}^T\hat{y}_z + \hat{z}_z \neq 0$ *can be generated by a function* $\psi(x,y,z)$ *from* (3.12) – (3.14) *with an arbitrary given* $C_0 = \begin{pmatrix} \alpha & 0 \\ \gamma & \delta \end{pmatrix}$.

However, the difficulty in the algorithmic implementation lies in the fact that, unlike $\bar{y}$ and $\bar{z}$, which are linear combinations of $\hat{y}, y$ and $\hat{z}, z$ with constant matrix coefficients, since $\bar{x} = d_1\hat{x} + d_2 x + d_0$ and d_1, d_2 are matrices with coefficients depending on $\bar{\psi}_z = \psi_z(\bar{x},\bar{y},\bar{z})$ which in turn depends on $\bar{x},\bar{y},\bar{z}$, the combination of $\bar{x}$ from $\hat{x}$ and x is not explicitly given, the entire equations for solving $\hat{x},\hat{y},\hat{z}$ in terms of x,y,z are highly implicit. The exceptional cases are the following.

(E1) $\alpha = 0, \delta = 0_n, \gamma = 0$.

$$\hat{\mu} = 1 - \psi_z(x,\hat{y},\hat{z}), \quad \bar{\mu} = 1, \quad (3.15)_1$$

$$\begin{cases} \hat{x} - x = -\psi_y(x,\hat{y},\hat{z}) + \hat{x}\psi_z(x,\hat{y},\hat{z}), \\ \hat{y} - y = \psi_x(x,\hat{y},\hat{z}), \\ \hat{z} - z = \psi_e(x,\hat{y},\hat{z}) = \psi(x,\hat{y},\hat{z}) - x^T\psi_x(x,\hat{y},\hat{z}). \end{cases} \quad (3.15)_2$$

(E2) $\alpha = 1, \delta = I_n, \gamma = 0$.

$$\hat{\mu} = \bar{\mu} = 1 + \psi_z(\hat{x},y,z), \quad (3.16)_1$$

$$\begin{cases} \hat{x} - x = -\psi_y(\hat{x},y,z) + x\psi_z(\hat{x},y,z), \\ \hat{y} - y = \psi_x(\hat{x},y,z), \\ \hat{z} - z = \psi_e(\hat{x},y,z) = \psi(\hat{x},y,z) - x^T\psi_x(\hat{x},y,z). \end{cases} \quad (3.16)_2$$

(E3) $\alpha = \dfrac{1}{2}, \delta = \dfrac{1}{2}I_n, \gamma = 0$.

$$\hat{\mu} = \frac{1+\frac{1}{2}\psi_z(\bar{x},\bar{y},\bar{z})}{1-\frac{1}{2}\psi_z(\bar{x},\bar{y},\bar{z})}, \quad \bar{\mu} = 1 + \frac{1}{2}\psi_z(\bar{x},\bar{y},\bar{z}), \quad (3.17)_1$$

$$\begin{cases} \hat{x} - x = -\psi_y(\bar{x},\bar{y},\bar{z}) + \psi_z(\bar{x},\bar{y},\bar{z}) \cdot \dfrac{\hat{x}+x}{2}, \\ \hat{y} - y = \psi_x(\bar{x},\bar{y},\bar{z}), \\ \hat{z} - z = \psi_e(\bar{x},\bar{y},\bar{z}) = \psi(\bar{x},\bar{y},\bar{z}) - \bar{x}^T\psi_x(\bar{x},\bar{y},\bar{z}), \end{cases} \quad (3.17)_2$$

with

$$\bar{x} = \frac{\hat{x}+x}{2} - \frac{1}{4}\psi_z(\bar{x},\bar{y},\bar{z})(\hat{x}-x), \quad \bar{y} = \frac{\hat{y}+y}{2}, \quad \bar{z} = \frac{\hat{z}+z}{2}. \tag{3.17$_3$}$$

For $\psi_z = \lambda =$ constant, the case (E3) reduces to

$$\hat{\mu} = \frac{1+\frac{1}{2}\lambda}{1-\frac{1}{2}\lambda}, \quad \bar{\mu} = 1 + \frac{1}{2}\lambda, \tag{3.18$_1$}$$

$$\begin{cases} \hat{x} - x = -\psi_y(\bar{x},\bar{y},\bar{z}) + \psi_z(\bar{x},\bar{y},\bar{z}) \cdot \dfrac{\hat{x}+x}{2}, \\ \hat{y} - y = \psi_x(\bar{x},\bar{y},\bar{z}), \\ \hat{z} - z = \psi_e(\bar{x},\bar{y},\bar{z}) = \psi(\bar{x},\bar{y},\bar{z}) - \bar{x}^T\psi_x(\bar{x},\bar{y},\bar{z}), \end{cases} \tag{3.18$_2$}$$

with

$$\bar{x} = \frac{\hat{x}+x}{2} - \frac{1}{4}\lambda(\hat{x}-x), \quad \bar{y} = \frac{\hat{y}+y}{2}, \quad \bar{z} = \frac{\hat{z}+z}{2}. \tag{3.18$_3$}$$

Note that the symplectic map induced by generating function ϕ from the relation (3.2) can be represented as the composition of the maps, non-symplectic generally, $z \to \bar{z}$ and $\bar{z} \to \hat{z}$

$$\begin{aligned} \bar{z} &= z + CJ\nabla\phi(\bar{z}), \\ \hat{z} &= \bar{z} + (I-C)J\nabla\phi(\bar{z}). \end{aligned}$$

Theorem 2 *Contact map $(x,y,z) \to (\hat{x},\hat{y},\hat{z})$ induced by contact generating function ψ from the relations (3.12)-(3.14) can be represented as the composition of the maps $(x,y,z) \to (\bar{x},\bar{y},\bar{z})$ and $(\bar{x},\bar{y},\bar{z}) \to (\hat{x},\hat{y},\hat{z})$ which are not contact generally and given, respectively, as follows*

$$\begin{cases} \bar{x} - x = -\delta\psi_y(\bar{x},\bar{y},\bar{z}) + \alpha\psi_z(\bar{x},\bar{y},\bar{z})x - \gamma\psi_z(\bar{x},\bar{y},\bar{z}), \\ \bar{y} - y = \left(I - \delta^T\right)\psi_x(\bar{x},\bar{y},\bar{z}), \\ \bar{z} - z = (1-\alpha)\psi_e(\bar{x},\bar{y},\bar{z}) - \gamma^T\psi_x(\bar{x},\bar{y},\bar{z}), \end{cases} \tag{3.19}$$

and

$$\begin{cases} \hat{x} - \bar{x} = -(I-\delta)\psi_y(\bar{x},\bar{y},\bar{z}) + (1-\alpha)\psi_z(\bar{x},\bar{y},\bar{z})\hat{x} + \gamma\psi_z(\bar{x},\bar{y},\bar{z}), \\ \hat{y} - \bar{y} = \delta^T\psi_x(\bar{x},\bar{y},\bar{z}), \\ \hat{z} - \bar{z} = \alpha\psi_e(\bar{x},\bar{y},\bar{z}) + \gamma^T\psi_x(\bar{x},\bar{y},\bar{z}). \end{cases} \tag{3.20}$$

(3.19) and (3.20) are the 2-stage form of the generating relation (3.12) of the contact map induced by generating function ψ under the type $C_0 = \begin{pmatrix} \alpha & 0 \\ \gamma & \delta \end{pmatrix}$. Corresponding to the exceptional cases (E1), (E2) and (E3), the above 2 -stage representation has simpler forms, we no longer write them here.

§4 Contact Algorithms for Contact Systems

Consider contact system (1.3) with the vector field a defined by contact Hamiltonian K according to Eq. (1.6). Take $\psi(x,y,z) = sK(x,y,z)$ in (3.12)-(3.14) as the generating function, we then obtain contact difference schemes with 1 st order of accuracy of the contact system (1.3) associated with all possible types $C_0 = \begin{pmatrix} \alpha & 0 \\ \gamma & \delta \end{pmatrix}$. The simplest and important cases are (write $\bar{K}_x = K_x(\bar{x},\bar{y},\bar{z})$, etc.) as follows.

$\widetilde{Q}$. Contact analog of symplectic method (p,Q)① $(\alpha = 0, \delta = 0_n, \gamma = 0)$.

$$\begin{aligned}
&\text{1-stage form}: && \begin{aligned} &\hat{x} = x + s\left(-K_y(x,\hat{y},\hat{z}) + \hat{x}K_z(x,\hat{y},\hat{z})\right) \\ &\hat{y} = y + sK_x(x,\hat{y},\hat{z}) \\ &\hat{z} = z + sK_e(x,\hat{y},\hat{z}). \end{aligned} \\
&\text{2-stage form}: && \begin{aligned} &\bar{x} = x, \bar{y} = y + s\bar{K}_x, \quad \bar{z} = z + s\bar{K}_e, \\ &\hat{x} = \bar{x} + s\left(-\bar{K}_y + \hat{x}\bar{K}_z\right), \quad \hat{y} = \bar{y}, \quad \hat{z} = \bar{z}. \end{aligned}
\end{aligned} \tag{4.1}$$

$\widetilde{P}$. Contact analog of symplectic method (P,q) $(\alpha = 1, \delta = I_n, \gamma = 0)$.

$$\begin{aligned}
&\text{1-stage form}: && \begin{aligned} &\hat{x} = x + s\left(-K_y(\hat{x},y,z) + xK_z(\hat{x},y,z)\right) \\ &\hat{y} = y + sK_x(\hat{x},y,z) \\ &\hat{z} = z + sK_e(\hat{x},y,z), \end{aligned} \\
&\text{2-stage form}: && \begin{aligned} &\bar{x} = x + s\left(-\bar{K}_y + x\bar{K}_z\right), \quad \bar{y} = y, \quad \bar{z} = z, \\ &\hat{x} = \bar{x}, \quad \hat{y} = \bar{y} + s\bar{K}_x, \quad \hat{z} = \bar{z} + s\bar{K}_e. \end{aligned}
\end{aligned} \tag{4.2}$$

$\widetilde{C}$. Contact version of centered Euler method $\left(\alpha = \frac{1}{2}, \delta = \frac{1}{2}I_n, \gamma = 0\right)$.

2-stage form:

$$\begin{aligned}
&\bar{x} = x + \frac{s}{2}\left(-\bar{K}_y + x\bar{K}_z\right), \quad \bar{y} = y + \frac{s}{2}\bar{K}_x, \quad \bar{z} = z + \frac{s}{2}\bar{K}_e, \\
&\hat{x} = \bar{x} + \frac{s}{2}\left(-\bar{K}_y + \hat{x}\bar{K}_x\right) = \left(\bar{x} - \frac{s}{2}\bar{K}_y\right)\left(1 - \frac{s}{2}\bar{K}_z\right)^{-1}, \\
&\hat{y} = \bar{y} + \frac{s}{2}\bar{K}_x = 2\bar{y} - y, \quad \hat{z} = \bar{z} + \frac{s}{2}\bar{K}_e = 2\bar{z} - z.
\end{aligned} \tag{4.3}$$

One might suggest, for example, the following scheme for (1.3):

$$\hat{x} = x + sa(\hat{x},y,z), \quad \hat{y} = y + sb(\hat{x},y,z), \quad \hat{z} = z + sc(\hat{x},y,z).$$

It differs from (4.2) only in one term for $\hat{x}$, i.e., $\hat{x}K(\hat{x},y,z)$ instead of $xK(\hat{x},y,z)$. This minute but delicate difference makes (4.2) contact and other non-contact!

① For Hamiltonian system $\dot{p} = -H_q(p,q), \dot{q} = H_p(p,q)$, the difference scheme $\hat{p} = p - sH_q(p,\hat{q}), \hat{q} = q + sH_p(p,\hat{q})$ is symplectic and we call it (p,Q) method because the pair $(p,\hat{q})$, composed by the old variables of p and the new variables of q, emerges in the Hamiltonian. The following (P,q) method has the similar meaning.

It should be noted that the $\widetilde{Q}$ and $\widetilde{P}$ methods are of order one of accuracy and the $\widetilde{C}$ method is of order two. The proof is similar to that for symplectic case. In principle, one can construct the contact difference schemes of arbitrarily high order of accuracy for contact systems, as was done for Hamiltonian systems, by suitably composing the $\widetilde{Q}, \widetilde{P}$ or $\widetilde{C}$ method and the respective reversible counterpart. Another general method for the construction of contact difference schemes is based on the generating functions for phase flows of contact systems which will be developed in the next section.

§5 Hamilton-Jacobi Equations for Contact Systems

We recall that a near identity contact map $g:(x,y,z)\to(\hat{x},\hat{y},\hat{z})$ can be generated from the so-called generating function $\psi(x,y,z)$, associated with a matrix $C_0=\begin{pmatrix}\alpha & 0\\ \gamma & \delta\end{pmatrix}$, by the relations (3.12)—(3.14) . Accordingly, to the phase flow e_K^t of a contact system with contact Hamiltonian K, there corresponds a time-dependent generating function $\psi^t(x,y,z)$ such that the map $e_K^t:(x,y,z)\to(\hat{x},\hat{y},\hat{z})$ is generated from ψ^t by the relations (3.12)-(3.14) in which ψ is replaced by ψ^t and C_0 is given in advance as above. The function ψ^t should be determined by K and C_0. Below we derive the relevant relations between them.

Let $H(p_0,p_1,q_0,q_1)=p_0K(p_1/p_0,q_1,q_0)$ for $p_0\neq 0$. With this conic Hamiltonian and with normal Darboux matrices $C=\begin{pmatrix}C_0 & 0\\ 0 & I-C_0\end{pmatrix}$ where $C_0=\begin{pmatrix}\alpha & 0\\ \gamma & \delta\end{pmatrix}$, we get the Hamilton-Jacobi equation

$$\frac{\partial}{\partial t}\phi^t(u)=H\left(u+(I-C)J\nabla\phi^t(u)\right)\qquad \text{with } u=(p_0,p_1,q_0,q_1)^T \tag{5.1}$$

satisfied by the generating function $\phi^t(u)$ of the phase flow g_H^t of the Hamiltonian system associated with the Hamiltonian H, while the phase flow g_H^t is generated from ϕ^t by the relation

$$g_H^t(u)-u=J\nabla\phi^t\left(Cg_H^t(u)+(I-C)u\right). \tag{5.2}$$

On the other hand, according to the discussions of Section 3, we have

$$\phi^t(p_0,p_1,q_0,q_1)=p_0\psi^t(x,y,z),\qquad \text{with } x=p_1/p_0, y=q_1, z=q_0.$$

So, by simple calculations,

$$u+(I-C)J\nabla\phi^t(u)=\begin{pmatrix}p_0-(1-\alpha)\phi_{q_0}\\ p_1+\gamma\phi_{q_0}-(I-\delta)\phi_{q_1}\\ q_0+\alpha\phi_{p_0}+\gamma^T\phi_{p_1}\\ q_1+\delta^T\phi_{p_1}\end{pmatrix}=\begin{pmatrix}p_0\left(1-(1-\alpha)\psi_z\right)\\ p_0\left(x+\gamma\psi_z-(I-\delta)\psi_y\right)\\ z+\alpha\psi_e+\gamma^T\psi_x\\ y+\delta^T\psi_x\end{pmatrix},$$

and

$$
\begin{aligned}
& H\left(u+(I-C)J\nabla\phi^t(u)\right) \\
& = p_0\left(1-(1-\alpha)\psi_z\right)K\left(\frac{x-(I-\delta)\psi_y+\gamma\psi_z}{1-(1-\alpha)\psi_z}, y+\delta^T\psi_x, z+\alpha\psi_e+\gamma^T\psi_x\right).
\end{aligned}
$$

Therefore, from Eq. (5.1), $\psi^t(x,y,z)$ satisfies

$$
\frac{\partial}{\partial t}\psi^t = \left(1-(1-\alpha)\psi_z\right)K\left(\frac{x-(I-\delta)\psi_y+\gamma\psi_z}{1-(1-\alpha)\psi_z}, y+\delta^T\psi_x, z+\alpha\psi_e+\gamma^T\psi_x\right). \tag{5.3}
$$

Now we claim [①] : for all u, the following equality is valid

$$
H\left(u+(I-C)J\nabla\phi^t(u)\right) = H\left(u-CJ\nabla\phi^t(u)\right). \tag{5.4}
$$

So, replacing C by $C-I$ in above discussions or, equivalently, replacing α and δ by $\alpha-1$ and $\delta-I$ with γ unchanging in (5.3), we obtain another equation satisfied by the ψ^t

$$
\frac{\partial}{\partial t}\psi^t = \left(1+\alpha\psi_z\right)K\left(\frac{x+\delta\psi_y+\gamma\psi_z}{1+\alpha\psi_z}, y+\left(\delta^T-I\right)\psi_x, z+(\alpha-1)\psi_e+\gamma^T\psi_x\right). \tag{5.5}
$$

(5.3) and (5.5) define the same function ψ^t. When $t=0$, e_K^t is identity, so we should impose the initial condition

$$
\psi^0(x,y,z)=0 \tag{5.6}
$$

for solving the first order partial differential equation (5.3) or (5.5). We call the both equations the Hamilton-Jacobi equations of the contact system associated with the contact Hamiltonian K and the matrix $C_0=\begin{pmatrix}\alpha & 0\\ \gamma & \delta\end{pmatrix}$.

Specifically, we have Hamilton-Jacobi equations for particular cases:

(E1) $\alpha=0, \delta=0, \gamma=0$.

$$
\frac{\partial}{\partial t}\psi^t = \left(1-\psi_z^t\right)K\left(\frac{x-\psi_y^t}{1-\psi_z^t}, y, z\right) = K\left(x, y-\psi_x^t, z-\psi_e^t\right); \tag{5.7}
$$

(E2) $\alpha=1, \delta=I_n, \gamma=0$.

$$
\frac{\partial}{\partial t}\psi^t = K\left(x, y+\psi_x^t, z+\psi_e^t\right) = \left(1+\psi_z^t\right)K\left(\frac{x+\psi_y^t}{1+\psi_z^t}, y, z\right); \tag{5.8}
$$

(E3) $\alpha=\dfrac{1}{2}, \delta=\dfrac{1}{2}I_n, \gamma=0$.

$$
\begin{aligned}
\frac{\partial}{\partial t}\psi^t &= \left(1-\frac{1}{2}\psi_z^t\right)K\left(\frac{x-\frac{1}{2}\psi_y^t}{1-\frac{1}{2}\psi_z^t}, y+\frac{1}{2}\psi_x^t, z+\frac{1}{2}\psi_e^t\right) \\
&= \left(1+\frac{1}{2}\psi_z^t\right)K\left(\frac{x+\frac{1}{2}\psi_y^t}{1+\frac{1}{2}\psi_z^t}, y-\frac{1}{2}\psi_x^t, z-\frac{1}{2}\psi_e^t\right).
\end{aligned} \tag{5.9}
$$

---

① Proof of the claim: let $\hat{u}=u+(I-C)J\nabla\phi^t(u)$ and $\bar{u}=u-CJ\nabla\phi^t(u)$. Then we have $u=C\hat{u}+(I-C)\bar{u}$. From (5.2), it follows that $\hat{u}=g_H^t(\bar{u})$. The claim is then proved since $H\left(g_H^t(\bar{u})\right)=H(\bar{u})$.

Remark on the construction of higher order contact difference schemes. If K is analytic, then one can solve $\psi^t(x,y,z)$ from the above Hamilton-Jacobi equations in the forms of power series in time t. Its coefficients are recursively determined by the K and the related matrix C_0. The power series are simply given from the corresponding conic Hamiltonian generating functions $\phi^t(p_0,p_1,q_0,q_1)$ by $\psi^t(x,y,z)=\phi^t(1,x,z,y)$, since the power series expressions of ϕ^t with respect to t from the conit Hamiltonian $H(p_0,p_1,q_0,q_1)=p_0K(p_1/p_0,q_1,q_0)$ have been well given in [7]. Taking a finite truncation of the power series up to order m, an arbitrary integer, with respect to the time t and replacing by the truncation the generating function ψ in (3.12)-(3.14), then one obtains a contact difference scheme of order m for the contact system defined by the contact Hamiltonian K. The proofs of these assertions are similar to those in the Hamiltonian system case and are omitted here.

References

[1] Arnold V I. Mathematical Methods of Classical Mechanics. New York: Springer, 1978

[2] Feng K. On difference schemes and symplectic geometry. In Feng K, ed. Proc 1984 Beijing Symp Diff Geometry and Diff Equations. Beijing: Science Press, 1985: 42-58

[3] Feng K. The Hamiltonian way for computing Hamiltonian dynamics. In Spigler R, ed. Applied and Industrial Mathematics. Netherlands: Kluwer, 1991: 17-35

[4] Feng K. Difference schemes for Hamiltonian formalism and symplectic geometry. J Comp Math. 1986 4(3): 279-289

[5] Feng K, Wang D L. Variations on a theme by Euler: [Preprint]. Beijing: CAS. Computing Center, 1991

[6] Feng K, Wang D L. Dynamical Systems and Geometric Construction of Algorithms. In Shi Z C, Yang C J, ed. Computational Mathematics of China. Contemporary Mathematics. 1994 163: 1-32

[7] Feng K, Wu H M, Qin M Z, Wang D L. Construction of canonical difference schemes for Hamiltonian formalism via generating functions. J Comp Math. 1989 7(1): 71-96

[8] Wang D L. Decomposition of vector fields and composition of algorithms. In Feng K, Shi Z C, ed. Proc Inter Conf on Computation of Differential Equations and Dynamical Systems. Singapore: World Scientific, 1993: 179-184

[9] Warner F W. Foundations of Differentiable Manifolds and Lie Groups, Springer, New York. 1983

[10] Feng K. How to compute properly Newton's equation of motion. In Ying L A, Guo B Y, ed. Proc of 2nd Conf on Numerical Methods for Partial Differential Equations. Singapore: World Scientific, 1992: 15-22

24 Volume-Preserving Algorithms for Source-Free Dynamical Systems[1][2]

无源动力系统的保体积算法

Abstract

In this paper, we first expound why the volume-preserving algorithms are proper for numerically solving source-free systems and then prove all the conventional methods are not volume-preserving. Secondly, we give a general method of constructing volume-preserving difference schemes for source-free systems on the basis of decomposing a source-free vector field as a finite sum of essentially 2-dimensional Hamiltonian fields and of composing the corresponding essentially symplectic schemes into a volume-preserving one. Lastly, we make some special discussions for so-called separable source-free systems for which arbitrarily high order explicit revertible volume-preserving schemes can be constructed.

§1 Introduction

Source-free dynamical systems on the Euclidean space $\mathbf{R}^n$ are defined by source-free (or divergence-free) vector fields $a : \mathbf{R}^n \to \mathbf{R}^n$,

$$(\operatorname{div} a)(x) = \sum_{i=1}^{n} \frac{\partial a_i}{\partial x_i}(x) = 0 \quad \text{identically for } x \in \mathbf{R}^n \tag{1.1}$$

through equations

$$\frac{dx}{dt} = \dot{x} = a(x). \tag{1.2}$$

Here and hereafter we use the coordinate description and matrix notation:

$$x = (x_1, \cdots, x_n)^T, \quad a(x) = (a_1(x), \cdots, a_n(x))^T, \tag{1.3}$$

where T denotes the transpose of matrix.

In this paper we mainly analyse and construct numerical algorithms proper for source-free systems. Such systems constitute one of the most important classical cases of dynamical systems preserving certain geometric structure and arise in many physical problems such as particle tracking in incompressible fluids and toroidal magnetic

---

① Joint with Shang Z J

② Numericshe Mathematik. 1995

surface-generating in stellarators. Since the difficulty and even impossibility of solving equations by quadrature, the numerical methods certainly play an important role in understanding the dynamic behavior of a system and in solving physical and engineering problems. On the other hand, the problem of whether a numerical algorithm is proper for a system is closely related to the problem of whether the algorithmic approximation to the corresponding phase flow approximates perfectly in some sense and even strictly preserve the structure of the system itself if the system has such structure. It has been evidenced with some typical examples in the Hamiltonian case that "nonproper" algorithms will result in essentially wrong approximations to the solutions of systems and "proper" algorithms may generate remarkably right ones.

But how does one evaluate a numerical algorithm to be proper for source-free systems? It is well known that intrinsic to all source-free systems, there is a volume form of the phase space $\mathbf{R}^n$, say

$$\alpha = dx_1 \wedge dx_2 \wedge \cdots \wedge dx_n, \tag{1.4}$$

such that the evolution of dynamics preserve this form. In other words, the phase flow e_a^t, of source-free system (1.2), satisfies the volume-preserving condition

$$\left(e_a^t\right)^* \alpha = \alpha, \tag{1.5}$$

or equivalently,

$$\det \frac{\partial e_a^t}{\partial x}(x) = 1 \quad \text{identically for } x \in \mathbf{R}^n \text{ and } t \in \mathbf{R}. \tag{1.5$'$}$$

In addition to this, e_a^t satisfies the group property in t

$$e_a^0 = \text{identity}, \quad e_a^{t+s} = e_a^t \circ e_a^s. \tag{1.6}$$

In fact, (1.5) and (1.6) completely describe the properties of the most general source-free dynamical systems. This fact suggests that a proper algorithmic approximation g_a^s to phase flow e_a^s for source-free vector field $a : \mathbf{R}^n \to \mathbf{R}^n$ should satisfy these two requirements. However, the group property (1.6) is too stringent in general for algorithmic approximations because only the phase flows satisfy it. Instead of it, a weaker requirement

$$g_a^0 = \text{identity}, \quad g_a^s \circ g_a^{-s} = \text{identity} \tag{1.7}$$

is reasonable and practicable for all vector fields $a : \mathbf{R}^n \to \mathbf{R}^n$ (see for example [4]). We call such algorithmic approximations revertible, that means g_a^s always generate coincident forward and backward orbits. As for the volume-preserving property (1.5), it characterizes the geometric structure——volume-preserving structure——of source-free systems. Our aim in this paper is just to construct difference schemes preserving this structure, which we call volume-preserving schemes, in the sense that the algorithmic approximations to the phase flows satisfy (1.5) for the most general source-free systems.

§2 Obstruction to Analytic Methods

We note that for $n = 2$, source-free vector fields = Hamiltonian fields, and area-preserving maps = symplectic maps, so the problem for area-preserving algorithms has been solved in principle.

But for $n \geqslant 3$, the problem is new, since all the conventional methods plus even the symplectic methods are generally not volume-preserving, even for linear source-free systems. As an illustration, solve on $\mathbf{R}^3$

$$\frac{dx}{dt} = a(x) = Ax, \quad trA = 0 \tag{2.1}$$

by the Euler centered method, we get algorithmic approximation G^s to $e_a^s = exp(sA)$ with

$$G^s = \left(I - \frac{s}{2}A\right)^{-1}\left(I + \frac{s}{2}A\right). \tag{2.2}$$

Simple calculations show that in 3 -dimensions, if tr A=0, then $\det G^s = 1 \Leftrightarrow \det A = 0$, which is exceptional. A more general conclusion in linear case is

Lemma 1 *Let $sl(n)$ denote the set of all $n \times n$ real matrices with trace equal to zero and $SL(n)$ the set of all $n \times n$ real matrices with determinant equal to one. Then for any real analytic function $\phi(z)$ defined in a neighbourhood of $z = 0$ in $\mathbf{C}$ satisfying the conditions:* 1) $\phi(0) = 1$ *and* 2) $\phi'(0) = 1$, *we have that $\phi(sl(n)) \subset SL(n)$ for some $n \geqslant 3$ if and only if $\phi(z) = exp(z)$.*

Proof. "If part" is a known conclusion. For the "only if part" it suffices to show it for $n = 3$. For this, we consider matrices of the diagonal form

$$D(s,t) = \begin{pmatrix} s & 0 & 0 \\ 0 & t & 0 \\ 0 & 0 & -(s+t) \end{pmatrix} \in sl(3), \quad s,t \in \mathbf{R}. \tag{2.3}$$

Since ϕ is analytic in a neighbourhood of the origin in $\mathbf{C}$, we have

$$\phi(D(s,t)) = \begin{pmatrix} \phi(s) & 0 & 0 \\ 0 & \phi(t) & 0 \\ 0 & 0 & \phi(-(s+t)) \end{pmatrix}, \quad s,t \sim 0. \tag{2.4}$$

By assumption, det $\phi(D(s,t)) = 1$ for $s, t \sim 0$. So,

$$\phi(s)\phi(t)\phi(-(s+t)) = 1, \quad s, t \sim 0. \tag{2.5}$$

This together with the condition $\phi(0) = 1$ yields

$$\phi(s)\phi(-s) = 1, \quad s \sim 0. \tag{2.6}$$

Multiplying the both sides of Eq. (2.5) by $\phi(s+t)$ and using (2.6), we get

$$\phi(s)\phi(t) = \phi(s+t), \quad s, t \sim 0. \tag{2.7}$$

This, together with the conditions 1) and 2) of the lemma, implies

$$\phi(z) = \exp(z)$$

which completes the proof.

Lemma 1 says that there are no consistent analytic approximations to the exponential function sending at the same time $sl(n)$ into $SL(n)$ other than the exponential itself. This shows that it is impossible to construct volume-preserving algorithms analytically depending on source-free vector fields. Thus we have

Theorem 1 *All the conventional methods including the well-known Runge-Kutta methods, linear multistep methods and Euler methods (explicit, implicit and centered) are non-volume-preserving.*

Consequently, to construct volume-preserving algorithms for source-free systems, we must break through the conventional model and explore new ways.

§3 "Essentially Hamiltonian Decompositions" of Source-free Vector Fields

In $\mathbf{R}^2$, every source-free field $a = (a_1, a_2)^T$ corresponds to a stream function or 2-dimensional Hamiltonian ψ, unique up to a constant:

$$a_1 = -\frac{\partial \psi}{\partial x_2}, \quad a_2 = \frac{\partial \psi}{\partial x_1}; \tag{3.1}$$

and in $\mathbf{R}^3$, every source-free field $a = (a_1, a_2, a_3)^T$ corresponds to a vector potential $b = (b_1, b_2, b_3)^T$, unique up to a gradient:

$$a = \operatorname{curl} b, \quad a_1 = \frac{\partial b_3}{\partial x_2} - \frac{\partial b_2}{\partial x_3}, a_2 = \frac{\partial b_1}{\partial x_3} - \frac{\partial b_3}{\partial x_1}, a_3 = \frac{\partial b_2}{\partial x_1} - \frac{\partial b_1}{\partial x_2}, \tag{3.2}$$

then we get source-free decomposition

$$a = \begin{pmatrix} a_1 \\ a_2 \\ a_3 \end{pmatrix} = \begin{pmatrix} 0 \\ \dfrac{\partial b_1}{\partial x_3} \\ -\dfrac{\partial b_1}{\partial x_2} \end{pmatrix} + \begin{pmatrix} -\dfrac{\partial b_2}{\partial x_3} \\ 0 \\ \dfrac{\partial b_2}{\partial x_1} \end{pmatrix} + \begin{pmatrix} \dfrac{\partial b_3}{\partial x_2} \\ -\dfrac{\partial b_3}{\partial x_1} \\ 0 \end{pmatrix} = a^{(1)} + a^{(2)} + a^{(3)}. \tag{3.3}$$

As a generalization of cases $n = 2, 3$, on $\mathbf{R}^n$, we have

Lemma 2 *To every source-free field* $a = (a_1, a_2, \cdots, a_n)^T$, *there corresponds a skew symmetric tensor field of order* $2, b = (b_{ik})_{1 \leqslant i,k \leqslant n}, b_{ik} = -b_{ki}$, *so that*

$$a_i = \sum_{k=1}^{n} \frac{\partial b_{ik}}{\partial x_k}, \quad i = 1, 2, \cdots, n. \tag{3.4}$$

Proof.[①] With the given $a = (a_1, \cdots, a_n)^T$, we define the 1-form on $\mathbf{R}^n$

$$\alpha = \sum_{i=1}^{n} a_i(x) dx_i. \tag{3.5}$$

Since a is source-free, we have $\delta\alpha = -\sum\limits_{i=1}^{n} \dfrac{\partial a_i}{\partial x_i} = -\operatorname{div} a = 0$, which means that α is δ-closed. By Poincaré's lemma, there exists a 2-form, say β, so that

$$\alpha = \delta\beta. \tag{3.6}$$

But for the 2-form β, there exists a skew symmetric tensor of order $2, b = (b_{ik})_{1 \leqslant i,k \leqslant n}$, $b_{ik} = -b_{ki}$, so that

$$\beta = \sum_{i,k=1}^{n} b_{ik} dx_i \wedge dx_k. \tag{3.7}$$

Seeing that

$$\delta\beta = \sum_{i=1}^{n} \left(\sum_{k=1}^{n} \frac{\partial b_{ik}}{\partial x_k} \right) dx_i \tag{3.8}$$

and noticing Eqs. (3.5) and (3.6), we get (3.4). The proof is completed.

By (3.4), we can decompose

$$a = \sum_{1 \leqslant i < k \leqslant n} a^{(ik)}, \quad a^{(ik)} = \left(0, \cdots, 0, \frac{\partial b_{ik}}{\partial x_k}, 0, \cdots, -\frac{\partial b_{ik}}{\partial x_i}, 0, \cdots, 0\right)^T, \quad i < k. \tag{3.9}$$

① For the definition of the operator δ, see[16],p.220; for the Poincaré's lemma for the exterior differential operator d, see[10], from which the Poincaré's lemma for the operator δ is easily derived.

Every vector field $a^{(ik)}$ in (3.9) is 2-dimensional Hamiltonian on the x_i-x_k plane and zero in other dimensions. We call such decompositions essentially Hamiltonian decompositions.

We note that the tensor potential $b=(b_{ik})_{1\leqslant i,k\leqslant n}$ is far from uniquely determined for a given source-free field $a=(a_1,\cdots,a_n)^T$ from Eq. (3.4). For uniqueness one may impose normalizing conditions in many different ways. One way is to impose, as was done by H. Weyl in [17] in 3-dimensional case,

$$N_0: b_{ik}=0,\quad |i-k|\geqslant 2, \tag{3.10}$$

(this condition is ineffective for $n=2$). The non-zero components are

$$b_{12}=-b_{21}, b_{23}=-b_{32},\cdots,b_{n-1,n}=-b_{n,n-1}. \tag{3.11}$$

$$N_k: b_{k,k+1}|_{x_{k+1}=0}=0, 1\leqslant k\leqslant n-2, \tag{3.12}$$

(this condition is ineffective for $n=2$).

$$N_{n-1}: b_{n-1,n}|_{x_{n-1}=x_n=0}=0. \tag{3.13}$$

Then simple calculations show that all $b_{k,k+1}$ are uniquely determined by quadrature

$$b_{12}=\int_0^{x_2} a_1 dx_2, \tag*{$(3.14)_1$}$$

$$b_{k,k+1}=\int_0^{x_{k+1}}\left(a_k+\frac{\partial b_{k-1,k}}{\partial x_{k-1}}\right)dx_{k+1}, 2\leqslant k\leqslant n-2, \tag*{$(3.14)_k$}$$

$$b_{n-1,n}=\int_0^{x_n}\left(a_{n-1}+\frac{\partial b_{n-2,n-1}}{\partial x_{n-2}}\right)dx_n-\int_0^{x_{n-1}} a_n|_{x_n=0}dx_{n-1}. \tag*{$(3.14)_{n-1}$}$$

So one gets an essentially Hamiltonian decomposition for a

$$a=\sum_{k=1}^{n-1}a^{(k)},\quad a^{(k)}=\left(0,\cdots,0,\frac{\partial b_{k,k+1}}{\partial x_{k+1}},-\frac{\partial b_{k,k+1}}{\partial x_k},0,\cdots,0\right)^T, \tag{3.15}$$

or in components,

$$\begin{aligned}
a_1&=\frac{\partial b_{12}}{\partial x_2},\\
a_2&=-\frac{\partial b_{12}}{\partial x_1}+\frac{\partial b_{23}}{\partial x_3},\\
&\cdots\cdots\\
a_{n-1}&=-\frac{\partial b_{n-2,n-1}}{\partial x_{n-2}}+\frac{\partial b_{n-1,n}}{\partial x_n},\\
a_n&=-\frac{\partial b_{n-1,n}}{\partial x_{n-1}}.
\end{aligned} \tag*{$(3.15)'$}$$

§4 Construction of Volume-preserving Difference Schemes

In this section, we give a general way to construct volume-preserving difference schemes for source-free systems by means of the essentially Hamiltonian decompositions of source-free vector fields and the symplectic difference schemes for 2-dimensional Hamiltonian systems. For this aim, we first prove

Lemma 3 *Let a be a smooth vector field on $\mathbf{R}^n$ and have decomposition*

$$a = \sum_{i=1}^{m} a^{(i)} \tag{4.1}$$

with smooth fields $a^{(i)} : \mathbf{R}^n \to \mathbf{R}^n, i = 1, \cdots, m$. Suppose that, for each $i = 1, \cdots, m, G_i^\tau$ is an approximation of order p to $e_{a(i)}^\tau$, the phase flow of the system associated to the field $a^{(i)}$, in the sense that $\lim_{\tau \to 0} \frac{1}{\tau^p}\left(G_i^\tau(x) - e_{a^{(i)}}^\tau(x)\right) = 0$ for all $x \in \mathbf{R}^n$ with some $p \geqslant 1$. Then we have

(1) for any permutation $(i_1 i_2 \cdots i_m)$ of $(12 \cdots m)$, the compositions

$$_1G_{i_1 i_2 \cdots i_m}^\tau := G_{i_m}^\tau \circ \cdots \circ G_{i_2}^\tau \circ G_{i_1}^\tau, \quad _1\hat{G}_{i_1 i_2 \cdots i_m}^\tau := \left(_1G_{i_1 i_2 \cdots i_m}^{-\tau}\right)^{-1} \tag{4.2}$$

are approximations, of order one, to e_a^τ; and the compositions

$$_2g_{i_1 i_2 \cdots i_m}^\tau :=_1 \hat{G}_{i_1 i_2 \cdots i_m}^{\tau/2} \circ_1 G_{i_1 i_2 \cdots i_m}^{\tau/2}, \quad _2\hat{g}_{i_1 i_2 \cdots i_m}^\tau =_1 G_{i_1 i_2 \cdots i_m}^{\tau/2} \circ_1 \hat{G}_{i_1 i_2 \cdots i_m}^{\tau/2} \tag{4.2$'$}$$

are revertible approximations, of order 2, to e_a^τ;

(2) if, for each $i = 1, 2, \cdots, m, G_i^\tau$ is an approximation, of order 2, to e_a^τ, then

$$_2G_{i_1 i_2 \cdots i_m}^\tau := G_{i_m}^{\tau/2} \circ \cdots \circ G_{i_2}^{\tau/2} \circ G_{i_1}^{\tau/2} \circ G_{i_1}^{\tau/2} \circ G_{i_2}^{\tau/2} \circ \cdots \circ G_{i_m}^{\tau/2} \tag{4.3}$$

is an approximation, of order 2, to e_a^τ; and it is revertible if each G_i^τ is revertible;

(3) if $_2G^\tau$ is a revertible approximation, of order 2, to e_a^τ, then the symmetric composition

$$_4G^\tau =_2 G^{\alpha_1 \tau} \circ_2 G^{\beta_1 \tau} \circ_2 G^{\alpha_1 \tau} \tag{4.4$_1$}$$

with

$$\alpha_1 = \left(2 - 2^{1/3}\right)^{-1}, \quad \beta_1 = 1 - 2\alpha_1 < 0 \tag{4.4$'_1$}$$

is a revertible approximation, of order 4, to e_a^τ; and generally, the symmetric composition, recursively defined as follows,

$$_{2(l+1)}G^\tau =_{2l} G^{\alpha_l \tau} \circ_{2l} G^{\beta_l \tau} \circ_{2l} G^{\alpha_l \tau} \tag{4.4$_l$}$$

with

$$\alpha_l = \left(2 - 2^{1/(2l+1)}\right)^{-1}, \quad \beta_l = 1 - 2\alpha_l < 0 \tag{4.4$'_l$}$$

is a revertible approximation, of order $2(l+1)$, to e_a^τ.

Proof. It is only needed to prove the lemma for $(i_1 i_2 \cdots i_m) = (12 \cdots m)$.

(1) It is easy to prove that the phase flow e_a^t has the series expansion

$$e_a^t(x) = x + \sum_{k=1}^{\infty} \frac{t^k}{k!} a^k(x), \quad x \in \mathbf{R}^n, t \sim 0, \tag{4.5}$$

where

$$a^1(x) = a(x), \quad a^2(x) = \frac{\partial a^1}{\partial x}(x) a(x), \quad a^k(x) = \frac{\partial a^{k-1}}{\partial x}(x) a(x), \quad k = 1, 2, \cdots. \tag{4.6}$$

The assumption that for $i = 1, 2, \cdots, m, G_i^\tau$ are approximations of order $p \geqslant 1$, to $e_{a(i)}^\tau$ implies that for all $x \in \mathbf{R}^n$,

$$G_i^\tau(x) = x + \tau a^{(i)}(x) + 0\left(\tau^2\right), \quad \tau \sim 0; \quad i = 1, 2, \cdots, m. \tag{4.7}$$

So, from Taylor expansion, we have that for $x \in \mathbf{R}^n$,

$$\left(G_2^\tau \circ G_1^\tau\right)(x) = G_2^\tau\left(G_1^\tau(x)\right) = x + \tau\left(a^{(1)}(x) + a^{(2)}(x)\right) + 0\left(\tau^2\right), \quad \tau \sim 0. \tag{4.8}$$

By induction for m, we get

$$\begin{aligned} {}_1G_{(12\cdots m)}^\tau(x) &= \left(G_m^\tau \circ \cdots \circ G_2^\tau \circ G_1^\tau\right)(x) \\ &= x + \tau\left(a^{(1)}(x) + a^{(2)}(x) + \cdots a^{(m)}(x)\right) + 0\left(\tau^2\right) \\ &= x + \tau a(x) + 0\left(\tau^2\right), \quad \tau \sim 0. \end{aligned} \tag{4.9}$$

This implies that ${}_1G_{(12\cdots m)}^\tau$ is an approximation, of order one, to e_a^τ. It was proved in [6] that ${}_2g_{i_1 i_2 \cdots i_m}^\tau$ and ${}_2\hat{g}_{i_1 i_2 \cdots i_m}^\tau$, defined by Eq. (4.2)′, are revertible approximations, of order 2, to e_a^τ. The conclusion (1) of the lemma is proved.

(2) By assumption, we have that for $x \in \mathbf{R}^n$ and $\tau \sim 0$,

$$G_i^\tau(x) = x + \tau a^{(i)}(x) + \frac{1}{2}\tau^2\left(a^{(i)}\right)^2(x) + 0\left(\tau^3\right), \quad i = 1, 2, \cdots, m. \tag{4.10}$$

Taylor expansion of the right hand side of Eq. (4.3) with $(i_1 i_2 \cdots i_m) = (12 \cdots m)$ yields

$${}_2G_{(12\cdots m)}^\tau(x) = x + \tau \sum_{i=1}^m a^{(i)}(x) + \frac{1}{2}\tau^2\left(\sum_{i,j=1}^m a^{(i)} a^{(j)}\right)(x) + 0\left(\tau^3\right), \quad \tau \sim 0. \tag{4.11}$$

Here we have used the convention

$$(ab)(x) = (a_* b)(x) = a_*(x) b(x), \quad a_*(x) = \frac{\partial a}{\partial x}(x) \tag{4.12}$$

for $a, b : \mathbf{R}^n \to \mathbf{R}^n$. On the other hand, we have

$$a^2 = a_* a = \left(\sum_{i=1}^m a^{(i)}\right)_* \left(\sum_{k=1}^m a^{(j)}\right) = \sum_{i,j=1}^m \left(a^{(i)}\right)_* a^{(j)} = \sum_{i,j=1}^m a^{(i)} a^{(j)}. \tag{4.13}$$

So,

$$e_a^\tau(x) = x + \tau a(x) + \frac{1}{2}\tau^2 a^2(x) + 0\left(\tau^3\right) ={}_2 G_{(12\cdots m)}^\tau(x) + 0\left(\tau^3\right), \quad \tau \sim 0.$$

This shows that ${}_2G_{(12\cdots m)}^\tau$ is an approximation, of order 2, to e_a^τ. By direct verification it is revertible if each component G_i^τ is revertible.

The conclusion (3) is Corollary 4.7 of [6].

Lemma 4 *Given system*

$$\dot{x} = a^{(k)}(x), \quad a^{(k)}(x) = \left(0, \cdots, 0, \frac{\partial b_{k,k+1}}{\partial x_{k+1}}(x), -\frac{\partial b_{k,k+1}}{\partial x_k}(x), 0, \cdots, 0\right)^T, \tag{4.14}$$

with $x = (x_1, \cdots, x_k, x_{k+1}, \cdots, x_n)^T$ *and smooth function* $b_{k,k+1} : \mathbf{R}^n \to \mathbf{R}^n$. *Then any symplectic difference scheme, of order* $p \geqslant 1$, *of the Hamiltonian system on the* x_k-x_{k+1} *plane*

$$\dot{x}_k = \frac{\partial b_{k,k+1}}{\partial x_{k+1}}, \quad \dot{x}_{k+1} = -\frac{\partial b_{k,k+1}}{\partial x_k} \tag{4.15}$$

with x_j, $j \neq k$, $k+1$ *as parameters naturally gives a volume-preserving difference scheme, of order* p, *of the source-free system* (4.14) *on the* n*-dimensional* $(x_1, \cdots, x_n)^T$*-space by simply freezing the coordinates* x_j, $j \neq k$, $k+1$ *and transforming* x_k *and* x_{k+1} *according to the symplectic difference scheme for* (4.15) *in which* x_j, $j \neq k$, $k+1$ *are considered as frozen parameters.*

Proof. It is obvious that the so-constructed difference scheme is of order p. As to the volume-preserving property, we easily prove that it is true by direct calculation of the determinant of the Jacobian of the step-transition map of the scheme, with the notice of the fact that the determinant of the Jacobian of a symplectic map is equal to one.

Now we construct volume-preserving difference schemes for source-free systems. Let $a = (a_1, \cdots, a_n)^T$ be a source-free field. As was proved in Section 3, we have an essentially Hamiltonian decomposition (3.15) for a with the functions $b_{k,k+1}$ given from a by (3.14). We denote by S_k^T the step-transition map of a volume-preserving difference scheme with step-size τ, as constructed in Lemma 4, associated to the vector field $a^{(k)} = \left(0, \cdots, 0, \frac{\partial b_{k,k+1}}{\partial x_{k+1}}, -\frac{\partial b_{k,k+1}}{\partial x_k}, 0, \cdots, 0\right)^T$ for $k = 1, 2, \cdots$. Then by Lemma 3, we have

Theorem 2 (1) *A simple composition of the* $n-1$ *components* $S_1^\tau, S_2^\tau, \cdots, S_{n-1}^\tau$, *say*

$${}_1G^\tau := S_{n-1}^\tau \circ \cdots \circ S_2^\tau \circ S_1^\tau, \tag{4.16}$$

is a volume-preserving algorithmic approximation, of order one, to e_a^τ; *and*

$${}_2g^\tau :={}_1 \hat{G}^{\tau/2} \circ_1 G^{\tau/2}, \quad {}_2\hat{g}^\tau ={}_1 G^{\tau/2} \circ_1 \hat{G}^{\tau/2} \tag{4.16$'$}$$

are revertible volume-preserving algorithmic approximations, of order 2, *to* e_a^τ.

(2) *If each* S_k^τ *is an approximation, of order* 2, *to* $e_{a(k)}^\tau$, *then the symmetric composition*

$$ {}_2G^\tau = S_{n-1}^{\tau/2} \circ \cdots \circ S_2^{\tau/2} \circ S_1^{\tau/2} \circ S_1^{\tau/2} \circ S_2^{\tau/2} \circ \cdots \circ S_{n-1}^{\tau/2} \tag{4.17}$$

is a volume-preserving approximation, of order 2, *to* e_a^τ.

(3) *If each* S_k^τ *is revertible, then the so-constructed* ${}_2G^\tau$ *is revertible too.*

(4) *From the above constructed revertible algorithmic approximation* ${}_2g^\tau$ *or* ${}_2G^\tau$, *we can further recursively constructed revertible approximations, of all even orders, to* e_α^τ *according to the process of Lemma* 3.

Remark 1 If a has essentially Hamiltonian decompositions other than (3.15) and (3.14), then one can construct volume-preserving difference schemes corresponding to these decompositions in a similar way to the above.

§5 Some Special Discussions for Separable Source-free Systems

For a source-free field $a = (a_1, \cdots, a_n)^T$ with essentially Hamiltonian decomposition (3.15), we take $S_k^\tau : x = (x_1, \cdots, x_n)^T \to \hat{x} = (\hat{x}_1, \cdots, \hat{x}_n)^T$ as determined from the following

$$\begin{cases} \hat{x}_j = x_j, \quad j \neq k, k+1 \\ \hat{x}_k = x_k + \tau \dfrac{\partial b_{k,k+1}}{\partial x_{k+1}} (x_1, \cdots, x_{k-1}, \hat{x}_k, x_{k+1}, \cdots, x_n) \\ \hat{x}_{k+1} = x_{k+1} - \tau \dfrac{\partial b_{k,k+1}}{\partial x_k} (x_1, \cdots, x_{k-1}, \hat{x}_k, x_{k+1}, \cdots, x_n). \end{cases} \tag{5.1}$$

Then simple calculations show that ${}_1G^\tau = S_{n-1}^\tau \circ \cdots \circ S_2^\tau \circ S_1^\tau$ is given from

$$\begin{cases} \hat{x}_1 = x_1 + \tau a_1 (\hat{x}_1, x_2, \cdots, x_n), \\ \hat{x}_j = x_j + \tau a_j (\hat{x}_1, \cdots, \hat{x}_j, x_{j+1}, \cdots, x_n) \\ \qquad + \tau \displaystyle\int_{x_j}^{\hat{x}_j} \sum_{l=1}^{j-1} \frac{\partial a_l}{\partial x_l} (\hat{x}_1, \cdots, \hat{x}_{j-1}, t, x_{j+1}, \cdots, x_n) \, dt, j = 2, \cdots, n-1, \\ \hat{x}_n = x_n + \tau a_n (\hat{x}_1, \cdots, \hat{x}_{n-1}, x_n) \end{cases} \tag{5.2}$$

and ${}_1\hat{G}^\tau = ({}_1G^{-\tau})^{-1}$ is given from

$$\begin{cases} \hat{x}_n = x_n + \tau a_n(x_1, \cdots, x_{n-1}, \hat{x}_n), \\ \hat{x}_j = x_j + \tau a_j(x_1, \cdots, x_j, \hat{x}_{j+1}, \cdots, \hat{x}_n) \\ \qquad - \tau \displaystyle\int_{x_j}^{\hat{x}_j} \sum_{l=1}^{j-1} \frac{\partial a_l}{\partial x_l}(x_1, \cdots, x_{j-1}, t, \hat{x}_{j+1}, \cdots, \hat{x}_n) dt, \quad j = 2, \cdots, n-1, \\ \hat{x}_1 = x_1 + \tau a_1(x_1, \hat{x}_2, \cdots, \hat{x}_n). \end{cases} \tag{5.3}$$

(5.2) and (5.3) are both volume-preserving difference scheme, of order one, of the source free system associated to the field a, with the step-transition maps ${}_1G^\tau$ and ${}_1\hat{G}^\tau$. They can be composed into revertible volume-preserving schemes of order 2, say, 2-stage scheme with step transition map ${}_2\hat{g}^\tau =_1 G^{\tau/2} \circ_1 \hat{G}^{\tau/2} : x = (x_1, \cdots, x_n)^T \to \hat{x} = (\hat{x}_1, \cdots, \hat{x}_n)^T$ as follows

$$\begin{cases} \hat{x}_n^{1/2} = x_n + \dfrac{\tau}{2} a_n \left(x_1, \cdots, x_{n-1}, \hat{x}_n^{1/2}\right), \\ \hat{x}_i^{1/2} = x_i + \dfrac{\tau}{2} a_i \left(x_1, \cdots, x_i, \hat{x}_{i+1}^{1/2}, \cdots, \hat{x}_n^{1/2}\right) \\ \qquad - \dfrac{\tau}{2} \displaystyle\int_{x_i}^{\hat{x}_i^{1/2}} \sum_{l=1}^{i-1} \frac{\partial a_l}{\partial x_l} \left(x_1, \cdots, x_{i-1}, t, \hat{x}_{i+1}^{1/2}, \cdots, \hat{x}_n^{1/2}\right) dt, \quad i = 2, \cdots, n-1, \\ \hat{x}_1^{1/2} = x_1 + \dfrac{\tau}{2} a_1 \left(x_1, \hat{x}_2^{1/2}, \cdots, \hat{x}_n^{1/2}\right); \\ \hat{x}_1 = \hat{x}_1^{1/2} + \dfrac{\tau}{2} a_1 \left(\hat{x}_1, \hat{x}_2^{1/2}, \cdots, \hat{x}_n^{1/2}\right), \\ \hat{x}_j = \hat{x}_j^{1/2} + \dfrac{\tau}{2} a_j \left(\hat{x}_1, \cdots, \hat{x}_j, \hat{x}_{j+1}^{1/2}, \cdots, \hat{x}_n^{1/2}\right) \\ \qquad + \dfrac{\tau}{2} \displaystyle\int_{\hat{x}_j^{1/2}}^{\hat{x}_j} \sum_{l=1}^{j-1} \frac{\partial a_l}{\partial x_l} \left(\hat{x}_1, \cdots, \hat{x}_{j-1}, t, \hat{x}_{j+1}^{1/2}, \cdots, \hat{x}_n^{1/2}\right) dt, \quad j = 2, \cdots, n-1, \\ \hat{x}_n = \hat{x}_n^{1/2} + \dfrac{\tau}{2} a_n \left(\hat{x}_1, \cdots, \hat{x}_{n-1}, \hat{x}_n^{1/2}\right). \end{cases} \tag{5.4}$$

Either (5.2) or (5.3) contains $n-1$ implicit equations generally. But for fields a with some specific properties, it will turn into explicit. For example, if $a = (a_1, \cdots, a_n)^T$ satisfies condition

$$\frac{\partial a_i}{\partial x_i} = 0, \quad i = 1, \cdots, n \tag{5.5}$$

(i.e., a_i does not depend on x_i), then (5.2) turns into

$$\begin{cases} \hat{x}_1 = x_1 + \tau a_1 \left(x_2, \cdots, x_n\right) \\ \hat{x}_j = x_j + \tau a_j \left(\hat{x}_1, \cdots, \hat{x}_{j-1}, x_{j+1}, \cdots, x_n\right), \quad j = 2, \cdots, n-1 \\ \hat{x}_n = x_n + \tau a_n \left(\hat{x}_1, \cdots, \hat{x}_{n-1}\right) \end{cases} \tag{5.6}$$

which is explicit. We note that, for $a=(a_1,\cdots,a_n)^T$,

$$a=\sum_{k=1}^{n}a^{\{k\}},\quad a^{\{k\}}=(0,\cdots,0,a_k,0,\cdots,0)^T,\quad k=1,2,\cdots,n. \tag{5.7}$$

It is easy to verify that if $a=(a_1,\cdots,a_n)^T$ satisfies the condition (5.5), then the scheme (5.6) is just the result of composing the Euler explicit schemes of the systems associated to the fields $a^{\{k\}}, k=1,\cdots,n$, i.e., we have

$${}_1G^\tau=E^\tau_{a\{n\}}\circ\cdots\circ E^\tau_{a\{2\}}\circ E^\tau_{a\{1\}}, \tag{5.8}$$

where

$$E^\tau_{a\{k\}}=1+\tau a^{\{k\}},\quad k=1,2,\cdots,n,\quad 1=\text{ identity}. \tag{5.9}$$

In fact, $E^\tau_{a\{k\}}$ are the phase flows $e^\tau_{a\{k\}}$, since $a_*^{\{k\}}a^{\{k\}}=0$ for $k=1,2,\cdots,n$, which is implied by the condition (5.5). According to Theorem 2, we then get a 2nd order explicit revertible volume-preserving scheme, with step transition map

$$\begin{aligned}{}_2G^\tau&=E^{\tau/2}_{a\{n\}}\circ\cdots\circ E^{\tau/2}_{a\{2\}}\circ E^{\tau/2}_{a\{1\}}\circ E^{\tau/2}_{a\{1\}}\circ E^{\tau/2}_{a\{2\}}\circ\cdots\circ E^{\tau/2}_{a\{n\}}\\&={}_1G^{\tau/2}\circ{}_1\hat{G}^{\tau/2}={}_2\hat{g}^\tau,\end{aligned} \tag{5.10}$$

which is given by (5.4) without the integral terms. Also, we can construct explicit revertible volume-preserving schemes of various even orders of the systems of the above type from the ${}_2G^\tau$ according to the constructing process stated in Theorem 2.

Systems satisfying the condition (5.5) are important in applications. For example, the well-known ABC flows and Jacobian elliptic curves are described by such systems. From Qin and Zhu's numerical computation for these two examples [7], one may see that volume-preserving algorithms are superior to the non-volume-preserving ones.

In [3,4,15], Feng and Wang introduced the concept of L-separability of vector fields in a Lie subalgebra L of the Lie algebra of all smooth vector fields on $\mathbf{R}^n$ in the sense that the fields can be decomposed as finite sums of vector fields which both belong to L and generate linear phase flows. In this sense, vector fields satisfying the condition (5.5) are source-free separable (all smooth source-free vector fields form a Lie algebra under the usual Lie bracket of vector fields), and so are finite sums of such fields. One easily verifies, in a similar way to that in the Hamiltonian case [3], that all polynomial source-free vector fields are source-free separable. Noting that the above discussions, one can construct explicit revertible volume-preserving difference schemes of various even orders for systems associated to source-free separable vector fields. We call such systems separable source-free ones.

Acknowledgements The authors would like to thank Qin Meng-zhao, Li Wang-yao, Wang Dao-liu, Zhang Mei-qing, Tang Yi-fa, Zhu Wen-jie, Shu Hai-bing and Jiang Li-xin for their helpful comments.

References

[1] Arnold V I. Mathematical Methods of Classical Mechanics. New York: Springer. 1978

[2] Feng K. The Hamiltonian way for computing Hamiltonian dynamics. In Spigler R, ed. Applied and Industrial Mathematics. Netherlands: Kluwer, 1991. 17-35

[3] Feng K, Wang D L. Variations on a theme by Euler: [Preprint]. Beijing: CAS. Computing Center. 1991

[4] Feng K, Wang D L. Dynamical systems and geometric construction of algorithms. In Shi Z C, Yang C C, eds. Comput Math in China. Contemporary Mathematics. AMS. 1994: 1-32

[5] Feng K, Wu H M, Q in M Z, Wang D L. Construction of canonical difference schemes for Hamiltonian formalism via generating functions. J Comp Math. 1989 7(1): 71-96

[6] Qin M Z, Zhu W J. Construction of higher order symplectic schemes by composition. Computing. 1992 47: 309-321

[7] Qin M Z, Zhu W J. Volume-preserving schemes and numerical experiments. Comp Math Appl. 1993 26: 33-42

[8] Shang Z J. Generating functions for volume-preserving mappings and Hamilton-Jacobi equations for source-free dynamical systems. Science in China (Series A). 1994 37(10): 1172-1188

[9] Shang Z J. Construction of volume-preserving difference schemes for source-free systems via generating functions. J Comp Math. 1994 12(3) 265-272

[10] Spivak M. Calculus on Manifolds. New York: Benjamin Inc. 1965

[11] Strang G. On the construction and the comparison of difference schemes. SIAM J Numer Anal. 1968 12: 506-517

[12] Strang G. Accurate partial difference methods I: Linear Cauchy problems. Arch Ration Mech Anal. 1963 12: 392-402

[13] Suzuki M. Fractal decomposition of exponential operators with applications to many-body theories and Monte Carlo simulations. Phys Letters A. 1990 146: 319-323

[14] Suzuki M. General theory of higher-order decomposition of exponential operators and symplectic integrators. Phys Letters A. 1992 165: 387-395

[15] Wang DL. Decomposition of vector fields and composition of algorithms. In Feng K, Shi Z C, ed. Proc Inter Conf on Computation of Differential Equations and Dynamical Systems. Singapore: World Scientific, 1993, 179-184

[16] Warner F W. Foundations of Differentiable Manifolds and Lie Groups. New York: Springer. 1983

[17] Weyl H. The method of orthogonal projection in potential theory. Duke Math J. 1940 7: 411-444

[18] Yoshida H. Construction of higher order symplectic integrators, Phys Letters A. 1990 150: 262-268

25 Symplectic Algorithms for Hamiltonian Systems[①]

哈密尔顿系统的辛算法

Abstract

In this paper we give a brief survey on the research, mainly undertaken by the authors and their research group, on various aspects of symplectic algorithms, such as construction of explicit and implicit symplectic algorithms of Hamiltonian systems, conservation laws, linear stability and formal energies.

§1 Introduction

Hamiltonian systems have a fundamental geometry — symplectic geometry. Their phase flow preserves the symplectic geometric structure, which is a one parameter group of symplectic transformations. Different from conventional numerical methods, symplectic algorithms simulate Hamiltonian systems in the same geometric framework, i.e., the step transition operators of symplectic algorithms are symplectic transformations. Hence, they have no artificial excitation and dissipation while most conventional numerical methods inevitably bring in artificial excitation and dissipation. Therefore, symplectic algorithms are pure and clean and suitable to long term tracking and qualitative simulations. The idea has been widely generalized to other systems with geometric structures or Lie algebraic structures, such as contact algorithms for contact systems, volume preserving algorithms for source free systems, which all are named structure preserving algorithms. Readers interested in these aspects can refer to [13, 15, 16, 19, 22, 47].

In this paper we only review symplectic algorithms for Hamiltonian systems, the construction of algorithms, conservation laws, linear stability and formal energies.

In section 2, we give some basic materials of ordinary differential equations and numerical methods. In section 3, we introduce symplectic algorithms for Hamiltonian systems. Section 4 is concerning with symplectically separable Hamiltonian systems and explicit symplectic algorithms. Symplectic algorithms are essentially implicit, but

---

① Joint with Wang D L

for symplectically separable systems, explicit schemes can be constructed. In section 5 we discuss unconditional symplectic algorithms and the generating function methods. Section 6 is for symplectic algorithms of perturbed Hamiltonian systems. In section 7 symplectic Runge-Kutta methods are given. In section 8, we consider the conservation laws of symplectic algorithms. In section 9, we discuss the linear stability of symplectic algorithms — H-stability. Section 10 concerns with formal and non-autonomous perturbed energies of symplectic algorithms.

This paper is based on the lectures given in Computational Mathematics Year held at the Institute of Mathematics of Nankai University. We would like to thank the organizers for their invitation and the Institute of Mathematics of Nankai University for their help in all aspects during the period of lectures.

§2 Preliminaries

2.1 Phase flow

Consider the autonomous system on $\mathbf{R}^m$

$$\dot{z}=a(z), \quad z=\left(z_1, \cdots, z_m\right)^T, \quad a(z)=\left(a_1(z), \cdots, a_m(z)\right)^T, \tag{2.1}$$

where $a(z)$ is a smooth vector field on $\mathbf{R}^m$. The superscript T denotes transpose of a matrix. The solution of System (2.1) with an initial point $z(0)=z_0$ is denoted by $z\left(t; z_0\right)=e_a^t z_0$. Usually, it exists for initial points in a domain $\mathcal{D}$ on $\mathbf{R}^m$ and for time in an interval (a, b) containing 0. For simplicity we assume $\mathcal{D}=\mathbf{R}^m$ and $(a, b)=\mathbf{R}$. The solution $z\left(t; z_0\right)$ smoothly depends not only on time t but also on initial points z_0. Hence for any fixed t, e_a^t is a transformation on $\mathbf{R}^m$ and is called the phase flow, or t-flow, of System (2.1).

The phase flow e_a^t of System (2.1) defined by the vector field a satisfies

$$\begin{aligned} \frac{d}{dt} e_a^t &= a \circ e_a^t, \\ e_a^0 &= \text{identity } := 1_m. \end{aligned}$$

e_a^t is a one parameter group in t

$$\begin{aligned} & e_a^{t+s}=e_a^t \circ e_a^s, \quad \forall t, s \in \mathbf{R}, \\ & e_a^0=1_m \end{aligned} \tag{2.2}$$

and expressible as a convergent, at least locally, power series

$$e_a^t=\sum_{k=0}^{\infty} t^k e_k, \tag{2.3}$$

where the coefficients are smooth vector functions from $\mathbf{R}^m$ to $\mathbf{R}^m$ and can be determined recursively

$$e_0 = e^0 = 1_m, e_1 = a, e_k = \frac{1}{k}(e_{k-1})_* e_1, \quad k = 1, 2, \cdots, \tag{2.4}$$

here Jacobian matrix of $u : \mathbf{R}^m \to \mathbf{R}^m$ is denoted by u_*.

2.2 Numerical methods

We now consider a one-step numerical method to approximate System (2.1). When $z^0 = z_0$ is given, we then can get $z^1, z^2, \cdots$ by the method (explicitly or implicitly). Obviously, z^0 can be taken as any initial point. Therefore, this method really defines a mapping on $\mathbf{R}^m$ with the time step-size, say s, as a parameter, denoted by g_a^s. Thus

$$z^0 \text{ given}, \; z^{k+1} = g_a^s z^k, \quad k \geqslant 0.$$

The method is said to be of order $r(\geqslant 1)$ if

$$g_a^s = e_a^s + O(s^{r+1}). \tag{2.5}$$

Therefore, it is equivalent to consider the approximation of $\{z^k\}$ to $z(ks)$ and of $(g_a^s)^k$ to e_a^{ks}. g_a^s is called the step transition operator (or mapping) of the method (or synonymously, algorithm). We always identify the step transition operator with a numerical method.

For example, the explicit and implicit Euler schemes for System (2.1) are

$$z \to \hat{z} = E_a^s z: \quad \hat{z} = z + sa(z),$$
$$z \to \hat{z} = I_a^s z: \quad \hat{z} = z + sa(\hat{z}).$$

Their step transition operators E_a^s and I_a^s are

$$E_a^s = 1_m + sa, \quad I_a^s = (1_m - sa)^{-1}. \tag{2.6}$$

Two-leg and one-leg weighted Euler schemes are

$$\begin{aligned} z \to \hat{z} = T_{a,c}^s z : \hat{z} = z + s(ca(\hat{z}) + (1-c)a(z)), \\ z \to \hat{z} = E_{a,c}^s z : \hat{z} = z + sa(c\hat{z} + (1-c)z), \end{aligned} \tag{2.7}$$

where c is a real number. Their step transition operators $T_{a,c}^s$ and $E_{a,c}^s$ are

$$T_{a,c}^s = (1_m - sca)^{-1} \circ (1_m + s(1-c)a) = I_a^{cs} \circ E_a^{(1-c)s},$$
$$E_{a,c}^s = (1_m + s(1-c)a) \circ (1_m - sca)^{-1} = E_a^{(1-c)s} \circ I_a^{sc}.$$

These show that $T^s_{a,c}$ and $E^s_{a,c}$ are really the composite schemes of the explicit and implicit schemes, $E_a^{(1-c)s}$ and I_a^{sc}, with different orders. Here the composite method $g^s = g_k^s \circ \cdots \circ g_1^s$ of methods $g_1^s, \cdots, g_k^s$ is implemented by the k-stage

$$z \to \hat{z} = g^s z: \quad z^1 = g_1^s z, z^2 = g_2^s z^1, \cdots, z^{k-1} = g_{k-1}^s z^{k-2}, \hat{z} = g_k^s z^{k-1}.$$

If $g_1^s, \cdots, g_k^s$ are explicit, then g^s is also explicit.

The transition operator g^s can be expressible as a power series

$$g_a^s = \sum_{k=0}^{\infty} s^k g_k, \quad g_0 = 1_m. \tag{2.8}$$

The order condition (2.5) is equivalent to

$$g_k = e_k, \quad k = 0, 1, \cdots, r; \quad g_{r+1} \neq e_{r+1}. \tag{2.9}$$

2.3 Revertibility

The group property (2.2) of the phase flow is a quite strong property. Usually, a numerical method can not be a group in step size s. Hence we consider a weak property of the phase flow e_a^t.

Let g^t be a one-parameter family of near-1 diffeomorphisms with $g^0 = 1_m$. g^t is called *revertible* if $g^t \circ g^{-t} = 1_m, \forall t$. We define the reversion of g^t as

$$\check{g}^t = \left(g^{-t}\right)^{-1}. \tag{2.10}$$

g^t is revertible if and only if $g^t = \check{g}^t$. Obviously, phase flow e_a^t is revertible. Moreover, we have

(1) $\check{g}^t \circ g^t$ and $g^t \circ \check{g}^t$ are always revertible for all g^t.

(2) $\check{\check{g}}^t = g^t, (f^t \circ g^t)^{\vee} = \check{g}^t \circ \check{f}^t$.

(3) If f^t and g^t are revertible, then $f^{\alpha t} \circ g^{\beta t} \circ f^{\alpha t}$ and $f^{\alpha t} \circ g^{\beta t} \circ g^{\beta t} \circ f^{\alpha t}$ are revertible for all $\alpha, \beta \in \mathbf{R}$.

(4) If f_i^t are revertible, $i = 1, \cdots, k$, then $f_1^{\alpha_1 t} \circ \cdots \circ f_k^{\alpha_k t} \circ f_k^{\alpha_k t} \circ \cdots \circ f_1^{\alpha_1 t}$ are revertible for all $\alpha_i \in \mathbf{R}, i = 1, \cdots, k$.

(5) When f^t and g^t are revertible, then $f^t \circ g^t$ is revertible if and only if $f^t \circ g^t = g^t \circ f^t$.

It can be seen from (2.6) and (2.7) that $I_a^s = \left(E_a^{-s}\right)^{-1} = \check{E}_a^s$, i.e., the reversion of the explicit Euler operator is the implicit Euler operator, and $\check{T}^s_{a,c} = T^s_{a,1-c}, \check{E}^s_{a,c} = E^s_{a,1-c}$, i.e. the reversion of the two- and one-leg Euler operators with the parameter c are the two- and one-leg Euler operators with the parameter $1 - c$.

Consequently, $T^s_{a,c}$ and $E^s_{a,c}$ are revertible for any vector field a if and only if $c = \dfrac{1}{2}$, i.e., they are trapezoidal and centered Euler schemes respectively.

2.4 Composition Methods

We write $g^s \approx e_a^s$ if $g^s = e_a^s + O(s^2)$, $g^s \approx e_a^s$, ord r if $g^s = e_a^s + O(s^{r+1})$, $g^s \approx e_a^s$, ord ∞ if $g^s = e_a^s$. A consistent revertible algorithm is always of even order $2l (l \geqslant 1)$. For $g^s \approx e_a^s$, ord 1,

$$g^{s/2} \circ \check{g}^{s/2} \quad \text{and} \quad \check{g}^{s/2} \circ g^{s/2} \approx e_a^s \tag{2.11}$$

are revertible and of order 2. If $g^s \approx e_a^s$ is revertible and of order 2, then the revertible composite of g^s :

$$g^{\alpha s} \circ g^{\beta s} \circ g^{\alpha s} \approx e_a^s \tag{2.12}$$

is of order 4 when

$$2\alpha + \beta = 1, \quad 2\alpha^3 + \beta^3 = 0, \tag{2.13}$$

i.e.,

$$\alpha = \frac{1}{2 - 2^{1/3}} > 0, \quad \beta = 1 - 2\alpha < 0.$$

Generally, if $g^s \approx e_a^s$ is revertible and of order $2l$, then the revertible composite (2.12) of g^s is of order $2(l+1)$, when

$$2\alpha + \beta = 1, \quad 2\alpha^{2l+1} + \beta^{2l+1} = 0, \tag{2.14}$$

i.e.,

$$\alpha = \frac{1}{2 - 2^{1/(2l+1)}} > 0, \quad \beta = 1 - 2\alpha < 0.$$

Since g^s is of order $2l$, there exists a vector function $\bar{e}$ such that

$$g^s = e^s + s^{2l+1}\bar{e} + O\left(s^{2l+2}\right).$$

Then

$$\begin{aligned} g^{\beta s} \circ g^{\alpha s} &= \left(e^{\beta s} + \beta^{2l+1} s^{2l+1} \bar{e}\right) \circ \left(e^{\alpha s} + \alpha^{2l+1} s^{2l+1} \bar{e}\right) + O\left(s^{2l+2}\right) \\ &= e^{(\beta+\alpha)s} + \left(\beta^{2l+1} + \alpha^{2l+1}\right) s^{2l+1} \bar{e} + O\left(s^{2l+2}\right). \end{aligned}$$

Similarly, we have

$$g^{\alpha s} \circ g^{\beta s} \circ g^{\alpha s} = e^{(\beta+2\alpha)s} + \left(\beta^{2l+1} + 2\alpha^{2l+1}\right) s^{2l+1} \bar{e} + O\left(s^{2l+2}\right).$$

Consequently, Condition (2.14) leads to that

$$g^{\alpha s} \circ g^{\beta s} \circ g^{\alpha s} = e^s + O\left(s^{2l+2}\right),$$

i.e., it is of order $2l+1$. Since it is revertible, it must be of order $2l+2$.

For more details, refer to [21, 33, 36, 52]

§3 Symplectic Algorithms — A Geometric View-point

We consider now Hamiltonian systems on $\mathbf{R}^{2n}$

$$\frac{dp}{dt} = -H_q, \quad \frac{dq}{dt} = H_p, \qquad p = \begin{pmatrix} p_1 \\ \vdots \\ p_n \end{pmatrix}, \quad q = \begin{pmatrix} q_1 \\ \vdots \\ q_n \end{pmatrix},$$

or in compact form,

$$\frac{dz}{dt} = J\nabla H(z), \quad z = \begin{pmatrix} p \\ q \end{pmatrix} \in \mathbf{R}^{2n}, \quad J = J_{2n} = \begin{pmatrix} 0 & -I_n \\ I_n & 0 \end{pmatrix}, \tag{3.1}$$

where $H(z)$ is a Hamiltonian function, $\nabla H(z) = H_z = (H_{z_1}(z), \cdots, H_{z_{2n}}(z))^T$ is the gradient of H with respect to z, 0 and I are the n by n zero and identity matrices. The phase flow of (3.1) is denoted by e_H^t which can be expanded as a power series

$$e_H^t = e_0 + te_1 + t^2e_2 + \cdots, \tag{3.2}$$

where

$$e_0 = 1_{2n}, e_1 = J\nabla H, e_2 = \frac{1}{2}(J\nabla H)_* J\nabla H,$$

$$e_k = \frac{1}{k}(e_{k-1})_* J\nabla H, k \geqslant 3. \tag{3.3}$$

The symplectic structure on $\mathbf{R}^{2n}$ is defined by the differential 2-form

$$\omega = \sum_{i=1}^{n} dp_i \wedge dq_i$$

which corresponds to the 2-dimensional oriented area on the tangent space of $\mathbf{R}^{2n}$. A *symplectic* transformation is a diffeomorphism $g : z \to \hat{z}$ on $\mathbf{R}^{2n}$ preserving the symplectic structure

$$g^*\omega = \omega \quad \text{i.e.,} \quad \sum d\hat{p}_i \wedge d\hat{q}_i = \sum dp_i \wedge dq_i,$$

or in matrix form,

$$g_*^T(z) J_{2n} g_*(z) \equiv J_{2n}.$$

Hence all symplectic transformations automatically preserve phase areas of even dimensions. All symplectic transformations on $\mathbf{R}^{2n}$ form a group under composition of transformations, denoted by SpD_{2n}.

Intrinsic to all Hamiltonian systems is that the phase flow of Hamiltonian systems preserves the symplectic structure, i.e.,

$$\left(e_H^t\right)^* \omega = \omega, \quad \text{or} \quad \left(e_H^t\right)_*^T (z) J \left(e_H^t\right)_* (z) = J, \quad \forall z \in \mathbf{R}^{2n}, t \in \mathbf{R}.$$

It follows that the phase areas of even-dimensions and the Hamiltonian (energy) are conserved in time evolution. It is natural and mandatory to require algorithms to be symplectic, such algorithms are called symplectic algorithms. More precisely, symplectic algorithms are numerical methods whose step transition operator g_H^s preserves the symplectic structure of the phase space as applied to Hamiltonian systems, i.e.,

$$(g_H^s)_*^T J (g_H^s)_* = J. \tag{3.4}$$

Hence symplectic algorithms also automatically preserve phase areas of even dimensions.

Conventional numerical methods only consider the accuracy of approximation of numerical solutions to exact solutions. They do not concern with the geometric properties that the systems possess. On the one hand, Hamiltonian systems are conversation systems, while most conventional numerical methods, such as Runge-Kutta methods, have an artificial excitation or dissipation which will completely destroy the behavior of the exact solutions in long term simulation. Symplectic algorithms are clean, they do not have such artificial excitation and dissipation and are able to preserve the behavior of the solution in long term simulation. On the other hand, a system of differential equations is a continuous dynamical system, a numerical method to approximate the system can be regarded as a discrete dynamical system for fixed time step-size. Therefore, when one uses a method to solve a system (problem), one really uses a discrete dynamical system to simulate a continuous dynamical system. In this sense, it is natural to require that the numerical simulation should be done in the same geometric framework, especially in long term tracking and qualitative numerical simulations. This requirement is easy to understand since in long term simulation, any numerical solution is far from the exact solution, only the qualitative properties remain. In this aspect, symplectic algorithms are overwhelmingly superior to conventional numerical methods. This idea has been widely generalized to other systems, such as contact systems, source free systems, Poisson systems, systems with Lie algebraic structures. Readers interested in this field can refer to [13, 15, 16, 19, 22, 47].

§4 Explicit Symplectic Algorithms

There are two kinds of symplectic algorithms. One is called unconditional symplectic algorithms which are symplectic for all Hamiltonian systems, another is symplectic for a kind of Hamiltonians. For example, explicit Euler scheme is generally non-symplectic, but it is symplectic for $H(z) = U(q)$ or $V(p)$. In this section we will construct explicit symplectic algorithms for symplectically separable Hamiltonian systems by composition of explicit Euler schemes.

Definition 1 A Hamiltonian function H is nilpotent of degree 2 if

$$(\nabla H)_* J \nabla H = 0 \quad \text{or} \quad H_{zz} J H_z = 0. \tag{4.1}$$

For Hamiltonians H nilpotent of degree 2, by the expressions (3.2) and (3.3), $e_k = 0$ for $k \geqslant 2$. Therefore the phase flow is

$$e_H^t(z) = z + t e_1(z) = z + tJ\nabla H(z),$$

which is just the explicit Euler scheme at $t = s : e_H^s(z) = E_H^s$. So E_H^s must be symplectic. In this case, E_H^s is also revertible since e_H^s is revertible.

Definition 2 $H(z)$ is *symplectically separable* if $H(z)$ can be decomposed as

$$H(z) = \sum_{i=1}^{k} H_i(z) \tag{4.2}$$

and every $H_i(z)$ is nilpotent of degree 2.

For symplectically separable Hamiltonians $H(z)$ with the decomposition (4.2), the following composite

$$z \to \hat{z} = g_1^s z, \quad g_1^s := E_{H_1}^s \circ E_{H_2}^s \circ \cdots \circ E_{H_k}^s \tag{4.3}$$

is explicit, symplectic and of order 1. The symmetric composite

$$\begin{aligned} &z \to \hat{z} = g_2^s z, \\ &g_2^s := g_1^{s/2} \circ \breve{g}_1^{s/2} = E_{H_1}^{s/2} \circ \cdots \circ E_{H_{k-1}}^{s/2} \circ E_{H_k}^s \circ E_{H_{k-1}}^{s/2} \circ \cdots \circ E_{H_1}^{s/2} \end{aligned} \tag{4.4}$$

is explicit, symplectic and revertible of order 2. Higher order revertible symplectic schemes can be constructed from these schemes by the procedure (2.12),(2.13) and (2.14).

If

$$H(z) = H(p, q) = U(p) \quad \text{or} \quad V(q),$$

where U and V are functions of n variables, then $H(z)$ is nilpotent of degree 2

$$\begin{pmatrix} H_p(p,q) \\ H_q(p,q) \end{pmatrix} = \begin{pmatrix} U_p(p) \\ 0 \end{pmatrix} \quad \text{or} \quad \begin{pmatrix} 0 \\ V_q(q) \end{pmatrix}.$$

The corresponding explicit Euler schemes

$$\begin{array}{lcl} \hat{p}=p & \text{or} & \hat{p}=p-sV_q(q) \\ \hat{q}=q+sU_p(p) & & \hat{q}=q \end{array}$$

are symplectic. If

$$H(p,q)=U(p)+V(q), \tag{4.5}$$

which is symplectically separable, then the composite schemes

$$\begin{array}{ll} E_V^s\circ E_U^s: & \hat{p}=p-sV_q(\hat{q}), \quad \hat{q}=q+sU_p(p), \\ E_U^s\circ E_V^s: & \hat{p}=p-sV_q(q), \quad \hat{q}=q+sU_p(\hat{p}) \end{array} \tag{4.6}$$

are explicit, symplectic and of order 1. The revertible scheme

$$\begin{gathered} g_2^s:=\left(E_V^{s/2}\circ E_U^{s/2}\right)^{\vee}\circ\left(E_V^{s/2}\circ E_U^{s/2}\right)=E_U^{s/2}\circ E_V^s\circ E_U^{s/2}: \\ q^1=q+\frac{s}{2}U_p(p), \quad \hat{p}=p-sV_q\left(q_1\right), \quad \hat{q}=q^1+\frac{s}{2}U_p(\hat{p}) \end{gathered} \tag{4.7}$$

is symplectic and of order 2. Its revertible composite

$$g_4^s:=g_2^{\alpha s}\circ g_2^{\beta_s}\circ g_2^{\alpha s}$$

gives an explicit symplectic scheme [24,32,33,36,52] of order 4 with the parameters (2.13), i.e.,

$$\begin{array}{ll} p^1=p-c_1sV_q(q), & q^1=q+d_1sU_p\left(p^1\right), \\ p^2=p^1-c_2sV_q\left(q^1\right), & q^2=q^1+d_2sU_p\left(p^2\right), \\ p^3=p^2-c_3sV_q\left(q^2\right), & q^3=q^2+d_3sU_p\left(p^3\right), \\ \hat{p}=p^3-c_4sV_q\left(q^3\right), & \hat{q}=q^3+d_4sU_p(\hat{p}), \end{array} \tag{4.8}$$

with the parameters $\alpha=\left(2-2^{1/3}\right)^{-1}, \beta=1-2\alpha$ and either

$$c_1=0, c_2=c_4=\alpha, c_3=\beta, d_1=d_4=\alpha/2, d_2=d_3=(\alpha+\beta)/2, \tag{4.9}$$

or

$$c_1=c_4=\alpha/2, \quad c_2=c_3=(\alpha+\beta)/2, \quad d_1=d_3=\alpha, \quad d_2=\beta, \quad d_4=0. \tag{4.10}$$

The 3rd order explicit symplectic scheme [37] for the separable Hamiltonian (4.5) is given by

$$\begin{array}{ll} p^1=p-c_1sV_q(q), & q^1=q+d_1sU_p\left(p^1\right), \\ p^2=p^1-c_2sV_q\left(q^1\right), & q^2=q^1+d_2sU_p\left(p^2\right), \\ \hat{p}=p^2-c_3sV_q\left(q^2\right), & \hat{q}=q^2+d_3sU_p(\hat{p}) \end{array} \tag{4.11}$$

with the parameters

$$c_1 = 7/24, c_2 = 3/4, c_3 = -1/24, d_1 = 2/3, d_2 = -2/3, d_3 = 1, \tag{4.12}$$

or

$$c_1 = 1, c_2 = -2/3, c_3 = 2/3, d_1 = -1/24, d_2 = 3/4, d_3 = 7/24. \tag{4.13}$$

Symplectically separable Hamiltonians have wide coverage in applications. It is easy to see that, the Hamiltonian of the form

$$H(p,q) = \phi(Ap + Bq), \quad AB^T = BA^T, \tag{4.14}$$

where $\phi(x)$ is a function of n variables, A and B are $n \times n$ matrices, is also nilpotent of degree 2. The explicit Euler scheme $E_H^s = E^s(\phi) = e_H^s$ is

$$E^s(\phi): \quad \begin{array}{l} \hat{p} = p - sB^T\phi_x(Ap + Bq), \\ \hat{q} = q + sA^T\phi_x(Ap + Bq). \end{array} \tag{4.15}$$

Hence we get a class of symplectically separable Hamiltonians [21]

$$H(p,q) = \sum_{i=1}^{k} H_i(p,q), \quad H_i(p,q) = \phi_i(A_i p + B_i q), \quad A_i B_i^T = B_i A_i^T, \tag{4.16}$$

where $\phi_i(x)$ are functions of n variables, A_i and B_i are $n \times n$ matrices. Moreover, all polynomial Hamiltonians belong to this class; in fact, it has been proved in [21] that every polynomial $H(p,q)$ in $2n$ variables p, q can be decomposed in the form of (4.16) with polynomials $\phi_i(x)$ of n variables x and diagonal matrices $A_i = \operatorname{diag}(a_1^i, \cdots, a_n^i)$ and $B_i = \operatorname{diag}(b_1^i, \cdots, b_n^i)$.

For symplectically separable Hamiltonians of the above class, all explicit revertible symplectic schemes formed by composition contain solely the Euler schemes $E^s(\phi_i)$ in the form (4.15) as basic components.

More generally, if we have a decomposition $H = H_1 + \cdots + H_k$, for which each H_i is integrable and its phase flow $e_{H_i}^s$ is algorithmically implementable, then the symmetrical composites

$$\begin{aligned} g_{2,H}^s &= e_{H_1}^{s/2} \circ \cdots \circ e_{H_{k-1}}^{s/2} \circ e_{H_k}^s \circ e_{H_{k-1}}^{s/2} \circ \cdots \circ e_{H_1}^{s/2} \\ g_{4,H}^s &= g_{2,H}^{a_s} \circ g_{2,H}^{\beta_s} \circ g_{2,H}^{a_2} \quad \text{with parameters (2.13)} \end{aligned}$$

give symplectic and revertible algorithms of order 2,4, etc. This general approach is widely applicable to *many body problems* in different physical contexts for which the *2-body problem* is *solvable*. Since in such problems the Hamiltonian usually admits a natural decomposition $H = \sum_{i<j} H_{ij}$, each H_{ij} accounts for a 2-body problem.

An interesting result from this approach is a construction[55] of explicit symplectic method for computing the Hamiltonian system of N vortices $z_i = (x_i, y_i)$ with intensities k_i

$$k_i \frac{dx_i}{dt} = \frac{\partial H}{\partial y_i}, \quad k_i \frac{dy_i}{dt} = -\frac{\partial H}{\partial x_i}, \quad i = 1, \cdots, N$$

with symplectic structure $\omega_k = \sum \frac{1}{k_i} dx_i \wedge dy_i$ and

$$H = \sum_{i<j} H_{ij}, \quad H_{ij} = -\frac{1}{2\pi} k_i k_j \ln r_{ij}, \quad r_{ij} = \left((x_i - x_j)^2 + (y_i - y_j)^2 \right)^{1/2}.$$

Each H_{ij} accounts for a solvable 2-vortex motion in which both vortices z_i, z_j rotate about their center of vorticity $z = (k_i z_i + k_j z_j)(k_i + k_j)^{-1}$ with the angular velocity $a = (k_i + k_j)/2\pi r_{ij}^2$ if $k_i + k_j \neq 0$ or translate with the linear velocity $(b(y_i - y_j), -b(x_i - x_j))$, $b = k_i/2\pi r_{ij}^2$ if $k_i + k_j = 0$. These simple phase flows $e^s_{H_{ij}}$ serve as the basic algorithmic components for successive compositions, resulting in efficient explicit revertible methods, symplectic for this specific system. They are promising in application to incompressible ideal flows for tracking the vortex particles.

§5 Generating Function Methods

Now we consider *unconditional* symplectic algorithms, i.e., they are symplectic for *all* Hamiltonian systems.

We first consider the simplest cases: one-leg weighted Euler scheme

$$\hat{z} = E^s_{H,c} z : \quad \hat{z} = z + sJH_z(c\hat{z} + (1-c)z) \tag{5.1}$$

with real number c. It is unconditionally symplectic if and only if $c = \frac{1}{2}$, which corresponds to the *centered Euler scheme*

$$\hat{z} = z + sJH_z\left(\frac{\hat{z} + z}{2}\right). \tag{5.2}$$

This illustrates a general situation: apart from some very rare exceptions, the vast majority of conventional schemes are non-symplectic. However, if we allow c in (5.1) to be a real matrix of order $2n$, we get a far-reaching generalization: (5.1) is symplectic if and only if

$$c = \frac{1}{2}(I_{2n} + J_{2n}B), \quad B^T = B, \quad c^T J + Jc = J. \tag{5.3}$$

The simplest and important cases are

$$
\begin{aligned}
&C: \quad c=\frac{1}{2}I_{2n}, \quad \hat{z}=z+sJH_z\left(\frac{\hat{z}+z}{2}\right),\\
&P: \quad c=\begin{pmatrix} I & 0\\ 0 & 0\end{pmatrix}, \quad \begin{aligned}\hat{p}&=p-sH_q(\hat{p},q),\\ \hat{q}&=q+sH_p(\hat{p},q),\end{aligned}\\
&Q: \quad c=\begin{pmatrix} 0 & 0\\ 0 & I\end{pmatrix}, \quad \begin{aligned}\hat{p}&=p-sH_q(p,\hat{q}),\\ \hat{q}&=q+sH_p(p,\hat{q}).\end{aligned}
\end{aligned} \tag{5.4}
$$

For $H(p,q)=U(p)+V(q)$, the above schemes P and Q reduce to explicit symplectic schemes (4.6).

Scheme (5.1) is revertible of order 2 for $c=\frac{1}{2}I$, this is (5.2). (5.1) is of order 1 for $c\neq\frac{1}{2}I$. Since $\check{E}^s_{H,c}=E^s_{H,I-c}$, the composites

$$
E^{s/2}_{H,c}\circ E^{s/2}_{H,I-c} \quad \text{and} \quad E^{s/2}_{H,I-c}\circ E^{s/2}_{H,c}
$$

are symplectic, revertible and of order 2. From these schemes we can get 4th and higher order revertible symplectic schemes by the procedure (2.12), (2.13) and (2.14).

So we get a great variety of simple symplectic schemes of order 1, 2, 4, etc., classified according to type matrices $B\in sm(2n):=$ space of symmetric matrices of order $2n$, which is a linear space of dimension $2n^2+n$.

A general methodology to construct unconditional symplectic algorithms is generating function method explained below.

A matrix α of order $4n$ is called a *Darboux matrix* if

$$
\alpha^T J_{4n}\alpha=\tilde{J}_{4n}, \quad J_{4n}=\begin{pmatrix} 0 & -I_{2n}\\ I_{2n} & 0\end{pmatrix}, \quad \tilde{J}_{4n}=\begin{pmatrix} J_{2n} & 0\\ 0 & -J_{2n}\end{pmatrix},
$$

$$
\alpha=\begin{pmatrix} a & b\\ c & d\end{pmatrix}, \quad \alpha^{-1}=\begin{pmatrix} a_1 & b_1\\ c_1 & d_1\end{pmatrix}.
$$

Every Darboux matrix induces a (linear) *fractional transform* between symplectic and symmetric matrices

$$
\begin{aligned}
\sigma_\alpha:\ & Sp(2n)\to sm(2n),\\
& \sigma_\alpha(S)=(aS+b)(cS+d)^{-1}=A, \quad \text{for} \quad |cS+d|\neq 0
\end{aligned}
$$

with the inverse transform $\sigma_\alpha^{-1}=\sigma_{\alpha^{-1}}$

$$
\begin{aligned}
\sigma_\alpha^{-1}:\ & sm(2n)\to Sp(2n),\\
& \sigma_\alpha^{-1}(A)=(a_1A+b_1)(c_1A+d_1)^{-1}=S, \quad \text{for } |c_1A+d_1|\neq 0,
\end{aligned}
$$

where $Sp(2n) = \{S \in GL(2n, \mathbf{R}) | S^T J_{2n} S = J_{2n}\}$ is the group of symplectic matrices.

The above machinery can be extended to generally non-linear operators on $\mathbf{R}^{2n}$. Denote $symm(2n)$ the space of symmetric operators on $\mathbf{R}^{2n}$ (not necessary one-one). Every $f \in symm(2n)$ corresponds, at least locally, to a real function ϕ (unique up to a constant) such that f is the gradient of ϕ : $f(w) = \nabla\phi(w)$, where $\nabla\phi(w) = (\phi_{w_1}(w), \cdots, \phi_{w_{2n}}(w))^T = \phi_w(w)$. Then we have

$$\sigma_\alpha : \quad S_p D_{2n} \to symm(2n),$$

$$\sigma_\alpha(g) = (a \circ g + b) \circ (c \circ g + d)^{-1} = \nabla\phi, \qquad \text{for} \qquad |cg_z + d| \neq 0,$$

or alternatively

$$ag(z) + bz = (\nabla\phi)(cg(z) + dz),$$

where ϕ is called the *generating function* of Darboux type α for the symplectic operator g. Then

$$\sigma_\alpha^{-1} : symm(2n) \to SpD_{2n},$$

$$\sigma_\alpha^{-1}(\nabla\phi) = (a_1 \circ \nabla\phi + b_1) \circ (c_1 \circ \nabla\phi + d_1)^{-1} = g, \text{ for } |c_1\phi_{ww} + d_1| \neq 0 \tag{5.5}$$

or alternatively

$$a_1\nabla\phi(w) + b_1 w = g\left(c_1\nabla\phi(w) + d_1 w\right), \tag{5.6}$$

where g is called the symplectic operator of Darboux type α for the generating function ϕ.

For the study of symplectic difference scheme we may narrow down the class of Darboux matrices to the subclass of *normal Darboux matrices*, i.e., those satisfying $a + b = 0, c + d = I_{2n}$. The normal Darboux matrices α can be characterized as

$$\alpha = \begin{pmatrix} a & b \\ c & d \end{pmatrix} = \begin{pmatrix} J & -J \\ c & I - c \end{pmatrix}, \quad c = \frac{1}{2}(I + JB), \quad B^T = B, \tag{5.7}$$

$$\alpha^{-1} = \begin{pmatrix} a_1 & b_1 \\ c_1 & d_1 \end{pmatrix} = \begin{pmatrix} (c - I)J & I \\ cJ & I \end{pmatrix}. \tag{5.8}$$

The fractional transform induced by a normal Darboux matrix establishes a 1-1 correspondence between *symplectic operators near identity* and *symmetric operators near nullity*.

For every Hamiltonian H with its phase flow e_H^t and for every normal Darboux matrix α, we get the *generating function* $\phi(w, t) = \phi_H^t(w) = \phi_{H,\alpha}^t(w)$ of *normal Darboux type* α for the *phase flow* of H by

$$\nabla\phi_{H,\alpha}^t = \left(Je_H^t - J\right) \circ \left(ce_H^t + I - c\right)^{-1}, \qquad \text{for small } |t|. \tag{5.9}$$

$\phi_{H,\alpha}^t$ satisfies the *Hamilton-Jacobi* equation

$$\frac{\partial}{\partial t}\phi(w,t) = -H\left(w + a_1\nabla\phi(w,t)\right) = -H\left(w + c_1\nabla\phi(w,t)\right) \tag{5.10}$$

and can be expressed by Taylor series in t

$$\phi(w,t) = \sum_{k=1}^{\infty}\phi^{(k)}(w)t^k, \quad |t| \text{ small.} \tag{5.11}$$

The coefficients can be determined recursively

$$\phi^{(1)}(w) = -H(w),$$

and for $k \geqslant 0,\ a_1 = (c-I)J:$

$$\phi^{(k+1)}(w) = \frac{-1}{k+1}\sum_{m=1}^{k}\frac{1}{m!}\sum_{\substack{j_1+\cdots+j_m=k \\ j_l\geqslant 1}} D^m H\left(a_1\nabla\phi^{(j_1)},\cdots,a_1\nabla\phi^{(j_m)}\right), \tag{5.12}$$

where we use the notation of the m-linear form

$$\begin{aligned} & D^m H(w)\left(a_1\nabla\phi^{(j_1)}(w),\cdots,a_1\nabla\phi^{(j_m)}(w)\right) \\ := & \sum_{i_1,\cdots,i_m=1}^{2n} H_{z_{i_1}\cdots z_{i_m}}(w)\left(a_1\nabla\phi^{(j_1)}(w)\right)_{i_1}\cdots\left(a_1\nabla\phi^{(j_m)}(w)\right)_{i_m}. \end{aligned}$$

By (5.9), the phase flow $\hat{z} := e_H^t z$ satisfies

$$\begin{aligned} \hat{z} - z &= -J\nabla\phi_{H,\alpha}^t(c\hat{z} + (I-c)z) \\ &= -\sum_{j=1}^{\infty} t^j J\nabla\phi^{(j)}(c\hat{z} + (I-c)z). \end{aligned} \tag{5.13}$$

Let ψ^s be a truncation of $\phi_{H,\alpha}^s$ up to a certain power s^m, say. Using inverse transform σ_α^{-1} we get the symplectic operator

$$g^s = \sigma_\alpha^{-1}\left(\nabla\psi^s\right), \quad |s| \text{ small}, \tag{5.14}$$

which depends on s, H, α (or equivalently B) and the mode of truncation. It is a symplectic approximation to the phase flow e_H^s and can serve as the transition operator of a symplectic difference scheme (for the Hamiltonian system (3.1))

$$z \to \hat{z} = g^s z: \quad \hat{z} = z - J\nabla\psi^s(c\hat{z} + (I-c)z), \quad c = \frac{1}{2}(I+JB). \tag{5.15}$$

Thus, using the machinery of phase flow generating functions we have constructed, for every H and every normal Darboux matrix α, a hierarchy of symplectic schemes by truncation. The simple symplectic schemes (5.4) correspond to the lowest truncation.

§6 Hamiltonian Algorithms for Hamiltonian Systems with a Perturbation Parameter

The machinery above can also be applied to construct symplectic algorithms for perturbed Hamiltonian systems defined by the perturbed Hamiltonian

$$H(z;\epsilon)=\sum_{k=0}^{\infty}\epsilon^k H_k(z)=H_0(z)+\sum_{k=1}^{\infty}\epsilon^k H_k(z), \tag{6.1}$$

where ϵ is the small perturbation parameter. $H_0(z)$ is usually an integrable Hamiltonian. The corresponding perturbed Hamiltonian system is

$$\frac{dz}{dt}=JH_z(z,\epsilon),\quad z\in\mathbf{R}^{2n}. \tag{6.2}$$

Its phase flow, denoted by $e_\epsilon^t=e_{H,\epsilon}^t$, depends on the parameter ϵ. The Hamilton-Jacobi equation and generating function are also parameterized by ϵ. That means, the parameterized generating function $\phi_\epsilon^t(w)=\phi^t(w,\epsilon)=\phi(w,t,\epsilon)$ satisfies the parameterized Hamilton-Jacobi equation

$$\frac{\partial}{\partial t}\phi(w,t,\epsilon)=-H\left(w+a_1\nabla\phi(w,t,\epsilon),\epsilon\right). \tag{6.3}$$

$\phi(w,t,\epsilon)$ can be expanded as a power series in ϵ in stead of t

$$\phi(w,t;\epsilon)=\sum_{k=0}^{\infty}\epsilon^k\phi^{(k)}(w,t). \tag{6.4}$$

The coefficients $\phi^{(k)}(w,t)$ satisfy the following equations:

$$\phi_t^{(0)}(w,t)=-\,H_0\left(w+a_1\nabla\phi^{(0)}(w,t)\right), \tag{6.5}$$

$$k\geqslant 1:\phi_t^{(k)}(w,t)=-\,H_k(w^*)-\sum_{i=1}^{k}\sum_{m=1}^{i}\frac{1}{m!}\sum_{\substack{i_1+\cdots+i_m=i\\ i_j\geqslant 1}}D^m H_{k-i}(w^*)$$
$$\times\left(a_1\nabla\phi^{(i_1)}(w,t),\cdots,a_1\nabla\phi^{(i_m)}(w,t)\right) \tag{6.6}$$

with the initial points $\phi^{(i)}(w,0)=0$, where $w^*=w+a_1\nabla\phi^{(0)}(w,t)$.

(6.5) is just the Hamilton-Jacobi equation for the unperturbed Hamiltonian system with Hamiltonian $H_0(z)$. The right hand side of (6.6) can be written as

$$-DH_0\left(w^*\right)\cdot a_1\nabla\phi^{(k)}(w,t)+R^{(k)}(w,t),$$

where the remainder $R^{(k)}(w,t)$ depends only on $H_i\left(w^*\right),i=0,1,\cdots,k$ and $\phi^{(i)}(w,t)$, $i=0,1,\cdots,k-1$. Once $\phi^{(0)}(w,t),\cdots,\phi^{(k-1)}(w,t)$ are known, $R^{(k)}(w,t)$ is also known. Therefore, if $\phi^{(0)}(w,t)$ can be solved from (6.5), then for $k\geqslant 1$,

$$\phi_t^{(k)}(w,t)=-DH_0\left(w^*\right)\cdot a_1\nabla\phi^{(k)}(w,t)+R^{(k)}(w,t)$$

are the linear partial differential equations for $\phi^{(k)}(w,t)$. They have the same coefficients, only those for $R^{(k)}(w,t)$ are different.

In some cases, we can solve (6.5) easily, refer to [45]. In general, it is difficult to solve (6.5) by analytical method. Nevertheless we can always give an approximative solution, for example, using the methods discussed above.

Let now $\psi_\epsilon^s(w)=\psi(w,s,\epsilon)$ be a truncation of $\phi(w,s,\epsilon)$ up to a certain power ϵ^m. Using the inverse transform σ_α^{-1} we get the symplectic operator

$$g_\epsilon^s=\sigma_\alpha^{-1}\left(\nabla\psi_\varepsilon^s\right),\quad |s| \text{ small.} \tag{6.7}$$

It is a symplectic approximation to the phase flow e_ϵ^s of order m in ϵ and can serve as the transition operator of a symplectic difference scheme for the perturbed Hamiltonian system (6.2)

$$z\rightarrow\hat{z}=g_\epsilon^s z:\quad \hat{z}=z-J\nabla\psi(c\hat{z}+(I-c)z,s,\epsilon),\quad c=\frac{1}{2}(I+JB). \tag{6.8}$$

For general perturbed Hamiltonians $H(z,\epsilon)$, these schemes are only consistent in the time stepsize s. But for the perturbed Hamiltonians with the form

$$H(z,\epsilon)=H_0(z)+\epsilon H_1(z), \tag{6.9}$$

the order of Scheme (6.8) in the time stepsize s is the same as in ϵ. More precisely, for the m-th order scheme g_ϵ^s for the perturbed Hamiltonian (6.9),

$$g_\epsilon^s=e_\epsilon^s+O\left((s\epsilon)^{m+1}\right). \tag{6.10}$$

Therefore, for small ϵ, the time stepsize s can be taken quite large. For more details, refer to [45].

§7 Symplectic Runge-Kutta Methods

For System (2.1) of differential equations, a r-stage Runge-Kutta method is defined by the following procedure

$$\begin{aligned} k_i&=z+s\sum_{j=1}^{r}d_{ij}a\left(k_j\right),\quad 1\leqslant i\leqslant r\\ \hat{z}&=z+s\sum_{j=1}^{r}b_j a\left(k_j\right), \end{aligned} \tag{7.1}$$

where d_{ij} and b_j are constants. The parameters d_{ij} and b_j characterize Runge-Kutta methods, which are chosen by order condition.

For Hamiltonian systems, Runge-Kutta methods are usually not symplectic. But if d_{ij} and b_j satisfy the condition

$$M_{ij} = b_i d_{ij} + b_j d_{ji} - b_i b_j \equiv 0 \quad (1 \leqslant i, j \leqslant r) \tag{7.2}$$

then they are symplectic [30,38]. In this case, the methods are implicit. The r -stage Gauss- Legendre methods satisfy Condition (7.2). They have the $2r$-th order of accuracy. The simplest one is

$$d_{11} = \frac{1}{2}, \quad b_1 = 1,$$

which gives

$$k_1 = J\nabla H\left(z + \frac{s}{2}k_1\right), \quad \hat{z} = z + sk_1.$$

This is just the centered Euler scheme (or midpoint rule)

$$\hat{z} = z + \frac{s}{2} J\nabla H\left(\frac{\hat{z}+z}{2}\right).$$

§8 Conservation Laws

Conservation laws we refer to here have two meanings. As is well known, the Hamiltonian system (3.1) itself has first integrals which are conserved in time evolution, e.g., the Hamiltonian is always a first integral. Hence, the first question is how many first integrals of Hamiltonian system (3.1) can be preserved by symplectic algorithms. The second question is whether or not there exist their own first integrals in case the original first integrals can not be preserved by symplectic algorithms.

8.1 Conservation laws

We first consider preservation of the first integrals of Hamiltonian systems by symplectic algorithms. For detailed discussion, refer to [4, 20, 28].

Consider the Hamiltonian system

$$\frac{dz}{dt} = J\nabla H(z). \tag{8.1}$$

Suppose

$$\hat{z} = g_H^s(z) \tag{8.2}$$

is a symplectic algorithm. Under a symplectic transformation $z = S(y)$, System (8.1) can be transformed into

$$\frac{dy}{dt} = J\nabla \tilde{H}(y), \tag{8.3}$$

where $\tilde{H}(y) = H(S(y))$ and Scheme (8.2) can be transformed into

$$\hat{y} = S^{-1} \circ g_H^s \circ S(y). \tag{8.4}$$

On the other hand, the algorithm g^s can apply to System (8.3) directly and the corresponding scheme is

$$\hat{y} = g_{\tilde{H}}^s(y). \tag{8.5}$$

Naturally, one can ask if (8.4) and (8.5) are the same. This introduces the following concept.

Definition 3 A symplectic algorithm g^s is invariant under the group $\mathcal{G}$ of symplectic transformations, or $\mathcal{G}$-invariant, for Hamiltonian H if

$$S^{-1} \circ g_H^s \circ S = g_{H\circ S}^s, \quad \forall S \in \mathcal{G};$$

g^s is symplectic invariant for Hamiltonian H if

$$S^{-1} g_H^s \circ S = g_{H\circ S}^s, \quad \forall S \in Sp(2n).$$

In practice, the second case is more common. Generally speaking, numerical algorithms depend on coordinates, i.e., they are locally represented. But many numerical algorithms may be independent of linear coordinate transformations.

Theorem 4 *Suppose F is a first integral of the Hamiltonian system* (8.1) *and e_F^t is the corresponding phase flow. Then F is conserved up to a constant by the symplectic algorithm g_H^s*

$$F \circ g_H^s = F + c \quad c \text{ is a constant} \tag{8.6}$$

if and only if g_H^s is e_F^t-invariant.

Proof. We first assume that the symplectic algorithm g_H^s is e_F^t -invariant, i.e.,

$$e_F^{-t} \circ g_H^s \circ e_F^t = g_{H\circ e_F^t}^s, \quad \forall t \in \mathbf{R}. \tag{8.7}$$

Since F is a first integral of the Hamiltonian system (8.1) with the Hamiltonian H, H is also the first integral of the Hamiltonian system with the Hamiltonian F, i.e.,

$$H \circ e_F^t = H. \tag{8.8}$$

It follows from (8.7) and (8.8) that

$$e_F^{-t} \circ g_H^s \circ e_F^t = g_H^s,$$

i.e.,

$$e_F^t = \left(g_H^s\right)^{-1} \circ e_F^t \circ g_H^s. \tag{8.9}$$

Differentiating (8.9) with respect to t at point 0 and noticing that

$$\left.\frac{de_F^t}{dt}\right|_{t=0} = J\nabla F,$$

we get

$$J\nabla F = \left(g_H^s\right)_*^{-1} J\nabla F \circ g_H^s. \tag{8.10}$$

Since g_H^s is symplectic, i.e.,

$$\left(g_H^s\right)_*^{-1} J = J\left(g_H^s\right)_*^T,$$

we have

$$J\nabla F = J\left(g_H^s\right)_*^T \nabla F \circ g_H^s = J\nabla\left(F \circ g_H^s\right).$$

Then

$$\nabla F = \left(g_H^s\right)_*^T \nabla F \circ g_H^s = \nabla\left(F \circ g_H^s\right).$$

It follows that

$$F \circ g_H^s = F + c. \tag{8.11}$$

We now assume that F is conserved by g_H^s, i.e., (8.6) is valid. Then noticing that the phase flows of the vector fields $J\nabla F$ and $\left(g_H^s\right)_*^{-1} J\nabla F \circ g_H^s$ are e_F^t and $\left(g_H^s\right)^{-1} \circ e_F^t \circ g_H^s$ respectively, we can get (8.7) similarly, i.e., g_H^s is e_F^t-invariant.

Symplectic invariant algorithms are invariant under the symplectic group $Sp(2n)$ and therefore invariant under the phase flow of any quadratic Hamiltonians.

Corolary 5 *Symplectic invariant algorithms for Hamiltonian systems preserve all quadratic first integrals of the original Hamiltonian systems up to a constant.*

If a symplectic scheme has a fixed point, i.e., there is a point z such that $g_H^s(z) = z$, then the constant $c = 0$ and the first integral is conserved exactly. Since linear schemes always have the fix point 0, we then have the following result.

Corolary 6 *Linear symplectic invariant algorithms for linear Hamiltonian systems preserve all quadratic first integrals of the original Hamiltonian systems.*

Example 1. Centered Euler scheme and symplectic Runge-Kutta methods are symplectic invariant. Hence they preserve all quadratic first integrals of System (8.1) up to a constant.

Example 2. Explicit symplectic schemes (4.6), (4.7), (4.11) and (4.8) are invariant under the linear symplectic transformations of the form $\operatorname{diag}\left(A^{-T}, A\right), A \in GL(n)$. Thus they preserve angular momentum $p^T Bq$ of the original Hamiltonian systems since their infinitesimal symplectic matrices are $\operatorname{diag}\left(-B^T, B\right), B \in gl(n)$.

In fact, these results can be improved. Symplectic Runge-Kutta methods preserve all quadratic first integrals of System (8.1) exactly[4]. For generating function methods, we have the following result[20].

Theorem 7 *Let $g^s_{H,\alpha}$ be a symplectic method constructed by the generating function method* (5.15) *with the Darboux type α. If $F(z) = \frac{1}{2} z^T Az, A \in sm(2n)$, is a quadratic first integral of the Hamiltonian system* (8.1) *and*

$$AJB - BJA = 0, \tag{8.12}$$

then $F(z)$ is conserved by $g^s_{H,\alpha}$, i.e.,

$$F(\hat{z}) = F(z), \qquad or \qquad F \circ g^s_{H,\alpha} = F. \tag{8.13}$$

For $B = 0$, i.e., the case of centered symplectic difference schemes, (8.12) is always valid. So all centered symplectic difference schemes preserve all quadratic first integrals of the Hamiltonian system (8.1) exactly.

Proof. Since $F(z)$ is the first integral of System (8.1),

$$\frac{1}{2}\hat{z}^T A\hat{z} = \frac{1}{2} z^T Az, \quad \hat{z} = e^t_H.$$

It can be rewritten as

$$\frac{1}{2}(\hat{z} + z)^T A(\hat{z} - z) = 0, \quad \hat{z} = e^t_H. \tag{8.14}$$

From (8.12) it follows that

$$\frac{1}{2}(JB(\hat{z} - z))^T A(\hat{z} - z) = \frac{1}{4}(\hat{z} - z)^T (AJB - BJA)(\hat{z} - z) = 0, \quad \forall \hat{z}, z \in \mathbf{R}^{2n}.$$

Combining it with (8.14), we have

$$(c\hat{z} + (I - c)z)^T A(z - z) = 0.$$

Using (5.13), it becomes

$$(c\hat{z} + (I - c)z)^T AJ \sum_{j=1}^{\infty} t^j \nabla\phi^{(j)}(c\hat{z} + (I - c)z) = 0.$$

From this we get

$$w^T AJ\nabla\phi^{(j)}(w) = 0, \quad \forall j \geqslant 1, \forall w \in \mathbf{R}^{2n}.$$

Taking $w = c\hat{z} + (I - c)z$, where

$$\hat{z} = g_{H,\alpha}^{s} z = z - J\nabla\psi^{(m)}(c\hat{z} + (I - c)z) = z - \sum_{j=1}^{m} s^j J\nabla\phi^{(j)}(c\hat{z} + (I - c)z),$$

we have

$$w^T A(\hat{z} - z) = -\sum_{j=1}^{m} s^j w^T AJ\nabla\phi^{(j)}(w) = -AJ\nabla\psi(w) = 0.$$

Since

$$\begin{aligned} w^T A(\hat{z} - z) &= \frac{1}{2}\hat{z}^T A\hat{z} - \frac{1}{2}z^T Az + \frac{1}{2}(\hat{z} - z)^T(AJB - BJA)(\hat{z} - z) \\ &= \frac{1}{2}\hat{z}^T A\hat{z} - \frac{1}{2}z^T Az, \end{aligned}$$

it leads to that $F(z)$ is the quadratic invariant of $g_{H,\alpha}^{s}$.

We list some of the most important normal Darboux's matrices c, the type matrices B, together with the corresponding form of symmetric matrices A of the conserved quadratic invariants $F(z) = \frac{1}{2}z^T Az$:

$c = I - c = \frac{1}{2}I, \quad B = 0, \quad A$ arbitrary .

$$c = \begin{pmatrix} I_n & 0 \\ 0 & 0 \end{pmatrix}, \quad B = \begin{pmatrix} 0 & -I_n \\ -I_n & 0 \end{pmatrix}, \qquad c = \begin{pmatrix} 0 & 0 \\ 0 & I_n \end{pmatrix}, \quad B = \begin{pmatrix} 0 & I_n \\ I_n & 0 \end{pmatrix}, \qquad A = \begin{pmatrix} 0 & b \\ b^T & 0 \end{pmatrix},$$

b arbitrary; angular momemtum type.

$$c = \frac{1}{2}\begin{pmatrix} I_n & \pm I_n \\ \mp I_n & I_n \end{pmatrix}, \quad B = \mp I_{2n}, \quad A = \begin{pmatrix} a & b \\ -b & a \end{pmatrix},$$

$a^T = a, b^T = -b$; Hermitian type.

$$c = \frac{1}{2}\begin{pmatrix} I & \pm I \\ \pm I & I \end{pmatrix}, \quad B = \pm\begin{pmatrix} I_n & 0 \\ 0 & -I_n \end{pmatrix}, \quad A = \begin{pmatrix} a & b \\ -b & -a \end{pmatrix},$$

$a^T = a$, $b^T = -b$.

8.2 Conservation laws of linear symplectic algorithms

Apart from the first integrals of the original Hamiltonian systems, a linear symplectic algorithm has its own quadratic first integrals[44]. For the linear Hamiltonian system

$$\frac{dz}{dt} = Lz, \quad L = JA \in sp(2n) \tag{8.15}$$

with a quadratic Hamiltonian $H(z) = \frac{1}{2}z^T Az, A^T = A$, denote its linear symplectic algorithm by

$$\hat{z} = g_H^s(z) = G(s, A)z, \quad G \in Sp(2n). \tag{8.16}$$

Assume that Scheme (8.16) is of order r. Then $G(s)$ has the form

$$G(s) = I + sL(s),$$
$$L(s) = L + \frac{s}{2!}L^2 + \frac{s^2}{3!}L^3 + \cdots + \frac{s^{r-1}}{r!}L^r + O\left(s^r\right).$$

For sufficiently small time step size $s, G(s)$ can be represented as

$$G(s) = e^{s\tilde{L}(s)}, \quad \tilde{L}(s) = L + O\left(s^r\right), \quad \tilde{L}(s) \in sp(2n).$$

So (8.16) becomes

$$\hat{z} = e^{s\tilde{L}(s)}z.$$

This is the solution $z(t)$ of the linear Hamiltonian system

$$\frac{dz}{dt} = \tilde{L}(s)z, \quad \tilde{L}(s) \in sp(2n) \tag{8.17}$$

with the initial value $z(0) = z^0$ evaluated at time s. The symplectic numerical solution

$$z^k = G^k(s)z^0 = e^{ks\tilde{L}(s)}z^0$$

is just the solution of System (8.17) at discrete points $ks, k = 0, \pm 1, \pm 2, \cdots$. Hence, for sufficiently small s, Scheme (8.16) corresponds to a perturbed linear Hamiltonian system (8.17) with the Hamiltonian

$$\widetilde{H}(z,s) = \frac{1}{2}\left(z, J^{-1}\tilde{L}(s)z\right) = \frac{1}{2}z^T J^{-1}Lz + O\left(s^r\right) = H(z) + O\left(s^r\right). \tag{8.18}$$

It is well-known that the linear Hamiltonian system (8.17) has n functionally independent quadratic first integrals. So does Scheme (8.16). The following

$$\widetilde{H}_i(z,s) = \frac{1}{2}z^T J^{-1}\widetilde{L}^{2i-1}(s)z, \quad i = 1, 2, \cdots, n \tag{8.19}$$

are the first integrals of the perturbed system (8.17), therefore, of Scheme (8.16), which approximate the first integrals of System (8.15)

$$H_i(z) = \frac{1}{2}z^T J^{-1}L^{2i-1}z, \quad i = 1, 2, \cdots, n$$

up to $O\left(s^r\right)$. Another group of first integrals of (8.16) is

$$\widehat{H}_i(z,s) = z^T J^{-1}G^i(s)z, \quad i = 1, 2, \cdots, n.$$

They can be checked easily. The first one is

$$\begin{aligned}\widehat{H}_1(z,s) &= z^T J^{-1}G(s)z = z^T J^{-1}(I + sL(s))z \\ &= sz^T J^{-1}L(s)z = 2sH(z) + O\left(s^3\right).\end{aligned}$$

§9 H-stability of Symplectic Algorithms

We know that Hamiltonian systems always appear in spaces of even dimensions. A more important fact is that there is no asymptotically stable linear Hamiltonian system. They are either Liapunov stable or unstable. So are linear symplectic algorithms. Therefore, usual stability concepts in numerical methods for ODE's are not suitable to symplectic algorithms for Hamiltonian systems, for example, A-stability and A(α)-stability, $\alpha < \pi/2$. Hence, usual A(α)-stability is useless for $\alpha < \pi/2$ and A-stability needs to be modified. In this section we discuss the linear stability of symplectic algorithms. Here we introduce a new test system and a new concept — H-stability (Hamiltonian stability) for symplectic algorithms and discuss the H-stability of symplectic invariant algorithms and the H-stability intervals of some explicit symplectic algorithms.

9.1 H-stability of symplectic algorithms

For the linear Hamiltonian system (8.15), a linear symplectic algorithm

$$z^{k+1} = g_H^s\left(z^k\right) = G(s, A)z^k, \quad k \geqslant 0 \tag{9.1}$$

is stable if $\exists C > 0$, such that

$$\left\|z^k\right\| = \left\|G^k(s, A)z^0\right\| \leqslant C\left\|z^0\right\|, \quad \forall k > 0,$$

where $\|\cdot\|$ is a well defined norm, such as Euclidean norm. Evidently, it is equivalent to $\left\|G^k(s)\right\|$ bounded, or, the eigenvalues of $G(s)$ are in the unit disk and its elementary divisors corresponding to the eigenvalues on the unit circle are linear. Since $G(s)$ is symplectic, then

$$G^{-1}(s) = J^{-1}G(s)^T J.$$

Hence if λ is an eigenvalue of $G(s)$, so is λ^{-1}, and they have the same elementary divisors. Therefore, the eigenvalue with the module less than 1 is always accompanied with the eigenvalue with the module great than 1. This implies that the linear symplectic method (9.1) can not be asymptotically stable. It also follows that the linear symplectic method (9.1) is stable if and only if the eigenvalues of $G(s)$ are unimodular and their elementary divisors are linear.

Herein we introduce the test Hamiltonian system

$$\frac{dz}{dt} = \alpha Jz, \quad z \in R^2, \quad \alpha \in R \tag{9.2}$$

with the test Hamiltonian

$$H(z) = H(p, q) = \frac{\alpha}{2}z^T z = \frac{\alpha}{2}\left(p^2 + q^2\right), \quad A = \alpha I. \tag{9.3}$$

Definition 8 A symplectic difference method is H-stable at $\mu = \alpha s$ if it is stable for the test Hamiltonian system (9.2) with the given μ. Such μ is called a stable point. The maximum interval in which every point is stable and which contains the original point is called the H-stability interval of the method. A symplectic difference method is H-stable if its H-stability interval is the whole real axis, $(-\infty, \infty)$. In this case, its numerical solutions are bounded for (9.2) with $\alpha \in \mathbf{R}$.

On the one hand, the test Hamiltonian (9.3) is the simplest in linear Hamiltonian systems; on the other hand, any stable linear Hamiltonian system has a normal form $H(p, q) = \frac{1}{2}\sum_{i=1}^{n} \alpha_i(p_i^2 + q_i^2)$ under symplectic transformations (see [2, 31, 48]). So the Hamiltonian (9.3) is also general enough. This shows that it is reasonable to take the Hamiltonian system (9.3) as a test system.

For the test system (9.2), (9.1) becomes

$$z^{k+1} = G(\mu)z^k, \tag{9.4}$$

where $G(\mu)$ is a 2×2 symplectic matrix. Denote

$$G(\mu) = \begin{pmatrix} a_1 & a_2 \\ a_3 & a_4 \end{pmatrix}.$$

Then $\det G(\mu) = a_1a_4 - a_2a_3 = 1$. Its characteristic polynomial is

$$|G(\mu) - \lambda I| = \begin{vmatrix} a_1 - \lambda & a_2 \\ a_3 & a_4 - \lambda \end{vmatrix} = \lambda^2 - (a_1 + a_4)\lambda + 1.$$

So its eigenvalues are

$$\lambda_{\pm} = \frac{a_1 + a_4}{2} \pm \sqrt{\left(\frac{a_1 + a_4}{2}\right)^2 - 1}.$$

Lemma 9 *(9.4) is stable at $\mu \neq 0$ if and only if*

$$\left(\frac{a_1 + a_4}{2}\right)^2 < 1 \quad or \quad -1 < \frac{a_1 + a_4}{2} < 1. \tag{9.5}$$

Applying the centered Euler scheme to the test system (9.3), it becomes

$$\hat{z} = z + \frac{1}{2}\mu J(\hat{z} + z), \quad \mu = \alpha s,$$

i.e.,

$$\hat{z} = \left(I + \frac{1}{2}\mu J\right)^{-1} \left(I - \frac{1}{2}\mu J\right) z,$$

where

$$G(\mu)=\left(I+\frac{1}{2}\mu J\right)^{-1}\left(I-\frac{1}{2}\mu J\right)=\frac{1}{1+\frac{1}{4}\mu^2}\begin{pmatrix}1-\frac{1}{4}\mu^2 & -\mu\\ \mu & 1-\frac{1}{4}\mu^2\end{pmatrix}.$$

So

$$\left(\frac{a_1+a_4}{2}\right)^2=\left(\frac{1-\frac{1}{4}\mu^2}{1+\frac{1}{4}\mu^2}\right)^2<1,\quad \text{for}\quad \mu\neq 0.$$

It follows from Lemma 9 that the centered Euler scheme is stable for all $\mu\neq 0$. Of course, it is stable at $\mu=0$. Hence the centered Euler scheme is H-stable.

9.2 H-stability of symplectic invariant methods

Theorem 10 *Symplectic invariant methods are H-stable.*

Proof. By Corolary 6, a symplectic invariant method preserves all quadratic first integrals of linear Hamiltonian systems. For the test Hamiltonian system (9.2), the Hamiltonian itself, $H(z)=\frac{\alpha}{2}z^Tz=\frac{\alpha}{2}\|z\|^2$, is a first integral. So $H(z)=\frac{1}{2}\alpha\|z\|^2$ is also a first integral of the symplectic method, i.e., if $z^k, k\geqslant 0$ is the solution of (9.1) applying to the test system (9.3), then

$$h\left(z^{k+1}\right)=h\left(z^k\right)=\cdots=h\left(z^0\right).$$

Therefore $\left\|z^k\right\|=\left\|z^0\right\|, k\geqslant 1$. This means that (9.1) is H-stable.

9.3 H-stability intervals of some explicit symplectic schemes

Explicit symplectic schemes constructed in section 4 are not symplectic invariant. We now consider the H-stability intervals of the 1st-4th order explicit symplectic schemes (4.6), (4.7), (4.11) and (4.8).

Applying the explicit symplectic schemes (4.6), (4.7), (4.11) and (4.8) to the test system (9.3), we get

$$z^{k+1}=G_i(\mu)z^k,\quad \mu=\alpha s,\quad i=1,2,3,4, \tag{9.6}$$

respectively, where G_i are their step transition matrices,

$$G_1(\mu) = \begin{pmatrix} 1 & -\mu \\ \mu & 1-\mu^2 \end{pmatrix},$$

$$G_2(\mu) = \begin{pmatrix} 1-\dfrac{1}{2}\mu^2 & -\mu \\ \mu\left(1-\dfrac{1}{4}\mu^2\right) & 1-\dfrac{1}{2}\mu^2 \end{pmatrix},$$

$$G_3(\mu) = \begin{pmatrix} 1-\dfrac{1}{2}\mu^2+\dfrac{1}{72}\mu^4 & -\mu\left(1-\dfrac{1}{6}\mu^2+\dfrac{7}{1728}\mu^4\right) \\ \mu\left(1-\dfrac{1}{6}\mu^2+\dfrac{1}{72}\mu^4\right) & 1-\dfrac{1}{2}\mu^2+\dfrac{5}{72}\mu^4-\dfrac{7}{1728}\mu^6 \end{pmatrix}.$$

$$G_4(\mu) = \begin{pmatrix} a_1 & a_2 \\ a_3 & a_4 \end{pmatrix},$$

$$a_1 = 1-\frac{1}{2}\mu^2+\frac{1}{24}\mu^4+\frac{1}{144}(1+\beta)^2\mu^6,$$

$$a_2 = -\mu\left(1-\frac{1}{6}\mu^2-\frac{1}{216}(2+\beta)(1+2\beta)\mu^4\right),$$

$$a_3 = \mu\left(1-\frac{1}{6}\mu^2+\frac{1}{216}(2+\beta)(1-\beta)\mu^4+\frac{1}{864}(2+\beta)(1+\beta)^2\mu^6\right),$$

$$a_4 = 1-\frac{1}{2}\mu^2+\frac{1}{24}\mu^4+\frac{1}{144}(1+\beta)^2\mu^6.$$

Hence, their H-stability intervals are $(-2,2)$, $(-2,2)$, $(-2.507, 2.507)$ and $(-1.573, 1.573)$ respectively.

§10 Formal Energy and Non-autonomous Perturbed Energy

10.1 Formal energy

In section 8, we mainly considered the conservation of the quadratic first integrals of Hamiltonian systems by symplectic algorithms and the quadratic first integrals of linear symplectic algorithms. Although the Hamiltonian H is a first integral, according to [29], in general, H can not be conserved by a symplectic algorithm. However, symplectic algorithms can have *formal* energies (Hamiltonians). For more details, refer to [11, 22, 53].

Suppose $H^s(z) = H(s,z)$ is a formal power series in s

$$H^s(z) = H(s,z) = \sum_{i=0}^{\infty} s^i H^{(i)}(z). \tag{10.1}$$

The formal phase flow $e^t_{H^s}$ satisfies

$$\frac{de^t_{H^s}}{dt} = J\nabla H^s \circ e^t_{H^s}, \quad e^0_{H^s} = 1_{2n}. \tag{10.2}$$

If

$$e_{H^s}^t\Big|_{t=s} = g_H^s, \tag{10.3}$$

then $H^s(z)$ is called a *formal* energy of the symplectic algorithm g_H^s. In this case,

$$H^s \circ g_H^s = H^s$$

in the convergent domain of (10.1). Hence, g_H^s formally preserves the formal energy H^s.

In the following, we want to determine H^s from g_H^s. Let the generating function of g_H^s with the normal Darboux matrix α be

$$\phi_{H^s,\alpha}^t(w) := \chi(t,s,w) = \sum_{k=1}^{\infty} t^k \chi^{(k)}(s,w). \tag{10.4}$$

The coefficients $\chi^{(k)}(s,w)$ can be recursively determined

$$\chi^{(1)}(s,w) = -H(s,w), \quad k \geqslant 1, \tag{10.5}$$

$$\begin{aligned}\chi^{(k+1)}(s,w) &= -\frac{1}{k+1}\sum_{m=1}^{k}\frac{1}{m!}\sum_{k_1+\cdots+k_m=k} D_w^m H\left(a_1\nabla\chi^{(k_1)},\cdots,a_1\nabla\chi^{(k_m)}\right)\\ &= \frac{1}{k+1}\sum_{m=1}^{k}\frac{1}{m!}\sum_{k_1+\cdots+k_m=k} D^m \chi^{(1)}\left(a_1\nabla\chi^{(k_1)},\cdots,a_1\nabla\chi^{(k_m)}\right),\end{aligned} \tag{10.6}$$

where the gradient ∇ is with respect to w. Let

$$\chi^{(k)}(s,w) = \sum_{i=0}^{\infty} s^i \chi^{(k,i)}(w), \tag{10.7}$$

then

$$\begin{aligned}\chi^{(k+1)}(s,w) &= \sum_{i=0}^{\infty} s^i \chi^{(k+1,i)}(w)\\ &= \sum_{i=0}^{\infty} s^i \frac{1}{k+1}\sum_{m=1}^{k}\frac{1}{m!}\sum_{\substack{i_0+\cdots+i_m=i\\ k_1+\cdots+k_m=k}} D_w^m \chi^{(1,i_0)}\left(a_1\nabla\chi^{(k_1,i_1)},\cdots,a_1\nabla\chi^{(k_m,i_m)}\right).\end{aligned}$$

Thus

$$\chi^{(k+1,i)}(w) = \frac{1}{k+1}\sum_{m=1}^{k}\frac{1}{m!}\sum_{\substack{i_0+\cdots+i_m=i\\ k_1+\cdots+k_m=k}} D_w^m \chi^{(1,i_0)}\left(a_1\nabla\chi^{(k_1,i_1)},\cdots,a_1\nabla\chi^{(k_m,i_m)}\right).$$

So the coefficient $\chi^{(k+1,i)}$ can be recursively determined. From (10.1), (10.5) and (10.7) it follows that

$$\chi^{(1,i)}(w) = -H^{(i)}(w), \quad i = 0,1,2,\cdots. \tag{10.8}$$

Assume that the generating function of g_H^s with the normal Darboux matrix α is

$$\psi_{g,\alpha}(s,w)=\sum_{k=1}^{\infty}s^k\psi^{(k)}(w). \tag{10.9}$$

Condition (10.3) can be rewritten in generating functions as

$$\chi(t,s,w)|_{t=s}=\chi(s,s,w)=\psi(s,w),$$

i.e.,

$$\sum_{k=1}^{\infty}s^k\sum_{i=0}^{\infty}s^i\chi^{(k,i)}(w)=\sum_{k=1}^{\infty}s^k\sum_{i=0}^{k-1}\chi^{(k-i,i)}(w)=\sum_{k=1}^{\infty}s^k\psi^{(k)}(w).$$

It follows

$$\sum_{i=0}^{k-1}\chi^{(k-i,i)}(w)=\psi^{(k)}(w),\quad k=1,2,\cdots.$$

Therefore,

$$\begin{aligned}
\chi^{(1,0)}&=\psi^{(1)},\\
\chi^{(1,1)}&=\psi^{(2)}-\chi^{(2,0)},\\
\chi^{(1,2)}&=\psi^{(3)}-\left(\chi^{(3,0)}+\chi^{(2,1)}\right),\\
&\cdots\cdots\\
\chi^{(1,k)}&=\psi^{(k+1)}-\left(\chi^{(k+1,0)}+\chi^{(k,1)}+\cdots+\chi^{(2,k-1)}\right),
\end{aligned} \tag{10.10}$$

Notice that $\chi^{(k+1,i)}$ can be determined by $\chi^{(k',i')}$, $k'\leqslant k, i'\leqslant i$. Then from $\psi^{(i)}$ we can get $\chi^{(1,i)}, i=0,1,2,\cdots$ by (10.10) and hence $H^{(i)}, i=0,1,2,\cdots$ by (10.8), i.e.,

$$H^{(i)}=-\chi^{(1,i)},\quad i=0,1,\cdots. \tag{10.11}$$

Finally we get the formal energy

$$H^s=\sum_{i=0}^{\infty}s^iH^{(i)}(z).$$

10.2 Non-autonomous perturbed energy

Formal energy of a symplectic algorithm is a good approximation to the original Hamiltonian. But usually, the convergence of the formal power series can not be guaranteed. There is another kind of non-autonomous perturbed energy of symplectic algorithms which can also exactly reproduce the symplectic algorithm but can not be conserved. For more details, refer to [44].

Since g_H^s is symplectic for all s, i.e.,

$$\left(g_*^s\right)^T J g_*^s = J, \quad \forall s, \tag{10.12}$$

differentiating it with respect to s, we then get

$$\left(\frac{dg_*^s}{ds}\right)^T J g_*^s + \left(g_*^s\right)^T J \frac{dg_*^s}{ds} = 0.$$

Therefore

$$J\frac{dg_*^s}{ds}\left(g_*^s\right)^{-1} = -\left(g_*^s\right)^{-T}\left(\frac{dg_*^s}{ds}\right)^T J = \left(J\frac{dg_*^s}{ds}\left(g_*^s\right)^{-1}\right)^T.$$

But

$$J\frac{dg_*^s}{dt}\left(g_*^s\right)^{-1} = \left(J\frac{dg^s}{dt}\circ\left(g^t\right)^{-1}\right)_*.$$

This shows that the Jacobian of $J\dfrac{dg^s}{ds}\circ\left(g^s\right)^{-1}$ is symmetric. By Poincaré Lemma, there is a function $\bar{H}(\cdot, s) : \mathbf{R}^{2n} \to \mathbf{R}$ such that

$$J\frac{dg^s}{ds}\circ\left(g^s\right)^{-1}(z) = \nabla\bar{H}(z, s),$$

i.e.,

$$\frac{dg^s}{ds} = J\nabla\bar{H}\circ g^s.$$

This shows that $\hat{z}(s) = g_H^s(z)$ is the solution of the non-autonomous Hamiltonian system

$$\frac{dz}{ds} = J\nabla\bar{H}(z, s).$$

More precisely,

$$\bar{H}(z, s) = \int_0^1 z^T J\frac{dg^s}{ds}\circ\left(g^s\right)^{-1}(\alpha z)d\alpha.$$

It follows that

$$\bar{H}(z, s) = H(z) + O\left(s^r\right).$$

Example 1. For the linear symplectic algorithm (8.15), the corresponding time-dependent Hamiltonian is

$$\bar{H}(z, s) = \frac{1}{2}z^T J\frac{dG(s)}{ds}G^{-1}(s)z,$$

which is different from (8.18)

Example 2. For the seperable Hamiltonian

$$H(p,q) = U(p) + V(q), \quad p, q \in \mathbf{R}^n,$$

a first order explicit symplectic difference scheme is

$$g^s: \quad \hat{p} = p - sV_q(q), \quad \hat{q} = q + sU_p(\hat{p}).$$

The perturbed Hamiltonian is

$$\bar{H}(p,q,s) = U(p) + V\left(q - sU_p(p)\right).$$

A second order explicit symplectic difference scheme is

$$g^s: \quad q^* = q + \frac{s}{2}U_p(p), \quad \hat{p} = p - sV_q\left(q^*\right), \quad \hat{q} = q^* + \frac{s}{2}U_p(\hat{p}).$$

The perturbed Hamiltonian is

$$\bar{H}(p,q,s) = \frac{1}{2}U(p) + \frac{1}{2}U\left(p + sV_q\left(q - \frac{s}{2}U_p(p)\right)\right) + V\left(q - \frac{s}{2}U_p(p)\right).$$

References

[1] Arnold V I. Mathematical Methods of Classical Mechanics. New York: Springer. 1978

[2] Bryuno A D. The normal form of a Hamiltonian system. Russian Math Surveys. 1988 41(1): 25-66

[3] Channell P J, Scovel J C. Symplectic Integration of Hamiltonian Systems. Nonlinearity. 1990 3: 213-259

[4] Cooper G. J. Stability of Runge-Kutta methods for Trajectory problems. IMA J Numer Anal. 1987 7: 1-13

[5] Feng K. On difference schemes and symplectic geometry. In Feng K, ed. Proc 1984 Beijing Symp Diff Geometry and Diff Equations. Beijing: Science Press, 1985: 42-58

[6] Feng K. Difference schemes for Hamiltonian formalism and symplectic geometry. J Comp Math. 1986 4(3): 279-289

[7] Feng K. The Hamiltonian way for computing Hamiltonian dynamics. In Spigler R, ed. Applied and Industrial Mathematics. Netherlands: Kluwer, 1991: 17-35

[8] Feng K. How to compute properly Newton's equation of motion. In Ying L A, Guo B Y, ed. Proc of 2nd Conf on Numerical Methods for Partial Differential Equations. Singapore: World Scientific, 1992: 15-22

[9] Feng K. Formal power series and numerical methods for differential equations. In Chan T, Shi Z C, ed. Proc of Inter Conf on Scientific Computation. Singapore: World Scientific, 1992: 28-35

[10] Feng K. The step transition operators for multi-step methods of ODE's: [Preprint]. Beijing: CAS. Computing Center, 1991

[11] Feng K. The calculus of generating functions and the formal energy for Hamiltonian algorithms: [Preprint]. Beijing: CAS. Computing Center, 1991

[12] Feng K. The calculus of formal power series for diffeomorphisms and vector fields: [Preprint]. Beijing: CAS. Computing Center, 1991

[13] Feng K. Symplectic, contact and volume-preserving algorithms. In Shi Z C, Ushijima T, ed. Proc 1st China-Japan Conf on Numer Math. Singapore: World Scientific, 1993. 1-28

[14] Feng K. Formal dynamical systems and numerical algorithms. In Feng K, shi Z C, ed. Proc Inter Conf on Computation of Differential Equations and Dynamical Systems. Singapore: World Scientific, 1993: 1-10

[15] Feng K. Contact algorithms for contact dynamical systems: [Preprint]. Beijing: CAS. Computing Center, 1992

[16] Feng K. Theory of contact generating functions: [Preprint]. Beijing: CAS. Computing Center, 1992

[17] Feng K, Qin M Z. The symplectic methods for computation of Hamiltonian equations. In Zhu Y L, Guo Ben-yu, ed. Proc Conf on Numerical Methods for PDE's. Berlin: Springer, 1987: 1-37. Lect Notes in Math 1297

[18] Feng K, Qin M Z. Hamiltonian algorithms and a comparative numerical study. Comput Phys Comm. 1991 65: 173-187

[19] Feng K, Shang Z J. Volume preserving algorithms for source-free dynamical systems. Numerische Mathematik. 1995

[20] Feng K, Wang D L. A note on conservation laws of symplectic difference schemes for Hamiltonian systems. J Comp Math. 1991 9(3): 229-237

[21] Feng K, Wang D L. Variations on a theme by Euler: [Preprint]. Beijing: CAS. Computing Center, 1991

[22] Feng K, Wang D L. Dynamical systems and geometric construction of algorithms. In Shi Z C, Yang C C. eds. Comput Math in China. Contemporary Mathematics. AMS. 1994: 1-32

[23] Feng K, Wu H M, Qin M Z, Wang D L. Construction of canonical difference schemes for Hamiltonian formalism via generating functions. J Comp Math. 1989 7(1): 71-96

[24] Forest E, Ruth R D. Fourth-order symplectic integration. Physica D. 1990 43: 105-117

[25] Ge Z. Generating functions for the Poisson map. [Preprint]. Beijing: CAS. Computing Center, 1987

[26] Ge Z. Equivariant symplectic difference schemes and generating functions. Physica D. 1991 49: 376-386

[27] Ge Z. Generating functions, Hamilton-Jacobi equations and symplectic groupsoids on Poisson manifolds. Indiana Univ Math J. 1990 39: 859-876

[28] Ge Z, Feng K. On the approximation of Hamiltonian systems. J Comp Math. 1988 6(1): 88-97

[29] Ge Z, Marsden J. Lie-Poisson Hamilton-Jacobi theory and Lie-Poisson integrators. Phys Lett A. 1988 133(3): 137-139

[30] Lasagni F. Canonical Runge-Kutta methods. ZAMP. 1988 39: 952-953

[31] Laub A J, Meyer K. Canonical forms for symplectic and Hamiltonian matrices. Celestial Mech. 1974 9: 213-238

[32] Neri F. Lie algebras and canonical integration: [Preprint]. Maryland: University of Maryland. Dept of Physics, 1987

[33] Qin M Z, Wang D L, Zhang M Q. Explicit symplectic difference schemes for separable Hamiltonian systems. J Comp Math. 1991 9(3): 211-221

[34] Qin M Z, Zhang M Q. Multi-stage symplectic schemes of two kind of Hamiltonian systems of wave equations. Computer Math Applic. 1990 19(10): 51-62

[35] Qin M Z, Zhang M Q. Explicit Runge-Kutta like unitary schemes to solve certain quantum operator equations of motion. J Stat Phys. 1990 60: 837-843

[36] Qin M Z, Zhu W J. Construction of higher order symplectic schemes by composition. Computing. 1992 47: 309-321

[37] Ruth R D. A canonical integration technique. IEEE Trans on Nuclear Sciences. 1983 NS-30 : 2669-2671

[38] Sanz-Serna J M. Runge-Kutta Schemes for Hamiltonian Systems. BIT. 1988 28: 877-883

[39] Shang Z J. On KAM theorem for symplectic algorithms for Hamiltonian systems: [PhD dissertation]. [Preprint]. Beijing: CAS. Computing Center, 1991

[40] Shu H B. A new approach to generating functions for contact systems. Computer Math Applic. 1993 25: 101-106

[41] Tang Y F. Hamiltonian systems and algorithms for geodesic flows on compact Riemannian manifolds: [Master Thesis]. [Preprint]. Beijing: CAS. Computing Center, 1990

[42] Wang D L. Semi-discrete Fourier spectral approximations of infinite dimensional Hamiltonian systems and conservation laws. Computer Math Applic. 1991 21(4): 63-75

[43] Wang D L. Poisson difference schemes for Hamiltonian systems on Poisson manifolds. J Comp Math. 1991 9(2): 115-124

[44] Wang D L. Some aspects of Hamiltonian systems and symplectic algorithms. Physica D. 1994 73: 1-16

[45] Wang D L. Symplectic difference schemes for perturbed Hamiltonian systems: [Preprint]. Berlin: ZIB, 1990

[46] Wang W L. Linear stability of symplectic methods: [Preprint]. Berlin: ZIB, 1991

[47] Wang D L. Decomposition of vector fields and composition of algorithms. In Feng K, Shi Z C, ed. Proc Inter Conf on Computation of Differential Equations and Dynamical Systems. Singapore: World Scientific, 1993: 179-184

[48] Williamson J. On the algebraic problem concerning the normal forms of linear dynamical systems. Amer J Math. 1936 58: 129-140

[49] Wisdom J, Holman M. Symplectic maps for N-body problems. Astron J . 1991 102(4): 1528-1538

[50] Wu Y H. Symplectic transformations and symplectic difference schemes. Chinese J Numer Math Applic. 1990 12(1): 23-31

[51] Wu Y H. Discrete variational principle to the Euler-Lagrange equation. Computer Math Applic. 1990 20(8): 61-75

[52] Yoshida H. Construction of higher order symplectic integrators. Phys Letters A. 1990 150: 262-268

[53] Yoshida H. Recent progress in the theory and application of symplectic integrators. Celestial Mechanics and Dynamical Astronomy. 1993 56: 27-43

[54] Zhang M Q. Explicit unitary schemes to solve quantum operator equations of motion. J Stat Phys. 1991 65(3/4): 793-799

[55] Zhang M Q, Qin M Z. Explicit symplectic schemes to solve vortex systems. Computer Math Applic. 1993 26: 51-56